LONDON MATHEMATICAL SOCIETY LECTURE NOTE SERIES

Managing Editor: Professor M. Reid, Mathematics Institute,
University of Warwick, Coventry CV4 7AL, United Kingdom

The titles below are available from booksellers, or from Cambridge University Press at
http://www.cambridge.org/mathematics

287 Topics on Riemann surfaces and Fuchsian groups, E. BUJALANCE, A.F. COSTA & E. MARTÍNEZ (eds)
288 Surveys in combinatorics, 2001, J.W.P. HIRSCHFELD (ed)
289 Aspects of Sobolev-type inequalities, L. SALOFF-COSTE
290 Quantum groups and Lie theory, A. PRESSLEY (ed)
291 Tits buildings and the model theory of groups, K. TENT (ed)
292 A quantum groups primer, S. MAJID
293 Second order partial differential equations in Hilbert spaces, G. DA PRATO & J. ZABCZYK
294 Introduction to operator space theory, G. PISIER
295 Geometry and integrability, L. MASON & Y. NUTKU (eds)
296 Lectures on invariant theory, I. DOLGACHEV
297 The homotopy category of simply connected 4-manifolds, H.-J. BAUES
298 Higher operads, higher categories, T. LEINSTER (ed)
299 Kleinian groups and hyperbolic 3-manifolds, Y. KOMORI, V. MARKOVIC & C. SERIES (eds)
300 Introduction to Möbius differential geometry, U. HERTRICH-JEROMIN
301 Stable modules and the D(2)-problem, F.E.A. JOHNSON
302 Discrete and continuous nonlinear Schrödinger systems, M.J. ABLOWITZ, B. PRINARI & A.D. TRUBATCH
303 Number theory and algebraic geometry, M. REID & A. SKOROBOGATOV (eds)
304 Groups St Andrews 2001 in Oxford I, C.M. CAMPBELL, E.F. ROBERTSON & G.C. SMITH (eds)
305 Groups St Andrews 2001 in Oxford II, C.M. CAMPBELL, E.F. ROBERTSON & G.C. SMITH (eds)
306 Geometric mechanics and symmetry, J. MONTALDI & T. RATIU (eds)
307 Surveys in combinatorics 2003, C.D. WENSLEY (ed.)
308 Topology, geometry and quantum field theory, U.L. TILLMANN (ed)
309 Corings and comodules, T. BRZEZINSKI & R. WISBAUER
310 Topics in dynamics and ergodic theory, S. BEZUGLYI & S. KOLYADA (eds)
311 Groups: topological, combinatorial and arithmetic aspects, T.W. MÜLLER (ed)
312 Foundations of computational mathematics, Minneapolis 2002, F. CUCKER *et al* (eds)
313 Transcendental aspects of algebraic cycles, S. MÜLLER-STACH & C. PETERS (eds)
314 Spectral generalizations of line graphs, D. CVETKOVIC, P. ROWLINSON & S. SIMIC
315 Structured ring spectra, A. BAKER & B. RICHTER (eds)
316 Linear logic in computer science, T. EHRHARD, P. RUET, J.-Y. GIRARD & P. SCOTT (eds)
317 Advances in elliptic curve cryptography, I.F. BLAKE, G. SEROUSSI & N.P. SMART (eds)
318 Perturbation of the boundary in boundary-value problems of partial differential equations, D. HENRY
319 Double affine Hecke algebras, I. CHEREDNIK
320 L-functions and Galois representations, D. BURNS, K. BUZZARD & J. NEKOVÁR (eds)
321 Surveys in modern mathematics, V. PRASOLOV & Y. ILYASHENKO (eds)
322 Recent perspectives in random matrix theory and number theory, F. MEZZADRI & N.C. SNAITH (eds)
323 Poisson geometry, deformation quantisation and group representations, S. GUTT *et al* (eds)
324 Singularities and computer algebra, C. LOSSEN & G. PFISTER (eds)
325 Lectures on the Ricci flow, P. TOPPING
326 Modular representations of finite groups of Lie type, J.E. HUMPHREYS
327 Surveys in combinatorics 2005, B.S. WEBB (ed)
328 Fundamentals of hyperbolic manifolds, R. CANARY, D. EPSTEIN & A. MARDEN (eds)
329 Spaces of Kleinian groups, Y. MINSKY, M. SAKUMA & C. SERIES (eds)
330 Noncommutative localization in algebra and topology, A. RANICKI (ed)
331 Foundations of computational mathematics, Santander 2005, L.M PARDO, A. PINKUS, E. SÜLI & M.J. TODD (eds)
332 Handbook of tilting theory, L. ANGELERI HÜGEL, D. HAPPEL & H. KRAUSE (eds)
333 Synthetic differential geometry (2nd Edition), A. KOCK
334 The Navier–Stokes equations, N. RILEY & P. DRAZIN
335 Lectures on the combinatorics of free probability, A. NICA & R. SPEICHER
336 Integral closure of ideals, rings, and modules, I. SWANSON & C. HUNEKE
337 Methods in Banach space theory, J.M.F. CASTILLO & W.B. JOHNSON (eds)
338 Surveys in geometry and number theory, N. YOUNG (ed)
339 Groups St Andrews 2005 I, C.M. CAMPBELL, M.R. QUICK, E.F. ROBERTSON & G.C. SMITH (eds)
340 Groups St Andrews 2005 II, C.M. CAMPBELL, M.R. QUICK, E.F. ROBERTSON & G.C. SMITH (eds)
341 Ranks of elliptic curves and random matrix theory, J.B. CONREY, D.W. FARMER, F. MEZZADRI & N.C. SNAITH (eds)
342 Elliptic cohomology, H.R. MILLER & D.C. RAVENEL (eds)
343 Algebraic cycles and motives I, J. NAGEL & C. PETERS (eds)
344 Algebraic cycles and motives II, J. NAGEL & C. PETERS (eds)
345 Algebraic and analytic geometry, A. NEEMAN
346 Surveys in combinatorics 2007, A. HILTON & J. TALBOT (eds)
347 Surveys in contemporary mathematics, N. YOUNG & Y. CHOI (eds)
348 Transcendental dynamics and complex analysis, P.J. RIPPON & G.M. STALLARD (eds)
349 Model theory with applications to algebra and analysis I, Z. CHATZIDAKIS, D. MACPHERSON, A. PILLAY & A. WILKIE (eds)
350 Model theory with applications to algebra and analysis II, Z. CHATZIDAKIS, D. MACPHERSON, A. PILLAY & A. WILKIE (eds)

351 Finite von Neumann algebras and masas, A.M. SINCLAIR & R.R. SMITH
352 Number theory and polynomials, J. MCKEE & C. SMYTH (eds)
353 Trends in stochastic analysis, J. BLATH, P. MÖRTERS & M. SCHEUTZOW (eds)
354 Groups and analysis, K. TENT (ed)
355 Non-equilibrium statistical mechanics and turbulence, J. CARDY, G. FALKOVICH & K. GAWEDZKI
356 Elliptic curves and big Galois representations, D. DELBOURGO
357 Algebraic theory of differential equations, M.A.H. MACCALLUM & A.V. MIKHAILOV (eds)
358 Geometric and cohomological methods in group theory, M.R. BRIDSON, P.H. KROPHOLLER & I.J. LEARY (eds)
359 Moduli spaces and vector bundles, L. BRAMBILA-PAZ, S.B. BRADLOW, O. GARCÍA-PRADA & S. RAMANAN (eds)
360 Zariski geometries, B. ZILBER
361 Words: Notes on verbal width in groups, D. SEGAL
362 Differential tensor algebras and their module categories, R. BAUTISTA, L. SALMERÓN & R. ZUAZUA
363 Foundations of computational mathematics, Hong Kong 2008, F. CUCKER, A. PINKUS & M.J. TODD (eds)
364 Partial differential equations and fluid mechanics, J.C. ROBINSON & J.L. RODRIGO (eds)
365 Surveys in combinatorics 2009, S. HUCZYNSKA, J.D. MITCHELL & C.M. RONEY-DOUGAL (eds)
366 Highly oscillatory problems, B. ENGQUIST, A. FOKAS, E. HAIRER & A. ISERLES (eds)
367 Random matrices: High dimensional phenomena, G. BLOWER
368 Geometry of Riemann surfaces, F.P. GARDINER, G. GONZÁLEZ-DIEZ & C. KOUROUNIOTIS (eds)
369 Epidemics and rumours in complex networks, M. DRAIEF & L. MASSOULIÉ
370 Theory of p-adic distributions, S. ALBEVERIO, A.YU. KHRENNIKOV & V.M. SHELKOVICH
371 Conformal fractals, F. PRZYTYCKI & M. URBANSKI
372 Moonshine: The first quarter century and beyond, J. LEPOWSKY, J. MCKAY & M.P. TUITE (eds)
373 Smoothness, regularity and complete intersection, J. MAJADAS & A. G. RODICIO
374 Geometric analysis of hyperbolic differential equations: An introduction, S. ALINHAC
375 Triangulated categories, T. HOLM, P. JØRGENSEN & R. ROUQUIER (eds)
376 Permutation patterns, S. LINTON, N. RUŠKUC & V. VATTER (eds)
377 An introduction to Galois cohomology and its applications, G. BERHUY
378 Probability and mathematical genetics, N. H. BINGHAM & C. M. GOLDIE (eds)
379 Finite and algorithmic model theory, J. ESPARZA, C. MICHAUX & C. STEINHORN (eds)
380 Real and complex singularities, M. MANOEL, M.C. ROMERO FUSTER & C.T.C WALL (eds)
381 Symmetries and integrability of difference equations, D. LEVI, P. OLVER, Z. THOMOVA & P. WINTERNITZ (eds)
382 Forcing with random variables and proof complexity, J. KRAJÍČEK
383 Motivic integration and its interactions with model theory and non-Archimedean geometry I, R. CLUCKERS, J. NICAISE & J. SEBAG (eds)
384 Motivic integration and its interactions with model theory and non-Archimedean geometry II, R. CLUCKERS, J. NICAISE & J. SEBAG (eds)
385 Entropy of hidden Markov processes and connections to dynamical systems, B. MARCUS, K. PETERSEN & T. WEISSMAN (eds)
386 Independence-friendly logic, A.L. MANN, G. SANDU & M. SEVENSTER
387 Groups St Andrews 2009 in Bath I, C.M. CAMPBELL *et al* (eds)
388 Groups St Andrews 2009 in Bath II, C.M. CAMPBELL *et al* (eds)
389 Random fields on the sphere, D. MARINUCCI & G. PECCATI
390 Localization in periodic potentials, D.E. PELINOVSKY
391 Fusion systems in algebra and topology, M. ASCHBACHER, R. KESSAR & B. OLIVER
392 Surveys in combinatorics 2011, R. CHAPMAN (ed)
393 Non-abelian fundamental groups and Iwasawa theory, J. COATES *et al* (eds)
394 Variational problems in differential geometry, R. BIELAWSKI, K. HOUSTON & M. SPEIGHT (eds)
395 How groups grow, A. MANN
396 Arithmetic dfferential operators over the p-adic Integers, C.C. RALPH & S.R. SIMANCA
397 Hyperbolic geometry and applications in quantum Chaos and cosmology, J. BOLTE & F. STEINER (eds)
398 Mathematical models in contact mechanics, M. SOFONEA & A. MATEI
399 Circuit double cover of graphs, C.-Q. ZHANG
400 Dense sphere packings: a blueprint for formal proofs, T. HALES
401 A double Hall algebra approach to affine quantum Schur-Weyl theory, B. DENG, J. DU & Q. FU
402 Mathematical aspects of fluid mechanics, J. ROBINSON, J.L. RODRIGO & W. SADOWSKI (eds)
403 Foundations of computational mathematics: Budapest 2011, F. CUCKER, T. KRICK, A. SZANTO & A. PINKUS (eds)
404 Operator methods for boundary value problems, S. HASSI, H.S.V. DE SNOO & F.H. SZAFRANIEC (eds)
405 Torsors, étale homotopy and applications to rational points, A.N. SKOROBOGATOV (ed)
406 Appalachian set theory, J. CUMMINGS & E. SCHIMMERLING (eds)
407 The maximal subgroups of the low-dimensional finite classical groups, J.N. BRAY, D.F. HOLT & C.M. RONEY-DOUGAL
408 Complexity science: the Warwick master's course, R. BALL, R.S. MACKAY & V. KOLOKOLTSOV (eds)
409 Surveys in combinatorics 2013, S. BLACKBURN, S. GERKE & M. WILDON (eds)
410 Representation theory and harmonic analysis of wreath products of finite groups, T. CECCHERINI-SILBERSTEIN, F. SCARABOTTI & F. TOLLI
411 Moduli spaces, L. BRAMBILA-PAZ, O. GARCÍA-PRADA, P. NEWSTEAD & R.P. THOMAS (eds)
412 Automorphisms and equivalence relations in topological dynamics, D.B. ELLIS & R. ELLIS
413 Optimal transportation, Y. OLLIVIER, H. PAJOT & C. VILLANI (eds)
414 Automorphic forms and Galois representations I, F. DIAMOND, P. KASSAEI & M. KIM (eds)
415 Automorphic forms and Galois representations II, F. DIAMOND, P. KASSAEI & M. KIM (eds)

London Mathematical Society Lecture Note Series: 417

Recent Advances in Algebraic Geometry

A Volume in Honor of Rob Lazarsfeld's 60th Birthday

Edited by

CHRISTOPHER D. HACON
University of Utah

MIRCEA MUSTAŢĂ
University of Michigan

MIHNEA POPA
Northwestern University

Shaftesbury Road, Cambridge CB2 8EA, United Kingdom

One Liberty Plaza, 20th Floor, New York, NY 10006, USA

477 Williamstown Road, Port Melbourne, VIC 3207, Australia

314–321, 3rd Floor, Plot 3, Splendor Forum, Jasola District Centre, New Delhi – 110025, India

103 Penang Road, #05–06/07, Visioncrest Commercial, Singapore 238467

Cambridge University Press is part of Cambridge University Press & Assessment, a department of the University of Cambridge.

We share the University's mission to contribute to society through the pursuit of education, learning and research at the highest international levels of excellence.

www.cambridge.org
Information on this title: www.cambridge.org/9781107647558

First published 2015

A catalogue record for this publication is available from the British Library

ISBN 978-1-107-64755-8 Paperback

Contents

Contributors

Thomas Bauer *Fachbereich Mathematik und Informatik, Philipps-Universität Marburg, Hans-Meerwein-Straße, D-35032 Marburg, Germany;* tbauer@mathematik.uni-marburg.de

Aaron Bertram *Department of Mathematics, University of Utah, Salt Lake City, UT 84112-0090, USA;* bertram@math.utah.edu

Sébastien Boucksom *CNRS - Institut de Mathématiques de Jussieu, 4 place Jussieu, 75252 Paris Cedex, France;* boucksom@math.jussieu.fr

Gregory Burnham *c/o Bridgewater Associates, 1 Glendinning plane, Westport, CT 06880, USA*

Frédéric Campana *Université de Lorraine, Institut Élie Cartan, UMR 7502 du CNRS, BP 70239, 54506 Vandœuvre-lès-Nancy Cedex, France;* frederic.campana@univ-lorraine.fr

Fabrizio Catanese *Lehrstuhl Mathematik VIII, Mathematisches Institut der Universität Bayreuth, NW II, Universitätsstr. 30, 95447 Bayreuth, Germany;* Fabrizio.Catanese@uni-bayreuth.de

Jungkai Alfred Chen *National Center for Theoretical Sciences, Taipei Office and Department of Mathematics, National Taiwan University, Taipei 106, Taiwan;* jkchen@math.ntu.edu.tw

Olivier Debarre *Département de Mathématiques et Applications, École Normale Supérieure et CNRS, 45 rue d'Ulm, 75230 Paris cedex 05, France;* olivier.debarre@ens.fr

Tommaso de Fernex *Department of Mathematics, University of Utah, Salt Lake City, UT 48112-0090, USA;* defernex@math.utah.edu

Jean-Pierre Demailly *Université de Grenoble I, Institut Fourier, UMR 5582 du CNRS, BP 74, 38402 Saint-Martin d'Hères, France;* demailly@fourier.ujf-grenoble.fr

Igor Dolgachev *Department of Mathematics, University of Michigan, 525 E. University Ave., Ann Arbor, MI 49109, USA;* idolga@umich.edu

David Eisenbud *Department of Mathematics, University of California, Berkeley, Berkeley, CA 94720;* eisenbud@math.berkeley.edu

Daniel Erman *Department of Mathematics, University of Wisconsin, Madison, WI 53706, USA;* derman@math.wisc.edu

Charles Favre *CNRS - Centre de Mathématiques Laurent Schwartz, École Polytechnique, 91128 Palaiseau Cedex, France;* favre@math.polytechnique.fr

Daniel Greb *Ruhr-Universität Bochum, Fakultät für Mathematik, Arbeitsgruppe Algebra/Topologie, 44780 Bochum, Germany;* daniel.greb@ruhr-uni-bochum.de

Christopher D. Hacon *Department of Mathematics, University of Utah, 155 South 1400 East, Salt Lake City, UT 48112-0090, USA;* hacon@math.utah.edu

Robin Hartshorne *Department of Mathematics, University of California, Berkeley, Berkeley, CA 94720;* robin@math.berkeley.edu

Benjamin Howard *Center for Communications Research, Institute for Defense Analysis, 805 Bunn Drive, Princeton, NJ 08540, USA;* bjhowa3@idaccr.org

Atanas Iliev *Department of Mathematics, Seoul National University, Gwanak Campus, Bldg. 27, Seoul 151-747, Korea;* ailiev2001@yahoo.com

János Kollár *Department of Mathematics, Princeton University, Princeton, NJ 08544-1000, USA;* kollar@math.princeton.edu

Sándor J Kovács *University of Washington, Department of Mathematics, Box 354350, Seattle, WA 98195-4350, USA;* skovacs@uw.edu

Alex Küronya *Budapest University of Technology and Economics, Department of Algebra. Address: Albert-Ludwigs-Universität Freiburg, Mathematisches Institut, Eckerstraße 1, D-79104 Freiburg, Germany;* alex.kueronya@math.uni-freiburg.de

Luigi Lombardi *Mathematisches Institut, Universität Bonn, Endenicher Allee 60, Bonn 53115, Germany;* lombardi@math.uni-bonn.de

Laurent Manivel *Institut Fourier, Université de Grenoble I et CNRS, BP 74, 38402 Saint-Martin d'Hères, France;* laurent.manivel@ujf-grenoble.fr

Shigeru Mukai *Research Institute for Mathematical Sciences, Kyoto University, Kyoto 606-8502, Japan;* mukai@kurims.kyoto-u.ac.jp

Hisanori Ohashi *Department of Mathematics, Faculty of Science and Technology, Tokyo University of Science, 2641 Yamazaki, Noda, Chiba 278-8510, Japan;* ohashi@ma.noda.tus.ac.jp, ohashi.hisanori@gmail.com

Roberto Paoletti *Dipartimento di Matematica e Applicazioni, Università degli Studi di Milano Bicocca, Via R. Cozzi 53, 20125 Milano, Italy;* `roberto.paoletti@unimib.it`

Giuseppe Pareschi *Dipartimento di Matematica, Università di Roma, Tor Vergata, Viale della Ricerca Scientifica, 00133 Roma, Italy;* `pareschi@axp.mat.uniroma2.it`

Thomas Peternell *Universität Bayreuth, Mathematisches Institut, D-95440 Bayreuth, Germany;* `thomas.peternell@uni-bayreuth.de`

Mihnea Popa *Department of Mathematics, Northwestern University, 2033 Sheridan Road, Evanston, IL 60208-2730, USA;* `mpopa@math.northwestern.edu`

Zvi Rosen *Department of Mathematics, University of California, Berkeley, Berkeley, CA 94720, USA;* `zhrosen@math.berkeley.edu`

Christian Schnell *Department of Mathematics, Stony Brook University, Stony Brook, NY 11794, USA;* `cschnell@math.sunysb.edu`

Frank-Olaf Schreyer *Mathematik und Informatik, Universität des Saarlandes, Campus E2 4, D-66123 Saarbrücken, Germany;* `schreyer@math.uni-sb.de`

Jessica Sidman *Department of Mathematics and Statistics, Mount Holyoke College, South Hadley, MA 01075, USA;* `jsidman@mtholyoke.edu`

Tomasz Szemberg *Instytut Matematyki UP, Podchorążych 2, PL-30-084 Kraków, Poland;* `tomasz.szemberg@gmail.com`

Stefano Urbinati *Università degli Studi di Padova, Via Trieste 63, 35121 Padova, Italy;* `urbinati.st@gmail.com`

Peter Vermeire *Department of Mathematics, Central Michigan University, Mount Pleasant, MI 48859, USA;* `p.vermeire@cmich.edu`

Claire Voisin *CNRS and École Polytechnique, Centre de mathématiques Laurent Schwartz, 91128 Palaiseau Cédex, France;* `voisin@math.polytechnique.fr`

Johnathan Wahl *University of North Carolina at Chapel Hill, Chapel Hill, NC 27599-3250, USA;* `jmwahl@email.unc.edu`

Preface

The conference "Recent Advances in Algebraic Geometry" was held between May 16 and 19, 2013, at the University of Michigan, Ann Arbor, to honor Robert K. Lazarsfeld (known as "Rob" among friends and colleagues) on the occasion of his 60th birthday. The conference honored Rob's outstanding contributions to algebraic geometry and the mathematical community, bringing together a large crowd, including many of his former students, collaborators, colleagues, and friends. It was a happy occasion for many of us who have known Rob and have been touched by his influence as students and peers, or simply as members of the algebraic geometry world.

From a personal point of view, we cannot even begin to discuss Rob's career without mentioning one of its most distinguished aspects, namely the unique influence he has had on the younger generations through teaching and mentoring. His style as a doctoral advisor and as an expositor is famous throughout the algebraic geometry community. He has been the advisor of more than 20 students, has numerous other mathematical descendants, and has mentored successful postdoctoral fellows. Many of these are now established mathematicians helping to expand the boundaries of Rob's mathematical vision. His generosity and ability to generate good problems, and his active support of the careers of his students, have been for many of us some of the most crucial aspects of our mathematical lives.

We highlight a few reference points in Rob's mathematical career. He received his B.A. from Harvard in 1975, and his Ph.D. from Brown in 1980, under the direction of William Fulton. He then went back to Harvard as a Benjamin Peirce Assistant Professor until 1983. During the 1981–82 academic year, Rob was awarded a postdoctoral fellowship from the American Mathematical Society, which he used to visit the Institute for Advanced Study in Princeton while on leave from Harvard. There he met Lawrence Ein, with whom he would later develop a long-lasting collaboration, resulting in over 25 joint papers. In 1983, Rob moved as an Assistant Professor to UCLA, where he became a Professor in 1987. He remained at UCLA until 1997, when he joined University of Michigan. There he was named Raymond L. Wilder Collegiate Professor of Mathematics in 2007. Starting in the Fall of 2013, Rob retired from Michigan and became a Professor at Stony Brook University. Over the years, Rob has received several honors, including a Sloan Fellowship (1984–87), the National Science Foundation Young Investigator Award (1985–90), and a Guggenheim Fellowship (1998–99); he was elected to the American Academy of Arts and Sciences in 2006.

While this is not the place to give a detailed account of Rob's work and accomplishments, it is inspiring to look back and give a brief overview of some of the highlights of his research that have had a profound impact on the field. His thesis was devoted to the study of low-degree ramified coverings of projective space. At the beginning of the 1980s, in joint work with William Fulton, Rob studied positivity properties of vector bundles, with applications to classical geometric questions, such as the connectedness of Brill–Noether loci. One of the fundamental results they proved describes the positive polynomials in the Chern classes of all ample vector bundles. Around the same time Rob began his work on the Castelnuovo–Mumford regularity of smooth projective varieties. Some landmark results in this direction that he obtained over the next decade concern sharp bounds for the regularity of curves (with Gruson and Peskine, generalizing a classical result of Castelnuovo) and surfaces, as well as a sharp bound in terms of the degrees of defining equations (with Bertram and Ein). In the mid-1980s, Rob began a collaboration with Mark Green that resulted in an extraordinarily influential series of papers, generating a large amount of research in algebraic geometry to this day. Some of these papers were devoted to the study of syzygies of smooth curves embedded in projective space. They contained important results and further conjectures on the precise connection between the algebraic invariants in the form of syzygies of the embedding, and the intrinsic geometry of the curve. Others were devoted to the study of cohomological support loci for topologically trivial line bundles, proving in particular their famous generic vanishing theorem. This led to a flurry of subsequent activity, involving both extensions and a wide array of applications, ranging from the study of singularities of theta divisors to that of the birational geometry of varieties with nontrivial holomorphic one-forms.

The most significant part of Rob's work since the beginning of the 1990s, largely done jointly with Lawrence Ein, but involving numerous other collaborators as well, revolved around geometric applications of vanishing theorems. Among the many fundamental results he obtained in this area, we mention only the proof of Fujita's conjecture for threefolds, an effective geometric version of Hilbert's Nullstellensatz, the fact that theta divisors have rational singularities, as well as various applications of asymptotic multiplier ideals to effective bounds in commutative algebra. Several new concepts, phenomena, or points of view in this circle of ideas, such as the notion of a graded sequence of ideals or the asymptotic study of linear series via the volume function, have their origin in Rob's work. Over the past few years, Rob has continued to ask fundamental questions and open new avenues of exploration, especially while studying Okounkov bodies, or the asymptotic behavior of syzygies. All of us influenced by Rob's work over the years are looking forward with excitement to Rob's future results and insights.

Rob's deep influence on the field of algebraic geometry and on how we think is not solely the outcome of his research papers and his teaching. When his book *Positivity in Algebraic Geometry* was published in 2004, it became an instant classic. It succeeded wonderfully in putting together under the same heading most of the areas of classical and modern complex algebraic geometry dedicated to, or influenced by, the study of positivity. It also developed for the first time the theory of multiplier ideals in textbook form, and introduced the theory of asymptotic multiplier ideals, tools that have since become of utmost importance in birational geometry. It is universally acknowledged that this will be one of a handful of fundamental references in the field of complex algebraic geometry for decades to come.

Before concluding, we would like to acknowledge the help we have received with funding and organizing the conference. We thank the National Science Foundation for support in the form of grant DMS-1262798 and the University of Michigan for financial and logistical assistance.

The papers collected in this volume are contributions from some of Rob's closest collaborators, students, and postdocs, as well as from some of the most prominent names in the subject. The reader will recognize in these contributions the extraordinary breadth of Rob's interests and influence. On behalf of the authors, all of those present at the conference, and the algebraic geometry community in general, we dedicate this volume to Rob with warmth and gratitude!

1

The effect of points fattening in dimension three

Th. Bauer[a]
Philipps-Universität Marburg

T. Szemberg[b]
Instytut Matematyki UP

Abstract

There has recently been increased interest in understanding the relationship between the symbolic powers of an ideal and the geometric properties of the corresponding variety. While a number of results are available for the two-dimensional case, higher dimensions are largely unexplored. In the present paper we study a natural conjecture arising from a result by Bocci and Chiantini. As a first step toward understanding the higher-dimensional picture, we show that this conjecture is true in dimension three. Also, we provide examples showing that the hypotheses of the conjecture may not be weakened.

Dedicated to Robert Lazarsfeld on the occasion of his sixtieth birthday

1 Introduction

The study of the effect of points fattening was initiated by Bocci and Chiantini [3]. Roughly speaking, they considered the radical ideal I of a finite set Z of points in the projective plane, its second symbolic power $I^{(2)}$, and deduced from the comparison of algebraic invariants of these two ideals various geometric properties of the set Z. Along these lines, Dumnicki *et al.* [7] studied higher symbolic powers of I. Similar problems were studied in [1] in the bi-homogeneous setting of ideals defining finite sets of points in $\mathbb{P}^1 \times \mathbb{P}^1$.

It is a natural task to try to generalize the result of Bocci and Chiantini [3, Theorem 1.1] to the higher-dimensional setting. Denoting by $\alpha(I)$ the *initial*

[a] Partially supported by DFG grant BA 1559/6-1.
[b] Partially supported by NCN grant UMO-2011/01/B/ST1/04875.

From *Recent Advances in Algebraic Geometry*, edited by Christopher Hacon, Mircea Mustaţă and Mihnea Popa

degree of a homogeneous ideal I, i.e., the least degree k such that $(I)_k \neq 0$, a natural generalization reads as follows:

Conjecture 1.1 *Let Z be a finite set of points in projective space $\mathbb{P}^n$ and let I be the radical ideal defining Z. If*

$$d := \alpha(I^{(n)}) = \alpha(I) + n - 1\,, \tag{1}$$

then either

$\alpha(I) = 1$, *i.e., Z is contained in a single hyperplane H in $\mathbb{P}^n$*

or

Z consists of all intersection points (i.e., points where n hyperplanes meet) of a general configuration of d hyperplanes in $\mathbb{P}^n$, i.e., Z is a star configuration. *For any polynomial in $I^{(n)}$ of degree d, the corresponding hypersurface decomposes into d such hyperplanes.*

The term *general configuration* in the conjecture means simply that no more than n hyperplanes meet in one point. This is equivalent to the *general linear position* for points in the dual projective space corresponding to the hyperplanes in the configuration. The result of Bocci and Chiantini is the case $n = 2$ of this conjecture. As a first step toward understanding the higher-dimensional picture, we show in the present paper:

Theorem 1.2 *The conjecture is true for $n = 3$.*

The assumption on the ideal I in the theorem amounts to the two equalities

$$\alpha(I^{(2)}) = \alpha(I) + 1$$
$$\alpha(I^{(3)}) = \alpha(I^{(2)}) + 1$$

and one might be tempted to relax the assumptions to only one of them. In Section 6 we provide examples showing, however, that neither is sufficient by itself to reach the conclusion of the theorem.

Star configurations are interesting objects of study in their own right. They are defined in [10] as unions of linear subspaces of fixed codimension c in projective space $\mathbb{P}^n$ that result as subspaces where exactly c of a fixed finite set of general hyperplanes in $\mathbb{P}^n$ intersect. The case described in Conjecture 1.1 corresponds thus to the $c = n$ situation. It is natural to wonder if the following further generalization of Conjecture 1.1 might be true: If Z is a finite collection of linear subspaces of codimension $c \leqslant n$ in $\mathbb{P}^n$ with the radical ideal I and such that

$$d = \alpha(I^{(c)}) = \alpha(I) + c - 1\,,$$

then Z is either contained in a hyperplane or forms a star configuration of codimension c subspaces. Janssen's recent preprint [12] deals with lines in $\mathbb{P}^3$ and shows that such a simple generalization would be too naive. Nevertheless we expect that there are some undiscovered patterns lurking beneath, and we hope to come back to this subject in the near future.

Throughout this paper we work over the complex numbers, and we use standard notation in algebraic geometry as in [13].

2 Initial degrees of symbolic powers

Definition 2.1 (Symbolic power) Let I be a homogeneous ideal in the polynomial ring $R = \mathbb{C}[\mathbb{P}^n]$. For a positive integer k, the ideal

$$I^{(k)} = R \cap \left(\bigcap_{\mathfrak{p} \in \mathrm{Ass}(I)} I^k R_{\mathfrak{p}} \right),$$

where the intersection is taken in the field of fractions of R, is the kth *symbolic power* of I.

Definition 2.2 (Differential power) Let I be a radical homogeneous ideal and let $V \subset \mathbb{P}^n$ be the corresponding subvariety. For a positive integer k, the ideal

$$I^{\langle k \rangle} = \bigcap_{P \in V} \mathfrak{m}_P^k,$$

where $\mathfrak{m}_P$ denotes the maximal ideal defining the point $P \in \mathbb{P}^n$, is the kth *differential power* of I.

In other words, the kth differential power of an ideal consists of all homogeneous polynomials vanishing to order at least k along the underlying variety. For radical ideals these two concepts fall together due to a result of Nagata and Zariski, see [8, Theorem 3.14] for prime ideals and [14, Corollary 2.9] for radical ideals:

Theorem 2.3 (Nagata–Zariski) *If I is a radical ideal in a polynomial ring over an algebraically closed field, then*

$$I^{\langle k \rangle} = I^{(k)}$$

for all $k \geqslant 1$.

We will make use of the following observation on symbolic powers:

Lemma 2.4 *Let I be an arbitrary radical homogeneous ideal. Then we have the inequality*

$$\alpha(I^{(k+1)}) - \alpha(I^{(k)}) \geqslant 1$$

for all $k \geqslant 1$.

Proof Let Z be the subscheme of $\mathbb{P}^n$ defined by I. The claim of the lemma follows immediately from the interpretation of symbolic powers as differential powers, Theorem 2.3, and the observation that if a polynomial f vanishes along Z to order at least $k+1$, then any of its partial derivatives vanishes along Z to order at least k. □

3 The $\mathbb{P}^2$ case revisited

As a warm-up, we give here a new proof of the result of Bocci and Chiantini. This proof has the advantage that it does not make use of the Plücker formulas.

Theorem 3.1 (Bocci–Chiantini) *Let Z be a finite set of points in the projective plane $\mathbb{P}^2$ and let I be its radical ideal. If*

$$d = \alpha(I^{(2)}) = \alpha(I) + 1\,,$$

then either Z consists of collinear points or Z is the set of all intersection points of a general configuration of d lines in $\mathbb{P}^2$.

Proof If $d = 2$, then we are done. So we assume $d \geqslant 3$.

By Lemma 4.2 below we may assume that Z consists of exactly $\binom{d}{2}$ points. Let X_2 be a divisor of degree d that is singular in all points of Z. Let P be one of the points in Z. Then there exists a divisor W_P of degree $d-2$ vanishing at all points in $Z \setminus \{P\}$ (and not vanishing at P).

We claim that W_P is contained in X_2. To see this, we begin by showing that they must have a common component. Indeed, this follows from Bézout's theorem, since otherwise we would get

$$d(d-2) = X_2 \cdot W_P \geqslant 2\left(\binom{d}{2} - 1\right) = d(d-1) - 2\,,$$

which is equivalent to $d \leqslant 2$ and contradicts our initial assumption in this proof.

Now let Γ be the greatest common divisor of X_2 and W_P, and let e be the degree of the divisor $W'_P = W_P - \Gamma$ (so that $\deg(\Gamma) = d-2-e$). There must be at least $\binom{e+2}{2} - 1$ points from $Z \setminus \{P\}$ on W'_P (otherwise there would be a pencil of such divisors W'_P and one could choose an element in this pencil passing through P, but then $W'_P + \Gamma$ would be an element of degree $d-2$ in I contradicting the assumption on $\alpha(I)$).

Then we intersect again $X_2 - \Gamma$ with W'_P and obtain

$$(e+2)e = (X_2 - \Gamma) \cdot W'_P \geqslant 2\left(\binom{e+2}{2} - 1\right) = (e+2)(e+1) - 2\,,$$

which gives $e = 0$.

It follows that $X_2 - W_P$ is a divisor of degree 2 with a double point at P. Hence, X_2 contains two lines intersecting in P. This holds for every point $P \in Z$. Since X_2 can contain at most d lines, we see that this is only possible if Z consists of the intersection points of a general configuration of d lines. □

4 A reduction result

We begin with a lemma concerning star configurations of points. We include the proof, since we were not able to trace it in the literature.

Lemma 4.1 *Let Z be a star configuration of points defined by hyperplanes $H_1, \ldots, H_d$ in $\mathbb{P}^n$. For $d \geqslant n+1$ the union*

$$H_1 \cup \ldots \cup H_d$$

is the only hypersurface F of degree d with the property

$$\operatorname{mult}_P F \geqslant n \quad \textit{for all} \quad P \in Z. \tag{2}$$

Proof We proceed by induction on the dimension $n \geqslant 2$. The initial case of $\mathbb{P}^2$ is dealt with simply by a Bézout-type argument. Indeed, assuming that there exists a curve F of degree d passing through all points in Z with multiplicity $\geqslant 2$ and taking a configuration line H_i, Bézout's theorem implies that H_i is a component of F. Since this holds for all lines in the configuration and $\deg(F) = d$, we are done.

For the induction step we assume that the lemma holds for dimension $n-1$ and all $d \geqslant n$. We want to conclude that it holds for $\mathbb{P}^n$ and all $d \geqslant n+1$. Of course we may assume that $n \geqslant 3$.

To this end let F by a hypersurface of degree d in $\mathbb{P}^n$ satisfying (2). Suppose that there exists a hyperplane H among $H_1, \ldots, H_d$, which is not a component of F. Then the restriction $G = F \cap H$ is a hypersurface of degree d in $H \simeq \mathbb{P}^{n-1}$ with $\operatorname{mult}_P G \geqslant n$ for all $P \in Z_H = Z \cap H$. Note that Z_H is itself a star configuration of points in H, defined by hyperplanes obtained as intersections $H_i \cap H$. So it is a star configuration of $d-1$ hyperplanes in $\mathbb{P}^{n-1}$. The polar system of G (i.e., the linear system defined by all first-order derivatives of the equation of G) is of dimension $n - 1 \geqslant 2$, and every element K in this system satisfies

$$\operatorname{mult}_P K \geqslant n - 1 \qquad \text{for all } P \in Z_H.$$

This contradicts the induction assumption. □

The following lemma allows us to assume $\#Z = \binom{d}{n}$ when proving the theorem (or when working on the conjecture).

Lemma 4.2 *Suppose that the set $Z \subset \mathbb{P}^n$ satisfies the assumptions of Conjecture 1.1 and that $\alpha(I) \geqslant 2$. Then there is a subset $W \subset Z$ with the following properties:*

(i) W is of cardinality $\binom{d}{n}$.
(ii) For the ideal J of W we have $\alpha(J^{(k)}) = \alpha(I^{(k)})$ for $k = 1, \ldots, n$.
(iii) If W is a star configuration, then $W = Z$.

Proof To begin with, note that the equality in (1) together with Lemma 2.4 implies $\alpha(I^{(k)}) = d - n + k$ for $k = 1, \ldots, n$.

The assumption $\alpha(I) \geqslant 2$ implies $d \geqslant n + 1$. Since there is no form of degree $\leqslant d - n$ vanishing along Z, there must be at least

$$s := \binom{d}{n} \tag{3}$$

points in Z.

Now we choose exactly s points $P_1, \ldots, P_s$ from Z that impose independent conditions on forms of degree $d - n$. (This can be done, since vanishing at each point in Z gives a linear equation on the coefficients of a form of degree $d - n$, so that we obtain a system of $\#Z$ linear equations of rank $s = \binom{d}{n}$ (which is the maximal possible rank). Then we choose a subsystem of s equations with maximal rank.) Let $W := \{P_1, \ldots, P_s\}$ and let J be the radical ideal of W. Since $W \subset Z$, we certainly have

$$\alpha(J^{(k)}) \leqslant \alpha(I^{(k)})$$

for all $k \geqslant 0$. On the other hand, we have

$$\alpha(J) = d - n + 1 = \alpha(I)$$

by the selection of W. Lemma 2.4 then implies that in fact

$$\alpha(J^{(k)}) = \alpha(I^{(k)})$$

for $k = 1, \ldots, n$. This shows that conditions (i) and (ii) are satisfied.

As for (iii): Suppose that W is a star configuration. By (ii) we have $\alpha(I^{(n)}) = \alpha(J^{(n)})$, hence it follows from Lemma 4.1 that $W = Z$. □

Further, we need the following elementary lemma on hypersurfaces that are obtained by taking derivatives.

Lemma 4.3 *Let $X \subset \mathbb{P}^n$ be a hypersurface defined by a polynomial f of degree d with a point P of multiplicity m such that $2 \leqslant m < d$. Then there exists a direction v such that the hypersurface Λ defined by the directional derivative of f in direction v has multiplicity m at P.*

Proof After a projective change of coordinates, we may assume $P = (1 : 0 : \ldots : 0)$. Then we can write

$$\begin{aligned} f(x_0 : x_1 : \ldots : x_n) = x_0^{d-m} g_m(x_1 : \ldots : x_n) \\ + x_0^{d-m-1} g_{m+1}(x_1 : \ldots : x_n) \\ + \ldots \\ + g_d(x_1 : \ldots : x_n) \end{aligned}$$

with homogeneous polynomials g_i of degree i for $i = m, \ldots, d$. Since $d > m$, the divisor defined by $\frac{\partial f}{\partial x_0} = 0$ has multiplicity m at P. □

5 Dimension three

In this section we give the

Proof of Theorem 1.2 We proceed by induction on d. For $d \leqslant 3$ the statement of the theorem is trivially satisfied, so we assume $d \geqslant 4$. By Lemma 4.2 we may assume that Z is of cardinality $\binom{d}{3}$. Let $X_3 \subset \mathbb{P}^3$ be the divisor defined by a polynomial of degree d in $I^{(3)}$. We assert that

$$X_3 \text{ is reducible.} \qquad (*)$$

To see this, we first note that thanks to $m = 3 < 4 \leqslant d$ there is, by Lemma 4.3, for any $P \in Z$ a directional derivative surface Λ_P of degree $d-1$ with multiplicity at least 3 at P. Arguing by contradiction, we assume that X_3 is irreducible, which implies that X_3 and Λ_P intersect properly, i.e., in a curve. Adapting the proof of [11, Proposition 3.1] to dimension 3, we see that the linear system of forms of degree $d-2$ vanishing along Z has only Z as its base locus. (This is due to the fact that the regularity of I is $d-2$.) We can therefore choose an element Y in this system that does not contain any component of the intersection curve of X_3 and Λ_P. Then the three surfaces X_3, Λ_P and Y intersect in points only, and we can apply Bézout's theorem to get

$$d(d-1)(d-2) = X \cdot \Lambda_P \cdot Y \geqslant 6\left(\binom{d}{3} - 1\right) + 9\,.$$

But this implies $0 \geqslant 3$, a contradiction. So (*) is established.

Now let Γ be an irreducible component of X_3 of smallest degree. Set $\gamma = \deg \Gamma$ and $X_3' = X_3 - \Gamma$. Our aim is to apply induction on X_3'. To this end we consider the set

$$Z' = Z \setminus \Gamma\,.$$

It is non-empty, as otherwise Z would be contained in Γ and then $\alpha(I)$ would be less than $d-2$. Indeed, we would have $\alpha(I) \leqslant \gamma \leqslant \left\lfloor \frac{d}{2} \right\rfloor$, which is less than $d-2$ if $d \geqslant 5$; and if $d = 4$, then $\gamma = 2$, so X_3 consists of two quadrics, which implies that it can have only two triple points – but then $\alpha(I) = 1$.

As Z' is non-empty, there is in particular a triple point on X_3', and hence $d - \gamma = \deg X_3' \geqslant 3$.

We claim that

$$\alpha(I_{Z'}) \geqslant d - \gamma - 2\,. \tag{4}$$

In fact, there is otherwise a surface S of degree $d - \gamma - 3$ passing through Z', and then $S + \Gamma$ is a divisor of degree $d-3$ passing through Z, which contradicts the assumption $\alpha(I) = d-2$.

Next, note that

$$\alpha(I_{Z'}^{(3)}) \leqslant \deg X_3' = d - \gamma\,. \tag{5}$$

In fact, as Γ does not pass through any of the points of Z', we know that X_3' has multiplicity at least 3 on Z'.

By Lemma 2.4, we obtain from (5) the inequality

$$\alpha(I_{Z'}) \leqslant d - \gamma - 2 \tag{6}$$

and this shows with (4) that equality holds in (6). From (5) we see, again with Lemma 2.4, that equality holds in (5) as well. We have thus established that the assumptions of the theorem are satisfied for the set Z'. By induction we conclude therefore that Z' is a star configuration and that X_3' decomposes into planes, or Z' is contained in a hyperplane and then the support of X_3' is that hyperplane. As Γ was chosen of minimal degree, it must be a plane as well, and hence X_3 decomposes entirely into planes. We can then run the above induction argument for *any* plane component Π of X_3 to see that the surface $X_3 - \Pi$ yields a star configuration. This shows immediately that there are no triple intersection lines among these planes and we conclude by just counting points with multiplicity at least 3 (in fact, exactly 3) that Z is a star configuration. □

6 Further results and examples

Recall that the *Waldschmidt constant* of a homogeneous ideal $I \subset \mathbb{P}^n$ is the asymptotic counterpart of the initial degree, defined as

$$\widehat{\alpha}(I) = \lim_{k\to\infty} \frac{\alpha(I^{(k)})}{k} = \inf_{k\geqslant 1} \frac{\alpha(I^{(k)})}{k} .$$

This invariant is indeed well defined, see [4] for this fact and some basic properties of $\widehat{\alpha}$. The Waldschmidt constants are interesting invariants that were recently rediscovered and studied by Bocci and Harbourne, see e.g. [5]. While Harbourne introduced the notation $\gamma(I)$, we propose here the notation $\widehat{\alpha}(I)$, as the Waldschmidt constant *is* the asymptotic version of the initial degree $\alpha(I)$, and the notation is then consistent with [9].

We state now a corollary of Theorem 1.2 dealing with the case where there is just one more α-jump by 1.

Corollary 6.1 *Let Z be a finite set of points in projective three-space $\mathbb{P}^3$ and let I be the radical ideal defining Z. If*

$$d := \alpha(I^{(4)}) = \alpha(I) + 3 , \tag{7}$$

then $\alpha(I) = 1$, i.e., Z is contained in a single plane in $\mathbb{P}^3$.

Proof Using Theorem 1.2 we need to exclude the possibility that Z forms a star configuration. To this end we apply in the case $n = 3$ the inequality

$$\widehat{\alpha}(I) \geqslant \frac{\alpha(I) + n - 1}{n} , \tag{8}$$

which was proved by Demailly [6, Proposition 6]. The assumptions of his result are satisfied by our Theorem 1.4. Combining (8) with the fact that $\widehat{\alpha}(I) \leqslant \frac{\alpha(I^{(n+1)})}{n+1}$ yields in our situation

$$\frac{\alpha(I) + n - 1}{n} \leqslant \frac{\alpha(I) + n}{n+1} ,$$

which gives immediately $\alpha(I) = 1$. □

Remark 6.2 (Waldschmidt constant of star configuration) Note that there is equality in (8) for star configurations of points by [4, Proof of Theorem 2.4.3].

Remark 6.3 If Conjecture 1.1 holds for any n, then the proof of the above corollary shows that if $\alpha(I^{(n+1)}) = \alpha(I) + n$, then $\alpha(I) = 1$.

We next provide examples showing that a single α-jump by 1 is not sufficient to reach the conclusion of Theorem 1.2.

Example 6.4 (Kummer surface) In this example we show that in general the assumption

$$\alpha(I^{(2)}) = \alpha(I) + 1 \tag{9}$$

for an ideal I of a set of points Z in $\mathbb{P}^3$ is not sufficient to conclude that the points in Z are coplanar or form a star configuration.

To this end, let $X \subset \mathbb{P}^3$ be the classical Kummer surface associated with an irreducible principally polarized abelian surface, and let Z be the set of the 16 double points on X. It is well known (see e.g. [2, Section 10.2]) that these 16 points form a 16_6 configuration, i.e., there are 16 planes Π_i in $\mathbb{P}^3$ such that each plane Π_i contains exactly 6 double points of X (and exactly 6 planes pass through every point in Z). We claim that

$$\alpha(I) = 3 \quad \text{and} \quad \alpha(I^{(2)}) = 4\,,$$

where I is the radical ideal of Z.

Granting this for a moment, we see immediately that the points in Z are neither coplanar nor form a star configuration, whereas the assumption in (9) is satisfied.

Turning to the proof, assume that there exists a surface S defined by an element of degree 3 in $I^{(3)}$. Let Π be one of the 16 planes Π_i. Then

$$1 \cdot 3 \cdot 4 = \Pi \cdot S \cdot X \geqslant 6 \cdot 1 \cdot 2 \cdot 2$$

implies that Π is a component of S. As the same argument works for all 16 planes, we get a contradiction. Hence $\alpha(I^{(2)}) = 4$.

A similar argument excludes the possibility that Z is contained in a quadric. We leave the details to the reader.

The following simpler example exhibiting the same phenomenon has been suggested by the referee.

Example 6.5 Let L_1, L_2, L_3 be mutually distinct and not coplanar lines in $\mathbb{P}^3$ intersecting at a point P. Let $A, B \in L_1$, $C, D \in L_2$ and $E \in L_3$ be points on these lines different from their intersection point P. Obviously the set $Z = \{A, B, C, D, E\}$ is not a star configuration. Let I be the radical ideal of Z. Then it is elementary to check that

$$\alpha(I) = 2 \text{ and } \alpha(I^{(2)}) = 3.$$

Note that 5 is the minimal number of non-coplanar points that can give $\alpha(I^{(2)}) = \alpha(I) + 1$ and not form a star configuration.

The next example has also been suggested by the referee and replaces a much more complicated example of our original draft.

Example 6.6 (Five general points in $\mathbb{P}^3$) In this example we show that in general the assumption

$$\alpha(I^{(3)}) = \alpha(I^{(2)}) + 1 \tag{10}$$

for an ideal I of a set of points Z in $\mathbb{P}^3$ is also not sufficient to conclude that the points in Z are coplanar or form a star configuration.

To this end let $Z = \{A, B, C, D, E\}$ consist of 5 points in general linear position in $\mathbb{P}^3$. For the radical ideal I of Z one then has

$$\alpha(I^{(3)}) = \alpha(I^{(2)}) + 1.$$

In fact, in this case one has

$$\alpha(I) = 2, \quad \alpha(I^{(2)}) = 4 \quad \text{and} \quad \alpha(I^{(3)}) = 5.$$

Acknowledgments We would like to thank Jean-Pierre Demailly for bringing reference [6] to our attention. Further, we thank Brian Harbourne for helpful discussions. We thank also the referee for detailed helpful remarks, comments and nice examples for Section 6.

References

[1] Baczyńska, M., Dumnicki, M., Habura, A., Malara, G., Pokora, P., Szemberg, T., Szpond, J., and Tutaj-Gasińska, H. Points fattening on $\mathbb{P}^1 \times \mathbb{P}^1$ and symbolic powers of bi-homogeneous ideals. *J. Pure Appl. Alg.* **218** (2014) 1555–1562.

[2] Birkenhake, C. and Lange, H. *Complex Abelian Varieties*. Berlin: Springer-Verlag, 2010.

[3] Bocci, C. and Chiantini, L. The effect of points fattening on postulation. *J. Pure Appl. Alg.* **215** (2011) 89–98.

[4] Bocci, C. and Harbourne, B. Comparing powers and symbolic powers of ideals. *J. Alg. Geom.* **19** (2010) 399–417.

[5] Bocci, C. and Harbourne, B. The resurgence of ideals of points and the containment problem. *Proc. Amer. Math. Soc.* **138** (2010) 1175–1190.

[6] Demailly, J.-P. Formules de Jensen en plusieurs variables et applications arithmétiques. *Bull. Soc. Math. Fr.* **110** (1982) 75–102.

[7] Dumnicki, M., Szemberg, T., and Tutaj-Gasińska, H. Symbolic powers of planar point configurations. *J. Pure Appl. Alg.* **217** (2013) 1026–1036.

[8] Eisenbud, D. *Commutative Algebra. With a view toward algebraic geometry.* New York: Springer-Verlag, 1995.

[9] de Fernex, T., Küronya, A., and Lazarsfeld, R. Higher cohomology of divisors on a projective variety. *Math. Ann.* **337** (2007) 443–455.

[10] Geramita, A., Harbourne, B., and Migliore, J. Star configurations in $\mathbb{P}^n$. *J. Alg.* **376** (2013) 279–299.

[11] Harbourne, B. and Huneke, C. Are symbolic powers highly evolved? *J. Ramanujan Math. Soc.* **28** (2013) 311–330.

[12] Janssen, M. On the fattening of lines in $\mathbb{P}^3$. arXiv:1306.4387, to appear in *J. Pure Appl. Alg.* (2014) http://dx.doi.org/10.1016/j.jpaa.2014.05.033.

[13] Lazarsfeld, R. *Positivity in Algebraic Geometry*. I.-II. Ergebnisse der Mathematik und ihrer Grenzgebiete, Vols 48–49. Berlin: Springer-Verlag, 2004.

[14] Sidman, J. and Sullivant, S. Prolongations and computational algebra. *Canad. J. Math.* **61** (2009) 930–949.

2

Some remarks on surface moduli and determinants

A. Bertram[a]
University of Utah

For Rob Lazarsfeld

1 Introduction

The slope of a vector bundle on a smooth projective curve C defines the families of stable and semi-stable bundles (of fixed degree and rank) that are coarsely represented by projective moduli spaces [17]. On a smooth projective polarized surface S, Mumford's H-slope defines H-stable and semi-stable torsion-free sheaves, but this is only the infinitessimal tip of the stability iceberg for S, in which slopes and compatible t-structures on the bounded **derived category** of coherent sheaves on S are the points of a **complex manifold** Stab(S) of stability conditions [14]. This gives rise to variations of determinant line bundles on moduli spaces of stable objects as the notion of stability varies. The stability conditions that are "closest" to the geometry of coherent sheaves on S are traditionally given in terms of a central charge (see e.g. [15]). Here I want to describe them in terms of a positive cohomology class α on S, my excuse being that the relationship with the determinant line bundles on moduli becomes *linear* in this coordinate system, and therefore the connection with moduli of (Gieseker) H-stable sheaves (and the classical moduli constructed by geometric invariant theory [16]) becomes more transparent.

The Chern character of a vector bundle E on C is the cohomology class $\mathrm{ch}(E) = \mathrm{rk}(E) + c_1(E) \in \mathrm{H}^0(C,\mathbb{Z}) \oplus \mathrm{H}^2(C,\mathbb{Z})$, and the slope of E can therefore be written in terms of the Chern character as

$$\mu(E) := \frac{\deg(E)}{\mathrm{rk}(E)} = \frac{\langle \mathrm{ch}(E), 1\rangle}{\langle \mathrm{ch}(E), H\rangle}$$

[a] Partially supported by NSF grant DMS-0901128.

From *Recent Advances in Algebraic Geometry*, edited by Christopher Hacon, Mircea Mustaţă and Mihnea Popa

where $H \in \mathrm{H}^2(C,\mathbb{Z})$ is the positive generator and $\langle\cdot,\cdot\rangle$ is the standard pairing on cohomology.

Mumford's H-slope on a surface S may similarly be defined as

$$\mu_H(E) := \frac{\langle \mathrm{ch}(E), H\rangle}{\langle \mathrm{ch}(E), H^2\rangle}$$

where H is an *ample* divisor class on S. The slopes we wish to consider here, however, have the form

$$\mu_\alpha(E) := \frac{\langle \mathrm{ch}(E), \alpha\rangle}{\langle \mathrm{ch}(E), \alpha\cdot H\rangle}$$

where $\alpha = \alpha_0 + \alpha_1 + \alpha_2 \in \mathrm{H}^{2*}(S,\mathbb{Q})$ is positive in a different sense:

$$\alpha_0 > 0 \text{ and } \langle\alpha_1,\alpha_1\rangle \geq 2\langle\alpha_0,\alpha_2\rangle.$$

By a theorem of Bogomolov, these are the inequalities that hold for the Chern characters of an H-stable vector bundle on S. The ample divisor H itself is not, of course, positive in this sense since it lacks the inequality $\alpha_0 > 0$. It is, however, a *limit* of positive classes $\epsilon\cdot\mathrm{td}(S)+H'$ as $\epsilon \to 0$ if $\mathrm{td}(S)$ is taken to be the Todd class. This is roughly the difference between Mumford and Gieseker stability.

Expressing the slope in terms of α amounts to a change of coordinates from Bridgeland's description of the slope in terms of a central charge defined via a divisor D and an ample class tH (as modified in [4]):

$$Z_{D,tH}(E) = -\int_S e^{-(D+itH)}\,\mathrm{ch}(E) \text{ and } \mu_Z(E) := -Re(Z(E))/Im(Z(E))$$

since with this definition

$$\mu_Z(E) = \frac{\langle \mathrm{ch}(E), 1 - D + \frac{1}{2}\left(D^2 - t^2H^2\right)\rangle}{\langle \mathrm{ch}(E), t(H - D\cdot H)\rangle} = \frac{\langle \mathrm{ch}(E), \alpha\rangle}{\langle \mathrm{ch}(E), \alpha\cdot(tH)\rangle}.$$

In the α coordinates, the orthogonal complement $c^\perp$ of a Chern class maps linearly via the determinant line bundle to $\mathrm{H}^2(M(c),\mathbb{Q})$, for moduli $M(c)$ of stable objects of class c. The recent positivity result of Bayer and Macrí [7] relates critical stability conditions (for which there exist strictly semi-stable objects) with critical values in the variation of the determinant line bundles. This relates the "wall and chamber structures" for the birational geometry of $M(c)$ with wall and chambers defined on the stability manifold [12]. The choice of a base point in $c^\perp \cap \{\alpha_0 = 1\}$ will allow us to identify rays in the ample cone of S with rays in the stability manifold that "point toward Gieseker stability." We will consider Chern classes of torsion-free sheaves, and then revisit the Serre map for surfaces in the context of torsion sheaves. This evokes fond memories for me, since Rob gave me my start with a question about the Serre map for curves.

2 Slopes and stabilities

A slope function on an abelian category $\mathcal{A}$, expressed as a ratio

$$\mu(A) = \frac{\lambda_d(A)}{\lambda_r(A)}; \ \lambda_d, \lambda_r : K(\mathcal{A})_{\mathbb{R}} \to \mathbb{R}$$

of $\mathbb{R}$-linear functions defines a pre-stability condition on $\mathcal{A}$ if

- $\lambda_r(A) \geq 0$ for all objects A of $\mathcal{A}$, and
- $\lambda_d(A) > 0$ for all objects (other than zero) for which $\lambda_r(A) = 0$.

If $\mathcal{A}$ is the heart of a bounded t-structure on a triangulated category $\mathcal{D}$, then the pre-stability condition extends to $\mathcal{D}$ in the obvious way.

Definition An object A of $\mathcal{A}$ is μ-stable if $\mu(B) < \mu(A)$ for all $B \subset A$.

It follows immediately from the two bullet points that:

Schur's Lemma 1 *Let A and B be μ-stable objects of $\mathcal{A}$. Then*

(i) Hom$(A, B) = 0$ if $\mu(A) > \mu(B)$.
(ii) Each nonzero $\phi \in \mathrm{Hom}(A, B)$ is an isomorphism if $\mu(A) = \mu(B)$.

If the abelian category is the heart of a bounded t-structure on a triangulated category $\mathcal{D}$, then the pre-slope extends to a Bridgeland pre-stability condition on $\mathcal{D}$. A different sort of boundedness is also required for the promotion of a pre-stability condition to a stability condition.

Definition Given a slope function μ on an abelian category $\mathcal{A}$, a bounded Harder–Narasimhan filtration on an object A has the form

$$0 = A_0 \subset A_1 \subset \cdots \subset A_n = A$$

where $\mu_i := \mu(A_i/A_{i-1})$ are strictly decreasing, and each $B_i = A_i/A_{i-1}$ admits a finite Jordan–Holder filtration

$$0 = B_i^0 \subset B_i^1 \subset \cdots \subset B_i^{m_i} = B_i$$

where each $C_i^j = B_i^j/B_i^{j-1}$ is stable of the *same* slope $\mu_i = \mu(B_i)$.

Objects B that admit Jordan–Holder filtrations are called *semi-stable*. It is a consequence of Schur's Lemma 1 that the Harder–Narasimhan filtration of each A is unique but that the Jordan–Holder filtrations, while not necessarily unique, do have associated graded objects $\oplus C_i^j$ that are unique up to isomorphism and reordering of the summands. Two semi-stable objects B and B' with isomorphic associated grades are said to be *s-equivalent*.

Definition A pre-stability condition is a *stability condition* if bounded Harder–Narasimhan filtrations exist for all nonzero objects of $\mathcal{A}$.

Our main example Let X be a smooth complex projective variety and $\mathcal{D}(X)$ be the bounded derived category of coherent sheaves on X. We will only consider stability conditions that factor through the Chern character, i.e., slope functions or central charges of the form

$$\mu(E) = \frac{\langle \mathrm{ch}(E), \alpha \rangle}{\langle \mathrm{ch}(E), \beta \rangle} \text{ or equivalently } Z(E) = -\langle \mathrm{ch}(E), \alpha \rangle + i\langle \mathrm{ch}(E), \beta \rangle$$

on the hearts $\mathcal{A}$ of compatible t-structures. Bridgeland proved that the locus of stability conditions has the structure of a complex manifold $\mathrm{Stab}(X)$ locally homeomorphic to a finite-dimensional complex vector space via the map

$$\text{stability condition } \sigma = (\mu_\sigma, \mathcal{A}_\sigma) \mapsto Z \in \mathrm{Hom}(K(X)/\equiv, \mathbb{C})$$

where $\equiv$ is *numerical equivalence*.

In this main example, we also have:

Schur's Lemma 2 *If E is stable, then $Hom(E, E) = \mathbb{C} \cdot id_E$.*

Consider the case $X = C$. Evidently, the slope function

$$\mu(E) = \frac{\langle \mathrm{ch}(E), 1 \rangle}{\langle \mathrm{ch}(E), H \rangle} = \frac{\deg(E)}{\mathrm{rk}(E)}$$

on the category $\mathcal{A}$ of coherent sheaves satisfies the bullet points, since

- the rank of a coherent sheaf on C is non-negative, and
- the length of a rank-zero coherent sheaf is its length

and the only sheaf of rank zero **and** length zero is the zero sheaf.

The stable coherent sheaves on a curve are either skyscraper sheaves (infinite slope) or stable vector bundles (finite slope). Geometric invariant theory can be used to show that s-equivalence classes of semi-stable bundles (of fixed rank and degree) have projective coarse moduli. The s-equivalence classes of semi-stable sheaves of rank zero also have projective moduli since they are points of the symmetric powers of C. Projectivity of moduli is certainly not a direct consequence of the definition of stability and begs the following:

Question Given a stability condition $\sigma = (\mu_\sigma, \mathcal{A}_\sigma)$ on $\mathcal{D}(X)$, when are the coarse moduli spaces of semi-stable objects of $\mathcal{A}_\sigma$ projective?

Definition A torsion-free sheaf E on a polarized smooth projective surface S is *H-stable* if $\mu_H(F) < \mu_H(E)$ for all subsheaves $F \subset E$ such that the support of E/F has codimension ≤ 1.

The problem with H-stability is that it violates the second bullet:

$$\mu_H(\mathbb{C}_x) = \frac{0}{0}$$

for skyscraper (and finite-length) sheaves.

In spite of this problem, Harder–Narasimhan filtrations exist, commencing with the torsion subsheaf $E_{tor} \subset E$ and continuing with a filtration on E/E_{tor} with H-semi-stable quotients B_i of strictly decreasing H-slope. As in the curve case the s-equivalence classes of H-semi-stable torsion-free sheaves B on S have projective coarse moduli (although the Gieseker slope is better suited for the contruction of moduli spaces by geometric invariant theory [16]).

There is one important new feature of H-stability:

Theorem (Bogomolov [13]) *The Chern classes of an H-stable sheaf E satisfy the following inequality:*

$$\langle c_1(E), c_1(E)\rangle \geq 2\langle \mathrm{rk}(E), ch_2(E)\rangle.$$

This inequality allows us to implement Bridgeland's tilting construction to produce t-structures that are compatible with slope functions of the form

$$\mu_\alpha(E) = \frac{\langle \mathrm{ch}(E), \alpha\rangle}{\langle \mathrm{ch}(E), \alpha \cdot H\rangle}$$

for $\alpha = \alpha_0 + \alpha_1 + \alpha_2 \in \mathrm{H}^*(S, \mathbb{Q})$ satisfying $\langle \alpha_1, \alpha_1\rangle > 2\langle \alpha_0, \alpha_2\rangle$.

Lemma 1 *If E is an H-stable sheaf and* $\langle ch(E), \alpha \cdot H\rangle = 0$, *then*

$$\langle ch(E), \alpha\rangle < 0.$$

Proof The inequalities at our disposal are

$$\alpha_0 > 0, \;\; \mathrm{rk}(E) > 0, \; \langle \alpha_1, \alpha_1\rangle > 2\alpha_0\alpha_2, \; \langle c_1(E), c_1(E)\rangle \geq 2\,\mathrm{rk}(E)\,\mathrm{ch}_2(E).$$

The vanishing assumption $\langle \mathrm{ch}(E), \alpha H\rangle = \langle \alpha_0 c_1(E) + \alpha_1 \,\mathrm{rk}(E), H\rangle = 0$ (i.e., the class $D = \alpha_0 c_1(E) + \alpha_1 \,\mathrm{rk}(E)$ is perpendicular to H), together with the inequalities, gives

$$\langle \mathrm{ch}(E), \alpha \rangle = \alpha_0 \,\mathrm{ch}_2(E) + \langle \alpha_1, \mathrm{ch}_1(E) \rangle + \alpha_2 \,\mathrm{rk}(E) < \frac{1}{2\alpha_0 \,\mathrm{rk}(E)} \langle D, D \rangle \leq 0$$

by the Hodge index theorem. □

Bridgeland's tilting construction requires the data of a *torsion pair.* This consists of a pair of full subcategories $(\mathcal{F}, \mathcal{T})$ of the category of coherent sheaves on S with the property that

(i) $\mathrm{Hom}(T, F) = 0$ for all objects T of $\mathcal{T}$ and F of $\mathcal{F}$.
(ii) Each coherent sheaf E fits into an exact sequence

$$0 \to T \to E \to F \to 0.$$

These produce a t-structure on the derived category $\mathcal{D}(S)$ whose heart consists of the complexes that have cohomologies only in two degrees: -1 (and belonging to $\mathcal{F}$) and 0 (and belonging to $\mathcal{T}$). That is, the tilted abelian category consists of objects $E^{\cdot}$ of the derived category that admit a cohomology sequence

$$0 \to F[1] \to E^{\cdot} \to T \to 0; \quad F \in \mathcal{F}, T \in \mathcal{T}.$$

These are determined by T and F and a *second* extension class $\epsilon \in \mathrm{Ext}^2(T, F)$ in the abelian category of coherent sheaves which is a first extension class $\epsilon \in \mathrm{Ext}^1(T, F[1])$ in the new tilted abelian category $\mathcal{A}^{\#}$.

The particular torsion pair associated with α is defined as follows:

- The objects of $\mathcal{F}_\alpha$ are all torsion-free sheaves F such that

$$\langle \mathrm{ch}(F'), \alpha \cdot H \rangle \leq 0 \text{ for all subsheaves } F' \subseteq F.$$

- The objects of $\mathcal{T}_\alpha$ are all coherent sheaves T such that

$$\langle \mathrm{ch}(T''), \alpha \cdot H \rangle > 0 \text{ or } \mathrm{len}(T'') < \infty \text{ for all quotients } T \to T'' \to 0.$$

Observation The pair $(\mathcal{F}_\alpha, \mathcal{T}_\alpha)$ satisfies (i) by definition, and (ii) by the Harder–Narasimhan filtrations defined above for a H-slope. The tilted abelian category with respect to this pair will be denoted by $\mathcal{A}_\alpha$.

Corollary 2 *The slope function μ_α satisfies the bullet points for a pre-stability condition on the tilted abelian category $\mathcal{A}_\alpha$.*

Proof The objects of $\mathcal{A}_\alpha$ are complexes of the form

$$0 \to F[1] \to E^{\cdot} \to T \to 0.$$

By definition, objects of T all satisfy $\langle \mathrm{ch}(T), \alpha \cdot H\rangle > 0$ *with the exception of torsion sheaves of finite length*. But these sheaves satisfy

$$\langle \mathrm{ch}(T), \alpha\rangle = \alpha_0 \,\mathrm{len}(T) > 0.$$

Objects of $F[1]$ satisfy $\langle \mathrm{ch}(F[1]), \alpha \cdot H\rangle = -\langle \mathrm{ch}(F), \alpha \cdot H\rangle > 0$ *with the exception of H-semi-stable torsion-free sheaves that pair to* 0. But by the key lemma (and linearity), such sheaves satisfy

$$\langle \mathrm{ch}(F[1]), \alpha\rangle = -\langle \mathrm{ch}(F), \alpha\rangle > 0$$

and hence the bullet points are satisfied for objects of $\mathcal{A}_\alpha$. □

Remark The finiteness of Harder–Narasimhan filtrations is not difficult to prove since we are restricting to rational coefficients [6]. It is curious that it is much more difficult to prove when the coefficients are real, even though the rational stability conditions are dense. Generalizations of the Bogomolov inequality to third Chern classes of stable complexes on threefolds have had some success [2, 9, 10, 18–21], although a useful such inequality for any projective Calabi–Yau threefold has yet to be found.

3 Determinants and moduli

In this section, B is always a "reasonable" base scheme (e.g., of finite type and quasi-projective over $\mathbb{C}$) and $S \times B$ is equipped with projections

$$p : S \times B \to S \text{ and } \pi : S \times B \to B.$$

A family of derived objects on S is a (not necessarily flat) coherent sheaf E_B on $S \times B$, or more generally an object of the derived category of $S \times B$. Families can be pushed forward to B (in the derived category), and the associated line bundle on the base

$$\Delta(E_B) := c_1(R\pi_* E_B)$$

is the determinant of the family. This will not, in general, define a line bundle on coarse moduli spaces of *isomorphism classes* of sheaves or derived objects. However, in the case where the Euler characteristic of the (derived) fibers over closed points vanishes:

$$\chi(S_b, E_b) := \chi(S \times \{b\}, Li_b^* E_B) = 0$$

then the determinant satisfies $\Delta(E_B) = \Delta(E_B \otimes \pi^* L)$ for any line bundle L on B and therefore descends to isomorphism classes of simple objects (i.e., objects with minimal automorphism group $\mathbb{C} \cdot id$).

In other words, in light of the Schur lemmas, the $\chi = 0$ condition on families is the precise condition that ought to result in a line bundle that descends to the coarse moduli spaces of σ-stable objects for any given stability condition σ.

A family of vector bundles E_B of rank r and degree d on a curve C of genus g can be transformed into a family of $\chi = 0$ vector bundles by choosing a vector F bundle on C with the property that

$$\mu(E_b) + \mu(F) = g - 1$$

and replacing E_B with the family $E_B \otimes p^*F$.

This not only gives a determinant line bundle, but also a pseudo-divisor on B associated with each family, with support

$$\Theta_F(E_B) := \{b \in B \mid \mathrm{H}^1(C_b, E_b \otimes F) \neq 0\}.$$

That is, the locus where the cohomology (in either degree) is nonzero.

The coarse moduli spaces $\mathcal{M}_C(s, L)$ of semi-stable vector bundles of rank s and fixed determinant $\wedge^s F = L$ are unirational and their Picard groups are generated by a single line bundle. It follows that the determinant line bundles Δ_F on $M_C(r, d)$ are independent of the choice of semi-stable bundle $F \in \mathcal{M}_C(s, L)$ (in each rank) and therefore that the pseudo-divisors Θ_F are linearly equivalent. Moreover, the line bundles Δ_F are ample, which can be proved directly or else by appealing to the fact that the Δ_F coincide with the ample line bundles (up to scaling) arising from the geometric invariant theory construction of moduli.

When we look for analogous polarizations on the moduli spaces of σ-stable objects on a surface S, we immediately run into the following:

Observation Orthogonal classes to $\mathrm{ch}(E)$ are not unique. Let

$$\widetilde{\mathrm{NS}}(S)_{\mathbb{Q}} = \mathrm{H}^0(S, \mathbb{Q}) \oplus \mathrm{NS}(S)_{\mathbb{Q}} \oplus \mathrm{H}^4(S, \mathbb{Q})$$

be the extended Néron–Severi vector space, of rank $\rho + 2$ over $\mathbb{Q}$, with the induced inner product from cohomology and let

$$c^{\perp} = \{\alpha \in \widetilde{\mathrm{NS}}(S)_{\mathbb{Q}} \mid \langle c, \alpha \rangle = 0\}$$

be the orthogonal complement of a Chern class $c = c_0 + c_1 + c_2$.

Now suppose $\alpha \in c^{\perp}$ and that F is an H-stable vector bundle with

$$\mathrm{ch}(F) \cdot \mathrm{td}(S) = \alpha \in \widetilde{\mathrm{NS}}(S)_{\mathbb{Q}}.$$

By the Hirzebruch–Riemann–Roch theorem, $\chi(S, E^{\cdot} \otimes F) = 0$ for any object $E^{\cdot}$ with $\mathrm{ch}(E^{\cdot}) = c$. Thus any *family* $E^{\cdot}_B$ of such objects gives rise to a determinant line bundle $\Delta_F(E^{\cdot}_B)$ on the base of the family with the desired invariance under tensoring by a line bundle from B. This is the candidate for the line bundle on coarse moduli of stable objects of class c, but it is important to notice that **both** the coarse moduli space (of α-stable objects of $\mathcal{A}_\alpha$) **and** the determinant line bundle on that moduli space *depend upon the choice of* $\alpha \in c^\perp$.

The dependence of the line bundles Δ_F on the class $\alpha \in c^\perp$ is linear. Indeed, the Grothendieck–Riemann–Roch theorem gives

$$c_1(R\pi_*(E_B^\cdot \otimes p^*F)) = \pi_*\left(\mathrm{ch}_1(E_B^\cdot)p^*\alpha_2 + \mathrm{ch}_2(E_B^\cdot)p^*\alpha_1 + \mathrm{ch}_3(E_B^\cdot)p^*\alpha_0\right)_1$$

where π_* is the push-forward on Chern classes.

Consider, for example, the **Hilbert schemes** $S^{[n]}$ with universal ideal sheaf $\mathcal{I}_Z \subset \mathcal{O}_{S\times S^{[n]}}$. This is a deceptively nice example. Since Hilbert schemes are smooth and projective of the right dimension $2n$, they are defined independently of the choice of ample class H and have a preferred universal family $\mathcal{I}_Z$.

The Chern character of the ideal sheaf $\mathcal{I}_Z$ of n points is $c = (1, 0, -n)$. Since $c^\perp = \{\alpha \mid n\alpha_0 = \alpha_2\}$, we have

$$c^\perp \cap \{\alpha_0 = 1\} = \{\alpha = (1, D, n) \mid D \in \mathrm{NS}(S)\}$$

and the positivity condition on α means that α determines a stability condition on S when

$$\langle D, D\rangle > 2n.$$

Let $D_t = -\frac{1}{2}K_S + tH'$ and $\alpha_t = (1, D_t, n)$ where H' is a second choice of ample divisor class (this will be important in later examples). For integer values of t, the resulting α is of the form

$$\mathrm{ch}(\mathcal{I}_W(tH'))\mathrm{td}(S) = \alpha_t$$

for a subscheme $W \subset S$ of the appropriate length. Thus, in this case, we may let $F = \mathcal{I}_W(tH')$ in the computation of the determinant line bundle (keeping in mind that this is torsion free and not locally free). Notice that for sufficiently large and small values of t, the class α_t defines stability conditions. When $t >> 0$, the second cohomology vanishes:

$$\mathrm{H}^2(S, \mathcal{I}_Z \otimes F) = \mathrm{H}^2(S, \mathcal{I}_Z \otimes \mathcal{I}_W \otimes \mathcal{O}_S(tH')) = 0$$

and it follows that for such values of t, the pseudo-divisors

$$\Theta_F = \{Z \mid \mathrm{H}^1(S, \mathcal{I}_Z \otimes F) \neq 0\}$$

represent the line bundle Δ_F. Its class is easily computed [3]:

$$\Delta_F = \pi_*\left(\mathrm{ch}_2(\mathcal{I}_Z)p^*D_t + \mathrm{ch}_3(\mathcal{I}_Z)\right) = -q^*(D_t^{(n)}) + \frac{1}{2}E$$

where $q : S^{[n]} \to S^n/\Sigma_n$ is the Hilbert–Chow morphism from the Hilbert scheme to the symmetric product, E is the exceptional divisor, and $D^{(n)}$ is the symmetric divisor on S^n/Σ_n associated with D.

Notice that this class is anti-ample for $t >> 0$. This is a consequence of the fact that the determinant line bundle is the divisor class given by the support of the sheaf $\mathrm{R}^1\pi_*(\mathcal{I}_\mathcal{Z} \otimes F)$, which has odd degree.

When $t << 0$, there is a similar result. The family of derived objects to consider in this case is the shift of the derived dual $\mathcal{I}_\mathcal{Z}^\vee[1]$ in the derived category of $S \times S^{[n]}$. This is a family of objects of the derived category of S with Chern character invariants $-c = (-1, 0, n)$. In this case, the determinant line bundle is the same as above but with opposite sign, reflecting the fact that in this case the pseudo-divisor gives the support of the sheaf $\mathrm{R}^2\pi_*(\mathcal{I}_\mathcal{Z}^\vee[1] \otimes F)$, which has even degree.

Turning to a more general example, suppose $c = (c_0, c_1, c_2) = \mathrm{ch}(E)$ where E is a Gieseker H-stable torsion-free sheaf on S. Gieseker stability is given in terms of the Hilbert polynomial, but can be defined for polarized surfaces as follows:

Definition A torsion-free sheaf E on S is *Gieseker H-unstable* if there is a subsheaf $F \subset E$ such that either

(i) $\mu_H(F) > \mu_H(E)$ (i.e., E is Mumford H-unstable), or else
(ii) $\mu_H(F) = \mu_H(E)$ and $\frac{\chi(S,F)}{\mathrm{rk}(F)} > \frac{\chi(S,E)}{\mathrm{rk}(E)}$.

It is *Gieseker H-stable* if for every subsheaf $F \subset E$, either

(i) $\mu_H(F) < \mu_H(E)$, or else
(ii) $\mu_H(F) = \mu_H(E)$ and $\frac{\chi(S,F)}{\mathrm{rk}(F)} < \frac{\chi(S,E)}{\mathrm{rk}(E)}$

and Gieseker H-semi-stable is defined in the usual way.

The moduli of Gieseker semi-stable equivalence classes have a natural construction via geometric invariant theory. This is also reflected in their naturalness from the point of view of α-stability conditions. As with the Hilbert scheme, consider the one-parameter family of elements of $c^\perp$, defined by another ample divisor class H' via

$$\alpha_t = (1, D_t, d_t), \text{ where } D_t = -\frac{K_S}{2} + tH', d_t = -\frac{1}{c_0}(D_t \cdot c_1 + c_2).$$

Remark For large t, we have the required positivity

$$\langle D_t, D_t \rangle > 2d_t$$

since the left side grows quadratically with t and the right grows linearly.

Lemma 3 *Suppose $ch(E^{\cdot}) = c$ and $E^{\cdot}$ is α_t-stable for all $t > t_0$. Then*

(i) $\mathrm{H}^{-1}(E^{\cdot}) = 0$, *i.e.,* $E^{\cdot} = E = \mathrm{H}^0(E^{\cdot})$ *is a coherent sheaf.*

(ii) E *is Mumford* H'*-semi-stable.*

(iii) E *is Gieseker* H'*-semi-stable.*

Proof Let $F = H^{-1}(E^{\cdot})$. For $F[1]$ to belong to the tilted category $\mathcal{A}_{\alpha_t}$, it is required that F be a torsion-free sheaf and that

$$\langle F, \alpha_t \cdot H\rangle = \langle c_1(F), H\rangle + \Big\langle \mathrm{ch}_0(F), \Big(-\frac{K_S}{2} + tH'\Big) \cdot H\Big\rangle \leq 0.$$

But this is positive when t is sufficiently large, proving (i).

Next, suppose E is H'-unstable, i.e., that there is an $F \subset E$ such that

$$\frac{\langle c_1(F), H'\rangle}{\mathrm{ch}_0(F)} > \frac{\langle c_1(E), H'\rangle}{\mathrm{ch}_0(E)}.$$

Then

$$\langle \mathrm{ch}(F), \alpha_t\rangle = t\left[-\frac{\mathrm{ch}_0(F)\langle H', c_1(E)\rangle}{\mathrm{ch}_0(E)} + \langle c_1(F), H'\rangle\right] + \text{constant}$$

and this is positive when t is large. But $\langle \mathrm{ch}(E), \alpha_t\rangle = 0$ by construction, so $\mu_{\alpha_t}(F) > 0 = \mu_{\alpha_t}(F)$ when t is large. Similarly, if E is Gieseker H'-unstable, then either it is Mumford unstable (already done) or else there is an $F \subset E$ with $\mu_H(F) = \mu_H(E)$ and $\frac{\chi(S,F)}{ch_0(F)} > \frac{\chi(S,E)}{ch_0(E)}$. In this case the linear term in $\langle \mathrm{ch}(F), \alpha_t\rangle$ vanishes, but the constant term is

$$-\frac{\mathrm{ch}_0(F)}{\mathrm{ch}_0(E)}\langle \mathrm{ch}(E), \mathrm{td}(S)\rangle + \langle \mathrm{ch}(F), \mathrm{td}(S)\rangle > 0$$

by the Riemann–Roch theorem. □

Remark There is little dependence here upon the choice of H. In fact, because $\langle \mathrm{ch}(E^{\cdot}), \alpha_t\rangle = 0$ for all t, it follows that $E^{\cdot}$ is α_t-stable if and only if $\langle \mathrm{ch}(F^{\cdot}), \alpha_t\rangle < 0$ for all $F^{\cdot} \subset E^{\cdot}$. The **only** dependence upon H is in the categories $\mathcal{A}_{\alpha_t}$ in which the inclusions $F^{\cdot} \subset E^{\cdot}$ take place. Since these categories $\mathcal{A}_{\alpha_t}$ eventually include any given coherent sheaf and exclude the shift of any given coherent sheaf, it is unsurprising that H disappears in the limit.

A converse to the lemma is also easy:

Lemma 4 *Suppose E is a Gieseker H'-semi-stable torsion-free sheaf with* $\mathrm{ch}(E) = c$ *and* $E \in \mathcal{A}_{\alpha_{t_0}}$. *Any α_{t_0}-destabilizing subobject $F^{\cdot} \subset E$ will fail to destabilize E for large t.*

Proof From the long exact sequence in cohomology, it follows that a subobject $F^{\cdot} \subset E$ is a coherent sheaf (though it may not be a subsheaf). If $Q^{\cdot} = E/F$ is the quotient in $\mathcal{A}_{t_0}$, then

$$0 \to \mathrm{H}^{-1}(Q^{\cdot}) \to F \to E \to \mathrm{H}^0(Q^{\cdot}) \to 0$$

is a long exact sequence of coherent sheaves. Now, as in the previous lemma, if $\mathrm{H}^{-1}(Q^{\cdot}) \neq 0$ then it is eventually not in the category $\mathcal{A}_{\alpha_t}$, and the inclusion $F \subset E$ is destroyed (even though both coherent sheaves are in the category $\mathcal{A}_{\alpha_t}$). On the contrary, if $\mathrm{H}^{-1}(Q^{\cdot}) = 0$ then $F \subset E$ is a subsheaf, and it follows that the Gieseker slope of F is either smaller than that of E, in which case the α_t-slope is also eventually smaller, or else they are equal, in which case $\langle \mathrm{ch}(F), \alpha_t \rangle = 0$ for **all** t and $F \subset E$ was not a destabilizing subobject at $t = t_0$. □

This is evidence that the coarse moduli spaces $\mathcal{M}_{H'}(c)$ of Gieseker semi-stable sheaves should be the moduli of α_t-semi-stable objects of $\mathcal{A}_{\alpha_t}$ for large t. More evidence is also available in the positivity of the determinant line bundle on the moduli space $\mathcal{M}_{H'}(c)$:

Theorem ([1]) *The determinant line bundles Δ_F for large t and F semi-stable with* $\mathrm{ch}(F)\cdot td(S) = \lambda\alpha_t$ *are positive on families of Gieseker H'-stable sheaves, nef on families of semi-stable sheaves and descend to a positive line bundle on the coarse moduli space $\mathcal{M}_{H'}(c)$.*

This is, at least, consistent with the theorem of Bayer–Macrí [7], which comes to the same conclusion for the determinant line bundle on the moduli of α_t-stable objects. It is hard to see how to make this into a proof, however, without uniform versions of the lemmas. That is, we are faced with the following:

Problems (a) How to show that there is a uniform bound T for all H'-semi-stable sheaves E such that for $t > T$, all "objections" to α_t-stability (in the form of nontrivial Harder–Narasimhan filtrations) disappear?

(b) How to show that there is a uniform bound T such that for all $t > T$, **all** objects other than semi-stable sheaves fail to be α_t-stable?

In special cases, these problems have been solved (e.g., [3, 4, 11] and notably [8], in which essentially everything is done for K3 surfaces), but there is not (to my knowledge) a one-size-fits-all-surfaces solution.

To sum up, we have an attractive picture in these α coordinates.

The picture For each Chern class $c = \mathrm{ch}(E)$, there is a base point

$$\alpha_0 = \left(1, -\frac{1}{2}K_S, d_0\right) \in c^\perp \cap \{\alpha_0 = 1\}$$

in the affine space, from which rays emanate in ample directions

$$\alpha_t := \left(1, -\frac{1}{2}K_S + tH', d_0 - t\left(\frac{c_1 \cdot H'}{c_0}\right)\right);\ t \geq 0$$

with the following properties:

(i) Each ray eventually enters the stability manifold.
(ii) Gieseker H'-semi-stable sheaves are eventually α_t-semi-stable.
(iii) Everything else is eventually α_t-unstable.

And it is expected that:

(iv) The moduli of α_t-stable sheaves for $t > T$ coincide with $\mathcal{M}_{H'}(c)$.

In any case, Gieseker H'-semi-stable sheaves are identified with rays in the affine space $c^\perp \cap \{\alpha_0 = 1\}$ emanating from the base point α_0. The choice of base point is actually quite important. If, for example, H' is a critical value for Gieseker moduli, in which there are strictly semi-stable sheaves although the invariants $c = (c_0, c_1, c_2)$ are primitive (meaning that for "nearby" ample classes H, semi-stability and stability coincide), then we expect to detect the intermediate variations in moduli, discovered by Matsuki and Wentworth [22], by varying the base point while continuing to point the ray in the direction H'.

4 The Serre map

One case remains, namely that of Gieseker-stable torsion sheaves on a surface S. Several examples have been studied recently in the context of Bridgeland stability, but I want to focus on one case, namely that of sheaves $\mathcal{E}$ with the following invariants:

$$c_1(\mathcal{E}) = 2H - K_S,\ \ \chi(S, \mathcal{E}) = 0$$

where H is an ample divisor class (for which $2H - K_S$ is effective).

One example of such sheaves are the quotients of maps

$$0 \to L^{-1} \otimes \omega_S \to L \to \mathcal{E} \to 0$$

when $L = \mathcal{O}_S(H)$ and $\omega_S = \mathcal{O}_S(K_S)$.

As before, we consider stability conditions α_t along a ray:

$$\alpha_t = \left(1, -\frac{1}{2}K_S, -t\right) \in c^\perp \cap \{\alpha_0 = 1\}$$

although in this case the variation is entirely concentrated in $H^4(S, \mathbb{Q})$. There are several interesting properties of these stability conditions α_t:

(i) The ray enters the stability manifold for some $t \le \frac{1}{8}K_S^2$.
(ii) The tilted categories $\mathcal{A}_{\alpha_t}$ are constant, since

$$\langle \mathrm{ch}(E), \alpha_t \cdot H\rangle = \mathrm{ch}_0(E)\left(-\frac{1}{2}K_S \cdot H\right) + \mathrm{ch}_1(E) \cdot H$$

is independent of t. Recall that extensions $0 \to F[1] \to E^{\cdot} \to T \to 0$ give the elements of $\mathcal{A}_{\alpha_t}$, where the semi-stable pieces of the Harder–Narasimhan filtration of F satisfy $\frac{c_1(E)\cdot H}{rk(E)} \le \frac{1}{2}K_S \cdot H$ and those of the sheaf T (other than the torsion) satisfy $\frac{c_1(E)\cdot H}{rk(F)} > \frac{1}{2}K_S \cdot H$.

In particular, there is a Serre map for surfaces, analogous to the family of vector bundles on a curve parametrized by

$$\mathbb{P}(\mathrm{Ext}^1(L, \omega_C)) = \mathbb{P}(\mathrm{H}^0(C, L)^*)$$

via the "lines" $\{\lambda\epsilon : 0 \to \omega_C \to E \to L \to 0; \lambda \in \mathbb{C}^*\}$ of extensions. Recall that the generic such extension gives a **stable** vector bundle of rank two provided that $\deg(L) > \deg(K_C)$.

On a surface, a family of objects of $\mathcal{A}_{\alpha_t}$ is parametrized by

$$\mathbb{P}(\mathrm{Ext}^2(L, L^{-1} \otimes \omega_S)) = \mathbb{P}(\mathrm{H}^0(S, L^{\otimes 2}))$$

via the "lines" $\{\lambda\epsilon : 0 \to L^{-1} \otimes \omega_S[1] \to E^{\cdot} \to L \to 0; \lambda \in \mathbb{C}^*\}$ of extensions provided that $\langle H, H\rangle > \langle \frac{1}{2}K_S, H\rangle$. Each value of t gives a *different* stability criterion for the objects $E^{\cdot}$. By the computation $\mu_{\alpha_t}(L) = 0 = \mu_{\alpha_t}(K_S - L) \Leftrightarrow t_1 = \langle H - K_S, H\rangle$ it follows that if L and $L^{-1} \otimes \omega_S[1]$ are both stable at t_1 (a nontrivial assertion! [5]), then

- each of the $E^{\cdot}$ parametrized by $\epsilon \in \mathbb{P}(\mathrm{Ext}^2(L, L^{-1} \otimes \omega_S))$ and
- each of the sheaves $\mathcal{E}$ parametrized by $\delta \in \mathbb{P}(\mathrm{Hom}(L^{-1} \otimes \omega_S, L))$

are strictly semi-stable objects, with Jordan–Hölder filtrations

$$0 \to L^{-1} \otimes \omega_S[1] \to E^{\cdot} \to L \to 0 \text{ and } 0 \to L \to \mathcal{E} \to L^{-1} \otimes \omega_S[1] \to 0$$

respectively. This suggests that for $t_2 < t < t_1$, a general extension ϵ will parametrize an α_t-stable object of $\mathcal{A}_{\alpha_t}$, while for $t > t_1$ one suspects (and it is proved in many cases) that already the moduli of Gieseker-stable sheaves coincides with the moduli of α_t-stable objects, i.e., that T is the uniform bound sought in the previous section. My student, Christian Martinez, has recently proven that the closure of the image of the Serre map has a particularly nice property:

Theorem 5 ([21]) *The involution on the derived category*

$$E^{\cdot} \mapsto (E^{\cdot})^{\vee} \otimes \omega_S[1]$$

induces an involution on the moduli of α_t-stable objects of Chern class $c = \mathrm{ch}(L) - \mathrm{ch}(L^{-1} \otimes \omega_S)$ and this involution fixes objects in the image of the Serre map.

References

[1] Álvarez-Cónsul, L. and King, A. A functorial construction of moduli of sheaves. *Invent. Math.* **168** (2007) 613–666.

[2] Arcara, E. and Bertram, A. Reider's theorem and Thaddeus pairs revisited. In *Grassmannians, Moduli Spaces and Vector Bundles*, D. Ellwood and E. Previato (eds). *Clay Proceedings*, Vol. 14 (2011), pp. 51–68.

[3] Arcara, D., Bertram, A., Coskun, I. and Huizenga, J. The minimal model program for the Hilbert schemes of points in $\mathbb{P}^2$ and Bridgeland stability. *Adv. Math.* **235** (2013) 580–626.

[4] Arcara, D. and Bertram, A. (with an appendix by Max Lieblich). Bridgeland-stable moduli spaces for K-trivial surfaces. *J. Eur. Mat. Soc. (JEMS)* **15** (2013) 1–38.

[5] Arcara, D. and Miles, E. Forthcoming.

[6] Bayer, A. and Macrí, E. The space of stability conditions on the local projective plane. *Duke Math. J.* **160** (2011) 263–322.

[7] Bayer, A. and Macrí, E. Projectivity and birational geometry of Bridgeland moduli spaces. arXiv:1203.4613, to appear in *J. Am. Math. Soc.*

[8] Bayer, A. and Macrí, E. MMP for moduli of sheaves on K3s via wall-crossing: nef and movable cones, Lagrangian fibrations. arXiv:1301.6968.

[9] Bayer, A., Macrì, E. and Toda, Y. Bridgeland stability conditions on threefolds I: Bogomolov–Gieseker type inequalities. *J. Alg Geom.* **23** (2014) 117–163.

[10] Bayer, A., Bertram, A., Macrì, E. and Toda, Y. Bridgeland stability conditions on threefolds II: An application to Fujita's conjecture. arXiv:1106.3430, to appear in *J. Alg. Geom.*

[11] Bertram, A. and Coskun, I. The birational geometry of the Hilbert scheme of points on surfaces. In *Birational Geometry, Rational Curves and Arithmetic*. Simons Symposia, Springer (2013), pp. 15–55.

[12] Bertram, A., Martinez, C. and Wang, J. The birational geometry of moduli spaces of sheaves on the projective plane. arXiv:1301.2011, to appear in *Geom. Ded.*

[13] Bogomolov, F. A. Holomorphic tensors and vector bundles on projective manifolds. *Izv. Akad. Nauk SSSR Ser. Mat.* **42**(6) (1978) 1227–1287, 1439 (in Russian).

[14] Bridgeland, T. Stability conditions on triangulated categories. *Ann. Math.* **166**(2) (2007) 317–345.

[15] Bridgeland, T. Stability conditions on K3 surfaces. *Duke Math. J.* **141**(2) (2008) 241–291.

[16] Huybrechts, D. and Lehn, M. *The Geometry of Moduli Spaces of Sheaves.* Cambridge: Cambridge University Press, 1997.

[17] LePotier, J. *Lectures on Vector Bundles.* Cambridge Studies in Advanced Mathematics, Vol. 54. Cambridge: Cambridge University Press, 1997.

[18] Maciocia, A. and Piyaratne, D. Fourier–Mukai transforms and Bridgeland stability conditions on abelian threefolds. arXiv:1304.3887.

[19] Maciocia, A. and Piyaratne, D. Fourier–Mukai transforms and Bridgeland stability conditions on abelian threefolds II. arXiv:1310.0299.

[20] Macrì, E. A generalized Bogomolov–Gieseker inequality for the three-dimensional projective space. rXiv:1207.4980.

[21] Martinez, C. Duality, Bridgeland wall crossing and flips of secant varieties. arXiv:1311.1183

[22] Matsuki, K. and Wentworth, R. Mumford–Thaddeus principle on the moduli space of vector bundles on an algebraic surface. *Int. J. Math.* **8** (1997) 97–148.

3

Valuation spaces and multiplier ideals on singular varieties

S. Boucksom[a]
CNRS – de Jussieu Institut de Mathématiques

T. de Fernex[b]
University of Utah

C. Favre[c]
CNRS – École Polytechnique

S. Urbinati
Università degli Studi di Padova

Abstract

We generalize to all normal complex algebraic varieties the valuative characterization of multiplier ideals due to Boucksom–Favre–Jonsson in the smooth case. To that end, we extend the log discrepancy function to the space of all real valuations, and prove that it satisfies an adequate properness property, building upon previous work by Jonsson and Mustaţă. We next give an alternative definition of the concept of numerically Cartier divisors previously introduced by the first three authors, and prove that numerically $\mathbb{Q}$-Cartier divisors coincide with $\mathbb{Q}$-Cartier divisors for rational singularities. These ideas naturally lead to the notion of numerically $\mathbb{Q}$-Gorenstein varieties, for which our valuative characterization of multiplier ideals takes a particularly simple form.

Dedicated to Robert Lazarsfeld on the occasion of his 60th birthday

[a] Supported by ANR projects MACK and POSITIVE.
[b] Supported by NSF CAREER grant DMS-0847059 and the Simons Foundation.
[c] Supported by the ERC-starting grant project "Nonarcomp" no. 307856.

1 Introduction

Multiplier ideal sheaves are a fundamental tool both in complex algebraic and complex analytic geometry. They provide a way to approximate a "singularity data," which can take the form of a (coherent) ideal sheaf, a graded sequence of ideal sheaves, a plurisubharmonic function, a nef b-divisor, etc., by a coherent ideal sheaf satisfying a powerful cohomology vanishing theorem. For the sake of simplicity, we will focus on the case of ideals and graded sequences of ideals in the present paper.

On a smooth (complex) algebraic variety X, the very definition of the multiplier ideal sheaf $\mathcal{J}(X, \mathfrak{a}^c)$ of an ideal sheaf $\mathfrak{a} \subset \mathcal{O}_X$ with exponent $c > 0$ is valuative in nature: a germ $f \in \mathcal{O}_X$ belongs to $\mathcal{J}(X, \mathfrak{a}^c)$ iff it satisfies

$$\nu(f) > c\nu(\mathfrak{a}) - A_X(\nu)$$

for all divisorial valuations ν, i.e., all valuations of the form $\nu = \mathrm{ord}_E$ (up to a multiplicative constant) with E a prime divisor on a birational model X', proper over X. Further, it is enough to test these conditions with X' a fixed log resolution of $\mathfrak{a}$ (which shows that $\mathcal{J}(X, \mathfrak{a}^c)$ is coherent, as the direct image of a certain coherent fractional ideal sheaf on X'). Here we have set as usual $\nu(\mathfrak{a}) := \min_{f \in \mathfrak{a}_x} \nu(f)$ with $x = c_X(\nu)$ the center of ν in X, and

$$A_X(\nu) := 1 + \mathrm{ord}_E\,(K_{X'/X})$$

is the *log discrepancy* (with respect to X) of the divisorial valuation ν.

The multiplier ideal sheaf $\mathcal{J}(X, \mathfrak{a}_\bullet^c)$ of a graded sequence of ideal sheaves $\mathfrak{a}_\bullet = (\mathfrak{a}_m)_{m \in \mathbb{N}}$ is defined as the stationary value of $\mathcal{J}(X, \mathfrak{a}_m^{c/m})$ for m large and divisible, but a direct valuative characterization was provided in [6] in the 2-dimensional case, in [3] for all non-singular varieties, and in [7, 8] for the general case of regular excellent noetherian $\mathbb{Q}$-schemes. More specifically, for each divisorial valuation ν, subadditivity of $m \mapsto \nu(\mathfrak{a}_m)$ allows us to define

$$\nu(\mathfrak{a}_\bullet) := \lim_{m \to \infty} m^{-1}\nu(\mathfrak{a}_m) = \inf_{m \geq 1} m^{-1}\nu(\mathfrak{a}_m)$$

in $[0, +\infty)$. By [3], a germ $f \in \mathcal{O}_X$ belongs to $\mathcal{J}(X, \mathfrak{a}_\bullet^c)$ iff there exists $0 < \varepsilon \ll 1$ such that

$$\nu(f) \geq (c + \varepsilon)\nu(\mathfrak{a}_\bullet) - A_X(\nu)$$

for all divisorial valuations ν.[1] In other words, the latter condition is shown to imply the existence of $m \gg 1$ such that $\nu(f) > cm^{-1}\nu(\mathfrak{a}_m) - A_X(\nu)$ for all divisorial valuations ν.

[1] In statements of this kind, if f is a germ at $x \in X$, it is implicitly understood that we are only considering those ν such that $x \in \overline{\{c_X(\nu)\}}$, so that we can make sense of $\nu(f)$.

The definition of multiplier ideals was extended to the case of an arbitrary normal algebraic variety X in [5]. If Δ is an effective $\mathbb{Q}$-Weil divisor on X such that $K_X + \Delta$ is $\mathbb{Q}$-Cartier (i.e., an *effective* $\mathbb{Q}$*-boundary* in MMP terminology), the log discrepancy function $A_{(X,\Delta)}$ is a by now classical object (see, e.g., [11]). It allows us to define the multiplier ideal sheaf $\mathcal{J}((X,\Delta);\mathfrak{a}^c)$ just as before for an ideal sheaf $\mathfrak{a} \subset \mathcal{O}_X$, and then $\mathcal{J}((X,\Delta);\mathfrak{a}_\bullet^c)$ for a graded sequence of ideals $\mathfrak{a}_\bullet$ as the largest element in the family $\mathcal{J}((X,\Delta);\mathfrak{a}_m^{c/m})$. It is proven in [5] that there is a unique maximal element in the family of ideals $\mathcal{J}((X,\Delta);\mathfrak{a}_\bullet^c)$ with Δ ranging over all effective $\mathbb{Q}$-boundaries, which coincides with the multiplier ideal $\mathcal{J}(X,\mathfrak{a}_\bullet^c)$ as defined in [5].

Note in particular that $\mathcal{J}(X,\mathcal{O}_X) = \mathcal{O}_X$ iff there exists an effective $\mathbb{Q}$-boundary Δ such that the pair (X,Δ) is klt – which simply means that X itself is log terminal when X is already $\mathbb{Q}$-Gorenstein (i.e., when K_X is $\mathbb{Q}$-Cartier).

In order to give a direct valuative description of these generalized multiplier ideals, one first needs to provide an adequate notion of log discrepancy for a divisorial valuation. As in [5], this is done by setting for $v = \mathrm{ord}_E$ with E a prime divisor on a birational model X' proper over X,

$$A_X(v) := 1 + \mathrm{ord}_E(K_{X'}) - \lim_{m\to\infty} m^{-1}\,\mathrm{ord}_E\,\mathcal{O}_X(-mK_X),$$

where K_X is now an actual canonical Weil divisor on X (as opposed to a linear equivalence class), $K_{X'}$ is the corresponding canonical Weil divisor on X', and $\mathcal{O}_X(-mK_X)$ is viewed as a fractional ideal sheaf on X. The definition is easily seen to be independent of the choices made.

Our first main result is as follows:

Theorem 1.1 *Let $\mathfrak{a}_\bullet$ be a graded sequence of ideal sheaves on a normal algebraic variety, and pick $c > 0$. For every closed subscheme $N \subset X$ containing both* Sing(X) *and the zero locus of $\mathfrak{a}_1$ (and hence of $\mathfrak{a}_m$ for all m) and every $0 < \varepsilon \ll 1$, we have*

$$\mathcal{J}(X,\mathfrak{a}_\bullet^c) = \{f \in \mathcal{O}_X \mid v(f) \geq c\,v(\mathfrak{a}_\bullet) - A_X(v) + \varepsilon v(\mathcal{I}_N)\quad \textit{for all divisorial valuations } v\},$$

with $\mathcal{I}_N \subset \mathcal{O}_X$ denoting the ideal sheaf defining N.

The key point in our approach is to construct an appropriate extension of A_X to the space Val_X of all *real* valuations on X, and to prove that it satisfies an adequate properness property, building upon [8]. Once this is done, the last ingredient is a variant of Dini's lemma. The argument will also prove:

Theorem 1.2 *Let $\mathfrak{a}_\bullet$ be a graded sequence of ideal sheaves on a normal algebraic variety, and pick $c > 0$. Then*

$$\mathcal{J}(X, \mathfrak{a}_\bullet^c) = \{f \in \mathcal{O}_X \mid v(f) > cv(\mathfrak{a}_\bullet) - A_X(v) \text{ for all real valuations } v\}.$$

In the special case $\mathfrak{a}_\bullet = \mathcal{O}_X$, this last result shows that there exists an effective $\mathbb{Q}$-boundary Δ with (X, Δ) klt iff $A_X(v) > 0$ for all nontrivial real valuations v. When X admits an effective $\mathbb{Q}$-boundary Δ with (X, Δ) log canonical, one easily sees that $A_X \geq 0$. However, the converse already fails in dimension three, as was shown recently by Yuchen Zhang for a normal isolated cone singularity [16].

In the last part of this paper we will provide an alternative approach to the notion of *numerically Cartier divisors* introduced in [2]. A Weil divisor on X is said to be numerically ($\mathbb{Q}$-)Cartier if it is the push-forward of a π-numerically trivial ($\mathbb{Q}$-)divisor for some (equivalently, any) resolution of singularities $\pi\colon X' \to X$. This naturally leads to the definition of a group of *numerical divisor classes* $\mathrm{Cl}_{\mathrm{num}}(X)$, defined as the quotient of the group of Weil divisors by numerically Cartier divisors. We prove that the abelian group $\mathrm{Cl}_{\mathrm{num}}(X)$ is always finitely generated. The $\mathbb{Q}$-vector space $\mathrm{Cl}_{\mathrm{num}}(X)_{\mathbb{Q}}$ is trivial when X is either $\mathbb{Q}$-factorial or has dimension two, thanks to Mumford's numerical pull-back. Building on an argument of Kawamata, we further prove that every numerically $\mathbb{Q}$-Cartier divisor is already $\mathbb{Q}$-Cartier when X has rational singularities.

We say that X is *numerically $\mathbb{Q}$-Gorenstein* when K_X is numerically $\mathbb{Q}$-Cartier. This means that for some (equivalently, any) resolution of singularities $\pi\colon X' \to X$, $K_{X'}$ is π-numerically equivalent to a π-exceptional $\mathbb{Q}$-divisor, which is necessarily unique, and denoted by $K^{\mathrm{num}}_{X'/X}$. It is related to the log discrepancy function by

$$A_X(\mathrm{ord}_E) = 1 + \mathrm{ord}_E\left(K^{\mathrm{num}}_{X'/X}\right)$$

for all prime divisors $E \subset X'$. As a consequence of Theorem 1.1, we show:

Theorem 1.3 *Assume that X is numerically $\mathbb{Q}$-Gorenstein, and let $\mathfrak{a} \subset \mathcal{O}_X$ be an ideal sheaf. Let also $\pi\colon X' \to X$ be a log resolution of $(X, \mathfrak{a})$, so that $\pi^{-1}\mathfrak{a} \cdot \mathcal{O}_{X'} = \mathcal{O}_{X'}(-D)$ with D an effective Cartier divisor. For each exponent $c > 0$ we then have*

$$\mathcal{J}(\mathfrak{a}^c) = \pi_*\mathcal{O}_{X'}\left(\lceil K^{\mathrm{num}}_{X'/X} - cD\rceil\right).$$

In dimension two, this result says that the multiplier ideals introduced in [5] agree with the numerical multiplier ideals defined using Mumford's numerical pull-back.

Since the underlying variety of any klt pair has rational singularities, Theorem 1.3 applied to $\mathfrak{a} = \mathcal{O}_X$ yields:

Corollary 1.4 *Let X be a normal algebraic variety. The following conditions are equivalent:*

(a) X is $\mathbb{Q}$-Gorenstein and log terminal;
(b) X is numerically $\mathbb{Q}$-Gorenstein and $A_X(\mathrm{ord}_E) > 0$ for all prime divisors E on some (equivalently, any) log resolution X' of X.

2 Valuation spaces

Throughout this paper we work over the field $\mathbb{C}$ of complex numbers. In this section we review some properties of valuation spaces, mostly following [8].

2.1 The space of real valuations

Let X be an algebraic variety. By a *valuation on X* we mean a *real-valued* valuation v on the function field of X that is trivial on the base field and admits a center on X. Recall that the latter is characterized as the unique scheme point $c_X(v) = \xi \in X$ such that $v \geq 0$ on $\mathcal{O}_{X,\xi}$ and $v > 0$ on its maximal ideal. We denote by Val_X the space of valuations on X, endowed with the topology of pointwise convergence.

The *trivial valuation*, which is identically zero on all nonzero rational functions, is the unique valuation on X centered at its generic point. Following [8], we denote by

$$\mathrm{Val}_X^* \subset \mathrm{Val}_X$$

the set of nontrivial valuations. Mapping a valuation to its center defines an *anticontinuous*[2] map

$$c_X\colon \mathrm{Val}_X \to X.$$

Every prime divisor E *over* X (i.e., in a normal birational model X', proper over X) determines a valuation $\mathrm{ord}_E \in \mathrm{Val}_X$ given by the order of vanishing at the generic point of E. A *divisorial valuation* is a valuation of the form $v = c\,\mathrm{ord}_E$ for some prime divisor E over X and some $c \in \mathbb{R}_+^*$. We denote by

$$\mathrm{DivVal}_X \subset \mathrm{Val}_X$$

the set of divisorial valuations.

[2] That is, the inverse image of an open subset is closed.

2.2 Normalized valuation spaces

For every (coherent) ideal sheaf $\mathfrak{a} \subset \mathcal{O}_X$ and every $v \in \mathrm{Val}_X$, we set as usual

$$v(\mathfrak{a}) := \min \{ v(f) \mid f \in \mathfrak{a}_{c_X(v)} \} \in [0, +\infty).$$

Definition 2.1 A *normalizing subscheme* is a (nontrivial) closed subscheme of X containing $\mathrm{Sing}(X)$. The *normalized valuation space* defined by N is

$$\mathrm{Val}_X^N := \{ v \in \mathrm{Val}_X \mid v(\mathcal{I}_N) = 1 \},$$

with $\mathcal{I}_N$ denoting the ideal sheaf defining N.

Note that

$$\mathbb{R}_+^* \cdot \mathrm{Val}_X^N = \{ v \in \mathrm{Val}_X \mid v(\mathcal{I}_N) > 0 \} = c_X^{-1}(N),$$

which is thus open in Val_X and only depends on the Zariski closed set N_{red}. We also clearly have

$$\mathrm{Val}_X^* = \bigcup_{N \subset X} \mathbb{R}_+^* \cdot \mathrm{Val}_X^N, \tag{2.1}$$

with N ranging over all normalizing subschemes.

The point of introducing this terminology is that the normalized valuation space Val_X^N admits a simple description as a limit of simplicial complexes.

Definition 2.2 A *good resolution* of a normalizing subscheme $N \subset X$ is a proper birational morphism $\pi \colon X_\pi \to X$ such that

- X_π smooth;
- π is an isomorphism over $X \setminus N$;
- $\pi^{-1}(N) \supset \mathrm{Exc}(\pi)$ both have pure codimension one, and $\pi^{-1}(N)_{\mathrm{red}}$ is a simple normal crossing divisor $\sum_{i \in I} E_i$ such that $E_J := \bigcap_{j \in J} E_j$ is irreducible (or empty) for all $J \subset I$.

Let π be a good resolution of N, and assume that E_J as above is non-empty. At its generic point η_J, the normal crossing condition guarantees that any choice of local equations $z_j \in \mathcal{O}_{X_\pi, \eta_J}$ for E_j, $j \in J$ yields a regular system of parameters. By Cohen's theorem we thus have

$$\widehat{\mathcal{O}}_{X_\pi, \eta_J} \cong \mathbb{C}(\eta_J)[[z_j, j \in J]],$$

with $\mathbb{C}(\eta_J)$ denoting the residue field at η_J. To every weight $w = (w_j)_{j \in J} \in \mathbb{R}_+^J$, we associate the monomial valuation v_w defined by

$$v_w \left(\sum_{\alpha \in \mathbb{N}^J} a_\alpha z^\alpha \right) := \min \left\{ \sum_i w_i \alpha_i \mid a_\alpha \neq 0 \right\}. \tag{2.2}$$

Viewed as a valuation on X, v_w is called a *quasi-monomial valuation*. The construction is independent of the choice of the local equations z_j, $j \in J$ (see [8, Sections 3–4] for more details).

Note that $v_w \in \mathrm{Val}_X^N$ iff $\sum_{j \in J} w_j \,\mathrm{ord}_{E_j}(N) = 1$. If we denote by $\Delta_\pi^N \subset \mathrm{Val}_X^N$ the set of all normalized quasi-monomial valuations so obtained, then Δ_π^N is a geometric realization of the *dual complex* of $\sum_i E_i$, i.e., the simplicial complex whose vertices are in bijection with I and that contains one simplicial face σ_J joining all vertices $j \in J$ for any subset $J \subset I$ such that $E_J \neq \emptyset$.

Further, there is a natural continuous retraction

$$r_\pi^N \colon \mathrm{Val}_X^N \to \Delta_\pi^N,$$

defined by letting $r_\pi^N(v)$ be the unique monomial valuation taking the value $v(E_j)$ on E_j. Note that $r_\pi^N(v)$ belongs to the relative interior of the face σ_J, with

$$J := \left\{ j \in I \mid c_{X_\pi}(v) \in E_j \right\}.$$

If π' factors through π (in which case we write $\pi' \geq \pi$), then there is a natural inclusion $\Delta_\pi^N \hookrightarrow \Delta_{\pi'}^N$. We then have:

Theorem 2.3 ([1, 15])

$$\mathrm{Val}_X^N = \overline{\bigcup_\pi \Delta_\pi^N},$$

where π runs over all good resolutions of N. More precisely, $\lim_\pi r_\pi^N(v) = v$ *for each $v \in$* Val_X.

A subset $\sigma \subset \mathrm{Val}_X^N$ is said to be a *face* if σ is a face of Δ_π^N for some π. A face of Δ_π^N is also called a π-*face*, and it can be endowed with a canonical affine structure induced from Δ_π^N. A real-valued function on Val_X^N is said to be *affine* (resp., *convex*) on a face if it is so in terms of the variable w as in (2.2). We say that a property holds *on small faces* if there exists π such that the property holds on the faces of $\Delta_{\pi'}$ for all $\pi' \geq \pi$.

2.3 Functions defined by ideals

Proposition 2.4 *Let $N \subset X$ be a normalizing subscheme and $\mathfrak{a} \subset \mathcal{O}_X$ be a nonzero coherent ideal sheaf. Then:*

(a) $v \mapsto v(\mathfrak{a})$ is a continuous function $\mathrm{Val}_X \to [0, +\infty)$*, and concave on each face of* Val_X^N*;*

(b) for each good resolution π of N we have $r_\pi^N(v)(\mathfrak{a}) \geq v(\mathfrak{a})$ on Val_X^N.

If N further contains the zero locus of $\mathfrak{a}$, and π is a good resolution of N dominating the blow-up of $\mathfrak{a}$, then $v \mapsto v(\mathfrak{a})$ is affine on the faces of Δ_π^N and $r_\pi^N(v)(\mathfrak{a}) = v(\mathfrak{a})$ on Val_X^N. In particular, $v \mapsto v(\mathfrak{a})$ is bounded on Val_X^N.

Proof The proof is basically contained in [3, 8]. We briefly recall the argument. Using the same notation introduced in Section 2.2, a valuation corresponding to a point in a face σ of some Δ_π^N is parametrized by $w = (w_j)_{j\in J}$ with $w_j \geq 0$ and $\sum_{j\in J} w_j \operatorname{ord}_{E_j}(N) = 1$. For every local function f on X we have $v_w(f) = \min\{\sum_i w_i\alpha_i \mid a_\alpha \neq 0\}$ with $f \circ \pi = \sum_\alpha a_\alpha z^\alpha$. Since $w \mapsto v_w(h)$ is the minimum of a collection of affine functions, it is concave. It follows that $v(\mathfrak{a}) = \min\{v(f) \mid f \in \mathfrak{a}\}$ is a concave function of $v \in \sigma$. Moreover, if N contains the zero locus of $\mathfrak{a}$ and π dominates the blow-up of $\mathfrak{a}$, then this function is affine on σ.

Finally, applying [8, Lemma 4.7] to a fixed resolution of singularities of X shows that $r_\pi^N(v)(\mathfrak{a}) \geq v(\mathfrak{a})$ for each valuation v, with equality if π dominates the blow-up of $\mathfrak{a}$. This concludes the proof. □

More generally, recall that a *graded sequence of ideals* $\mathfrak{a}_\bullet = (\mathfrak{a}_m)_{m\geq 0}$ is a sequence of coherent ideal sheaves such that $\mathfrak{a}_m \cdot \mathfrak{a}_n \subset \mathfrak{a}_{m+n}$ for all m, n. We will always assume that $\mathfrak{a}_1 \neq 0$, and hence $\mathfrak{a}_m \neq 0$ for all $m \geq 1$. Since $v(\mathfrak{a}_m)$ is a subadditive sequence for each $v \in \mathrm{Val}_X$, we can set

$$v(\mathfrak{a}_\bullet) = \lim_{m\to\infty} m^{-1} v(\mathfrak{a}_m) = \inf_{m\geq 1} m^{-1} v(\mathfrak{a}_m).$$

Proposition 2.4 generalizes to

Proposition 2.5 *Let $N \subset X$ be a normalizing subscheme and let π be a good resolution of N. For any graded sequence of ideal sheaves $\mathfrak{a}_\bullet = (\mathfrak{a}_m)_{m\geq 0}$, $v \mapsto v(\mathfrak{a}_\bullet)$ defines an upper semicontinuous function $\mathrm{Val}_X \to [0, +\infty)$ such that*

(a) $v \mapsto v(\mathfrak{a}_\bullet)$ is concave and continuous on each face of Val_X^N;
(b) $r_\pi^N(v)(\mathfrak{a}_\bullet) \geq v(\mathfrak{a}_\bullet)$ on Val_X^N for each good resolution π of N.

Furthermore, if N contains the zero locus of $\mathfrak{a}_1$ (or, equivalently, of $\mathfrak{a}_m$ for all m), then $v \mapsto v(\mathfrak{a}_\bullet)$ is also bounded on Val_X^N.

Proof Only the continuity on the faces is not a direct consequence of Proposition 2.4. But since each face σ of Val_X^N is a simplex, it follows from an elementary fact in convex analysis [13, Theorem 10.2] that $v \mapsto v(\mathfrak{a})$, being concave and usc, is automatically continuous on σ. □

Remark 2.6 As a consequence of [4, Theorem B], one can show the following uniform Lipschitz property: assume that a given face σ of Val_X^N has the property that the closure $Z \subset X$ of the center of some (equivalently, any) valuation of the relative interior of σ is proper over $\mathbb{C}$. Then there exists $C > 0$ such that for any graded sequence of ideal sheaves $\mathfrak{a}_\bullet$ the function $\nu \mapsto \nu(\mathfrak{a}_\bullet)$ is Lipschitz continuous on σ with Lipschitz constant $\leq C \operatorname{ord}_Z(\mathfrak{a}_\bullet)$.

3 The log discrepancy function

Throughout this section, X denotes a normal algebraic variety.

3.1 The log discrepancy of a divisorial valuation

Let K_X be a canonical Weil divisor on X, i.e., the closure in X of the divisor of a given rational form of top degree on X_{reg}. The choice of K_X induces on the one hand a graded sequence of fractional ideal sheaves $(\mathcal{O}_X(-mK_X))_{m\in\mathbb{N}}$, and on the other hand a canonical Weil divisor $K_{X'}$ for each birational model X' of X.

Following [2, 5], we define for all $m \geq 1$ the *m-limiting log discrepancy function* as the unique homogeneous function $A_X^{(m)}\colon \operatorname{DivVal}_X \to \mathbb{R}$ such that

$$A_X^{(m)}(\operatorname{ord}_E) = 1 + \operatorname{ord}_E(K_{X'}) - m^{-1} \operatorname{ord}_E \mathcal{O}_X(-mK_X)$$

for each prime divisor E on a birational model X'. Here $\operatorname{ord}_E(K_{X'})$ simply means the coefficient of E in $K_{X'}$ viewed as a cycle of codimension one. The definition is independent of the choices made, and the subadditivity of the sequence $\operatorname{ord}_E \mathcal{O}_X(-mK_X)$ shows that $A_X^{(m)}$ converges pointwise to a function $A_X\colon \operatorname{DivVal}_X \to \mathbb{R}$, the *log discrepancy function*, with $A_X = \sup_{m\geq 1} A_X^{(m)}$.

3.2 The log discrepancy of a real valuation

Theorem 3.1 *There is a unique way to extend A_X and $A_X^{(m)}$ ($m \geq 1$) to homogeneous, lower semicontinuous functions $\operatorname{Val}_X \to \mathbb{R} \cup \{+\infty\}$ such that the following properties hold for each normalizing subscheme $N \subset X$:*

(i) $A_X^{(m)}$ and A_X are convex and continuous on all faces of Val_X^N, and $A_X^{(m)}$ is even affine on small faces;

(ii) on Val_X^N we have

$$A_X^{(m)} = \sup_\pi A_X^{(m)} \circ r_\pi^N \quad \textit{and} \quad A_X = \lim_\pi A_X \circ r_\pi^N,$$

where π runs over all good resolutions of N and $r_\pi^N\colon \operatorname{Val}_X^N \to \Delta_\pi^N$ is the corresponding retraction;

(iii) for each $a \in \mathbb{R}$, $\left\{A_X^{(1)} \le a\right\}$ *is a compact subset of* Val_X, *and* $A_X^{(m)}$ *converges uniformly to* A_X *on this set.*

Further, we have $A_X = \sup_{m \ge 1} A_X^{(m)}$ *on* Val_X.

Remark 3.2 Combined with Remark 2.6, our proof will show that A_X is in fact Lipschitz continuous on any face of Val_X^N containing valuations with proper center in X.

Remark 3.3 We do not know whether $A_X \ge A_X \circ r_\pi^N$ holds on Val_X^N for π large enough in general.

Theorem 3.1 will be proved by reduction to the smooth case. The next result summarizes the required properties for X smooth, all of which are contained in [8].

Lemma 3.4 *Assume that X is smooth. Then Theorem 3.1 holds; further, if* $\mathfrak{a}_\bullet$ *is a graded sequence of ideal sheaves on X and N is a normalizing subscheme containing the zero locus of* $\mathfrak{a}_1$ *(and hence of* $\mathfrak{a}_m$ *for all m),* $m^{-1}v(\mathfrak{a}_m) \to v(\mathfrak{a}_\bullet)$ *uniformly for* $v \in \{A_X \le a\} \cap \mathrm{Val}_X^N$, *for each* $a \in \mathbb{R}$.

Proof When X is smooth, properties (i) and (ii) of Theorem 3.1 follow from [8, Proposition 5.1, Corollary 5.8]. The compactness of $\{A_X \le a\}$ is a consequence of the Skoda–Izumi inequality, just as in the proof of [8, Proposition 5.9]. By Dini's lemma and the subadditivity of $(v(\mathfrak{a}_m))_{m \in \mathbb{N}}$, the uniform convergence is equivalent to the continuity of $v \mapsto v(\mathfrak{a}_\bullet)$ on $\{A_X \le a\}$, which is [8, Corollary 6.4]. □

Proof of Theorem 3.1 Uniqueness is clear: since the rational points of each dual complex Δ_π^N consist of divisorial valuations, A_X and $A_X^{(m)}$ are uniquely determined on Δ_π^N by (i), and hence on Val_X^N by (ii). By homogeneity, they are also uniquely determined on

$$\mathrm{Val}_X^* = \bigcup_N \mathbb{R}_+^* \cdot \mathrm{Val}_X^N .$$

In order to prove existence, we fix the choice of a projective birational morphism $\mu \colon X' \to X$ such that X' is smooth and μ is an isomorphism over X_{reg}.

We claim that there exists a μ-exceptional effective divisor D on X' such that the graded sequence of fractional ideal sheaves

$$\mathfrak{a}_m := \mu^{-1}\mathcal{O}_X(-mK_X) \cdot \mathcal{O}_{X'}(mK_{X'}) \cdot \mathcal{O}_{X'}(-mD)$$

is a sequence of actual ideal sheaves. To see this, it is enough to choose D such that $\mathfrak{a}_1 \subset \mathcal{O}_{X'}$. But we may add a Cartier divisor Z to K_X so that $K_X + Z$ is effective. The divisorial part of the ideal sheaf

$$\mu^{-1}\mathcal{O}_X(-K_X - Z) \cdot \mathcal{O}_{X'}$$

coincides with $\mathcal{O}_{X'}(-K_{X'} - \mu^* Z)$ up to a μ-exceptional divisor D, and we get $\mathfrak{a}_1 \subset \mathcal{O}_{X'}$ as desired for this choice of D.

Note that we have by definition

$$A_X^{(m)}(v) = A_{X'}(v) + v(D) - m^{-1} v(\mathfrak{a}_m)$$

and

$$A_X(v) = A_{X'}(v) + v(D) - v(\mathfrak{a}_\bullet)$$

for all $v \in \mathrm{DivVal}_X \simeq \mathrm{DivVal}_{X'}$. Using the canonical homeomorphism $\mathrm{Val}_X \simeq \mathrm{Val}_{X'}$, we can now use these formulas to *define* A_X and $A_X^{(m)}$ on Val_X. Propositions 2.4 and 2.5 already show that A_X and $A_X^{(m)}$ are homogeneous and lsc on Val_X. It remains to see that they satisfy (i), (ii) and (iii) of Theorem 3.1.

Let $N \subset X$ be a given normalizing subscheme. Each good resolution $\pi' : X'_{\pi'} \to X'$ of $N' := \mu^{-1}(N)$ induces a good resolution $\pi := \mu \circ \pi'$ of N such that $\Delta_{\pi'}^{N'} = \Delta_\pi^N$, and the retractions $r_\pi^N : \mathrm{Val}_X^N \to \Delta_\pi^N$ and $r_{\pi'}^{N'} : \mathrm{Val}_{X'}^{N'} \to \Delta_{\pi'}^{N'}$ identify modulo the canonical homeomorphism $\mathrm{Val}_{X'} \simeq \mathrm{Val}_X$.

Since N' contains the support of D and the zero locus of $\mathfrak{a}_1$, Proposition 2.4 shows that $v \mapsto v(D) - v(\mathfrak{a}_\bullet)$ is bounded and lower semicontinuous on $\mathrm{Val}_{X'}^{N'}$, and continuous and convex on the faces of $\mathrm{Val}_{X'}^{N'}$, while $v \mapsto v(D) - m^{-1} v(\mathfrak{a}_m)$ is affine on small faces. It follows that A_X and $A_X^{(m)}$ satisfy (i).

Now pick $a \in \mathbb{R}$, and set for simplicity

$$K := \{A_X^{(1)} \leq a\} \cap \mathrm{Val}_X^N .$$

Since $v(D) - v(\mathfrak{a}_1)$ is bounded for $v \in \mathrm{Val}_X^N$, K is contained in $\{A_{X'} \leq a'\} \cap \mathrm{Val}_{X'}^{N'}$ for some $a' \in \mathbb{R}$, and hence is compact by Lemma 3.4 and the lower semicontinuity of $A_X^{(1)}$. Lemma 3.4 also shows that $m^{-1} v(\mathfrak{a}_m) \to v(\mathfrak{a}_\bullet)$ uniformly for $v \in K$, which proves (iii).

Let $v \in \mathrm{Val}_X^N$. If $A_X(v)$ is finite, then so is $A_{X'}(v)$, and $r_{\pi'}^{N'}(v)$ stays in the compact set

$$\{A_{X'} \leq A_{X'}(v)\} \cap \mathrm{Val}_{X'}^{N'}$$

since $A_{X'} \geq A_{X'} \circ r_{\pi'}^{N'}$. For each m fixed we have, by Proposition 2.4,

$$v(\mathfrak{a}_m) = r_{\pi'}^{N'}(v)(\mathfrak{a}_m)$$

for π' large enough, which proves that $A_X^{(m)}$ satisfies (ii). By uniform convergence of $m^{-1}\nu'(\mathfrak{a}_m)$ to $\nu'(\mathfrak{a}_\bullet)$ for $\nu' \in \{A_{X'} \leq A_{X'}(\nu)\} \cap \mathrm{Val}_{X'}^{N'}$, we infer

$$\nu(\mathfrak{a}_\bullet) = \lim_{\pi'} r_{\pi'}^{N'}(\nu)(\mathfrak{a}_\bullet),$$

so that A_X satisfies (ii) on the locus of Val_X^N where it is finite. If now $\nu \in \mathrm{Val}_X^N$ has $A_X(\nu) = +\infty$, then

$$\lim_{\pi'} A_{X'}(r_{\pi'}^{N'}(\nu)) = A_{X'}(\nu) = +\infty,$$

while $r_{\pi'}^{N'}(\nu)(\mathfrak{a}_\bullet)$ remains bounded, and we thus get (ii) at ν as well. The same argument also proves the last assertion of Theorem 3.1. □

4 Valuative characterization of multiplier ideals

4.1 Multiplier ideal sheaves

We briefly recall the definition of multiplier ideals in the context of a normal variety, as introduced in [5]. We follow the presentation of [2, Section 3], which is phrased in the language of b-divisors, and therefore closer to our present valuative point of view. Indeed, it suffices to recall that a b-divisor is nothing but a homogeneous function on DivVal_X, with the extra property that it is non-zero on only finitely many prime divisors of X (and hence on every model X' over X, since $X' \to X$ has only finitely many exceptional prime divisors).

If $\mathfrak{a} \subset \mathcal{O}_X$ is a coherent ideal sheaf and c is a positive real number, the *m-limiting multiplier ideal sheaf* of $\mathfrak{a}^c$ is defined as

$$\mathcal{J}_m(X, \mathfrak{a}^c) := \left\{ f \in \mathcal{O}_X \mid \nu(f) > c\nu(\mathfrak{a}) - A_X^{(m)}(\nu) \text{ for all } \nu \in \mathrm{DivVal}_X \right\},$$

where $A_X^{(m)}$ is the m-limiting log discrepancy function from Section 3. This is a reformulation of [2, Definition 3.7], which is phrased in the equivalent language of b-divisors. It is proved in [2] that $\mathcal{J}_m(X, \mathfrak{a}^c)$ is actually coherent.

We have $\mathcal{J}_m(X, \mathfrak{a}^c) \subset \mathcal{J}_l(X, \mathfrak{a}^c)$ whenever m divides l, and the *multiplier ideal sheaf* $\mathcal{J}(X, \mathfrak{a}^c)$ can thus be defined as the unique maximal element of the family $(\mathcal{J}_m(X, \mathfrak{a}^c))_{m \in \mathbb{N}}$. By [2, Theorem 3.8], $\mathcal{J}(X, \mathfrak{a}^c)$ is also the largest element in the family of "classical" multiplier ideals $\mathcal{J}((X, \Delta); \mathfrak{a}^c)$, where Δ runs over all effective $\mathbb{Q}$-Weil divisors on X such that $K_X + \Delta$ is $\mathbb{Q}$-Cartier (so that (X, Δ) is a pair in the sense of Mori theory).

More generally, when $\mathfrak{a}_\bullet = (\mathfrak{a}_m)_{m \in \mathbb{N}}$ is a graded sequence of (coherent) ideal sheaves, the multiplier ideal $\mathcal{J}(X, \mathfrak{a}_\bullet^c)$ is defined as the maximal element of the family $\mathcal{J}(X, \mathfrak{a}_m^{c/m})$ [2, Definition 3.12]. Equivalently, $\mathcal{J}(X, \mathfrak{a}_\bullet^c)$ is also the largest element in the family $\mathcal{J}_m\left(X, \mathfrak{a}_m^{c/m}\right)$, cf. [2, Lemma 3.13].

4.2 Valuative characterization

Theorems 1.1 and 1.2 are restated together in the following result:

Theorem 4.1 *Let X be a normal variety. Let $\mathfrak{a}_\bullet$ be a graded sequence of (coherent) ideal sheaves on X, and let $c > 0$ be a real number. Then the following two characterizations of the multiplier ideal sheaf $\mathcal{J}(X, \mathfrak{a}_\bullet^c) \subset \mathcal{O}_X$ hold:*

(a)

$$\mathcal{J}(X, \mathfrak{a}_\bullet^c) = \{f \in \mathcal{O}_X \mid v(f) > cv(\mathfrak{a}_\bullet) - A_X(v) \text{ for all } v \in \mathrm{Val}_X\};$$

(b) for every normalizing subscheme $N \subset X$ containing the zero locus of $\mathfrak{a}_1$ (and hence of $\mathfrak{a}_m$ for all m) and every $0 < \varepsilon \ll 1$ we have

$$\mathcal{J}(X, \mathfrak{a}_\bullet^c) = \{f \in \mathcal{O}_X \mid v(f) \geq cv(\mathfrak{a}_\bullet) - A_X(v) + \varepsilon v(\mathcal{I}_N) \text{ for all } v \in \mathrm{DivVal}_X\}.$$

A key ingredient in the proof is the following simple variant of Dini's lemma:

Lemma 4.2 *Let Z be a Hausdorff topological space, and let $\phi_m \colon Z \to \mathbb{R} \cup \{+\infty\}$ be a non-decreasing sequence of lower semicontinuous functions converging to ϕ. Assume also that each sublevel set $\{\phi_1 \leq a\}$ with $a \in \mathbb{R}$ is compact. Then $\inf_Z \phi_m \to \inf_Z \phi$.*

Proof Note that $\inf_Z \phi_m \leq \inf_Z \phi$, just because ϕ_m is non-decreasing. Assume first that $\inf_Z \phi < +\infty$ (which is the only case we shall actually use), and let $\varepsilon > 0$. Setting

$$K_m := \left\{\phi_m \leq \inf_Z \phi - \varepsilon\right\}$$

defines a decreasing sequence of compact sets, by lower semicontinuity of ϕ_m and the compactness of sublevel sets of ϕ_1. Since

$$\bigcap_{m \in \mathbb{N}} K_m = \left\{\phi \leq \inf_Z \phi - \varepsilon\right\} = \emptyset,$$

it follows that $K_m = \emptyset$ for all $m \gg 1$, i.e., $\inf_Z \phi_m > \inf_Z \phi - \varepsilon$ for all $m \gg 1$. In case $\phi \equiv +\infty$, the same argument applies, fixing $A > 0$ instead of ε and replacing K_m with $K'_m := \{\phi_m \leq A\}$. □

Proof of Theorem 4.1 Let $N \subset X$ be a normalizing subscheme containing the zero locus of $\mathfrak{a}_1$. Let also $U \subset X$ be an affine open set and pick $f \in \mathcal{O}(U)$. The theorem will follow from the equivalence between the following properties:

(i) $f \in \mathcal{J}(X, \mathfrak{a}_\bullet^c)(U)$;
(ii) $v(f) > cm^{-1}v(\mathfrak{a}_m) - A_X^{(m)}(v)$ on DivVal_U^N for all m large and divisible;
(iii) $v(f) > cm^{-1}v(\mathfrak{a}_m) - A_X^{(m)}(v)$ on Val_U^N for all m large and divisible;
(iv) $v(f) > cv(\mathfrak{a}_\bullet) - A_X(v)$ on Val_U^N;
(v) $v(f) \geq cv(\mathfrak{a}_\bullet) - A_X(v) + \varepsilon$ on Val_U^N for some $0 < \varepsilon \ll 1$;
(vi) $v(f) \geq cv(\mathfrak{a}_\bullet) - A_X(v) + \varepsilon$ on DivVal_U^N for some $0 < \varepsilon \ll 1$;
(vii) $v(f) \geq cv(\mathfrak{a}_\bullet) - A_X(v) + \varepsilon v(\mathcal{I}_N)$ on DivVal_U for some $0 < \varepsilon \ll 1$.

Let us first check (i)$\Longleftrightarrow$(ii). Since U is affine, $\mathcal{J}(X, \mathfrak{a}_\bullet^c)(U)$ is the largest element in the family of ideals $\mathcal{J}_m(X, \mathfrak{a}_m^{c/m})(U)$ of $\mathcal{O}(U)$, and (i) thus amounts to $v(f) > cm^{-1}v(\mathfrak{a}_m) - A_X^{(m)}(v)$ on DivVal_U for all m large and divisible, which implies (ii). Conversely, (ii) implies (i) since for any $v \in \mathrm{DivVal}_U$ centered outside $N \supset \mathrm{Sing}(X)$ we have $A_X^{(m)}(v) = A_X(v) > 0$ (since U is smooth at the center of v) while $v(\mathfrak{a}_m) = 0$.

Next, consider the functions $\phi, \phi_m \colon \mathrm{Val}_U^N \to \mathbb{R} \cup \{+\infty\}$ defined by

$$\phi(v) := v(f) - cv(\mathfrak{a}_\bullet) + A_X(v)$$

and

$$\phi_m(v) := v(f) - cm^{-1}v(\mathfrak{a}_m) + A_X^{(m)}(v).$$

For each m fixed, Proposition 2.4 and Theorem 3.1 show that ϕ_m is lower semicontinuous, affine on small faces of Val_U^N, and satisfies $\phi_m \geq \phi_m \circ r_\pi^N$ for all π large enough. This shows that $\phi_m > 0$ on DivVal_U^N iff $\phi_m > 0$ on Val_U^N, i.e., (ii)$\Longleftrightarrow$(iii).

Further, each sublevel set $\{\phi_1 \leq a\}$ with $a \in \mathbb{R}$ is compact by Theorem 3.1, and Lemma 4.2 thus yields (iv)$\Rightarrow$(iii), while the converse follows from $A_X \geq A_X^{(m)}$.

Since ϕ is lower semicontinuous and has compact sublevel sets, it achieves its infimum on Val_U^N, which proves (iv)$\Longleftrightarrow$(v). Next, (v) trivially implies (vi), while the converse holds since ϕ is continuous on each dual complex Δ_π^N and satisfies $\phi = \lim_\pi \phi \circ r_\pi^N$ on Val_U^N, again by Theorem 3.1.

As to (vi)$\Longleftrightarrow$(vii), it holds because $v(\mathcal{I}_N) = -1$ on Val_U^N by definition of the latter, while we have just as above $v(f) - cv(\mathfrak{a}_\bullet) + A_X(v) - \varepsilon v(\mathcal{I}_N) \geq A_X(v) \geq 0$ on any $v \in \mathrm{DivVal}_U$ centered outside N.

To get (b) in Theorem 4.1 from (vii), note that the ideals

$$\{f \in \mathcal{O}(U) \mid v(f) \geq cv(\mathfrak{a}_\bullet) - A_X(v) + \varepsilon v(\mathcal{I}_N) \text{ on } \mathrm{DivVal}_U\}$$

are independent of $0 < \varepsilon \ll 1$, by the Noetherian property of $\mathcal{O}(U)$. □

5 Numerically Cartier divisors

5.1 The group of numerical divisor classes

In this section, we provide an alternative and more concrete approach to the notion of numerically Cartier divisors introduced in [2, Section 2].

As a matter of notation, we respectively denote by $\mathrm{Car}(X)$ and $Z^1(X)$ the groups of Cartier and Weil divisors of a normal variety X. We define the *local class group of* X as

$$\mathrm{Cl}_{\mathrm{loc}}(X) := \mathrm{Z}^1(X)/\,\mathrm{Car}(X).$$

By definition, $\mathrm{Cl}_{\mathrm{loc}}(X)$ is trivial iff X is (locally) factorial. Since the usual divisor class group $\mathrm{Cl}(X)$ is defined as the quotient of $Z^1(X)$ by the subgroup of principal divisors, we have an exact sequence

$$0 \to \mathrm{Pic}(X) \to \mathrm{Cl}(X) \to \mathrm{Cl}_{\mathrm{loc}}(X) \to 0.$$

Remark 5.1 When X only has an isolated singularity at $0 \in X$, $\mathrm{Cl}_{\mathrm{loc}}(X)$ coincides with the divisor class group of the local ring $\mathcal{O}_{X,0}$.

Definition 5.2 Let X be a normal variety.

(i) A Weil divisor $D \in \mathrm{Z}^1(X)$ is *numerically Cartier* if there exists a resolution of singularities $\mu\colon X' \to X$ (i.e., a projective birational morphism with X' smooth) and a μ-numerically trivial Cartier divisor D' on X' such that $D = \mu_* D'$.

(ii) We denote by $\mathrm{NumCar}(X) \subset \mathrm{Z}^1(X)$ the subgroup of numerically Cartier divisors, and elements of $\mathrm{NumCar}(X)_{\mathbb{Q}} \subset Z^1(X')_{\mathbb{Q}}$ are called *numerically $\mathbb{Q}$-Cartier*.

(iii) The *group of numerical divisor classes* of X is defined as the quotient

$$\mathrm{Cl}_{\mathrm{num}}(X) := Z^1(X)/\,\mathrm{NumCar}(X).$$

(iv) We say that X is *numerically factorial* (resp. *numerically $\mathbb{Q}$-factorial*) if $\mathrm{Cl}_{\mathrm{num}}(X) = 0$ (resp. $\mathrm{Cl}_{\mathrm{num}}(X)_{\mathbb{Q}} = 0$).

By definition, $\mathrm{Cl}_{\mathrm{num}}(X)$ is a quotient of $\mathrm{Cl}_{\mathrm{loc}}(X)$, and it is in fact much smaller in general. Indeed, as we shall see shortly, $\mathrm{Cl}_{\mathrm{num}}(X)$ is always finitely generated as an abelian group.

In order to analyze further numerically Cartier divisors, we first show that it is enough to work with a *fixed* resolution of singularities.

Proposition 5.3 *Let $\mu\colon X' \to X$ be a projective birational morphism.*

(i) If X' is factorial, every $D \in \mathrm{NumCar}(X)$ can be written as $D = \mu_* D'$ for a unique μ-numerically trivial $D' \in \mathrm{Car}(X')$.

(ii) *If X' is $\mathbb{Q}$-factorial, every $D \in \mathrm{NumCar}(X)_{\mathbb{Q}}$ is of the form $D = \mu_* D'$ for a unique μ-numerically trivial $D' \in \mathrm{Car}(X')_{\mathbb{Q}}$.*

In both cases, we set $\mu^*_{\mathrm{num}} D := D'$ and call it the numerical pull-back *of D.*

Proof The kernel of $\mu_* \colon Z^1(X') \to Z^1(X)$ is exactly the space of μ-exceptional divisors. By the negativity lemma, there is no nontrivial divisor on X' that is both μ-numerically trivial and μ-exceptional, which proves the uniqueness part in both cases.

Now pick $D \in \mathrm{NumCar}(X)$. By definition, there exists a resolution $\mu'' \colon X'' \to X$ such that $D = \mu''_* D''$ for some μ''-numerically trivial $D'' \in \mathrm{Car}(X'')$. Since the pull-back of D'' to a higher resolution remains relatively numerically trivial, we may assume that μ'' dominates μ, i.e., $\mu'' = \mu \circ \rho$ for a birational morphism $\rho \colon X'' \to X'$. Since X' is factorial (resp. $\mathbb{Q}$-factorial), $D' := \rho_* D''$ belongs to $\mathrm{Car}(X')$ (resp. $\mathrm{Car}(X')_{\mathbb{Q}}$), and $D'' - \rho^* D'$ is both ρ-exceptional and ρ-numerically trivial, hence trivial. By the projection formula, it follows that D' is μ'-numerically trivial and $D = \mu'_* D'$. □

Corollary 5.4 *With the same assumption as in Proposition 5.3, $\mu_* \colon Z^1(X') \to Z^1(X)$ induces:*

(i) *an exact sequence of abelian groups*

$$0 \to \mathrm{Exc}^1(\mu) \to N^1(X'/X) \to \mathrm{Cl}_{\mathrm{num}}(X) \to 0$$

if X' is factorial, where $\mathrm{Exc}^1(\mu)$ is the (free abelian) group of μ-exceptional divisors and $N^1(X'/X)$ is the group of μ-numerical equivalence classes;

(ii) *an exact sequence of $\mathbb{Q}$-vector spaces*

$$0 \to \mathrm{Exc}^1(\mu)_{\mathbb{Q}} \to N^1(X'/X)_{\mathbb{Q}} \to \mathrm{Cl}_{\mathrm{num}}(X)_{\mathbb{Q}} \to 0$$

if X' is $\mathbb{Q}$-factorial.

In particular, $\mathrm{Cl}_{\mathrm{num}}(X)$ is a finitely generated abelian group.

Proof The exact sequences in (i) and (ii) follow immediately from Proposition 5.3. The last assertion is a consequence of the relative version of the theorem of the base [10, p. 334, Proposition 3], which guarantees that $N^1(X'/X)$ is finitely generated. □

Remark 5.5 As a special case of (ii) above, if X' is $\mathbb{Q}$-factorial and (E_i) denotes the μ-exceptional prime divisors, then X is numerically $\mathbb{Q}$-factorial iff for every $D' \in \mathrm{Car}(X')$ there exist $a_i \in \mathbb{Q}$ such that

$$\left(D' + \sum_i a_i E_i\right) \cdot C = 0 \tag{5.1}$$

holds for all curves $C \subset X'$ contained in a μ-fiber.

Example 5.6 If X is an affine cone over a smooth projective polarized variety (Y, L), then

$$\mathrm{Cl}_{\mathrm{loc}}(X) \simeq \mathrm{Pic}(Y)/\mathbb{Z}L$$

and

$$\mathrm{Cl}_{\mathrm{num}}(X) \simeq \mathrm{NS}(Y)/\mathbb{Z}c_1(L).$$

In particular, X is numerically $\mathbb{Q}$-factorial iff $\rho(Y) = 1$.

Example 5.7 Every surface is numerically $\mathbb{Q}$-factorial. This is directly related to the existence of Mumford's numerical pull-back. Indeed, let $\mu\colon X' \to X$ be a resolution of singularities, with exceptional divisor $\sum_i E_i$. Since the intersection matrix $(E_i \cdot E_j)$ is negative definite, for $D' \in \mathrm{Car}(X')$ we can find $a_i \in \mathbb{Q}$ such that (5.3) holds for each curve $C = E_j$.

Note that $\mathrm{Cl}_{\mathrm{num}}(X)$ is however nontrivial in general, even for a surface. For instance, it follows from Example 5.6 that $\mathrm{Cl}_{\mathrm{num}}(X) = \mathbb{Z}/2\mathbb{Z}$ for an A_1-singularity.

Example 5.8 If X is log terminal in the sense of [5], i.e., if (X, Δ) is klt for some effective $\mathbb{Q}$-Weil divisor Δ, it follows from the current knowledge in the minimal model program that there exists a small projective birational morphism $\mu\colon X' \to X$ such that X' is $\mathbb{Q}$-factorial. By (ii) of Corollary 5.4, we then have $N^1(X'/X)_{\mathbb{Q}} \simeq \mathrm{Cl}_{\mathrm{num}}(X)_{\mathbb{Q}}$, which is thus trivial iff μ is an isomorphism. In other words, X is numerically $\mathbb{Q}$-factorial iff X is $\mathbb{Q}$-factorial. Since X has rational singularities, the previous conclusion will also follow from Theorem 5.11 below.

Let us now check that Definition 5.2 is indeed compatible with [2, Definition 2.26, Remark 2.27].[3] Since this result is not strictly necessary in the rest of the paper, we simply refer to [2, Section 2] for details about the notions used.

Proposition 5.9 *A Weil divisor D on X is numerically $\mathbb{Q}$-Cartier in the sense of Definition 5.2 iff*

$$\nu(\mathcal{O}_X(-mD)) = -\nu(\mathcal{O}_X(mD)) + o(m) \tag{5.2}$$

for all $\nu \in \mathrm{DivVal}_X$.

For each projective birational morphism $\mu\colon X' \to X$ with X' $\mathbb{Q}$-factorial, we then have

$$\lim_{m\to\infty} m^{-1}\nu(\mathcal{O}_X(-mD)) = \nu(\mu^*_{\mathrm{num}}D) \tag{5.3}$$

[3] More precisely, numerically $\mathbb{Q}$-Cartier divisors in the present sense correspond to numerically Cartier divisors in the sense of [2].

for all $\nu \in \mathrm{DivVal}_X$. *In particular, the limit on the left-hand side is rational.*

Proof In the terminology of [2, Section 2], (5.2) reads

$$\mathrm{Env}_X(-D) = -\,\mathrm{Env}_X(D),$$

where $\mathrm{Env}_X(D)$ is the *nef envelope* of D, i.e., the b-divisor over X characterized by

$$\nu(\mathrm{Env}_X(D)) = \lim_{m\to\infty} m^{-1}\nu(\mathcal{O}_X(mD))$$

for all $\nu \in \mathrm{DivVal}_X$. Assume first that $D \in Z^1(X)$ is numerically $\mathbb{Q}$-Cartier. Let $\mu\colon X' \to X$ be a projective birational morphism with X' $\mathbb{Q}$-factorial and set $D' := \mu^*_{\mathrm{num}}D$. The Cartier b-divisor $\overline{D}'$ induced by pulling back D' is then relatively nef over X and satisfies $\overline{D}'_X = D$, and hence

$$\overline{D}' \le \mathrm{Env}_X(D)$$

by [2, Proposition 2.12]. Since $-\overline{D}'$ is also relatively nef, we similarly get

$$-\overline{D}' \le \mathrm{Env}_X(-D).$$

Summing up these two inequalities and using the trivial inequality

$$\mathrm{Env}_X(D) + \mathrm{Env}_X(-D) \le \mathrm{Env}_X(D - D) = 0,$$

we infer

$$\mathrm{Env}_X(D) = \overline{D}'$$

and

$$\mathrm{Env}_X(-D) = -\overline{D}',$$

which proves (5.2) and (5.3).

Conversely, assume that $D \in Z^1(X)$ satisfies $\mathrm{Env}_X(-D) = -\,\mathrm{Env}_X(D)$. By [2, Lemma 2.10], it follows that $D' := \mathrm{Env}_X(D)_{X'} \in \mathrm{Car}(X')_{\mathbb{R}}$ is μ-numerically trivial. Since $\mu_*D' = D$ belongs to $Z^1(X)_{\mathbb{Q}}$ and μ_* is defined over $\mathbb{Q}$, the injectivity of μ_* on μ-numerically trivial divisors implies that D' is in fact a $\mathbb{Q}$-divisor, and hence that D is numerically $\mathbb{Q}$-Cartier with $D' = \mu^*_{\mathrm{num}}D$. □

Remark 5.10 In particular, this result shows that the envelope $\mathrm{Env}_X(D)$ of a Weil divisor $D \in Z^1(X)$ such that $\mathrm{Env}_X(-D) = -\,\mathrm{Env}_X(D)$ is a $\mathbb{Q}$-Cartier b-divisor (the rationality of the coefficients being in particular not obvious from the definition). In fact, the whole point of the present view is to highlight the fact that the $\mathbb{R}$-vector space of $\mathbb{R}$-Weil divisors $D \in Z^1(X)_{\mathbb{R}}$ with $\mathrm{Env}_X(-D) = -\,\mathrm{Env}_X(D)$ is in fact defined over $\mathbb{Q}$.

5.2 The case of rational singularities

In this section we prove:

Theorem 5.11 *Let X be a normal variety with at most rational singularities. Then*

$$\mathrm{NumCar}(X)_{\mathbb{Q}} = \mathrm{Car}(X)_{\mathbb{Q}},$$

i.e., a Weil divisor is numerically $\mathbb{Q}$-Cartier iff it is $\mathbb{Q}$-Cartier. In particular, X is numerically $\mathbb{Q}$-factorial iff X is $\mathbb{Q}$-factorial.

The proof is inspired by that of [9, Lemma 1.1], which states that the $\mathbb{Q}$-vector space $Z^1(X)_{\mathbb{Q}}/\mathrm{Car}(X)_{\mathbb{Q}}$ is finite dimensional when X has rational singularities. We will need the following two results:

Lemma 5.12 ([14, Proposition 1]) *A Weil divisor on a normal variety X is locally Cartier at a point $x \in X$ iff its restriction to the formal completion of X at x is Cartier.*

Lemma 5.13 *If Y is a (possibly reducible) projective complex variety, a line bundle L on Y is numerically trivial, i.e., $L \cdot C = 0$ for all curves $C \subset Y$, iff $c_1(L) = 0$ in $H^2(Y, \mathbb{Q})$.*

Proof When Y is nonsingular, the result is well known and amounts to the Hodge conjecture for 1-dimensional cycles (which follows from the 1-codimensional case via the Hard Lefschetz theorem). However, we haven't been able to locate a reference in the literature in the general singular case; we are very grateful to Claire Voisin for having shown us the following argument. Let $\pi\colon Y' \to Y$ be a resolution of singularities. Since π^*L is also numerically trivial, we have $\pi^*c_1(L) = 0$ in $H^2(Y', \mathbb{Q})$, by the result in the smooth case. By [12, Corollary 5.42], this means that $c_1(L) \in W_1H^2(Y, \mathbb{Q})$, where $W_\bullet$ denotes the weight filtration of the mixed Hodge structure. The problem is thus to show that $W_1H^2(Y, \mathbb{Q})$ only meets the image of $\mathrm{Pic}(Y)$ at 0.

To see this, note that there exists a morphism $f\colon Y \to Z$ to a smooth projective variety Z such that $L = f^*M$ for some line bundle M on Z; indeed, this is true with Z a projective space when L is very ample, and writing L as a difference of very ample line bundles gives the general case, with Z a product of two projective spaces.

Since $f^*\colon H^2(Z, \mathbb{Q}) \to H^2(Y, \mathbb{Q})$ is a morphism of mixed Hodge structures, it is *strict* with respect to weight filtrations, and we get

$$c_1(L) \in f^*H^2(Z, \mathbb{Q}) \cap W_1H^2(Y, \mathbb{Q}) = f^*\left(W_1H^2(Z, \mathbb{Q})\right),$$

which is zero since Z is smooth. □

Proof of Theorem 5.11 Let $\mu\colon X' \to X$ be a resolution of singularities and let $D' \in \mathrm{Car}(X')$ be μ-numerically trivial. Our goal is to show that $D := \mu_* D'$ is $\mathbb{Q}$-Cartier. By Lemma 5.12, it is enough to show that every (closed) point $x \in X$ has an analytic neighborhood U on which D^{an} is $\mathbb{Q}$-Cartier.

The exponential exact sequence on the associated complex analytic variety X'^{an} yields an exact sequence

$$R^1\mu_*^{\mathrm{an}}O \to R^1\mu_*^{\mathrm{an}}O^* \to R^2\mu_*\mathbb{Z} \to R^2\mu_*^{\mathrm{an}}O$$

of sheaves on X^{an}, where the two extreme terms coincide by GAGA with the analytifications of $R^q\mu_*O$ for $q = 1, 2$, and hence vanish since X has rational singularities. We thus have an isomorphism

$$\left(R^1\mu_*^{\mathrm{an}}O^*\right)_x \simeq \left(R^2\mu_*^{\mathrm{an}}\mathbb{Z}\right)_x = H^2\left(\mu^{-1}(x), \mathbb{Z}\right),$$

where the right-hand equality holds by properness of μ^{an} (which again follows from (the easy direction of) GAGA). Since D' has degree 0 on each projective curve $C \subset \mu^{-1}(x)$, its image in $H^2\left(\mu^{-1}(x), \mathbb{Q}\right)$ is trivial by Lemma 5.13. By the above isomorphism, the image of D'^{an} in $\left(R^1\mu_*^{\mathrm{an}}O^*\right)_x \otimes \mathbb{Q}$ is also trivial, which means that D'^{an} is $\mathbb{Q}$-linearly equivalent to 0 on $(\mu^{\mathrm{an}})^{-1}(U)$ for a small enough analytic neighborhood U of x. Since the morphism $(\mu^{\mathrm{an}})^{-1}(U) \to U$ is a proper modification, it follows as desired that D^{an} is $\mathbb{Q}$-Cartier on U. □

5.3 Multiplier ideals in the numerically $\mathbb{Q}$-Gorenstein case

Definition 5.14 A normal variety X is *numerically $\mathbb{Q}$-Gorenstein* if K_X is numerically $\mathbb{Q}$-Cartier.

Given a resolution of singularities $\mu\colon X' \to X$, Corollary 5.4 shows that X is numerically $\mathbb{Q}$-Gorenstein iff $K_{X'}$ is μ-numerically equivalent to a μ-exceptional $\mathbb{Q}$-divisor, which is then uniquely determined and denoted by $K^{\mathrm{num}}_{X'/X}$. In other words, we set

$$K^{\mathrm{num}}_{X'/X} := K_{X'} - \mu^*_{\mathrm{num}} K_X.$$

By Proposition 5.9, for each prime divisor $E \subset X'$ we then have

$$A_X(\mathrm{ord}_E) = 1 + \mathrm{ord}_E\left(K^{\mathrm{num}}_{X'/X}\right). \tag{5.4}$$

Lemma 5.15 *Assume that X is numerically $\mathbb{Q}$-Gorenstein, and let $N \subset X$ be a normalizing subscheme. For each good resolution π of N, the log discrepancy function $A\colon \mathrm{Val}_X \to \mathbb{R} \cup \{+\infty\}$ is then affine on the faces of the dual complex Δ^N_π, and*

$$A_X = \sup_\pi A_X \circ r^N_\pi$$

on Val_X^N*, where* π *ranges over all good resolutions of* N*.*

Proof Write

$$K_{X'/X}^{\mathrm{num}} = K_{X'} - \pi_{\mathrm{num}}^* K_X = \sum_i a_i E_i$$

with $a_i \in \mathbb{Q}$ and E_i π-exceptional and prime. Modulo the canonical homeomorphism $\mathrm{Val}_{X'} \simeq \mathrm{Val}_X$, (5.4) yields

$$A_X(v) = A_{X'}(v) + \sum_i a_i v(E_i)$$

on Val_X. Since X' is smooth, $A_{X'}$ is affine on the faces of Δ_π^N and satisfies $A_{X'} \geq A_{X'} \geq r_\pi^N$. Since Proposition 2.4 shows that $v \mapsto v(E_i)$ is also affine on the faces Δ_π^N, and satisfies $r_\pi^N(v)(E_i) = v(E_i)$, the result follows. □

The next result is Theorem 1.3 from the Introduction:

Theorem 5.16 *Assume that* X *is numerically* $\mathbb{Q}$*-Gorenstein, and let* $\mathfrak{a} \subset \mathcal{O}_X$ *be an ideal sheaf. Let also* $\mu \colon X' \to X$ *be a log resolution of* $(X, \mathfrak{a})$*, so that* $\mu^{-1}\mathfrak{a} \cdot \mathcal{O}_{X'} = \mathcal{O}_{X'}(-D)$ *with* D *an effective Cartier divisor. For each exponent* $c > 0$ *we then have*

$$\mathcal{J}(X, \mathfrak{a}^c) = \mu_* \mathcal{O}_{X'}\left(\lceil K_{X'/X}^{\mathrm{num}} - cD \rceil\right).$$

Proof Let N be a normalizing subscheme containing the zero locus of $\mathfrak{a}$, and pick a good resolution π of N factoring as $\pi = \mu \circ \rho$. Using Lemma 5.15 and arguing as in the proof of Theorem 4.1, we easily get

$$\mathcal{J}(X, \mathfrak{a}^c) = \pi_* \mathcal{O}_{X_\pi}\left(\lceil K_{X_\pi/X}^{\mathrm{num}} - cD \rceil\right).$$

Also, we have

$$K_{X_\pi/X}^{\mathrm{num}} = \rho^* K_{X'/X}^{\mathrm{num}} + K_{X_\pi/X'},$$

since both sides of the equality are π-exceptional and π-numerically equivalent to K_{X_π}. Since $K_{X_\pi/X'}$ is effective and π-exceptional, we obtain as desired

$$\pi_* \mathcal{O}_{X_\pi}\left(\lceil K_{X_\pi/X}^{\mathrm{num}} - cD \rceil\right) = \mu_* \mathcal{O}_{X'}\left(\lceil K_{X'/X}^{\mathrm{num}} - cD \rceil\right).$$

□

Corollary 5.17 *Assume that* X *is numerically* $\mathbb{Q}$*-Gorenstein. Then* X *has log terminal singularities (in the usual sense,* i.e., *with* K_X $\mathbb{Q}$*-Cartier) iff* $A_X > 0$ *on* DivVal_X.

Proof By Theorem 5.16, we have $A_X > 0$ on DivVal_X iff $\mathcal{J}(X, \mathcal{O}_X) = \mathcal{O}_X$, which is the case iff there exists an effective $\mathbb{Q}$-Weil divisor Δ such that the pair (X, Δ) is klt [5] (see also [2]). But this implies that X has rational singularities, and Theorem 5.11 thus shows that K_X is $\mathbb{Q}$-Cartier. Since (X, Δ) is klt, so is $(X, 0)$, which means that X is log terminal in the classical sense. □

Acknowledgments We are very grateful to Claire Voisin for providing a proof of Lemma 5.13. We would also like to thank Mattias Jonsson for a key observation that helped us simplify the statements of the main results.

References

[1] Berkovich, V. G. 1990. Spectral theory and analytic geometry over non-Archimedean fields. *Mathematical Surveys and Monographs*, Vol. 33. Providence, RI: American Mathematical Society.

[2] Boucksom, S., de Fernex, T., and Favre, C. 2012. The volume of an isolated singularity. *Duke Math. J.*, **161**, 1455–1520.

[3] Boucksom, S., Favre, C., and Jonsson, M. 2012. Valuations and plurisubharmonic singularities. *Publ. Res. Inst. Math. Sci.*, **44**, 449–494.

[4] Boucksom, S., Favre, C., and Jonsson, M. 2012. A refinement of Izumi's Theorem. `arXiv:1209.4104`.

[5] de Fernex, T. and Hacon, C. 2009. Singularities on normal varieties. *Compos. Math.*, **145**, 393–414.

[6] Favre, C. and Jonsson, M. 2005. Valuations and multiplier ideals. *J. Amer. Math. Soc.*, **18**, 655–684.

[7] Jonsson, M. and Mustaţă, M. 2012. An algebraic approach to the openness conjecture of Demailly and Kollár. `arXiv:1205.4273`.

[8] Jonsson, M. and Mustaţă, M. 2010. Valuations and asymptotic invariants for sequences of ideals. To appear in *Ann. Inst. Fourier*. `arXiv:1011.3699`.

[9] Kawamata, Y. 1988. Crepant blowing-up of 3-dimensional canonical singularities and its application to degenerations of surfaces. *Ann. Math.*, **127**, 93–163.

[10] Kleiman, S. 1966. Toward a numerical theory of ampleness. *Ann. Math.*, **84**, 293–344.

[11] Kollár, J. 1997. Singularities of pairs. In *Algebraic Geometry—Santa Cruz 1995*. Proceedings of the Symposium on Pure Mathematics, Vol. 62, pp. 221–287. Providence, RI: American Mathematical Society.

[12] Peters, C. and Steenbrink, J. 2008. *Mixed Hodge Structures*. Ergebnisse der Mathematik und ihrer Grenzgebiete. 3. Folge. A Series of Modern Surveys in Mathematics [Results in Mathematics and Related Areas. 3rd Series. A Series of Modern Surveys in Mathematics], Vol. 52. Berlin: Springer-Verlag.

[13] Rockafellar, R. T. 1970. *Convex Analysis*. Princeton Mathematical Series, No. 28. Princeton, NJ: Princeton University Press.

[14] Samuel, P. 1961. Sur les anneaux factoriels. *Bull. Soc. Math. France*, **89**, 155–173.
[15] Thuillier, A. 2007. Géométrie toroïdale et géométrie analytique non archimédienne. Application au type d'homotopie de certains schémas formels. *Manuscripta Math.*, **123**, 381–451.
[16] Zhang, Y. 2013. On the volume of isolated singularities. To appear in *Compos. Math.* `arXiv:1301.6121`.

4

Line arrangements modeling curves of high degree: Equations, syzygies, and secants

G. Burnham
Bridgewater Associates

Z. Rosen
University of California at Berkeley

J. Sidman
Mount Holyoke College

P. Vermeire
Central Michigan University

Abstract

We study curves consisting of unions of projective lines whose intersections are given by graphs. Under suitable hypotheses on the graph, these so-called *graph curves* can be embedded in projective space as line arrangements. We discuss property N_p for these embeddings and are able to obtain products of linear forms that generate the ideal in certain cases. We also briefly discuss questions regarding the higher-dimensional subspace arrangements obtained by taking the secant varieties of graph curves.

1 Introduction

An arrangement of linear subspaces, or subspace arrangement, is the union of a finite collection of linear subspaces of projective space. In this paper we study arrangements of lines called graph curves with high degree relative to genus. We are particularly interested in the defining equations and syzygies of these subspace arrangements. We will assume an algebraically closed ground field of characteristic zero throughout.

From *Recent Advances in Algebraic Geometry*, edited by Christopher Hacon, Mircea Mustaţă and Mihnea Popa

Let $G = (V, E)$ be a simple, connected graph with vertex set V and edge set E. Following [9], we assume that G is *subtrivalent*, meaning that each vertex has degree at most three. The (abstract) *graph curve* C_G associated with G is constructed by taking the union of $\{L_v \mid v \in V\}$, where each L_v is a copy of $\mathbb{P}^1$ and lines L_u and L_v intersect in a node if and only if there is an edge between u and v in G. (Note that if we think of the nodes of C_G as vertices and the lines L_v as edges, then C_G is the graph dual to G.) Since we are assuming that each vertex has degree less than or equal to three, C_G is specified by purely combinatorial data; we may assume that on each component of C_G the nodes are at 0, 1 or ∞. Note that if each vertex of G is trivalent, then each copy of $\mathbb{P}^1$ in C_G contains three nodes, and C_G is stable (see [4, 9]).

The motivation for the work presented here was to see if the syzygies of a high-degree graph curve and its secant varieties would behave as they are expected to when the curve is smooth. The kth secant variety, Σ_k, of a smooth curve in $\mathbb{P}^r$ has expected dimension $\min\{2k + 1, r\}$. Thus, we expect the kth secant variety of C_G to be an arrangement of subspaces of dimension $2k + 1$.

Many authors [10, 16, 22, 25] have given generalizations of the results for smooth curves to higher-dimensional varieties, showing that embeddings via line bundles satisfying various positivity conditions will also satisfy property N_p. However, recent work of Ein and Lazarsfeld [10] shows that these results describe only a small portion of the minimal free resolution of a higher-dimensional variety, and what happens in the remaining piece is quite complicated, contrary to the belief that positivity of an embedding simplifies syzygies.

One can view the conjecture of [29], which says that we should expect property $N_{k+2,p}$ (ideal generators of degree $k + 2$ and linear syzygies through stage p) for the kth secant variety of a smooth curve of genus g embedded via a complete linear series of degree at least $2g + 2k + 1 + p$, as an alternate way of generalizing property N_p for curves to higher-dimensional varieties. Some progress was made for first secant varieties, using geometric methods in [27], but the recursive nature of these methods makes generalizing those techniques to higher secant varieties daunting. If a similar result were true for the secant varieties of graph curves, the proof methods would necessarily be very different and the hope is that they would shed new light on understanding secant varieties of smooth curves.

Based on many examples computed with Macaulay2 [18], the situation looks promising. However, when $g > 2$, the combinatorics can be intricate even if we only consider curves and not secant varieties. We will focus on curves in Sections 2 and 3 and turn to a discussion of the syzygies of secant varieties of graph curves in Section 4.

We begin by setting some assumptions and notation. Let d be the number of vertices in G. The topology of G determines the arithmetic genus of C_G as we may view G as a 1-dimensional simplicial complex, from which it follows that $p_a(C_G) = h^1(G, k)$ if G is connected (see Proposition 1.1 in [4]). We refer to this quantity as the genus g of G, and $|E| = d + g - 1$. Note that g is not the genus of G in the usual graph-theoretic sense.

The story that we wish to generalize to the setting of graph curves began with Green and Lazarsfeld [19] in the early 1980s, who showed that if C is a smooth and irreducible curve of genus g embedded in projective space via a complete linear series of degree $d \geq 2g+1+p$, then C satisfies property N_p. In other words, its ideal is generated by quadrics with syzygy modules generated by linear forms through the pth stage of the resolution.

We conjecture that if G satisfies Assumption 1.1, then property N_p will hold for C_G embedded as a line arrangement in $\mathbb{P}^{d-g}$.

Assumption 1.1 Fix $p \geq 0$ and let G be a simple, connected, subtrivalent graph with $d \geq 2g + 1 + p$. Assume that if G' is a connected subgraph induced on $V' \subset V$, $d' = |V'|$, and g' is the genus of G', then $d' \geq 2g' + 1 + p$ if $g' \geq 1$.

To see that the recursive hypotheses are necessary, note that a graph may satisfy $d \geq 2g + 2$, but if it contains a triangle, then the ideal of C_G cannot be generated by quadrics.

If Assumption 1.1 is satisfied for some $p \geq 0$, then C_G embeds in $\mathbb{P}^{d-g}$ as a line arrangement via [7] and is arithmetically Cohen–Macaulay by [15]. If C_G is arithmetically Cohen–Macaulay, we may proceed as in [19] and property N_p for C_G will follow if property N_p holds for a general hyperplane section. In [19], Green and Lazarsfeld deduce property N_p for points in linearly general position, and conjecture that the failure of property N_p for a set of $2r + 1 + p$ points implies the existence of a subset of $2k + 2 - p$ points on a $\mathbb{P}^k$. As shown in [14, 20], this conjecture for point sets is a consequence of the linear syzygy conjecture of Eisenbud, Koh, and Stillman [13]. Green proved the linear syzygy conjecture in [20], and for graph curves of degree $g \leq 2$ we can show that an embedding of C_G as a line arrangement via a complete linear series must satisfy N_p if Assumption 1.1 is satisfied.

Graph curves associated with graphs in which every vertex is trivalent are canonical curves, and have been studied in several different contexts. For example, Ciliberto, Harris, and Miranda [8] used graph curves to understand the surjectivity of the Wahl map, Ciliberto and Miranda [9] related graph curves to graph colorings, and Bayer and Eisenbud [4] studied graph curves in connection with Green's conjecture. In fact, Proposition 3.1 in [4] gives an explicit description of generators of the ideal of a canonical graph curve using the combinatorics of G. More recently, Ballico has written several papers about graph curves [1, 2].

We present an explicit embedding of C_G into projective space in Section 3. If the ideal of C_G is generated by quadrics, this allows us to show that I_{C_G} may be generated by products of linear forms (Theorem 3.7).

Although a subspace arrangement may always be cut from products of linear forms set-theoretically, we do not generally expect the ideal of a subspace arrangement to be generated by products of linear forms, cf. Propositions 5.4 and 5.7 in [5]. The most interesting examples of subspace arrangements with ideals generated by products of linear forms occur when the intersections among the subspaces have a rich combinatorial structure [5, 23, 24]. If G is a path or a cycle, then C_G can be embedded in projective space so that its ideal is generated by square-free monomials. In both cases, the ideals of the nontrivial secant varieties of these curves are also generated by square-free monomials and are examples of "combinatorial secant varieties" [28].

In addition to viewing graph curves and their secant varieties as combinatorial models of smooth curves and their secant varieties, we can also think of them as a new way of generating arrangements of linear subspaces with interesting interactions between the combinatorics of the arrangements, the geometry of the embeddings, and their defining equations. We present conjectures and questions for further work in this direction in Section 5.

2 Regularity and property N_p

In this section we will show that if $g \leq 2$, then the ideal of a linearly normal embedding of C_G as a line arrangement satisfies property N_p if G satisfies Assumption 1.1 for some $p \geq 0$, following the idea of the "quick" proof that a smooth and irreducible curve of degree $d \geq 2g+1+p$ satisfies N_p given in [19].

A key assumption in [19] is that a hyperplane section of a smooth curve of degree $d \geq 2g + 1 + p$ will consist of points in linearly general position. This fact is used to show that the points in a hyperplane section of the curve impose independent conditions on quadrics.

Using Lemma 2.1 we can show that this is not the case for a graph curve if G contains a cycle as a proper subgraph.

Lemma 2.1 *If G is a cycle on d vertices, then a hyperplane section of C_G has a 1-dimensional space of linear dependence relations and all of the points are contained in the support of the relation.*

Proof A cycle of length d embeds into $\mathbb{P}^{d-1}$, so the hyperplane section consists of d points in $\mathbb{P}^{d-2}$. A set of d points spanning $\mathbb{P}^{d-2}$ must satisfy a unique relation up to scalar. □

Therefore, if we have a cycle as a proper subset of a graph G, the points of a hyperplane section must fail to be in linearly general position. Because N_p fails if G contains a cycle of length $p + 2$, it will often be impossible to reproduce the graded Betti diagrams of a smooth curve with the graded Betti diagrams of a graph curve. For instance, for genus $g = 2$ and degree d, the length of the smallest cycle has an upper bound of $\lfloor \frac{2d-1}{3} \rfloor + 1$.

Nevertheless, we will show that if G satisfies Assumption 1.1 for $g \leq 2$, then a general hyperplane section of C_G imposes independent conditions on quadrics. This follows from the weaker assumption that no $2k + 2$ of the points lie on a $\mathbb{P}^k$ using ideas from [14].

Theorem 2.2 *Suppose that G satisfies Assumption 1.1 for some $p \geq 0$, $g \leq 2$, and C_G is embedded in $\mathbb{P}^{d-g}$ as a line arrangement via a complete linear series. If H is a general hyperplane and $X = H \cap C_G$, then there is no set of $2k + 2 - p$ dependent points of X lying on a $\mathbb{P}^k$.*

Proof Let $Y \subset X$. Suppose for contradiction that $|Y| = 2k + 2 - p$ and Y spans a $\mathbb{P}^k$. This means that there is a $2k + 2 - p - (k + 1) = k + 1 - p = m$-dimensional space of dependence relations on Y. Since $g \leq 2$, we know that $m \leq 2$. If $m = 0$, then the points are independent, which contradicts our hypotheses.

If $m = 1$, then $k = p$. Either the support of the unique dependence relation on Y contains a cycle of points, or the relation is a linear combination of dependence relations on two cycles in which at least one point has been eliminated from their support. If $\{\gamma_i\}$ form a basis for $H_1(G; \mathbb{R})$, the corresponding dependence relations $\{R_i\}$ form a basis for the space of linear relations on X, and Assumption 1.1 implies that $\gamma_1 \cup \gamma_2$ contains at least $5 + p$ points. The cycles γ_1 and γ_2 can be combined in $H_1(G; \mathbb{R})$ to form a distinct cycle γ_3, which also supports a unique linear dependence. Therefore, if we fix the coefficient of R_1 there is a unique multiple of R_2 that eliminates the shared points in the interior of their common path to create a dependence relation with support on γ_3. Consequently, we see that we cannot simultaneously eliminate the endpoints of this path and the points between them from the support. Therefore, if a linear combination of R_1 and R_2 is not supported on a full cycle, it contains at least $2 \cdot 2 + 1 + p - 2 = 2 + 1 + p = 3 + p$ points, implying that Y spans a projective space of dimension at least $p + 1$, which is a contradiction as $k = p$.

If $m = 2$, then $k = p + 1$. In this case $g = 2$, and Y must contain the support of both cycles of G, in which case $2k + 2 - p \geq 2 \cdot 2 + 1 + p$, or $2k \geq 2p + 3$, which contradicts $k = p + 1$. □

Remark We conjecture that if G satisfies Assumption 1.1 and C_G is embedded via a complete linear series, then Theorem 2.2 holds for all g. The idea is that if there is an m-dimensional space of dependence relations on Y, then we need at least m independent cycles of G to span this space. The support of m cycles contains at least $2m+1+p$ points. If more than m cycles are needed to span the space of dependence relations of Y, then we may have eliminated some points from the support, but we will always have at least $2m+1+p$ points remaining.

Theorem 2.3 *If G satisfies Assumption 1.1 for some $p \geq 0$, and no $2k+2-p$ points of X lie on a $\mathbb{P}^k$, then a general hyperplane section of C_G has a 3-regular ideal and satisfies property N_p.*

Proof The proposition on p. 169 of [14] states that X imposes independent conditions on quadrics if X does not contain a subset of $2k + 2$ points on a projective k-plane. This implies that the ideal of X is 3-regular by Lemma 2 of [14]. The ideal of X satisfies N_p as a consequence of Theorem 2.1 in [20]. □

Theorem 2.4 *Suppose that G satisfies Assumption 1.1 for some $p \geq 0$, $d \geq 2g + 1 + p$, and C_G is embedded in $\mathbb{P}^{d-g}$ as a line arrangement via a complete linear series. If no $2k + 2 - p$ points of a general hyperplane section lie on a $\mathbb{P}^k$, then this embedding is arithmetically Cohen–Macualay, 3-regular, and satisfies N_p.*

Proof For 3-regularity we need $H^1(\mathcal{I}_{C_G}(2)) = H^2(\mathcal{I}_{C_G}(1)) = 0$. We know that $H^1(\mathcal{O}_{C_G}(1)) = 0$ by Serre duality and our hypothesis that $d \geq 2g + 2$. This implies that $H^2(\mathcal{I}_{C_G}(1)) = 0$. To see the vanishing of $H^1(\mathcal{I}_{C_G}(2))$, note via Theorem 2.3 the regularity of the ideal of a general hyperplane section X of C_G is 3, which implies that $H^1(\mathcal{I}_X(2)) = 0$. Since C_G is embedded via a complete linear series, $H^1(\mathcal{I}_{C_G}(1)) = 0$, and we conclude that $H^1(\mathcal{I}_{C_G}(2)) = 0$.

The curve $C_G \subset \mathbb{P}^{d-g}$ is arithmetically Cohen–Macaulay if its homogeneous coordinate ring is Cohen–Macaulay. Equivalently, the hypersurfaces of degree m are a complete linear series, which holds if and only if $H^1(\mathcal{I}_{C_G}(m)) = 0$ for all $m \geq 0$ (see Section 8A of [11]). When $m = 0$, this follows because C_G is connected. We know that $H^1(\mathcal{I}_{C_G}(1)) = 0$ from the linear normality of the embedding, and $H^1(\mathcal{I}_{C_G}(k)) = 0$ for all $k \geq 2$ by the 3-regularity of the ideal. □

Corollary 2.5 *If G satisfies Assumption 1.1 for some $p \geq 0$, and $g \leq 2$, then an embedding of C_G as a line arrangement via a complete linear series is arithmetically Cohen–Macualay, 3-regular, and satisfies N_p.*

Proof Theorem 2.2 implies that the hypotheses of Theorems 2.3 and 2.4 are satisfied. □

Note that by Theorem 4.2 of [15], we know that if Assumption 1.1 holds for some $p \geq 0$, then an embedding of C_G as a line arrangement is always Cohen–Macaulay, as our singularities are planar. Moreover, Ballico and Franciosi [3] proved that a line bundle L on a reduced curve C satisfies property N_p under certain numerical conditions on the positivity of L with respect to subcurves constructed from an ordering of the irreducible components of C. Their hypothesis on the degree of L restricted to an irreducible component fails if G contains a cycle or if $p > 0$, and L has degree 1 on each line. However, if G is a tree, then Assumption 1.1 is automatically satisfied, so we expect that the ideal of C_G is 2-regular in this case. In fact, this follows from [12] because the lines in C_G can be ordered in such a way that the ith line intersects the span of the previous lines in a single point.

3 Line arrangements generated by products of linear forms

In this section we present an embedding of C_G into projective space if its edges can be labeled according to certain rules described below. If the ideal of C_G is generated by quadrics, then we identify conditions on the labeling that guarantee the existence of generators of the ideal of C_G that are monomial and binomial products of linear forms.

Given a graph G satisfying Assumption 1.1, construct $\tilde{G}$ from G by adding a loop to each vertex of degree 1 so that vertices of degree 1 in G are incident to two edges in $\tilde{G}$. For the induction in Theorem 3.7 we also need to allow the possibility of the addition of a loop at vertices with degree 2 in G. We describe the embedding of $C_G \subset \mathbb{P}^{d-g}$ by labeling the edges of $\tilde{G}$ with monomial and binomial linear forms in $S[x_0, \ldots, x_{d-g}]$ that indicate how coordinates of $\mathbb{P}^{d-g}$ parameterize each line L_v.

Label each edge of $\tilde{G}$ with a monomial x_i or a binomial $x_i - x_j$ subject to the following rules:

1. We require that each variable x_i appears as a monomial edge label exactly once.
2. Binomials only appear on non-loop edges.

3. Each edge labeled with a binomial is incident to a vertex with three incident edges.
4. If v has three incident edges, then they are labeled x_j, x_k, and $x_j - x_k$, where $j \neq k \in \{0, \dots, d-g\}$.

For a fixed graph G, it may be the case that some $\tilde{G}$ can be labeled according to these rules and others can not.

To define the ideal of L_v, let Ω_v be the set defined by deleting all of the variables appearing on the edges incident to v from the set of variables of S and then adding in the binomial edge label incident to v if v has only two incident edges in $\tilde{G}$. We let $I_v = \langle \Omega_v \rangle$ be the ideal of L_v. Thus, the line L_v is parameterized by the coordinates on the incident edges, with coordinates i and j equal if $x_i - x_j$ appears at v but x_i and x_j do not.

Example 3.1 The graph G below has $g = 2$ and $d = 5$.

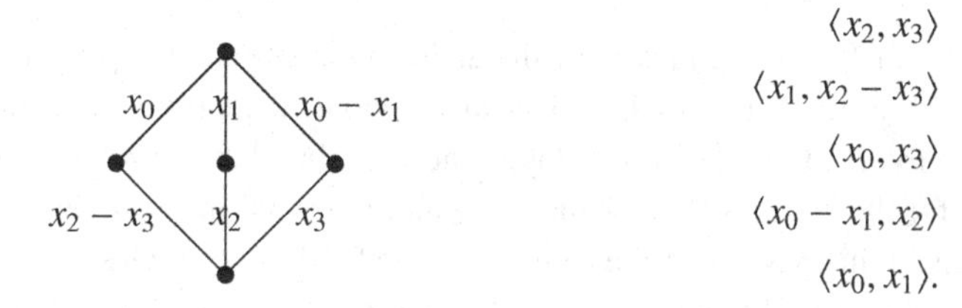

The ideals of the five lines are:

$$\langle x_2, x_3 \rangle$$
$$\langle x_1, x_2 - x_3 \rangle$$
$$\langle x_0, x_3 \rangle$$
$$\langle x_0 - x_1, x_2 \rangle$$
$$\langle x_0, x_1 \rangle.$$

Via Macaulay2 [18], the ideal of the arrangement is

$$\langle x_0x_2 - x_0x_3 + x_1x_3, x_1x_2x_3, x_0x_1x_3 - x_1^2x_3 \rangle.$$

The labeling gives rise to an embedding of C_G, but the ideal of this embedding is not generated by products of linear forms and is not generated by quadrics.

Theorem 3.2 *If G satisfies Assumption 1.1 for $p \geq 0$ and $\tilde{G}$ is labeled as described above, then $I = \cap_{v \in V} I_v$ is the ideal of an embedding of C_G into $\mathbb{P}^{d-g}$.*

Proof If u and v are connected by an edge in G and ℓ is the linear form on the edge that joins them, then the lines L_u and L_v intersect at the point of $\mathbb{P}^{d-g}$ that has coordinates appearing in ℓ set to 1 and all other coordinates set to 0.

To see that a labeling defines an embedding of C_G we must show that if u and v are not connected by an edge, then L_u and L_v do not intersect. If they intersect then a variable appearing on an edge incident to u must also appear on an edge incident to v. If x_i is an edge label at u, and v is incident to an edge with a label containing x_i, we must have the configuration on the left in Figure 1. But then the only coordinates of L_v with x_i nonzero also have x_j nonzero, and L_u does not contain any points with x_j nonzero unless w is trivalent and

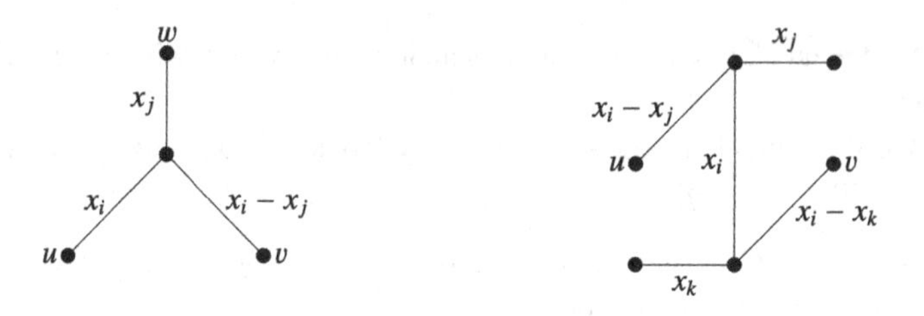

Figure 1

there is an edge labeled $x_j - x_k$ incident to u. However, this is not possible, because G does not contain any triangles. Hence, the lines cannot intersect.

The only other possibility is that x_i appears in a binomial at u and at v as in the diagram on the right in Figure 1. But then if the lines L_u and L_v intersect, the three coordinates x_i, x_j, x_k must all be nonzero and equal. This means that the edge labeled x_k must be incident to u and the edge labeled x_j must be incident to v. But this is forbidden because $d \geq 2g + 1$ for all subgraphs of genus g. □

Our method of labeling edges with linear forms is similar in spirit to the description of the generators of the ideal of a canonical graph curve (corresponding to a trivalent graph) in [4]. Bayer and Eisenbud label the edges in G with a basis for the space of 1-cochains of G and intersect an ideal generated by monomials in this basis with the ring generated by the 1-cocycles.

In order to describe the generators of I_{C_G} explicitly, we must make some further assumptions on the relative placement of labels.

Assumption 3.3 The labeling on $\tilde{G}$ satisfies the following conditions:

1. Incident edges never both have binomial labels. In other words, the labeling below never appears:

   $x_i - x_j$ $x_k - x_\ell$

   v

2. If v is a vertex of degree 2 as depicted below (with i, j, k distinct), then there are no other edges with labels containing x_i that are incident to edges with labels containing x_j or x_k:

   x_i $x_j - x_k$

   u v w

3. The vertices of G are ordered $v_1, \ldots, v_d$, G_i is the graph induced on $v_1, \ldots, v_i$, and $\tilde{G}_{i-1}$ is obtained from $\tilde{G}_i$ by removing v_i and replacing any non-loop edge uv_i labeled with a monomial by a loop at u labeled with the same monomial:

(a) G_i is connected;
(b) v_i has at most two incident edges in $\tilde{G}_i$;
(c) if v_i is connected to a vertex u in G_{i-1} via an edge labeled with a binomial, then u is incident to three edges in G_i. (i.e., L_u has a monomial ideal).

In what follows, let $G_{\hat{v}}$ denote the subgraph of G obtained by removing v and all of its incident edges. If C_G is embedded in $\mathbb{P}^{d-g}$, we let $C_{G_{\hat{v}}}$ be the corresponding subcurve. Note that if $\deg v = 1$ in G and we remove the line L_v from C_G embedded as above, then $C_{G_{\hat{v}}}$ is embedded as a line arrangement via a complete linear series in a hyperplane. If $\deg v = 2$ in G and v is contained in a cycle, then $G_{\hat{v}}$ is still connected, the genus drops by 1, and the remaining subcurve is embedded via a complete linear series. We do not allow the removal of vertices of degree 3 because if $\deg v = 3$, and $G_{\hat{v}}$ is connected, then the genus drops by 2 and $C_{G_{\hat{v}}}$ is not embedded via a complete linear series.

Lemma 3.4 *Suppose that Assumption 1.1 holds for some $p \geq 1$ and Assumption 3.3 also holds. If the configuration in part 2 of Assumption 3.3 appears in a labeling of $\tilde{G}$, then $x_i(x_j - x_k)$ is in the ideal of C_G and $x_i x_j$, $x_i x_k$ are in the ideal of $C_{G_{\hat{v}}}$.*

Proof To see that $x_i(x_j - x_k)$ is in the ideal of C_G, note that $x_j - x_k$ is in the ideal of L_v. Our hypotheses imply that for any vertex $v' \neq v$, if x_i appears on an incident edge, neither x_j nor x_k do, so $x_j - x_k$ is in the ideal of $L_{v'}$. Otherwise, x_i is in the ideal of $L_{v'}$. Hence, $x_i(x_j - x_k)$ is in the ideal of each line.

It is easy to see that neither $x_i x_j$ nor $x_i x_k$ vanish on L_v. Since this is the only line where the coordinate x_i is paired with x_j or x_k, it follows that these two monomials are contained in the ideal of $C_{G_{\hat{v}}}$. □

Example 3.5 If $g = 2$, G has precisely two trivalent vertices, and it satisfies Assumption 1.1 for some $p \geq 1$, then it can be labeled according to Assumption 3.3. If the cycles are disjoint, then G must consists of two cycles and a bridge between them. Putting one binomial label in each cycle satisfies Assumption 3.3 because each cycle has length at least 4.

If the cycles overlap, then we have three paths between trivalent vertices u and v. Label the shortest path with monomials and put one binomial label on each of the remaining paths. For example, the graph in Figure 2 satisfies Assumption 3.3 and has defining ideal $\langle x_3x_4, x_0x_4 - x_2x_4, x_0x_3 - x_1x_3, x_1x_2 \rangle$.

Theorem 3.6 *Suppose that G satisfies Assumption 1.1 with $p = 1$. Fix a $\tilde{G}$ and a labeling that gives an embedding of C_G into $\mathbb{P}^{d-g}$ as a line arrangement. If Assumption 3.3 is satisfied, and $I_{C_{G_i}}$ is generated by quadrics for all*

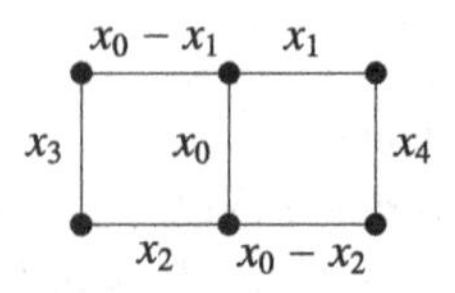

Figure 2

$i \geq 2$, then I_{C_G} is generated by elements of the form x_ix_j, $x_i(x_j - x_k)$, where the variables in each product are distinct.

Proof of Theorem 3.7 We proceed by induction on d. The result is easy to check when $d = 2$. Assume the result for all graphs on $d - 1$ vertices satisfying our hypotheses. Our hypotheses hold for G_i and $\tilde{G}_i$ for all $i \geq 2$. Let $v = v_{i+1}$. We may assume that $G = G_{i+1}$ and $G_i = G_{\hat{v}}$.

Case 1 *v has degree 1 in G.* The vertex v is incident to exactly one vertex $u \in G_{\hat{v}}$ with $u \neq v$. We may assume that L_v is spanned by a point p in $C_{G_{\hat{v}}}$ and the point $[0 : \cdots : 0 : 1]$. (By Assumption 3.3 part 3, all loops are labeled by monomials.) Then $I_{C_{G_{\hat{v}}}} = Q + \langle x_{d-g} \rangle$, where Q is generated by elements of the form x_ix_j and $x_i(x_j - x_k)$ in which no term is divisible by x_{d-g}.

We argue that $Q \subset I_{L_v}$. Let $q = fh$ be one of the generators of Q fixed above where f and h are linear forms. Since q must vanish at p, without loss of generality we may assume that f vanishes at p. Since x_{d-g} does not appear in f, then f must also vanish on $[0 : \cdots : 1]$. Thus, f is a linear form vanishing at two points of L_v; hence it must vanish on all of L_v. Therefore, $Q \subset I_{L_v}$. Thus, we see that $I_{C_G} = Q + \langle x_{d-g} \rangle \cdot I_{L_v}$. Moreover, we see that I_{C_G} is generated by the generators of Q and elements of the form $x_{d-g}x_i$ and $x_{d-g}(x_j - x_k)$.

Case 2 *v has degree 2 in G.* By Assumption 3.3 part 3, there cannot be a loop at v. Then there are two cases: without loss of generality, either the labels on the edges incident to v have the form x_0 and x_1 or they have the form x_0 and $x_1 - x_2$.

In the first case, we claim that $x_0x_1 \in I_{C_{G_{\hat{v}}}}$. Indeed, we have the configuration

x_0 x_1
u v w

If z is a vertex in $G_{\hat{v}}$ such that x_0 does not vanish on L_z, then x_0 must appear in a label on an edge incident at z. If $z = u$, then x_1 cannot appear in a binomial on any edge incident at z via Assumption 3.3 part 2, and so x_1 vanishes on L_z. If $z \neq u$, then $x_0 - x_j$ must appear on an edge incident at z. Again, if x_1 does not vanish on L_z, then it must appear on an edge incident at z. It cannot appear in a binomial by Assumption 3.3 part 1, in which case z must be equal to w, which creates a triangle. We conclude that either x_0 or x_1 vanishes on every irreducible component in $C_{G_{\hat{v}}}$, and hence that $x_0x_1 \in I_{C_{G_{\hat{v}}}}$.

Define a binomial minimal generator of $I_{C_{G_{\hat{v}}}}$ to be a binomial quadric in the ideal such that neither of its monomials is in $I_{C_{G_{\hat{v}}}}$. If $x_0(x_1 - x_i)$ is in $I_{C_{G_{\hat{v}}}}$, then so is x_0x_i. Hence we may assume that we have no minimal binomial generators of the form $x_0(x_1 - x_i)$. Similarly, we may assume that we have no generators of the form $x_1(x_0 - x_i)$.

The ideal of I_{C_G} is the intersection of $I_{C_{G_{\hat{v}}}}$ with $I_v = \langle x_2, \ldots, x_{d-g} \rangle$, and it is generated by quadrics. The only monomial quadrics not contained in $\langle x_2, \ldots, x_{d-g} \rangle$ are x_0^2, x_1^2, x_0x_1. The monomials x_0^2, x_1^2 do not appear in any minimal generator of $I_{C_{G_{\hat{v}}}}$. Since $x_0(x_1 - x_i)$ and $x_1(x_0 - x_i)$ are not generators of $I_{C_{G_{\hat{v}}}}$, every generator of the form x_ix_j and $x_i(x_j - x_k)$ must be in I_v, except for x_0x_1. Therefore, since I_{C_G} is generated by a space of quadrics whose dimension must be less than the dimension of the space of quadrics generating $I_{C_{G_{\hat{v}}}}$, we conclude that I_{C_G} is generated by the generators of $I_{C_{G_{\hat{v}}}}$ minus x_0x_1.

In the second case, x_0x_1 and x_0x_2 are in $I_{C_{G_{\hat{v}}}}$ but not I_{C_G} by Lemma 3.4 and I_{C_G} is the intersection of $I_{C_{G_{\hat{v}}}}$ with $I_v = \langle x_1 - x_2, \ldots, x_{d-g} \rangle$. We can find generators of $I_{C_{G_{\hat{v}}}}$ that have the form x_ix_j and $x_i(x_j - x_k)$. Note that all square-free monomials x_ix_j are in I_v except for x_0x_1, x_0x_2, and x_1x_2. The monomial x_1x_2 cannot be in the ideal of $C_{G_{\hat{v}}}$ because it contains the line parameterized by x_1 and x_2.

We claim that all of the binomial minimal generators of $I_{C_{G_{\hat{v}}}}$ are contained in I_v. If $x_i(x_j - x_k)$ is not contained in I_v, then i must be 0, 1, or 2. If it is 0, then exactly one of j and k is in the set $\{1, 2\}$. But then x_0x_1 and x_0x_2 are already in $I_{C_{G_{\hat{v}}}}$, so $x_0(x_j - x_k)$ is not a binomial minimal generator.

So, without loss of generality, assume $i = 1$. Let w be the trivalent vertex with L_w parameterized by x_1 and x_2, and note that $w \in G_{\hat{v}}$. If neither of j or k is in the set $\{0, 2\}$, then $x_i(x_j - x_k)$ is in I_v. If one of them is equal to 0, then $x_1(x_0 - x_k)$ is not a binomial minimal generator because $x_0x_1 \in I_{C_{G_{\hat{v}}}}$. So assume that we have $x_1(x_2 - x_k)$ with $k \neq 0, 1, 2$. Then x_1x_k is in the ideal of L_w. If $x_1(x_2 - x_k)$ were in $I_{C_{G_{\hat{v}}}}$ it would also have to be in the ideal of L_w. But $x_1x_k, x_1(x_2 - x_k)$ in the ideal of L_w would imply that x_1x_2 would also be in the ideal of L_w, which is a contradiction. Therefore, we have no generators of the form $x_1(x_2 - x_k)$.

We conclude that all of the monomial and binomial minimal generators of $I_{C_{G_{\hat{v}}}}$ are in I_{C_G} except for x_0x_1 and x_0x_2. But $x_0(x_1 - x_2) \in I_{C_G}$, and the conclusion follows as in the first case since we have identified a space of quadrics in I_{C_G} of dimension exactly one less than the dimension of the space of quadrics in $I_{C_{G_{\hat{v}}}}$. □

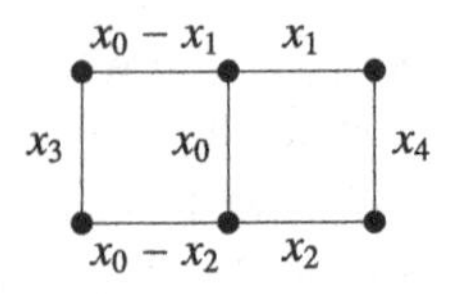

Figure 3

Via Corollary 2.5, if $g \leq 2$, we know that each $I_{C_{G_i}}$ is generated by quadrics and we obtain the following:

Corollary 3.7 *Suppose that G satisfies Assumption 1.1 for* $p = 1$*. Fix a* $\tilde{G}$ *and a labeling that gives an embedding of* C_G *into* $\mathbb{P}^{d-g}$ *as a line arrangement. If Assumption 3.3 is satisfied, and* $g \leq 2$*, then* I_{C_G} *is generated by elements of the form* $x_i x_j, x_i(x_j - x_k)$*, where the variables in each product are distinct.*

The result in Corollary 3.7 is sharp, as witnessed by the following example.

Example 3.8 Let G be the graph in Figure 3, where $d = 6$ and $g = 2$. Both of the vertices on the left fail part 1 of Assumption 3.3.

The ideal of the embedding corresponding to this labeling is

$$(x_3x_4, x_0x_4, x_0x_3 - x_1x_3 - x_2x_3, x_1x_2).$$

The terms in the trinomial do not appear in any other generators of the ideal. Therefore, it is impossible to find a set of minimal generators that does not contain an element with at least three terms.

Corollary 3.9 *If G satisfies Assumption 1.1 for* $p = 1$*,* $g \leq 2$*, and G has at most two trivalent vertices, then there exists an embedding of* $C_G \subset \mathbb{P}^{d-g}$ *so that* I_{C_G} *is generated by elements of the form* $x_i x_j$ *and* $x_i(x_j - x_k)$*.*

Proof Let the vertices of degree 2 with an incident edge labeled by a binomial be the last vertices in the order (so the first to get stripped off in the induction). Combine Corollary 2.5 with Example 3.5 and Corollary 3.7. □

4 Secant varieties and property $N_{k,p}$

In this section we show when $N_{3,p}$ must fail for the secant line variety Σ_1. The key idea of the proof comes from [12], whose authors state that their Theorem 1.1 has a natural generalization for higher-degree forms. We give a precise statement of a special case below:

Theorem 4.1 *Suppose that $X \subset \mathbb{P}^n$ is a variety that satisfies $N_{k,p}$. Let W be a linear subspace of dimension p with $Z = X \cap W$. If* $\dim Z = 0$, *then Z contains at most $\binom{p+k-1}{p}$ points.*

Proof It follows from the proof of Theorem 1.1 from [12] that the ideal of Z in the homogeneous coordinate ring of W is k-regular. Via Theorem 4.2 in [11], the degree in which the Hilbert function and the Hilbert polynomial of S_Z agree is the regularity of S_Z. We know that the Hilbert polynomial of S_Z is constant, equal to the number of points in Z. If $I(Z)$ is k-regular then S_Z is $k-1$-regular.

If $\dim(S_Z)_{k-1}$ is equal to the size of Z, then $\dim S_{k-1}$ must be at least the size of Z. Hence, $|Z| \leq \binom{p+k-1}{p}$. □

Corollary 4.2 *If C_G contains a cycle of m lines, then $N_{3,m-4}$ fails for the secant variety of C_G.*

Proof Since the m lines in the cycle are contained in a $\mathbb{P}^{m-1}$, so is the span of any subset of these lines. Thus, each 3-plane obtained by taking the span of non-adjacent lines in the cycle is contained in this $\mathbb{P}^{m-1}$. There are $\binom{m}{2} - m = \frac{1}{2}m(m-3)$ such 3-planes.

A general plane of dimension $m-4$ intersects a 3-plane in $\mathbb{P}^{m-1}$ in a point. Therefore, a general $(m-4)$-plane in this $\mathbb{P}^{m-1}$ intersects the secant variety of X in $\frac{1}{2}m(m-3)$ points. However, $\binom{m-4+3-1}{m-4} = \binom{m-2}{2} = \frac{1}{2}(m-2)(m-3)$. Thus, $N_{3,m-4}$ fails. □

5 Questions and conjectures

Computations with Macaulay 2 [18] were essential in all of our computations of embeddings of graph curves. In addition to the results proved in this paper we have several questions and conjectures regarding the defining equations and syzygies of graph curves and their secant varieties motivated by the examples that we have seen.

In Section 3 we saw that under certain hypotheses I_{C_G} is generated by products of linear forms that can be described explicitly in terms of the combinatorics of the graph G. The combinatorics of the kth secant variety of C_G is encoded in an *intersection lattice* whose elements are constructed by intersecting subsets of the subspaces. From the intersection lattice of an arrangement, we get a partially ordered set ordered by reverse inclusion of subspaces.

Question 1 Does the partially ordered set associated with the kth secant variety have any interesting combinatorial features? We conjecture that Σ_k is Cohen–Macaulay, so will the corresponding poset be shellable?

It is also natural to ask if there is an analogue of Theorem 3.7 for secant varieties, perhaps requiring additional hypotheses on the intersection lattice of the secant varieties of C_G.

Question 2 Are the secant varieties of C_G defined by products of linear forms?

Finding generators of I_{Σ_k} that are products of linear forms is equivalent to finding an explicit and special basis for the ideal that may have combinatorial interest. Of course, a module does not typically have a unique generating set or a unique minimal free graded resolution. However, the number of minimal generators of degree j of the ith syzygy module is invariant under a change of basis. Given a finitely generated graded module M, the graded Betti number $\beta_{i,j}$ is the number of minimal generators of degree j required at the ith stage of a minimal free graded resolution of M. A standard way of displaying the graded Betti numbers of a module is with a graded Betti diagram organized as follows:

	0	1	2	
0	$\beta_{0,0}$	$\beta_{1,1}$	$\beta_{2,2}$	$\cdots$
1	$\beta_{0,1}$	$\beta_{1,2}$	$\beta_{2,3}$	$\cdots$

Bounds on the number of rows and columns of the graded Betti diagram of a module give a rough sense of how complicated it is. Specifically, recall that the *regularity* of a finitely generated graded module M is equal to $\sup\{j - i \mid \beta_{i,j} \neq 0 \text{ for some } i\}$, and thus regularity gives a bound on the number of rows of the graded Betti diagram of M. Additionally, by the Auslander–Buchsbaum formula, a variety $X \subset \mathbb{P}^n$ is arithmetically Cohen–Macaulay if $\sup\{i \mid \beta_{i,j} \neq 0 \text{ for some } i\} = \operatorname{codim} X$, which bounds the number of columns of the graded Betti diagram off M. The following conjecture is the graph curve analogue of Conjecture 1.4 in [27], which refines conjectures from [29].

Conjecture 1 If Assumption 1.1 holds for some $p \geq 2k$, then the kth secant variety of C_G has regularity equal to $2k + 1$ and is arithmetically Cohen–Macaulay.

Note that as the secant varieties of C_G are not normal, we cannot expect projective normality.

In addition to bounding the length and width of the graded Betti diagram, we conjecture that under certain conditions one particular graded Betti number counts the number of cycles of minimal length in the graph. Recall that the *girth* of a graph is the length of its smallest cycle.

Conjecture 2 Let G be a graph on d vertices, embedded as in Theorem 1.3. Let n denote the girth of G. Assume that $d = 2g + 1 + p$ and $n - 2 \leq p$. Then property N_p fails and $\beta_{n-2,n}$ is equal to the number of cycles of length n in G.

Example 5.1 gives an illustration of the properties discussed in Conjectures 1, 2.

Example 5.1 ($g = 2, d = 10$) Let G be as given in Figure 4.

The ideal of C_G corresponding to this labeling is given below:

$$I_{C_G} = \begin{matrix} (x_5x_8, & x_4x_8, & x_3x_8, & x_2x_8, & x_1x_8, & x_6x_7, & x_5x_7, \\ x_2x_7, & x_1x_7, & x_0x_7, & x_4x_6, & x_3x_6, & x_2x_6, & x_1x_6, \\ x_3x_5, & x_2x_5, & x_1x_5, & x_0x_5, & x_2x_4, & x_1x_4, & x_0x_4, \\ x_1x_3, & x_0x_3, & x_0x_2, & x_3x_7 - x_4x_7, & x_0x_8 - x_6x_8). & & \end{matrix}$$

The graded Betti diagram of S/I_{C_G} shows that $N_{2,5}$ fails as $\beta_{5,7} = 2$. As Conjecture 2 predicts, the girth of G is 7, and G contains precisely two cycles of length 7:

	0	1	2	3	4	5	6	7
total:	1	26	98	168	154	72	15	2
0:	1	.	.	.	.	.	.	..
1:	.	26	98	168	154	70	8	.
2:	.	.	.	.	.	2	7	2

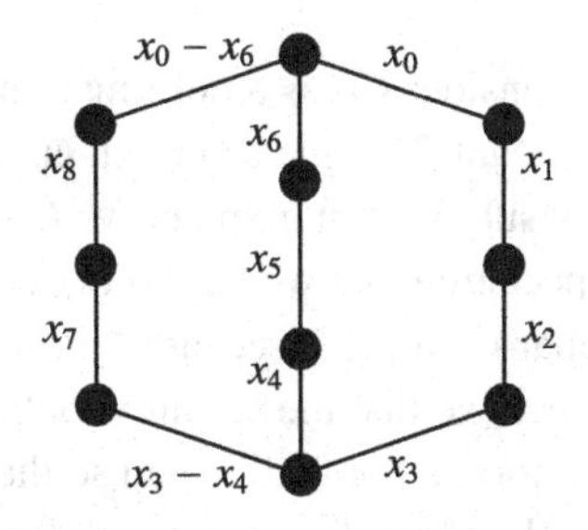

Figure 4

We can also compute the ideal of Σ:

$$I(\Sigma) = \begin{array}{llll} (x_3x_5x_8, & x_2x_5x_8, & x_1x_5x_8, & x_0x_4x_8 - x_4x_6x_8, \\ x_2x_4x_8, & x_1x_4x_8, & x_1x_3x_8, & x_0x_3x_8 - x_3x_6x_8, \\ x_2x_6x_7, & x_1x_6x_7, & x_2x_5x_7, & x_0x_2x_8 - x_2x_6x_8, \\ x_1x_5x_7, & x_0x_5x_7, & x_0x_2x_7, & x_3x_6x_7 - x_4x_6x_7, \\ x_2x_4x_6, & x_1x_4x_6, & x_1x_3x_6, & x_1x_3x_7 - x_1x_4x_7, \\ x_1x_3x_5, & x_0x_3x_5, & x_0x_2x_5, & x_0x_3x_7 - x_0x_4x_7, \\ x_0x_2x_4) & & & \end{array}$$

and its graded Betti diagram

	0	1	2	3	4	5
total:	1	25	58	43	12	3
0:	1	.	.	.	.	.
1:	.	.	.	.	.	.
2:	.	25	58	41	.	.
3:	.	.	.	.	7	.
4:	.	.	.	2	5	3

We see that $N_{3,3}$ fails for Σ and that $\beta_{3,7} = 2$, which is the number of cycles of length equal to the girth of G. We can also see from the graded Betti diagram that Σ is arithmetically Cohen–Macaulay and that $I(\Sigma)$ has regularity 5.

It is natural to ask if combinatorics can be used to compute other values of the $\beta_{i,j}$. One result that gives the flavor of what might be possible is due to Gasharov, Peeva, and Welker [17], who used the lcm lattice of a monomial ideal to compute graded Betti numbers of monomial ideals.

Question 3 Is there an analogue of the lcm lattice for graph curves and their secant varieties that would allow us to compute (or estimate) the graded Betti numbers of graph curves?

Further work on understanding the graded Betti numbers of graph curves has been done in [6].

It is also interesting to consider C_G as a deformation of a smooth curve. In Example 5.1, C_G has a 7-secant $\mathbb{P}^5$ while a smooth curve of genus 2 in $\mathbb{P}^8$ has no such $\mathbb{P}^5$. As any strictly subtrivalent graph curve $C_G \subset \mathbb{P}^n$ is smoothable in $\mathbb{P}^n$ [21, 29.9], it is our expectation that we have a family of seven 6-secant $\mathbb{P}^5$s to smooth curves that collapse to the 7-secant $\mathbb{P}^5$ in the singular limit C_G. It also seems reasonable to believe that the secant varieties to embedded curves in a flat family themselves form a flat family, and so the secant varieties to C_G should, in particular, have the same dimension and degree as those to smooth curves. In fact, since each pair of disjoint lines in C_G spans a $\mathbb{P}^3$, we have a

3-dimensional secant plane for each edge in the complement of the graph G. If C_G has degree d and genus g, then G has d vertices and $d + g - 1$ edges. Thus, the number of edges in the complement of G is $\binom{d}{2} - d - g + 1 = \binom{d-1}{2} - g$, which is the degree of the secant variety of a smooth curve of degree d and genus g.

Acknowledgments The third author thanks the Abel Symposium "Combinatorial Aspects of Commutative Algebra and Algebraic Geometry" from 2009, where David Eisenbud and Mike Stillman suggested the problem of considering graph curves and their secant varieties. We thank Ian Barnett, David Cox, David Eisenbud, Marco Franciosi, and Jenia Tevelev for helpful correspondences and conversations. We are very grateful to an anonymous referee who led us to make significant and necessary changes to the original version of this paper[1]. This work began at the Mount Holyoke College Summer Mathematics Institute, funded by NSF grant DMS-0849637. The third author was supported by NSF grant DMS-0600471. Finally, the third author wishes to thank Rob Lazarsfeld for introducing her to algebraic geometry, syzygies, and subspace arrangements.

References

[1] Ballico, E. On the gonality of graph curves. *Manuscripta Math.*, **129**(2): 169–180, 2009.

[2] Ballico, E. Graph curves with gonality one. *Rend. Semin. Mat. Univ. Politec. Torino*, **68**(1): 17–28, 2010.

[3] Ballico, E. and Franciosi, M. On property N_p for algebraic curves. *Kodai Math. J.*, **23**(3): 432–441, 2000.

[4] Bayer, D. and Eisenbud, D. Graph curves. *Adv. Math.*, **86**(1): 1–40, 1991. With an appendix by Sung Won Park.

[5] Björner, A., Peeva, I., and Sidman, J. Subspace arrangements defined by products of linear forms. *J. London Math. Soc. (2)*, **71**(2): 273–288, 2005.

[6] Bruce, D., Kao, P.-H., Nash, E., and Vermeire, P. Betti tables of reducible algebraic curves. *Proc. Amer. Math. Soc.*, in press.

[7] Catanese, F., Franciosi, M., Hulek, K., and Reid, M. Embeddings of curves and surfaces. *Nagoya Math. J.*, **154**: 185–220, 1999.

[8] Ciliberto, C. Harris, J., and Miranda, R. On the surjectivity of the Wahl map. *Duke Math. J.*, **57**(3): 829–858, 1988.

[9] Ciliberto, C. and Miranda, R. Graph curves, colorings, and matroids. In *Zero-Dimensional Schemes (Ravello, 1992)*, pp. 89–111. Berlin: de Gruyter, 1994.

[10] Ein, L. and Lazarsfeld, R. Asymptotic syzygies of algebraic varieties. *Invent. Math.*, **190**(3): 603–646, 2012.

[1] Readers of [26] should note that references there refer to the original version of this paper, and results are not in the same place or form here.

[11] Eisenbud, D. *The Geometry of Syzygies*: Volume 229 of *Graduate Texts in Mathematics*. New York: Springer-Verlag, 2005. A second course in commutative algebra and algebraic geometry.

[12] Eisenbud, D., Green, M., Hulek, K., and Popescu, S. Restricting linear syzygies: algebra and geometry. *Compos. Math.*, **141**(6): 1460–1478, 2005.

[13] Eisenbud, D., Koh, J., and Stillman, M. Determinantal equations for curves of high degree. *Amer. J. Math.*, **110**(3): 513–539, 1988.

[14] Eisenbud, D. and Koh, J.-H. Remarks on points in a projective space. In *Commutative Algebra (Berkeley, CA, 1987)*. Vol. 15 of Mathematical Sciences Research Institute Publications, pp. 157–172. New York: Springer-Verlag, 1989.

[15] Franciosi, M. and Elisa, T. The canonical ring of a 3-connected curve*. arXiv:1107.5535v2.

[16] Gallego, F. J. and Purnaprajna, B. P. Projective normality and syzygies of algebraic surfaces. *J. Reine Angew. Math.*, **506**: 145–180, 1999.

[17] Gasharov, V., Peeva, I., and Welker, V. The lcm-lattice in monomial resolutions. *Math. Res. Lett.*, **6**(5 & 6): 521–532, 1999.

[18] Grayson, D. R. and Stillman, M. E. Macaulay2, a software system for research in algebraic geometry. Available at http://www.math.uiuc.edu/Macaulay2/.

[19] Green, M. and Lazarsfeld, R. Some results on the syzygies of finite sets and algebraic curves. *Compos. Math.*, **67**(3): 301–314, 1988.

[20] Green, M. L. The Eisenbud–Koh–Stillman conjecture on linear syzygies. *Invent. Math.*, **136**(2): 411–418, 1999.

[21] Hartshorne, R. *Deformation Theory*. Vol. 257 of *Graduate Texts in Mathematics*. New York: Springer-Verlag, 2010.

[22] Hering, M., Schenck, H., and Smith, G. G. Syzygies, multigraded regularity and toric varieties. *Compos. Math.*, **142**(6): 1499–1506, 2006.

[23] Li, W. C. W. and Li, S.-Y. R. On generators of ideals associated with unions of linear varieties. *Bull. London Math. Soc.*, **13**(1): 59–65, 1981.

[24] Lovász, L. Stable sets and polynomials. *Discrete Math.*, **124**(1–3): 137–153, 1994. *Graphs and Combinatorics (Qawra, 1990)*.

[25] Ottaviani, G. and Paoletti, R. Syzygies of Veronese embeddings. *Compositio Math.*, **125**(1): 31–37, 2001.

[26] Schenck, H. and Sidman, J. Commutative algebra of subspace and hyperplane arrangements. In *Commutative Algebra*, pp. 639–665. New York: Springer-Verlag, 2013.

[27] Sidman, J. and Vermeire, P. Syzygies of the secant variety of a curve. *Algebra Number Theory*, **3**(4): 445–465, 2009.

[28] Sturmfels, B. and Sullivant, S. Combinatorial secant varieties. *Pure Appl. Math. Q.*, **2**(3, part 1): 867–891, 2006.

[29] Vermeire, P. Regularity and normality of the secant variety to a projective curve. *J. Algebra*, **319**(3): 1264–1270, 2008.

5

Rationally connected manifolds and semipositivity of the Ricci curvature

F. Campana
Université de Lorraine

J.-P. Demailly
Université de Grenoble I

Th. Peternell
Universität Bayreuth

Abstract

This paper establishes a structure theorem for compact Kähler manifolds with semipositive anticanonical bundle. Up to finite étale cover, it is proved that such manifolds split holomorphically and isometrically as a product of Ricci flat varieties and of rationally connected manifolds. The proof is based on a characterization of rationally connected manifolds through the non-existence of certain twisted contravariant tensor products of the tangent bundle, along with a generalized holonomy principle for pseudoeffective line bundles. A crucial ingredient for this is the characterization of uniruledness by the property that the anticanonical bundle is not pseudoeffective.

Dedicated to Rob Lazarsfeld on the occasion of this 60th birthday

1 Main results

The goal of this work is to understand the geometry of compact Kähler manifolds with semipositive Ricci curvature, and especially to study the relations that tie Ricci semipositivity with rational connectedness. Many of the ideas are borrowed from [DPS96] and [BDPP]. Recall that a compact complex manifold X is said to be rationally connected if any two points of X can be joined by a chain of rational curves. A line bundle L is said to be

From *Recent Advances in Algebraic Geometry*, edited by Christopher Hacon, Mircea Mustaţă and Mihnea Popa

hermitian semipositive if it can be equipped with a smooth hermitian metric of semipositive curvature form. A sufficient condition for hermitian semipositivity is that some multiple of L is spanned by global sections; on the other hand, the hermitian semipositivity condition implies that L is numerically effective (nef) in the sense of [DPS94], which, for X projective algebraic, is equivalent to saying that $L \cdot C \geq 0$ for every curve C in X. Examples contained in [DPS94] show that all three conditions are different (even for X projective algebraic). The Ricci curvature is the curvature of the anticanonical bundle $K_X^{-1} = \det(T_X)$, and by Yau's solution of the Calabi conjecture (see [Aub76], [Yau78]), a compact Kähler manifold X has a hermitian semipositive anticanonical bundle K_X^{-1} if and only if X admits a Kähler metric ω with $\mathrm{Ricci}(\omega) \geq 0$. A classical example of a projective surface with K_X^{-1} nef is the complex projective plane $\mathbb{P}^2_{\mathbb{C}}$ blown-up in nine points, no three of which are collinear and no six of which lie on a conic; in that case Brunella [Bru10] showed that there are configurations of the nine points for which K_X^{-1} admits a smooth (but non-real analytic) metric with semipositive Ricci curvature; depending on some diophantine condition introduced in [Ued82], there are also configurations for which some multiple K_X^{-m} of K_X^{-1} is generated by sections and others for which K_X^{-1} is nef without any smooth metric. Finally, let us recall that a line bundle $L \to X$ is said to be pseudoeffective if there exists a singular hermitian metric h on L such that the Chern curvature current $T = i\Theta_{L,h} = -i\partial\overline{\partial}\log h$ is non-negative; equivalently, if X is projective algebraic, this means that the first Chern class $c_1(L)$ belongs to the closure of the cone of effective $\mathbb{Q}$-divisors.

We first give a criterion characterizing rationally connected manifolds by the non-existence of sections in certain twisted tensor powers of the cotangent bundle; this is only a minor variation of Theorem 5.2 in [Pet06], cf. also Remark 5.3 therein.

1.1 Criterion for rational connectedness *Let X be a projective algebraic n-dimensional manifold. The following properties are equivalent:*

- *(a) X is rationally connected.*
- *(b) For every invertible subsheaf $\mathcal{F} \subset \Omega_X^p := \mathcal{O}(\Lambda^p T_X^*)$, $1 \leq p \leq n$, $\mathcal{F}$ is not pseudoeffective.*
- *(c) For every invertible subsheaf $\mathcal{F} \subset \mathcal{O}((T_X^*)^{\otimes p})$, $p \geq 1$, $\mathcal{F}$ is not pseudo-effective.*
- *(d) For some (resp. for any) ample line bundle A on X, there exists a constant $C_A > 0$ such that*

$$H^0(X, (T_X^*)^{\otimes m} \otimes A^{\otimes k}) = 0 \qquad \text{for all } m, k \in \mathbb{N}^* \text{ with } m \geq C_A k.$$

1.2 Remark The proof follows easily from the uniruledness criterion established in [BDPP]: a non-singular projective variety X is uniruled if and only if

K_X is not pseudoeffective. A conjecture attributed to Mumford asserts that the weaker assumption

(*d'*) $H^0(X, (T^*_X)^{\otimes m}) = 0$

for all $m \geq 1$ should be sufficient to imply rational connectedness. Mumford's conjecture can actually be proved by essentially the same argument if one uses the abundance conjecture in place of the more demanding uniruledness criterion from [BDPP] – more specifically that $H^0(X, K_X^{\otimes m}) = 0$ for all $m \geq 1$ would imply uniruledness.

1.3 Remark By [DPS94], Criteria 1.1 (b) and (c) make sense on an arbitrary compact complex manifold and imply that $H^0(X, \Omega^2_X) = 0$. If X is assumed to be compact Kähler, then X is automatically projective algebraic by Kodaira [Kod54], therefore, 1.1 (b) or (c) also characterizes rationally connected manifolds among all compact Kähler ones. □

The following structure theorem generalizes the Bogomolov–Kobayashi–Beauville structure theorem for Ricci-flat manifolds ([Bog74a], [Bog74b], [Kob81], [Bea83]) to the Ricci semipositive case. Recall that a *holomorphic symplectic manifold* X is a compact Kähler manifold admitting a holomorphic symplectic 2-form ω (of maximal rank everywhere); in particular, $K_X = \mathcal{O}_X$. A *Calabi–Yau* manifold is a simply connected projective manifold with $K_X = \mathcal{O}_X$ and $H^0(X, \Omega^p_X) = 0$ for $0 < p < n = \dim X$ (or a finite étale quotient of such a manifold).

1.4 Structure theorem *Let X be a compact Kähler manifold with K_X^{-1} hermitian semipositive. Then*

(a) The universal cover $\widetilde{X}$ admits a holomorphic and isometric splitting

$$\widetilde{X} \simeq \mathbb{C}^q \times \prod Y_j \times \prod S_k \times \prod Z_\ell$$

where Y_j, S_k, Z_ℓ are compact simply connected Kähler manifolds of respective dimensions n_j, n'_k, n''_ℓ with irreducible holonomy, Y_j being Calabi–Yau manifolds (holonomy $\mathrm{SU}(n_j)$), S_k holomorphic symplectic manifolds (holonomy $\mathrm{Sp}(n'_k/2)$), and Z_ℓ rationally connected manifolds with $K_{Z_\ell}^{-1}$ semipositive (holonomy $\mathrm{U}(n''_\ell)$).

(b) There exists a finite étale Galois cover $\widehat{X} \to X$ such that the Albanese variety $\mathrm{Alb}(\widehat{X})$ is a q-dimensional torus and the Albanese map $\alpha\colon \widehat{X} \to \mathrm{Alb}(\widehat{X})$ is an (isometrically) locally trivial holomorphic fiber bundle whose fibers are products $\prod Y_j \times \prod S_k \times \prod Z_\ell$ of the type described in (a). Even more holds after possibly another finite étale cover: $\hat{X}$ is a fiber bundle with fiber $\prod Z_\ell$ on $\prod Y_j \times \prod S_k \times \mathrm{Alb}(\widehat{X})$.

(c) *We have* $\pi_1(\widehat{X}) \simeq \mathbb{Z}^{2q}$ *and* $\pi_1(X)$ *is an extension of a finite group* Γ *by the normal subgroup* $\pi_1(\widehat{X})$. *In particular, there is an exact sequence*

$$0 \to \mathbb{Z}^{2q} \to \pi_1(X) \to \Gamma \to 0,$$

and the fundamental group $\pi_1(X)$ *is almost abelian.*

The proof relies on the holonomy principle, De Rham's splitting theorem [DR52], and Berger's classification [Ber55]. Foundational background can be found in papers by Lichnerowicz [Lic67], [Lic71] and Cheeger–Gromoll [CG71], [CG72]. The restricted holonomy group of a hermitian vector bundle (E, h) of rank r is by definition the subgroup $H \subset \mathrm{U}(r) \simeq U(E_{z_0})$ generated by parallel transport operators with respect to the Chern connection ∇ of (E, h), along loops based at z_0 that are contractible (up to conjugation, H does not depend on the base point z_0). We need here a generalized "pseudoeffective" version of the holonomy principle, which can be stated as follows.

1.5 Generalized holonomy principle *Let* E *be a holomorphic vector bundle of rank* r *over a compact complex manifold* X. *Assume that* E *is equipped with a smooth hermitian structure* h *and* X *with a hermitian metric* ω, *viewed as a smooth positive* $(1,1)$*-form* $\omega = i \sum \omega_{jk}(z) dz_j \wedge d\overline{z}_k$. *Finally, suppose that the* ω*-trace of the Chern curvature tensor* $\Theta_{E,h}$ *is semipositive, that is*

$$i\Theta_{E,h} \wedge \frac{\omega^{n-1}}{(n-1)!} = B \frac{\omega^n}{n!}, \qquad B \in \mathrm{Herm}(E, E), \quad B \geq 0 \text{ on } X,$$

and denote by H *the restricted holonomy group of* (E, h).

(a) *If there exists an invertible sheaf* $\mathcal{L} \subset \mathcal{O}((E^*)^{\otimes m})$ *which is pseudoeffective as a line bundle, then* $\mathcal{L}$ *is flat and* $\mathcal{L}$ *is invariant under parallel transport by the connection of* $(E^*)^{\otimes m}$ *induced by the Chern connection* ∇ *of* (E, h); *in fact,* H *acts trivially on* $\mathcal{L}$.

(b) *If* H *satisfies* $H = \mathrm{U}(r)$, *then none of the invertible subsheaves* $\mathcal{L}$ *of* $\mathcal{O}((E^*)^{\otimes m})$ *can be pseudoeffective for* $m \geq 1$.

The generalized holonomy principle is based on an extension of the Bochner formula as found in [BY53], [Kob83]: for (X, ω) Kähler, every section u in $H^0(X, (T_X^*)^{\otimes m})$ satisfies

$$\Delta(\|u\|^2) = \|\nabla u\|^2 + Q(u), \tag{1.6}$$

where $Q(u) \geq m\lambda_1 \|u\|^2$ is bounded from below by the smallest eigenvalue λ_1 of the Ricci curvature tensor of ω. If $\lambda_1 \geq 0$, the equality $\int_X \Delta(\|u\|^2)\omega^n = 0$ implies $\nabla u = 0$ and $Q(u) = 0$. The generalized principle consists essentially of considering a general vector bundle E rather than $E = T_X^*$, and replacing

$\|u\|^2_\omega$ with $\|u\|^2_\omega e^\varphi$ where u is a local trivializing section of $\mathcal{L}$, where φ is the corresponding local plurisubharmonic weight representing the metric of $\mathcal{L}$ and ω a Gauduchon metric, cf. (3.2).

1.7 Remark If one makes the weaker assumption that K_X^{-1} is nef, then Qi Zhang [Zha96, Zha05] proved that the Albanese mapping $\alpha\colon X \to \mathrm{Alb}(X)$ is surjective in the case where X is projective, and Păun [Pau12] recently extended this result to the general Kähler case (cf. also [CPZ03]). One may wonder whether there still exists a holomorphic splitting

$$\widetilde{X} \simeq \mathbb{C}^q \times \prod Y_j \times \prod S_k \times \prod Z_\ell$$

of the universal covering as above. However, the example where $X = \mathbb{P}(E)$ is the ruled surface over an elliptic curve $C = \mathbb{C}/(\mathbb{Z} + \mathbb{Z}\tau)$ associated with a nontrivial rank 2 bundle $E \to C$ with

$$0 \to \mathcal{O}_C \to E \to \mathcal{O}_C \to 0$$

shows that $\widetilde{X} = \mathbb{C} \times \mathbb{P}^1$ cannot be an isometric product for a Kähler metric ω on X. Actually, such a situation would imply that $K_X^{-1} = \mathcal{O}_{\mathbb{P}(E)}(1)$ is semipositive, but we know by [DPS94] that $\mathcal{O}_{\mathbb{P}(E)}(1)$ is nef and non-semipositive. Under the mere assumption that K_X^{-1} is nef, it is unknown whether the Albanese map $\alpha\colon X \to \mathrm{Alb}(X)$ is a submersion, unless X is a projective threefold [PS98], and even if it is supposed to be so, it seems to be unknown whether the fibers of α may exhibit nontrivial variation of the complex structure (and whether they are actually products of Ricci flat manifolds by rationally connected manifolds). The main difficulty is that, a priori, the holonomy argument used here breaks down – a possibility would be to consider some sort of "asymptotic holonomy" for a sequence of Kähler metrics satisfying $\mathrm{Ricci}(\omega_\varepsilon) \geq -\varepsilon\omega_\varepsilon$, and dealing with the Gromov–Hausdorff limit of the variety. □

2 Proof of the criterion for rational connectedness

In this section we prove Criterion 1.1. Observe first that if X is rationally connected, then there exists an immersion $f\colon \mathbb{P}^1 \subset X$ passing through any given finite subset of X such that f^*T_X is ample, see e.g. [Kol96, Theorem 3.9, p. 203]. In other words $f^*T_X = \bigoplus \mathcal{O}_{\mathbb{P}^1}(a_j)$, $a_j > 0$, while $f^*A = \mathcal{O}_{\mathbb{P}^1}(b)$, $b > 0$. Hence

$$H^0(\mathbb{P}^1, f^*((T_X^*)^{\otimes m} \otimes A^{\otimes k})) = 0 \qquad \text{for } m > kb/\min(a_i).$$

As the immersion f moves freely in X, we immediately see from this that 1.1 (a) implies 1.1 (d) with any constant value $C_A > b/\min(a_j)$.

To see that 1.1 (d) implies 1.1 (c), assume that $\mathcal{F} \subset (T_X^*)^{\otimes p}$ is a pseudoeffective line bundle. Then there exists $k_0 \gg 1$ such that

$$H^0(X, \mathcal{F}^{\otimes m} \otimes A^{k_0}) \neq 0$$

for all $m \geq 0$ (for this, it is sufficient to take k_0 such that $A^{k_0} \otimes (K_X \otimes G^{n+1})^{-1} > 0$ for some very ample line bundle G). This implies $H^0(X, (T_X^*)^{\otimes mp} \otimes A^{k_0}) \neq 0$ for all m, contradicting assumption 1.1 (d).

The implication 1.1 (c) $\Rightarrow$ 1.1 (b) is trivial.

It remains to show that 1.1 (b) implies 1.1 (a). First note that K_X is not pseudoeffective, as one sees by applying the assumption 1.1 (b) with $p = n$. Hence X is uniruled by [BDPP]. We consider the quotient with maximal rationally connected fibers (rational quotient or MRC fibration, see [Cam92], [KMM92])

$$f : X \dashrightarrow W$$

to a smooth projective variety W. By [GHS01], W is not uniruled, otherwise we could lift the ruling to X and the fibers of f would not be maximal. We may further assume that f is holomorphic. In fact, assumption 1.1 (b) is invariant under blow-ups. To see this, let $\pi : \hat{X} \to X$ be a birational morphism from a projective manifold $\hat{X}$ and consider a line bundle $\hat{\mathcal{F}} \subset \Omega^p_{\hat{X}}$. Then $\pi_*(\hat{\mathcal{F}}) \subset \pi_*(\Omega^p_{\hat{X}}) = \Omega^p_X$, hence we introduce the line bundle

$$\mathcal{F} := (\pi_*(\hat{\mathcal{F}}))^{**} \subset \Omega^p_X.$$

Now, if $\hat{\mathcal{F}}$ were pseudoeffective, so would $\mathcal{F}$ be. Thus 1.1 (b) is invariant under π and we may suppose f holomorphic. In order to show that X is rationally connected, we need to prove that $p := \dim W = 0$. Otherwise, $K_W = \Omega^p_W$ is pseudoeffective by [BDPP] and we obtain a pseudoeffective invertible subsheaf $\mathcal{F} := f^*(\Omega^p_W) \subset \Omega^p_X$, in contradiction to 1.1 (b). □

3 Bochner formula and generalized holonomy principle

Let (E, h) be a hermitian holomorphic vector bundle over an n-dimensional compact complex manifold X. The semipositivity hypothesis on $B = \mathrm{Tr}_\omega \Theta_{E,h}$ is invariant by a conformal change of metric ω. Without loss of generality we can assume that ω is a Gauduchon metric, i.e., that $\partial\overline{\partial}\omega^{n-1} = 0$ (cf. [Gau77]). We consider the Chern connection ∇ on (E, h) and the corresponding parallel transport operators. At every point $z_0 \in X$, there exists a local coordinate system $(z_1, \ldots, z_n)$ centered at z_0 (i.e., $z_0 = 0$ in coordinates), and a holomorphic frame $(e_\lambda(z))_{1\le\lambda\le r}$ such that

$$\langle e_\lambda(z), e_\mu(z)\rangle_h = \delta_{\lambda\mu} - \sum_{1\le j,k\le n} c_{jk\lambda\mu} z_j \bar z_k + O(|z|^3),\ 1 \le \lambda, \mu \le r, \tag{3.1}$$

$$\Theta_{E,h}(z_0) = \sum_{1\le j,k,\lambda,\mu\le n} c_{jk\lambda\mu} dz_j \wedge d\bar z_k \otimes e_\lambda^* \otimes e_\mu,\ c_{kj\mu\lambda} = \overline{c_{jk\lambda\mu}}, \tag{3.1'}$$

where $\delta_{\lambda\mu}$ is the Kronecker symbol and $\Theta_{E,h}(z_0)$ is the curvature tensor of the Chern connection ∇ of (E, h) at z_0.

Assume that we have an invertible sheaf $\mathcal{L} \subset O((E^*)^{\otimes m})$ that is pseudoeffective. There exist a covering U_j by coordinate balls and holomorphic sections f_j of $\mathcal{L}_{|U_j}$ generating $\mathcal{L}$ over U_j. Then $\mathcal{L}$ is associated with the Čech cocycle g_{jk} in O_X^* such that $f_k = g_{jk} f_j$, and the singular hermitian metric $e^{-\varphi}$ of $\mathcal{L}$ is defined by a collection of plurisubharmonic functions $\varphi_j \in \mathrm{PSH}(U_j)$ such that $e^{-\varphi_k} = |g_{jk}|^2 e^{-\varphi_j}$. It follows that we have a globally defined bounded measurable function

$$\psi = e^{\varphi_j} \|f_j\|^2 = e^{\varphi_j} \|f_j\|^2_{h^{*m}}$$

over X, which can be viewed also as the hermitian metric ratio $(h^*)^m/e^{-\varphi}$ along $\mathcal{L}$, i.e., $\psi = (h^*)^m_{|\mathcal{L}} e^{\varphi}$. We are going to compute the Laplacian $\Delta_\omega \psi$. For simplicity of notation, we omit the index j and consider a local holomorphic section f of $\mathcal{L}$ and a local weight $\varphi \in \mathrm{PSH}(U)$ on some open subset U of X. In a neighborhood of an arbitrary point $z_0 \in U$, we write

$$f = \sum_{\alpha\in\mathbb{N}^m} f_\alpha\, e_{\alpha_1}^* \otimes \ldots \otimes e_{\alpha_m}^*, \qquad f_\alpha \in O(U),$$

where (e_λ^*) is the dual holomorphic frame of (e_λ) in $O(E^*)$. The hermitian matrix of (E^*, h^*) is the transpose of the inverse of the hermitian matrix of (E, h), hence (3.1) implies

$$\langle e_\lambda^*(z), e_\mu^*(z)\rangle_h = \delta_{\lambda\mu} + \sum_{1\le j,k\le n} c_{jk\mu\lambda} z_j \bar z_k + O(|z|^3), \qquad 1 \le \lambda, \mu \le r.$$

On the open set U the function $\psi = (h^*)^m_{|\mathcal{L}} e^{\varphi}$ is given by

$$\psi = \Big(\sum_{\alpha\in\mathbb{N}^m} |f_\alpha|^2 + \sum_{\alpha,\beta\in\mathbb{N}^m,\,1\le j,k\le n,\,1\le\ell\le m} f_\alpha \overline{f_\beta}\, c_{jk\beta_\ell\alpha_\ell} z_j \bar z_k + O(|z|^3)|f|^2\Big) e^{\varphi(z)}.$$

By taking $i\partial\bar\partial(\ldots)$ of this at $z = z_0$ in the sense of distributions (that is, for almost every $z_0 \in X$), we find

$$i\partial\bar\partial\psi = e^{\varphi}\Big(|f|^2 i\partial\bar\partial\varphi + i\langle \partial f + f\partial\varphi, \partial f + f\partial\varphi\rangle + \sum_{\alpha,\beta,j,k,1\le\ell\le m} f_\alpha \overline{f_\beta}\, c_{jk\beta_\ell\alpha_\ell}\, i dz_j \wedge d\bar z_k\Big).$$

Since $i\partial\overline{\partial}\psi \wedge \frac{\omega^{n-1}}{(n-1)!} = \Delta_\omega\psi\,\frac{\omega^n}{n!}$ (we actually take this as a definition of Δ_ω), a multiplication by ω^{n-1} yields the fundamental inequality

$$\Delta_\omega\psi \geq |f|^2 e^{\varphi}(\Delta_\omega\varphi + m\lambda_1) + |\nabla_h^{1,0} f + f\partial\varphi|^2_{\omega,h^{*m}}\, e^{\varphi}, \tag{3.2}$$

where $\lambda_1(z) \geq 0$ is the lowest eigenvalue of the hermitian endomorphism $B = \mathrm{Tr}_\omega\,\Theta_{E,h}$ at an arbitrary point $z \in X$. As $\partial\overline{\partial}\omega^{n-1} = 0$, we have

$$\int_X \Delta\psi\frac{\omega^n}{n!} = \int_X i\partial\overline{\partial}\psi \wedge \frac{\omega^{n-1}}{(n-1)!} = \int_X \psi \wedge \frac{i\partial\overline{\partial}(\omega^{n-1})}{(n-1)!} = 0$$

by Stokes' formula. Since $i\partial\overline{\partial}\varphi \geq 0$, (3.2) implies $\Delta_\omega\varphi = 0$, i.e., $i\partial\overline{\partial}\varphi = 0$, and $\nabla_h^{1,0} f + f\partial\varphi = 0$ almost everywhere. This means in particular that the line bundle $(\mathcal{L}, e^{-\varphi})$ is flat. In each coordinate ball U_j the pluriharmonic function φ_j can be written $\varphi_j = w_j + \overline{w}_j$ for some holomorphic function $w_j \in \mathcal{O}(U_j)$, hence $\partial\varphi_j = dw_j$ and the condition $\nabla_h^{1,0} f_j + f_j\partial\varphi_j = 0$ can be rewritten $\nabla_h^{1,0}(e^{w_j} f_j) = 0$ where $e^{w_j} f_j$ is a local holomorphic section. This shows that $\mathcal{L}$ must be invariant by parallel transport and that the local holonomy of the Chern connection of (E, h) acts trivially on $\mathcal{L}$. Statement 1.5 (a) follows.

Finally, if we assume that the restricted holonomy group H of (E, h) is equal to $\mathrm{U}(r)$, there cannot exist any holonomy-invariant invertible subsheaf $\mathcal{L} \subset \mathcal{O}((E^*)^{\otimes m})$, $m \geq 1$, on which H acts trivially, since the natural representation of $\mathrm{U}(r)$ on $(\mathbb{C}^r)^{\otimes m}$ has no invariant line on which $\mathrm{U}(r)$ induces a trivial action. Property 1.5 (b) is proved. □

4 Proof of the structure theorem

We suppose here that X is equipped with a Kähler metric ω such that $\mathrm{Ricci}(\omega) \geq 0$, and we set $n = \dim_{\mathbb{C}} X$. We consider the holonomy representation of the tangent bundle $E = T_X$ equipped with the hermitian metric $h = \omega$. Here

$$B = \mathrm{Tr}_\omega\,\Theta_{E,h} = \mathrm{Tr}_\omega\,\Theta_{T_X,\omega} \geq 0$$

is nothing but the Ricci operator.

Proof of 1.4 (a) Let

$$(\widetilde{X}, \omega) \simeq \prod (X_i, \omega_i)$$

be the De Rham decomposition of $(\widetilde{X}, \omega)$, induced by a decomposition of the holonomy representation in irreducible representations. Since the holonomy is contained in $\mathrm{U}(n)$, all factors (X_i, ω_i) are Kähler manifolds with irreducible holonomy and holonomy group $H_i \subset \mathrm{U}(n_i)$, $n_i = \dim X_i$. By Cheeger–Gromoll [CG71], there is possibly a flat factor $X_0 = \mathbb{C}^q$ and the other factors X_i, $i \geq 1$, are compact and simply connected. Also, the product structure shows that each $K_{X_i}^{-1}$ is hermitian semipositive. By Berger's classification of holonomy groups [Ber55] there are only three possibilities, namely $H_i = \mathrm{U}(n_i)$, $H_i = \mathrm{SU}(n_i)$, or $H_i = \mathrm{Sp}(n_i/2)$. The case $H_i = \mathrm{SU}(n_i)$ leads to X_i being a Calabi–Yau manifold, and the case $H_i = \mathrm{Sp}(n_i/2)$ implies that X_i is holomorphic symplectic (see, e.g., [Bea83]). Now, if $H_i = \mathrm{U}(n_i)$, the generalized holonomy principle 1.5 shows that none of the invertible subsheaves $\mathcal{L} \subset O((T_{X_i}^*)^{\otimes m})$ can be pseudoeffective for $m \geq 1$. Therefore, X_i is rationally connected by Criterion 1.1. □

Proof of 1.4 (b) Set $X' = \prod_{i \geq 1} X_i$. The group of covering transformations acts on the product $\widetilde{X} = \mathbb{C}^q \times X'$ by holomorphic isometries of the form $x = (z, x') \mapsto (u(z), v(x'))$. At this point, the argument is slightly more involved than in Beauville's paper [Bea83], because the group G' of holomorphic isometries of X' need not be finite (e.g., X' may be a projective space); instead, we imitate the proof of ([CG72], Theorem 9.2) and use the fact that X' and $G' = \mathrm{Isom}(X')$ are compact. Let $E_q = \mathbb{C}^q \rtimes U(q)$ be the group of unitary motions of $\mathbb{C}^q$. Then $\pi_1(X)$ can be seen as a discrete subgroup of $E_q \times G'$. As G' is compact, the kernel of the projection map $\pi_1(X) \to E_q$ is finite and the image of $\pi_1(X)$ in E_q is still discrete with compact quotient. This shows that there is a subgroup Γ of finite index in $\pi_1(X)$ which is isomorphic to a crystallographic subgroup of $\mathbb{C}^q$. By Bieberbach's theorem, the subgroup $\Gamma_0 \subset \Gamma$ of elements which are translations is a subgroup of finite index. Taking the intersection of all conjugates of Γ_0 in $\pi_1(X)$, we find a normal subgroup $\Gamma_1 \subset \pi_1(X)$ of finite index, acting by translations on $\mathbb{C}^q$. Then $\widehat{X} = \widetilde{X}/\Gamma_1$ is a fiber bundle over the torus $\mathbb{C}^q/\Gamma_1$, with X' as fiber and $\pi_1(X') = 1$. Therefore, $\widehat{X}$ is the desired finite étale covering of X. □

For the second assertion we consider fiberwise the rational quotient and obtain a factorization

$$\widehat{X} \xrightarrow{\beta} W \xrightarrow{\gamma} \mathrm{Alb}(\widehat{X})$$

with fiber bundles β (fiber $\prod Z_\ell$) and γ (fiber $\prod Y_j \times \prod S_k$). Since clearly $K_W \equiv 0$, the claim follows from the Beauville–Bogomolov decomposition theorem.

Proof of 1.4 (c) The statement is an immediate consequence of 1.4 (b), using the homotopy exact sequence of a fibration. □

5 Further remarks

We finally point out two direct consequences of Theorem 1.4. Since the property

$$H^0(X, (T_X^*)^{\otimes m}) = 0 \qquad (m \geq 1)$$

is invariant under finite étale covers, we obtain immediately from Theorem 1.4:

5.1 Corollary *Let X be a compact Kähler manifold with K_X^{-1} hermitian semipositive. Assume that $H^0(X, (T_X^*)^{\otimes m}) = 0$ for all positive m. Then X is rationally connected.*

This establishes Mumford's conjecture in case X has semipositive Ricci curvature.

Theorem 1.4 also gives strong implications for small deformations of a manifold with semipositive Ricci curvature:

5.2 Corollary *Let X be a compact Kähler manifold with K_X^{-1} hermitian semipositive. Let $\pi\colon \mathcal{X} \to \Delta$ be a proper submersion from a Kähler manifold $\mathcal{X}$ to the unit disk $\Delta \subset \mathbb{C}$. Assume that $X_0 = \pi^{-1}(0) \simeq X$. Then there exists a finite étale cover $\widehat{\mathcal{X}} \to \mathcal{X}$ with projection $\hat{\pi}\colon \widehat{\mathcal{X}} \to \Delta$ such that – after possibly shrinking Δ – the following holds:*

(a) The relative Albanese map $\alpha\colon \widehat{\mathcal{X}} \to \mathrm{Alb}(\mathcal{X}/\Delta)$ is a surjective submersion, thus the Albanese map $\alpha_t\colon \widehat{X}_t = \widehat{\pi}^{-1}(t) \to \mathrm{Alb}(X_t)$ is a surjective submersion for all t.

(b) Every fiber of α_t is a product of Calabi–Yau manifolds, irreducible symplectic manifolds, and irreducible rationally connected manifolds.

(c) There exists a factorization of α,

$$\widehat{\mathcal{X}} \xrightarrow{\beta} \mathcal{Y} \xrightarrow{\gamma} \mathrm{Alb}(\mathcal{X}/\Delta),$$

such that $\beta_t = \beta_{|\widehat{X}_t}$ is a submersion and a rational quotient of $\widehat{X}_t$ for all t, and $\gamma_t = \gamma_{|Y_t}$ is a trivial fiber bundle.

Corollary 5.2 is an immediate consequence of Theorem 1.4 and the following proposition:

5.3 Proposition *Let $\pi\colon \mathcal{Y} \to \Delta$ be a proper Kähler submersion over the unit disk. Assume that $Y_0 \simeq \prod X_i \times \prod Y_j \times \prod Z_k$ with X_i Calabi–Yau, Y_j irreducible symplectic, and Z_k irreducible rationally connected. Then (possibly after shrinking Δ) every Y_t has a decomposition*

$$Y_t \simeq \prod X_{i,t} \times \prod Y_{j,t} \times \prod Z_{k,t}$$

with factors of the same type as above, and the factors form families $\mathcal{X}_i$, $\mathcal{Y}_j$, and $\mathcal{Z}_k$.

Proof It suffices to treat the case of two factors, say $Y_0 = A_1 \times A_2$ where the A_i are Calabi–Yau, irreducible symplectic, or rationally connected. Since $H^1(A_j, \mathcal{O}_{A_j}) = 0$, the factors A_j deform to the neighboring Y_t. By the properness of the relative cycle space, we obtain families $q_i\colon U_i \to S_i$ over Δ with projections $p_i\colon U_i \to \mathcal{Y}$. Possibly after shrinking Δ, this yields holomorphic maps $f_i\colon \mathcal{Y} \to S_i$. Then the map

$$f_1 \times f_2\colon \mathcal{Y} \to S_1 \times S_2$$

is an isomorphism, since $A_t \cdot B_t = A_0 \cdot B_0 = 1$. This gives the families $(A_i)_t$ we are looking for. □

A Appendix (by Jean-Pierre Demailly) A flag variety version of the holonomy principle

Our goal here is to derive a related version of the holonomy principle over flag varieties, based on a modified Bochner formula which we hope to be useful in other contexts (especially since no assumption on the base manifold is needed). If E is as before a holomorphic vector bundle of rank r over an n-dimensional complex manifold, we denote by $F(E)$ the flag manifold of E, namely the bundle $F(E) \to X$ whose fibers consist of flags

$$\xi:\ E_x = V_0 \supset V_1 \supset \ldots \supset V_r = \{0\}, \quad \dim E_x = r, \quad \operatorname{codim} V_\lambda = \lambda$$

in the fibers of E, along with the natural projection $\pi\colon F(E) \to X$, $(x, \xi) \mapsto x$. We let Q_λ, $1 \le \lambda \le r$ be the tautological line bundles over $F(E)$ such that

$$Q_{\lambda,\xi} = V_{\lambda-1}/V_\lambda,$$

and for a weight $a = (a_1, \ldots, a_r) \in \mathbb{Z}^r$ we set

$$Q^a = Q_1^{a_1} \otimes \ldots \otimes Q_r^{a_r}.$$

In additive notation, viewing the Q_j as divisors, we also denote by

$$a_1 Q_1 + \ldots + a_r Q_r$$

any real linear combination ($a_j \in \mathbb{R}$). Our goal is to compute explicitly the curvature tensor of the line bundles Q^a with respect to the tautological metric induced by h. For convenience of notation, we prefer to work on the dual flag manifold $F(E^*)$, although there is a biholomorphism $F(E) \simeq F(E^*)$ given by

$$(E_x = W_0 \supset W_1 \supset \ldots \supset W_r = \{0\}) \mapsto (E_x^* = V_0 \supset V_1 \supset \ldots \supset V_r = \{0\}),$$

where $V_\lambda = W_{r-\lambda}^\dagger$ is the orthogonal subspace of $W_{r-\lambda}$ in E_x^*. In this context, we have an isomorphism

$$V_{\lambda-1}/V_\lambda = W_{r-\lambda+1}^\dagger / W_{r-\lambda}^\dagger \simeq (W_{r-\lambda}/W_{r-\lambda+1})^*.$$

This shows that $Q^a \to F(E^*)$ is isomorphic to $Q^b \to F(E)$ where $b_\lambda = -a_{r-\lambda+1}$, that is

$$(b_1, b_2, \ldots, b_{r-1}, b_r) = (-a_r, -a_{r-1}, \ldots, -a_2, -a_1).$$

We now proceed to compute the curvature of $Q^a \to F(E^*)$, using the same notation as in Section 3. In a neighborhood of every point $z_0 \in X$, we can find a local coordinate system $(z_1, \ldots, z_n)$ centered at z_0 and a holomorphic frame $(e_\lambda)_{1\le\lambda\le r}$ such that

$$\langle e_\lambda(z), e_\mu(z)\rangle = \mathbf{1}_{\{\lambda=\mu\}} - \sum_{1\le j,k\le n} c_{jk\lambda\mu} z_j \bar{z}_k + O(|z|^3),\ 1 \le \lambda, \mu \le r, \tag{A.1}$$

$$\Theta_{E,h}(z_0) = \sum_{1\le j,k,\lambda,\mu\le n} c_{jk\lambda\mu} dz_j \wedge d\bar{z}_k \otimes e_\lambda^* \otimes e_\mu,\ c_{kj\mu\lambda} = \overline{c_{jk\lambda\mu}}, \tag{A.1'}$$

where $\mathbf{1}_S$ denotes the characteristic function of the set S. For a given point $\xi_0 \in F(E_{z_0}^*)$ in the flag variety, one can always adjust the frame (e_λ) in such a way that the flag corresponding to ξ_0 is given by

$$V_{\lambda,0} = \mathrm{Vect}(e_1, \ldots, e_\lambda)^\dagger \subset E_{z_0}^*. \tag{A.2}$$

A point (z, ξ) in a neighborhood of (z_0, ξ_0) is likewise represented by the flag associated with the holomorphic tangent frame $(\widetilde{e}_\lambda(z,\xi))_{1\le\lambda\le r}$ defined by

$$\widetilde{e}_\lambda(z,\xi) = e_\lambda(z) + \sum_{\lambda<\mu\le r} \xi_{\lambda\mu} e_\mu(z), \qquad (\xi_{\lambda\mu})_{1\le\lambda<\mu\le r} \in \mathbb{C}^{r(r-1)/2}. \tag{A.3}$$

We obtain in this way a local coordinate system $(z_j, \xi_{\lambda\mu})$ near (z_0, ξ_0) on the total space of $F(E^*)$, where the $(\xi_{\lambda\mu})$ are the fiber coordinates. The frame $\widetilde{e}(z,\xi)$ is not orthonormal, but by the Gram–Schmidt orthogonalization process the flag ξ is also induced by the (non-holomorphic) orthonormal frame $(\widehat{e}_\lambda(z,\xi))$ obtained inductively by putting $\widehat{e}_1 = \widetilde{e}_1/|\widetilde{e}_1|$ and

$$\widehat{e}_\lambda = \Big(\widetilde{e}_\lambda - \sum_{1\le\mu<\lambda} \langle \widetilde{e}_\lambda, \widehat{e}_\mu\rangle \widehat{e}_\mu\Big)/(\text{norm of numerator}).$$

Straightforward calculations imply that the hermitian inner products involved are $O(|\xi| + |z|^2)$ and the norms equal to $1 + O((|\xi| + |z|)^2)$, hence we get

$$\widetilde{e}_\lambda(z,\xi) = e_\lambda(z,\xi) + \sum_{\lambda<\mu\leq r} \xi_{\lambda\mu} e_\mu(z) - \sum_{1\leq\mu<\lambda} \overline{\xi}_{\mu\lambda} e_\mu(z) + O((|\xi| + |z|)^2)$$

and more precisely (omitting variables for simplicity of notation)

$$\begin{aligned}\widetilde{e}_\lambda = \Big(1 - \frac{1}{2}\sum_{1\leq\mu<\lambda} |\xi_{\mu\lambda}|^2 - \frac{1}{2}\sum_{\lambda<\mu\leq r} |\xi_{\lambda\mu}|^2 + \frac{1}{2}\sum_{1\leq j,k\leq n} c_{jk\lambda\lambda} z_j \overline{z}_k\Big) e_\lambda \\ + \sum_{\lambda<\mu\leq r} \xi_{\lambda\mu} e_\mu \\ - \sum_{1\leq\mu<\lambda} \Big(\overline{\xi}_{\mu\lambda} + \sum_{\lambda<\nu\leq r} \xi_{\lambda\nu}\overline{\xi}_{\mu\nu} - \sum_{1\leq j,k\leq n} c_{jk\lambda\mu} z_j\overline{z}_k\Big) e_\mu \\ + O((|\xi| + |z|)^3).\end{aligned} \tag{A.4}$$

The curvature of the tautological line bundle $Q_\lambda = V_{\lambda-1}/V_\lambda$ can be evaluated by observing that the dual line bundle

$$Q_\lambda^* = V_\lambda^\dagger / V_{\lambda-1}^\dagger = \mathrm{Vect}(\widetilde{e}_1, \ldots, \widetilde{e}_\lambda)/\,\mathrm{Vect}(\widetilde{e}_1, \ldots, \widetilde{e}_{\lambda-1})$$

admits a holomorphic section given by

$$v_\lambda(z,\xi) = \widetilde{e}_\lambda(z,\xi) \bmod \mathrm{Vect}(\widetilde{e}_1, \ldots, \widetilde{e}_{\lambda-1}).$$

The tautological norm of this section is

$$\begin{aligned}|v_\lambda|^2 &= |\widetilde{e}_\lambda|^2 - \sum_{1\leq\mu<\lambda} |\langle\widetilde{e}_\lambda, \widetilde{e}_\mu\rangle|^2 \\ &= 1 - \sum_{1\leq j,k\leq n} c_{jk\lambda\lambda} z_j\overline{z}_k \\ &\qquad + \sum_{\lambda<\mu\leq r} |\xi_{\lambda\mu}|^2 - \sum_{1\leq\mu<\lambda} |\xi_{\mu\lambda}|^2 + O((|z| + |\xi|)^3).\end{aligned}$$

Therefore we obtain the formula

$$\begin{aligned}\Theta_{Q_\lambda}(z_0,\xi_0) &= \partial\overline{\partial} \log |v_\lambda|^2_{|(z_0,\xi_0)} \\ &= -\sum_{1\leq j,k\leq n} c_{jk\lambda\lambda}\, dz_j \wedge d\overline{z}_k \\ &\qquad + \sum_{\lambda<\mu\leq r} d\xi_{\lambda\mu} \wedge d\overline{\xi}_{\lambda\mu} - \sum_{1\leq\mu<\lambda} d\xi_{\mu\lambda} \wedge d\overline{\xi}_{\mu\lambda},\end{aligned}$$

$$\begin{aligned}\Theta_{Q^a}(z_0,\xi_0) &= \sum_{1\le\lambda\le r} a_\lambda \Theta_{Q_\lambda}(z_0,\xi_0)\\ &= -\sum_{1\le j,k\le n,\,1\le\lambda\le r} a_\lambda c_{jk\lambda\lambda} dz_j\wedge d\bar z_k\\ &\quad+\sum_{1\le\lambda<\mu\le r}(a_\lambda-a_\mu)d\xi_{\lambda\mu}\wedge d\bar\xi_{\lambda\mu}.\end{aligned}$$

This calculation holds true only at (z_0,ξ_0), but it shows that we have at every point a decomposition of Θ_{Q^a} in horizontal and vertical parts given by

$$\Theta_{Q^a} = \theta^H_a + \theta^V_a, \tag{A.5}$$

$$\left\{\begin{aligned}\theta^H_a(z_0,\xi_0) &= -\textstyle\sum_{j,k,\lambda} a_\lambda c_{jk\lambda\lambda} dz_j\wedge d\bar z_k\\ &= -\textstyle\sum_{1\le\lambda\le r} a_\lambda\, \pi^*\langle\Theta_{T_X,\omega}(e_\lambda),e_\lambda\rangle,\end{aligned}\right. \tag{A.6^H}$$

$$\theta^V_a(z_0,\xi_0) = \sum_{1\le\lambda<\mu\le r}(a_\lambda-a_\mu)d\xi_{\lambda\mu}\wedge d\bar\xi_{\lambda\mu}. \tag{A.6^V}$$

The decomposition is taken here with respect to the C^∞ splitting of the exact sequence

$$0\to T_{Y/X}\to T_Y\to \pi^*T_X\to 0,\qquad Y:=F(E^*) \tag{A.7}$$

provided by the Chern connection ∇ of (E,h) ; horizontal directions are those coming from flags associated with ∇-parallel frames. In order to express (A.6^H) in a more intrinsic way at an arbitrary point $(z,\xi)\in Y$, we have to replace $(e_\lambda(z))$ by the orthonormal frame $(\widehat{e}_\lambda(z,\xi))$ associated with the flag ξ. Such frames are not unique, actually they are defined up to the action of $(S^1)^r$, but such a change does not affect the expression of θ^H_a. We then get the intrinsic formula

$$\begin{aligned}\theta^H_a(z,\xi) &= -\sum_{1\le\lambda\le r} a_\lambda\,\pi^*\langle\Theta_{T_X,\omega}(\widehat{e}_\lambda(z,\xi)),\widehat{e}_\lambda(z,\xi)\rangle\\ &= -\sum_{1\le\lambda\le r} a_\lambda \sum_{1\le j,k\le n,\,1\le\sigma,\tau\le r} c_{jk\sigma\tau}(z)\,\widehat{e}_{\lambda\sigma}(z,\xi)\overline{\widehat{e}_{\lambda\tau}(z,\xi)}\,dz_j\wedge d\bar z_k,\end{aligned} \tag{A.8}$$

where we put

$$\widehat{e}_\lambda(z,\xi) = \sum_{1\le\sigma\le r}\widehat{e}_{\lambda\sigma}(z,\xi)\,e_\sigma(z)$$

(the coefficients $\widehat{e}_{\lambda\sigma}(z,\xi)$ can be computed from (A.4)). Moreover, since θ^V_a and Θ_{Q^a} have the same restriction to the fibers of $Y\to X$, we conclude that θ^V_a is in fact unitary invariant along the fibers (the tautological metric of Q^a clearly has this property). Let us consider the vertical and normalized unitary invariant relative volume form η of $Y \to X$ given by

$$\text{(A.9)} \qquad \eta(z_0, \xi_0) = \bigwedge_{1 \le \lambda < \mu \le r} i\, d\xi_{\lambda\mu} \wedge d\overline{\xi}_{\lambda\mu} \qquad \text{at } (z_0, \xi_0).$$

Let $N = r(r-1)/2$ be the fiber dimension. For a strictly dominant weight a, i.e., $a_1 > a_2 > \ldots > a_r$, the line bundle Q^a is relatively ample with respect to the projection $\pi\colon Y = F(E^*) \to X$, and $i\theta^V_a$ induces a Kähler form on the fibers. Formula (A.6^V) shows that the corresponding volume form is

$$(i\theta^V_a)^N = N! \prod_{1 \le \lambda < \mu \le r} (a_\lambda - a_\mu)\, \eta.$$

A.10 Curvature formulas *Consider as above $Q^a \to Y := F(E^*)$. Then*

(a) The curvature form of Q^a is given by $\Theta_{Q^a} = \theta^H_a + \theta^V_a$, where the horizontal part is given by

$$\theta^H_a = - \sum_{1 \le \lambda \le r} a_\lambda\, \pi^* \langle \Theta_{T_X,\omega}(\widehat{e}_\lambda), \widehat{e}_\lambda \rangle$$

and the vertical part by

$$\theta^V_a(z_0, \xi_0) = \sum_{1 \le \lambda < \mu \le r} (a_\lambda - a_\mu) d\xi_{\lambda\mu} \wedge d\overline{\xi}_{\lambda\mu}$$

in normal coordinates at any point (z_0, ξ_0).

(b) The relative canonical bundle $K_{Y/X}$ is isomorphic with Q^ρ for the (anti-dominant) canonical weight $\rho_\lambda = 2\lambda - r - 1$, $1 \le \lambda \le r$. For any positive definite $(1,1)$-form ω on X we have

$$\begin{aligned} i\partial\overline{\partial}\eta \wedge \pi^*\omega^{n-1} &= -i\theta^H_\rho \wedge \eta \wedge \pi^*\omega^{n-1} \\ &= \sum_{1 \le \lambda \le r} \rho_\lambda\, \pi^* \langle i\Theta_{T_X,\omega}(\widehat{e}_\lambda), \widehat{e}_\lambda \rangle \wedge \eta \wedge \pi^*\omega^{n-1}. \end{aligned}$$

Proof (a) follows entirely from the previous discussion.
(b) The formula for the canonical weight is a classical result in the theory of flag varieties. As $(i\theta^V_a)^N$ and η are proportional for a strictly dominant, we compute instead

$$\partial\overline{\partial}(\theta^V_a)^N = N\, (\theta^V_a)^{N-1} \wedge \partial\overline{\partial}\theta^V_a + N(N-1)\, (\theta^V_a)^{N-2} \wedge \partial\theta^V_a \wedge \overline{\partial}\theta^V_a,$$

and for this, we use a Taylor expansion of order 2 at (z_0, ξ_0). Since Θ_{Q^a} is closed, we have $\partial\overline{\partial}\theta^V_a = -\partial\overline{\partial}\theta^H_a$, hence

$$\partial\overline{\partial}\theta^V_a = \partial\overline{\partial} \sum_{1 \le \lambda \le r} a_\lambda \sum_{1 \le j,k \le n,\, 1 \le \sigma,\tau \le r} c_{jk\sigma\tau}(z) \widehat{e}_{\lambda\sigma}(z,\xi) \overline{\widehat{e}_{\lambda\tau}(z,\xi)}\, dz_j \wedge d\overline{z}_k,$$

and we have similar formulas for $\partial(\theta_a^V)$ and $\overline{\partial}(\theta_a^V)$. When taking ∂, $\overline{\partial}$ and $\partial\overline{\partial}$ we need only consider the differentials in ξ, otherwise we get terms $\Lambda^{\geq 3}(dz, d\overline{z})$ of degree at least three in the dz_j or $d\overline{z}_k$ and the wedge product of these with $\pi^*\omega^{n-1}$ is zero. For the same reason, $\partial\theta_a^V \wedge \overline{\partial}\theta_a^V$ will not contribute to the result since it produces terms of degree four or more in dz_j, $d\overline{z}_k$. Formula (A.4) gives

$$\begin{aligned}\widehat{e}_{\lambda\sigma} = \mathbf{1}_{\{\lambda=\sigma\}}\Big(1 - \frac{1}{2}\sum_{1\leq\mu<\lambda}|\xi_{\mu\lambda}|^2 - \frac{1}{2}\sum_{\lambda<\mu\leq r}|\xi_{\lambda\mu}|^2\Big)\\ + \mathbf{1}_{\{\lambda<\sigma\}}\xi_{\lambda\sigma} - \mathbf{1}_{\{\sigma<\lambda\}}\Big(\overline{\xi}_{\sigma\lambda} + \sum_{\mu>\lambda}\xi_{\lambda\mu}\overline{\xi}_{\sigma\mu}\Big) + O(|z|^2 + |\xi|^3).\end{aligned}$$

Notice that we do not need to look at the terms $O(|z|)$, $O(|z|^2)$ as they will produce no contribution at (z_0, ξ_0). From this we infer

$$\begin{aligned}\widehat{e}_{\lambda\sigma}\overline{\widehat{e}_{\lambda\tau}} =& \mathbf{1}_{\{\lambda=\sigma=\tau\}}\Big(1 - \sum_{1\leq\mu<\lambda}|\xi_{\mu\lambda}|^2 - \sum_{\lambda<\mu\leq r}|\xi_{\lambda\mu}|^2\Big)\\ &+ \mathbf{1}_{\{\lambda=\tau<\sigma\}}\xi_{\lambda\sigma} - \mathbf{1}_{\{\sigma<\lambda=\tau\}}\overline{\xi}_{\sigma\lambda} + \mathbf{1}_{\{\lambda=\sigma<\tau\}}\overline{\xi}_{\lambda\tau} - \mathbf{1}_{\{\tau<\lambda=\sigma\}}\xi_{\tau\lambda}\\ &+ \mathbf{1}_{\{\sigma,\tau>\lambda\}}\xi_{\lambda\sigma}\overline{\xi}_{\lambda\tau} + \mathbf{1}_{\{\sigma,\tau<\lambda\}}\xi_{\tau\lambda}\overline{\xi}_{\sigma\lambda}\\ &+ \mathbf{1}_{\{\tau<\lambda<\sigma\}}\xi_{\lambda\sigma}\xi_{\tau\lambda} + \mathbf{1}_{\{\sigma<\lambda<\tau\}}\overline{\xi}_{\sigma\lambda}\overline{\xi}_{\lambda\tau}\\ &- \sum_{1\leq\mu\leq r}\mathbf{1}_{\{\sigma<\lambda=\tau<\mu\}}\xi_{\lambda\mu}\overline{\xi}_{\sigma\mu} + \mathbf{1}_{\{\tau<\lambda=\sigma<\mu\}}\xi_{\tau\mu}\overline{\xi}_{\lambda\mu} \mod(|z|^2, |\xi|^3).\end{aligned}$$

By virtue of (A.7), only "diagonal terms" of the form $d\xi_{\lambda\mu} \wedge d\overline{\xi}_{\lambda\mu}$ in the $\partial\overline{\partial}$ of this expression can contribute to $(\theta_a^V)^{N-1} \wedge \partial\overline{\partial}\theta_a^V$, all others vanish at $z = \xi = 0$. The useful terms are thus

$$\begin{aligned}\partial\overline{\partial}\Big(\sum_{1\leq\lambda\leq r}a_\lambda\sum_{1\leq\sigma,\tau\leq r}c_{jk\sigma\tau}\widehat{e}_{\lambda\sigma}\overline{\widehat{e}_{\lambda\tau}}dz_j \wedge d\overline{z}_k\Big) = (\text{unneeded terms})\\ + \sum_{1\leq\lambda<\mu\leq r}(-a_\mu c_{jk\mu\mu} - a_\lambda c_{jk\lambda\lambda} + a_\lambda c_{jk\mu\mu} + a_\mu c_{jk\lambda\lambda})\,d\xi_{\lambda\mu}\wedge d\overline{\xi}_{\lambda\mu}\wedge dz_j\wedge d\overline{z}_k\\ = \sum_{1\leq\lambda<\mu\leq r}(a_\lambda - a_\mu)(c_{jk\mu\mu} - c_{jk\lambda\lambda})\,d\xi_{\lambda\mu} \wedge d\overline{\xi}_{\lambda\mu} \wedge dz_j \wedge d\overline{z}_k + (\text{unneeded}).\end{aligned}$$

From this we infer

$$\partial\overline{\partial}(\theta_a^V)^N \wedge \pi^*\omega^{n-1} = (\theta_a^V)^N \wedge \sum_{1\leq j,k\leq n,\ 1\leq\lambda\leq r}(2\lambda - 1 - r)\,c_{jk\lambda\lambda}\,dz_j \wedge d\overline{z}_k \wedge \pi^*\omega^{n-1}.$$

In fact, the coefficient of $c_{jk\lambda\lambda}$ is the number $(\lambda - 1)$ of indices $< \lambda$ (coming from the term $(a_\lambda - a_\mu)c_{jk\mu\mu}$ above) minus the number $r - \lambda$ of indices $> \lambda$ (coming from the term $-(a_\lambda - a_\mu)c_{jk\lambda\lambda}$). Formula A.10 (b) follows. □

A.11 Bochner formula *Assume that X is a compact complex manifold possessing a balanced metric. That is, a positive smooth $(1,1)$-form $\omega = i\sum_{1\le j,k\le n}\omega_{jk}(z)\,dz_j\wedge d\overline{z}_k$ such that $d\omega^{n-1} = 0$. Assume also that for some dominant weight a $(a_1\ge\ldots\ge a_r\ge 0)$, the $\mathbb{R}$-line bundle Q^a is pseudoeffective on $Y := F(E^*)$,* i.e., *that there exists a quasi-plurisubharmonic function φ such that $i(\Theta_{Q^a}+\partial\overline{\partial}\varphi)\ge 0$ on Y. Then*

$$\int_Y (i\partial\varphi\wedge\overline{\partial}\varphi - i\theta^H_{a-\rho})e^{\varphi}\,\eta\wedge\pi^*\omega^{n-1}\le 0,$$

or equivalently

$$\int_Y \Big(i\partial\varphi\wedge\overline{\partial}\varphi + \sum_{1\le\lambda\le r}(a_\lambda-\rho_\lambda)\langle i\Theta_{TX,\omega}(\widehat{e}_\lambda),\widehat{e}_\lambda\rangle\Big)e^{\varphi}\,\eta\wedge\pi^*\omega^{n-1}\le 0.$$

Proof The idea is to use the $\partial\overline{\partial}$-formula

$$\int_Y i\partial\overline{\partial}(e^{\varphi})\wedge\eta\wedge\pi^*\omega^{n-1} - e^{\varphi}\wedge i\partial\overline{\partial}\eta\wedge\pi^*\omega^{n-1}$$
$$= \int_Y d\Big(i\overline{\partial}(e^{\varphi})\wedge\eta\wedge\pi^*\omega^{n-1} + e^{\varphi}\,i\partial\eta\wedge\pi^*\omega^{n-1}\Big) = 0,$$

which follows immediately from Stokes. We get

$$\text{(A.12)}\qquad \int_Y (i\partial\overline{\partial}\varphi + i\partial\varphi\wedge\overline{\partial}\varphi)e^{\varphi}\wedge\eta\wedge\pi^*\omega^{n-1} - e^{\varphi}\,i\partial\overline{\partial}\eta\wedge\pi^*\omega^{n-1} = 0.$$

Now, $i\partial\overline{\partial}\varphi\ge -i\Theta_{Q^a}$ in the sense of currents, and therefore by A.10 (a) and (b) we obtain

$$\text{(A.13)}\quad i\partial\overline{\partial}\varphi\wedge\eta\wedge\pi^*\omega^{n-1} - i\partial\overline{\partial}\eta\wedge\pi^*\omega^{n-1}\ge(-i\theta^H_a + i\theta^H_\rho)\wedge\eta\wedge\pi^*\omega^{n-1}.$$

The combination of (A.12) and (A.13) yields the inequality of Corollary A.11. □

The parallel transport operators of (E,h) can be considered to operate on the global flag variety $Y = F(E^*)$ as follows. For any piecewise smooth path $\gamma\colon[0,1]\to X$, we get a (unitary) hermitian isomorphism $\tau_\gamma\colon E_p\to E_q$ where $p=\gamma(0)$, $q=\gamma(1)$. Therefore τ_γ induces an isomorphism $\widetilde{\tau}_\gamma\colon F(E^*_p)\to F(E^*_q)$ of the corresponding flag varieties, and an isomorphism over $\widetilde{\tau}_\gamma$ of the tautological line bundles Q^a. Given a local C^∞ vector field v on an open set $U\subset X$, there is a unique horizontal lifting $\widetilde{v}$ of v to a C^∞ vector field on $\pi^{-1}(U)\subset Y$, where horizontality refers again to $\nabla = \nabla_{E,h}$. Now, the flow of $\widetilde{v}$ consists of parallel transport operators along the trajectories of v. By definition, h is invariant by parallel transport, therefore the associated hermitian metric h_a on each line bundle Q^a is also invariant. Another metric $h_{a,\varphi} = h_a e^{-\varphi}$ is invariant if and only if the weight function φ is invariant by the flows of all such vector fields $\widetilde{v}$ on Y, that is if $d\varphi(\zeta) = 0$ for all horizontal vector fields $\zeta \in T_Y$.

A.14 Theorem *Let $E \to X$ be a holomorphic vector bundle of rank r over a compact complex manifold X. Assume that X is equipped with a hermitian metric ω and E with a hermitian structure h such that $B := \mathrm{Tr}_\omega(i\Theta_{E,h}) \geq 0$. At each point $z \in X$, let*

$$0 \leq b_1(z) \leq \ldots \leq b_r(z)$$

be the eigenvalues of $B(z)$ with respect to $h(z)$. Finally, let Q^a be a pseudo-effective $\mathbb{R}$-line bundle on $Y := F(E^*)$ associated with a dominant weight $a_1 \geq \ldots \geq a_r \geq 0$, and let φ be a quasi-plurisubharmonic function on Y such that $i(\Theta_{Q^a} + \partial\overline{\partial}\varphi) \geq 0$. Then

(a) The function $\psi(z) = \sup_{\xi \in F(E_z^*)} \varphi(z,\xi)$ is constant and $b_\lambda \equiv 0$ as soon as $a_\lambda > 0$, and in particular $B \equiv 0$ if $a_r > 0$.

(b) Assume that $B \equiv 0$. Then the function φ must be invariant by parallel transport on Y.

Proof Since our hypotheses are invariant by a conformal change on the metric ω, we can assume by Gauduchon [Gau77] that $\partial\overline{\partial}\omega^{n-1} = 0$.

(a) Notice that if a is integral and φ is given by a holomorphic section of Q^a, then e^φ is the square of the norm of that section with respect to h, and e^ψ is the sup of that norm on the fibers of $Y \to X$. In general, formula A.10 (a) shows that

$$i\partial\overline{\partial}\varphi(z,\xi) \geq -i\theta_a^H(z,\xi) - i\theta_a^V(z,\xi),$$

hence

$$i\partial\overline{\partial}^H\varphi(z,\xi) \wedge \omega^{n-1}(z) \geq \sum_{1\leq\lambda\leq r} a_\lambda \langle i\Theta_{TX,\omega}(\widetilde{e}_\lambda), \widetilde{e}_\lambda\rangle(z,\xi) \wedge \omega^{n-1}(z), \tag{A.15}$$

where $i\partial\overline{\partial}^H\varphi$ means the restriction of $i\partial\overline{\partial}\varphi$ to the horizontal directions in T_Y. By taking the supremum in ξ, we conclude from standard arguments of subharmonic function theory that

$$\Delta_\omega\psi(z) \geq \sum_{1\leq\lambda\leq r} a_\lambda b_\lambda(z),$$

since the RHS is the minimum of the coefficient of the (n,n)-form occurring in the RHS of (A.15). Therefore, ψ is ω-subharmonic and so must be constant on

X by Aronszajn [Aro57]. It follows that $b_\lambda \equiv 0$ whenever $a_\lambda > 0$, in particular $B \equiv 0$ if $a_r > 0$.

(b) Under the assumption $B = \mathrm{Tr}_\omega \Theta_{E,h} \equiv 0$, the calculations made in the course of the proof of A.10 (b) imply that

$$\partial\eta \wedge \pi^*\omega^{n-1} = 0, \qquad \partial\overline{\partial}\eta \wedge \pi^*\omega^{n-1} = 0.$$

By the proof of the Bochner formula A.11 (the fact that $\partial\overline{\partial}\omega^{n-1} = 0$ is enough here), we get

$$0 \leq \int_Y i\partial\varphi \wedge \overline{\partial}\varphi \wedge \eta \wedge \pi^*\omega^{n-1} \leq 0,$$

and we conclude from this that the horizontal derivatives $\partial^H\varphi$ vanish. Therefore, φ is invariant by parallel transport. □

In the vein of Criterion 1.1, we have the following additional statement:

A.16 Proposition *Let X be a compact Kähler manifold. Then X is projective and rationally connected if and only if none of the $\mathbb{R}$-line bundles Q^a over $Y = F(T_X^*)$ is pseudoeffective for weights $a \neq 0$ with $a_1 \geq \ldots \geq a_r \geq 0$.*

Proof If X is projective rationally connected and some Q^a, $a \neq 0$, is pseudoeffective, we obtain a contradiction to Theorem A.14 by pulling back T_X and Q^a via a map $f : \mathbb{P}^1 \to X$ such that $E = f^*T_X$ is ample on $\mathbb{P}^1$ (as $B > 0$ in this circumstance).

Conversely, if the $\mathbb{R}$-line bundles Q^a, $a \neq 0$, are not pseudoeffective on $Y = F(T_X^*)$, we obtain by taking $a_1 = \ldots = a_p = 1$, $a_{p+1} = \ldots = a_n = 0$ that $\pi_*Q^a = \Omega_X^p$. Therefore, $H^0(X, \Omega_X^p) = 0$ and all invertible subsheaves $\mathcal{F} \subset \Omega_X^p$ are not pseudoeffective for $p \geq 1$. Hence X is projective (take $p = 2$ and apply Kodaira [Kod54]) and rationally connected by Criterion 1.1 (b). □

Acknowledgments This work was completed while the three authors were visiting the Mathematisches Forschungsinstitut Oberwolfach in September 2012. They wish to thank the Institute for its hospitality and the exceptional quality of the environment.

References

[Aro57] Aronszajn, N. A unique continuation theorem for solutions of elliptic partial differential equations or inequalities of second order. *J. Math. Pures Appl.* **36**, 235–249 (1957).

[Aub76] Aubin, T. Equations du type Monge–Ampère sur les variétés kähleriennes compactes. *C. R. Acad. Sci. Paris Ser. A* **283**, 119–121 (1976); *Bull. Sci. Math.* **102**, 63–95 (1978).

[Bea83] Beauville, A. Variétés kähleriennes dont la première classe de Chern est nulle. *J. Diff. Geom.* **18**, 775–782 (1983).

[Ber55] Berger, M. Sur les groupes d'holonomie des variétés à connexion affine des variétés riemanniennes. *Bull. Soc. Math. France* **83**, 279–330 (1955).

[BY53] Bochner, S. and Yano, K. Curvature and Betti numbers. *Annals of Mathematics Studies*, No. 32. Princeton, NJ: Princeton University Press, 1953.

[Bog74a] Bogomolov, F.A. On the decomposition of Kähler manifolds with trivial canonical class. *Math. USSR Sbornik* **22**, 580–583 (1974).

[Bog74b] Bogomolov, F.A. Kähler manifolds with trivial canonical class. *Izvestija Akad. Nauk* **38**, 11–21 (1974).

[BDPP] Boucksom, S., Demailly, J.-P., Paun, M., and Peternell, T. The pseudo-effective cone of a compact Kähler manifold and varieties of negative Kodaira dimension. *J. Alg. Geom.* **22**, 201–248 (2013).

[Bru10] Brunella, M. On Kähler surfaces with semipositive Ricci curvature. *Riv. Math. Univ. Parma (N.S.)*, **1**, 441–450 (2010).

[Cam92] Campana, F. Connexité rationnelle des variétés de Fano. *Ann. Sci. Ec. Norm. Sup.* **25**, 539–545 (1992).

[CPZ03] Campana, F., Peternell, Th., and Zhang, Q. On the Albanese maps of compact Kähler manifolds. *Proc. Amer. Math. Soc.* **131**, 549–553 (2003).

[CG71] Cheeger, J. and Gromoll, D. The splitting theorem for manifolds of nonnegative Ricci curvature. *J. Diff. Geom.* **6**, 119–128 (1971).

[CG72] Cheeger, J. and Gromoll, D. On the structure of complete manifolds of nonnegative curvature. *Ann. Math.* **96**, 413–443 (1972).

[DPS94] Demailly, J.-P., Peternell, T., and Schneider, M. Compact complex manifolds with numerically effective tangent bundles. *J. Alg. Geom.* **3**, 295–345 (1994).

[DPS96] Demailly, J.-P., Peternell, T., and Schneider, M. Compact Kähler manifolds with hermitian semipositive anticanonical bundle. *Compos. Math.* **101**, 217–224 (1996).

[DR52] de Rham, G. Sur la reductibilité d'un espace de Riemann. *Comment. Math. Helv.* **26**, 328–344 (1952).

[Gau77] Gauduchon, P. Le théorème de l'excentricité nulle. *C. R. Acad. Sci. Paris* **285**, 387–390 (1977).

[GHS01] Graber, T., Harris, J., and Starr, J. Families of rationally connected varieties. *J. Amer. Math. Soc.* **16**, 57–67 (2003).

[KMM92] Kollár, J., Miyaoka, Y., and Mori, S. Rationally connected varieties. *J. Alg.* **1**, 429–448 (1992).

[Kob81] Kobayashi, S. Recent results in complex differential geometry. *Jber. dt. Math.-Verein.* **83**, 147–158 (1981).

[Kob83] Kobayashi, S. Topics in complex differential geometry. In *DMV Seminar*, Vol. 3. Berlin: Birkhäuser, 1983.

[Kod54] Kodaira, K. On Kähler varieties of restricted type. *Ann. Math.* **60**, 28–48 (1954).

[Kol96] Kollár, J. *Rational Curves on Algebraic Varieties*. Ergebnisse der Mathematik und ihrer Grenzgebiete, 3. Folge, Band 32. Berlin: Springer-Verlag, 1996.

[Lic67] Lichnerowicz, A. Variétés kähleriennes et première classe de Chern. *J. Diff. Geom.* **1**, 195–224 (1967).

[Lic71] Lichnerowicz, A. Variétés Kählériennes à première classe de Chern non négative et variétés riemanniennes à courbure de Ricci généralisée non négative. *J. Diff. Geom.* **6**, 47–94 (1971).

[Pau12] Păun, M. Relative adjoint transcendental classes and the Albanese map of compact Kähler manifolds with nef Ricci classes. arXiv: 1209.2195.

[Pet06] Peternell, Th. Kodaira dimension of subvarieties II. *Int. J. Math.* **17**, 619–631 (2006).

[PS98] Peternell, Th. and Serrano, F. Threefolds with anti canonical bundles. *Coll. Math.* **49**, 465–517 (1998).

[Ued82] Ueda, T. On the neighborhood of a compact complex curve with topologically trivial normal bundle. *J. Math. Kyoto Univ.* **22**, 583–607 (1982/83).

[Yau78] Yau, S. T. On the Ricci curvature of a complex Kähler manifold and the complex Monge–Ampère equation I. *Comm. Pure Appl. Math.* **31**, 339–411 (1978).

[Zha96] Zhang, Q. On projective manifolds with nef anticanonical bundles. *J. Reine Angew. Math.* **478**, 57–60 (1996).

[Zha05] Zhang, Q. On projective varieties with nef anticanonical divisors. *Math. Ann.* **332**, 697–703 (2005).

6

Subcanonical graded rings which are not Cohen–Macaulay

F. Catanese[a]
Universität Bayreuth

Abstract

We answer a question by Jonathan Wahl, giving examples of regular surfaces (so that the canonical ring is Gorenstein) with the following properties:

(1) the canonical divisor $K_S \equiv rL$ is a positive multiple of an ample divisor L;
(2) the graded ring $\mathcal{R} := \mathcal{R}(X, L)$ associated to L is not Cohen–Macaulay.

In the Appendix, Wahl shows how these examples lead to the existence of Cohen–Macaulay singularities with K_X $\mathbb{Q}$-Cartier which are not $\mathbb{Q}$-Gorenstein, since their index one cover is not Cohen–Macaulay.

Dedicated to Rob Lazarsfeld on the occasion of his 60th birthday

1 Introduction

The situation that we consider in this paper is the following: L is an ample divisor on a complex projective manifold X of complex dimension n, and we assume that L is subcanonical, i.e., there exists an integer h such that we have the linear equivalence $K_X \equiv hL$, where $h \neq 0$. There are then two cases: $h < 0$ and X is a Fano manifold, or $h > 0$ and X is a manifold with ample canonical divisor (in particular X is of general type). Assume that X is a Fano manifold and that $-K_X = rL$, with $r > 0$: then, by Kodaira vanishing,

$$H^j(mL) := H^j(\mathcal{O}_X(mL)) = 0, \qquad \forall m \in \mathbb{Z}, \forall 1 \leq j \leq n-1.$$

[a] The present work took place in the realm of the DFG Forschergruppe 790 "Classification of algebraic surfaces and compact complex manifolds."

From *Recent Advances in Algebraic Geometry*, edited by Christopher Hacon, Mircea Mustaţă and Mihnea Popa

For $m < 0$ this follows from Kodaira vanishing (and holds for $j \geq 1$), while for $m \geq 0$ Serre duality gives $h^j(mL) = h^{n-j}(K - mL) = h^{n-j}((-r - m)L) = 0$. At the other extreme, if K_X is ample and $K_X \equiv rL$ (thus $r > 0$), by the same argument we get vanishing outside of the interval

$$0 \leq m \leq r.$$

We associate to L, as usual, the finitely generated graded $\mathbb{C}$-algebra

$$\mathcal{R}(X, L) := \oplus_{m \geq 0} H^0(X, \mathcal{O}_X(mL)).$$

Therefore in the Fano case, the divisor L is arithmetically Cohen–Macaulay (see [Hart77]) and the above graded ring is a Gorenstein ring. The question is whether, in the case where K_X is ample, one may also hope for such a good property.

The above graded ring is integral over the canonical ring $\mathcal{A} := \mathcal{R}(X, K_X)$, which is a Gorenstein ring if and only if we have *pluri-regularity*, i.e., vanishing

$$H^j(\mathcal{O}_X) = 0, \qquad \forall\, 1 \leq j \leq n - 1.$$

Jonathan Wahl asked the following question (which makes sense only for $n \geq 2$):

Question 1 (J. Wahl) *Are there examples of subcanonical pluri-regular varieties X such that the graded ring $\mathcal{R}(X, L)$ is not Cohen–Macaulay?*

We show that the answer is positive, also in the case of regular subcanonical surfaces with K_X ample, where by the assumption we have the vanishing

$$H^1(mL) = 0, \qquad \forall m \leq 0, \text{ or } r \leq m$$

and the question boils down to requiring the vanishing also for $1 \leq m \leq r - 1$.

The following theorem answers the question by J. Wahl:

Theorem 2 *For each $r = n - 3$, where $n \geq 7$ is relatively prime to 30, and for each m, $1 \leq m \leq r - 1$, there are Beauville-type surfaces S with $q(S) = 0$ ($q(S) := \dim H^1(S, \mathcal{O}_S)$) such that $K_S = rL$ and $H^1(mL) \neq 0$.*

We therefore get examples of the following situation: $\mathcal{A} := \mathcal{R}(S, K_S)$ is a Gorenstein graded ring, and a subring of the ring $\mathcal{R} := \mathcal{R}(S, L)$, which is not arithmetically Cohen–Maculay. Hence, we have constructed examples of non-Cohen–Macaulay singularities (Spec($\mathcal{R}$)) with K_Y Cartier which are cyclic quasi-étale covers of a Gorenstein singularity (Spec($\mathcal{A}$)). In the Appendix, J. Wahl uses these to construct Cohen–Macaulay singularities with K_X $\mathbb{Q}$-Cartier whose index one cover is not Cohen–Macaulay.

In fact, we can consider three graded rings, two of which are subrings of the third, and which are cones associated to line bundles on the surface S:

- $Y := \mathrm{Spec}(\mathcal{R})$, the cone associated to L, which is not Cohen–Macaulay, while K_Y is Cartier;
- $Z := \mathrm{Spec}(\mathcal{A})$, the cone associated to K_S, which is Gorenstein;
- $X := \mathrm{Spec}(\mathcal{B})$, the cone associated to $K_S + L$ (for instance), which is Cohen–Macaulay with K_X $\mathbb{Q}$-Cartier, but whose index 1 (or canonical) cover $Y = \mathrm{Spec}(\mathcal{R})$ is not Cohen–Macaulay.

2 The special case of even surfaces

Recall: a smooth projective surface S is said to be **even** if there is a divisor L such that $K_S \equiv 2L$. This is a topological condition; it means that the second Stiefel Whitney class $w_2(S) = 0$, or, equivalently, the intersection form

$$H^2(S, \mathbb{Z}) \to \mathbb{Z}$$

is even (takes only even values). In particular, an even surface is a minimal surface. In particular, if S is of general type and even, the self-intersection

$$K_S^2 = 4L^2 = 8k$$

for some integer $k \geq 1$. The first numerical case is therefore the case $K_S^2 = 8$.

Proposition 3 *Assume that S is an even surface of general type with $K_S^2 = 8$ and $p_g(S) = h^0(K_S) = 0$. Then, if $K_S \equiv 2L$, we have $H^1(L) = 0$.*

Proof We have made the assumption that S is even, $K \equiv 2L$, and $p_g = 0$. Since the intersection form is even, and $K^2 \leq 9$ by the Bogomolov–Miyaoka–Yau inequality, we obtain that $L^2 = 2$. The Riemann–Roch theorem tells us that $\chi(L) = 1 + \frac{1}{2}L(L - K) = 1 + \frac{1}{2}L(-L) = 0$. On the other hand, by Serre duality $\chi(L) = 2h^0(L) - h^1(L)$, so if $H^1(L)$ is different from zero, then $H^0(L) \neq 0$, contradicting $p_g = 0$. □

Our construction for $n = 5$ shows in particular that the "Beauville surface," constructed in [Bea78] (see also [BPHV]) is an even surface with $K_S^2 = 8, q(S) = p_g(S) = 0$, but with $H^1(L) = 0$.

3 Canonical linearization on Fermat curves

Fix a positive integer $n \geq 5$, and let C be the degree-n Fermat curve

$$C := \{(x, y, z) \in \mathbb{P}^2 \mid f(x, y, z) := x^n + y^n + z^n = 0\}.$$

Let, as usual, μ_n be the group of n-roots of unity. Then the group

$$G := \mu_n^2 = \mu_n^3/\mu_n$$

acts on C, and we obtain a natural linearization of $\mathcal{O}_C(1)$ by letting $(\zeta, \eta) \in \mu_n^2$ act as follows:

$$z \mapsto z, x \mapsto \zeta x, y \mapsto \eta y.$$

In other words, $H^0(\mathcal{O}_C(1))$ splits as a direct sum of 1-dimensional eigenspaces (respectively generated by x, y, z) corresponding to the characters $(1,0), (0,1), (0,0) \in (\mathbb{Z}/n)^2 \cong \mathrm{Hom}(G, \mathbb{C}^*)$. Similarly, for $m \le n-1$, the monomial $x^a y^b z^{m-a-b} \in H^0(\mathcal{O}_C(m))$ generates the unique eigenspace for the character (a, b) (we identify here $\mathbb{Z}/n \cong \{0, 1, \ldots, n-1\}$ and we obviously require $a + b \le m$). However, any two linearizations differ (see [Mum70]) by a character of the group.

Definition 4 Assume that n is not divisible by 3. We call the **canonical** linearization on $H^0(\mathcal{O}_C(1))$ the one obtained from the natural one by twisting with the character $(n-3)^{-1}(1,1)$. Thus x corresponds to the character $v_1 := (1,0) + (n-3)^{-1}(1,1) = (-3)^{-1}(-2,1)$, y corresponds to the character $v_2 := (0,1) + (n-3)^{-1}(1,1) = (-3)^{-1}(1,-2)$, and z corresponds to the character $v_3 := (-3)^{-1}(1,1)$.

Remark 5 (I) Observe that v_1, v_2 are a basis of $(\mathbb{Z}/n)^2$ as soon as n is not divisible by 3. Indeed, $v_1 + v_2 = \frac{1}{3}(1,1) = 3^{-1}(1,1)$, hence

$$(1,0) = v_1 + 3^{-1}(1,1) = 2v_1 + v_2, \quad (0,1) = 2v_2 + v_1.$$

(II) Observe that the above linearization induces a linearization on all multiples of L, and, in the case where $m = (n-3)$, we obtain the natural linearization on the canonical divisor of C, $\mathcal{O}_C(n-3) \cong \Omega^1_C$. Since, if we take affine coordinates where $z = 1$, and we let f be the equation of C, we have

$$H^0(\Omega^1_C) = \left\{ P(x,y)\frac{dx}{f_y} = -P(x,y)\frac{dy}{f_x} \right\}$$

and the monomial $P = x^a y^b$ corresponds under this isomorphism to the character $(a+1, b+1)$.

(III) In particular, Serre duality

$$H^0(\mathcal{O}_C(m)) \times H^1(\Omega^1_C(-m)) \to H^1(\Omega^1_C) \cong \mathbb{C},$$

where $\mathbb{C}$ is the trivial G-representation, is G-invariant.

From the previous discussion it follows that:

Lemma 6 *The monomial* $x^a y^b z^c \in H^0(\mathcal{O}_C(m))$ *(here* $a, b, c \geq 0,\ a+b+c = m$*) corresponds to the character* χ, *equal to*

$$(a, b) + (-3)^{-1}(m, m) = (a - c)v_1 + (b - c)v_2.$$

Proof $v_1 + v_2 = \frac{1}{3}(1, 1)$, hence $(a, b) + (-3)^{-1}(m, m) = av_1 + bv_2 + (-m + a + b)(3)^{-1}(1, 1) = (a - c)v_1 + (b - c)v_2$. □

4 Abelian Beauville surfaces and their subcanonical divisors

We recall now the construction (see also [Cat00], [BCG05], or [Cat08]) of a Beauville surface with Abelian group $G \cong (\mathbb{Z}/n)^2$, where n is not divisible by 2 and by 3.

Definition 7 (1) Let $\Sigma \subset G$ be the union of the three respective subgroups generated by $(1, 0), (0, 1), (1, 1)$.

(2) Let $\psi : G \to G$ be a homomorphism such that, setting $\Sigma^* := \Sigma \setminus \{(0, 0)\}$, $\psi(\Sigma^*) \cap \Sigma^* = \emptyset$ (equivalently, $\psi(\Sigma) \cap \Sigma = \{(0, 0)\}$).

(3) Let C be the degree-n Fermat curve and let

$$S = (C \times C)/(\mathrm{Id} \times \psi)(G),$$

i.e., the quotient of $C \times C$ by the action of G s.t. $g(P_1, P_2) = (g(P_1), \psi(g)(P_2))$.

Remark 8 (i) By property (2), G acts freely and S is a projective smooth surface with ample canonical divisor.

(ii) The line bundle $\mathcal{O}_{C\times C}(1, 1)$ is $G \times G$ linearized, in particular it is $G \cong (\mathrm{Id} \times \psi)(G)$-linearized, therefore it descends to S, and we get a divisor L on S such that the pull-back of $\mathcal{O}_S(L)$ is the above G-linearized bundle.

(iii) By the previous remarks, we have a linear equivalence

$$K_S \equiv (n - 3)L.$$

5 Cohomology of multiples of the subcanonical divisor L

We consider now an integer m with

$$1 \leq m \leq n - 4$$

and determine the space $H^1(\mathcal{O}_S(mL))$.

Observe first that $H^1(\mathcal{O}_S(mL)) \cong H^1(\mathcal{O}_{C\times C}(m,m))^G$. By the Künneth formula

$$(9) \quad H^1(\mathcal{O}_{C\times C}(m,m)) \cong [H^0(\mathcal{O}_C(m)) \otimes H^1(\mathcal{O}_C(m))] \bigoplus [H^1(\mathcal{O}_C(m)) \otimes H^0(\mathcal{O}_C(m))].$$

We want to decompose the right hand side as a representation of $G \cong (\mathrm{Id} \times \psi)(G)$.

Explicitly, $H^0(\mathcal{O}_C(m)) = \oplus_\chi V_\chi$, where if we write the character $\chi = (a,b) + (-3)^{-1}(m,m)$ ($\chi = (a-(m-a-b))v_1 + (b-(m-a-b))v_2$ as we saw) then V_χ has dimension equal to one and corresponds to the monomial $x^a y^b z^{m-a-b}$, where $a, b \geq 0,\ a+b \leq m$. By Serre duality, $H^1(\mathcal{O}_C(m)) = \oplus_{\chi'} V_{-\chi'}$, where if we write as above $\chi' = (a',b') + (-3)^{-1}(m',m')$, then $V_{-\chi'}$ is the dual of $V_{\chi'}$, corresponding to the monomial $x^{a'} y^{b'} z^{m'-a'-b'}$, where $m' = n-3-m$, so $1 \leq m' \leq n-4$ also, and where $a', b' \geq 0,\ a'+b' \leq m'$.

Now, the homomorphism $\psi\colon G \to G$ induces a dual homomorphism $\phi := \psi^\vee\colon G^\vee \to G^\vee$, therefore we can finally write $H^1(\mathcal{O}_{C\times C}(m,m))$ as a representation of $G \cong (\mathrm{Id} \times \psi)(G)$:

$$H^1(\mathcal{O}_{C\times C}(m,m)) = \bigoplus_{\chi,\chi'} [(V_\chi \otimes V_{-\phi(\chi')}) \oplus (V_{-\chi'} \otimes V_{\phi(\chi)})].$$

We have proven therefore:

Lemma 10 $H^1(\mathcal{O}_S(mL)) \neq 0$ *if and only if there are characters* $\chi = (a-c)v_1 + (b-c)v_2$ *and* $\chi' = (a'-c')v_1 + (b'-c')v_2$ *with* $a, b \geq 0,\ a+b \leq m$, $a', b' \geq 0,\ a'+b' \leq m' = n-3-m$ *such that*

$$\chi = \phi(\chi') \quad \text{or} \quad \chi' = \phi(\chi).$$

Proof of Theorem 2 We now take ϕ to be given by a diagonal matrix in the basis v_1, v_2, i.e., such that

$$\phi(v_j) = \lambda_j v_j,\ j = 1, 2,\ \lambda_j \in (\mathbb{Z}/n)^*.$$

For further use we also set $\lambda := \lambda_1, \mu := \lambda_2$.

Given n relatively prime to 30 and $1 \leq m \leq n-4$, we want to find λ_1 and μ such that the equations

$$(a-c) = \lambda(a'-c')$$

$$(b-c) = \mu(b'-c')$$

have solutions with $a, b, c \geq 0, a+b+c = m$, and $a', b', c' \geq 0, a'+b'+c' = m'$.

The first idea is simply to take $b = c$ and $b' = c'$, so that μ can be taken arbitrarily. For the first equation some care is needed, since we want that λ be a unit: for this it suffices that $(a - c), (a' - c')$ are both units, for instance they could be chosen to be equal to one of the three numbers $1, 2, 3$, according to the congruence class of m, respectively m', modulo 3. With this proviso we have to verify that we have a free action on the product.

Lemma 11 *If $n \geq 7$, given λ a unit, there exists a unit μ such that $\psi = \phi^\vee$ satisfies the condition $\psi(\Sigma) \cap \Sigma = \{(0,0)\}$.*

Proof Since $(1,0) = 2v_1 + v_2$ and $(0,1) = v_1 + 2v_2$, the matrix of ϕ in the standard basis is the matrix

$$\phi = \frac{1}{3}\begin{pmatrix} 4\lambda - \mu & 2(\lambda - \mu) \\ 2(\mu - \lambda) & 4\mu - \lambda \end{pmatrix}$$

while the matrix of ψ is the matrix

$$\psi = \frac{1}{3}\begin{pmatrix} A := 4\lambda - \mu & B := 2(\mu - \lambda) \\ C := 2(\lambda - \mu) & D := 4\mu - \lambda \end{pmatrix}.$$

The conditions for a free action boil down to

$$A, B, C, D, A + B, C + D \text{ are units in } \mathbb{Z}/n$$

and moreover $A \neq B$, $C \neq D, A + B \neq C + D$. These are in turn equivalent to the condition that

$$\lambda, \mu, \lambda - 4\mu, \lambda - \mu, \mu - 4\lambda, \lambda + 2\mu, 2\lambda + \mu \in (\mathbb{Z}/n)^*.$$

Given $\lambda \in (\mathbb{Z}/n)^*$, consider its direct-sum decomposition given by the Chinese remainder theorem and the primary factorization of n. For each prime p dividing n, the residue classes modulo p which are excluded by the above condition are at most five values inside $(\mathbb{Z}/p)^*$, hence we are done if $(\mathbb{Z}/p)^*$ has at least six elements.

Now, since n is relatively prime to 30, each prime number dividing it is greater than or equal to $p = 7$. □

Proposition 12 *Consider the Beauville surface S constructed in [Bea78], corresponding to the case $n = 5$. Then S is an even surface and $K_S \equiv 2L$, where $H^1(L) = 0$.*

Proof We observe that L is unique, because the torsion group of S is of exponent 5 (see [BC04]). The existence of L follows exactly as in the proof of the main theorem, where the condition $n \geq 7$ was not used. That $H^1(L) = 0$ follows directly from Proposition 3. □

Acknowledgments I would like to thank Jonathan Wahl for asking the above question. In the Appendix below he describes a construction based on our main result.

References

[BC04] Bauer, I. C. and Catanese, F. Some new surfaces with $p_g = q = 0$. The Fano Conference, pp. 123–142. University of Torino, Turin, 2004.

[BCG05] Bauer, I., Catanese, F., and Grunewald, F. Beauville surfaces without real structures. In, *Geometric Methods in Algebra and Number Theory*. Progress in Mathematics, Vol. 235. Berlin: Birkhäuser, 2005, pp. 1–42.

[BPHV] Barth, W., Peters, C., Van de Ven, A. *Compact Complex Surfaces*. Ergebnisse der Mathematik und ihrer Grenzgebiete (3), 4. Berlin: Springer-Verlag, 1984. Second edition by Barth, W., Hulek, K., Peters, C. and Van de Ven, A. Ergebnisse der Mathematik und ihrer Grenzgebiete. 3. Folge. A, 4. Berlin: Springer-Verlag, 2004.

[Bea78] Beauville, A. *Surfaces Algébriques Complexes*. Astérisque No. 54. Paris: Société Mathématique de France, 1978.

[Cat00] Catanese, F. Fibred surfaces, varieties isogenous to a product and related moduli spaces. *Amer. J. Math.* **122**(1) (2000), 1–44.

[Cat08] Catanese, F. *Differentiable and deformation type of algebraic surfaces, real and symplectic structures*. In Symplectic 4-Manifolds and Algebraic Surfaces, 55–167. Lecture Notes in Mathematics, 1938. Springer, Berlin, (2008).

[Hart77] Hartshorne, R. *Algebraic Geometry*. Graduate Texts in Mathematics, No. 52. New York: Springer-Verlag, 1977.

[Mum70] Mumford, D. *Abelian Varieties*. Tata Institute of Fundamental Research Studies in Mathematics, No. 5. Published for the Tata Institute of Fundamental Research, Bombay. London: Oxford University Press, 1970.

Appendix by J. Wahl: A non-$\mathbb{Q}$-Gorenstein Cohen–Macaulay cone X with K_X $\mathbb{Q}$-Cartier

A germ $(X, 0)$ of an isolated normal complex singularity of dimension $n \geq 2$ is called $\mathbb{Q}$-*Gorenstein* if:

1. $(X, 0)$ is Cohen–Macaulay.
2. The dualizing sheaf K_X is $\mathbb{Q}$-Cartier (i.e., the invertible sheaf $\omega_{X-\{0\}}$ has finite order r).

3. The corresponding cyclic *index one* (or *canonical*) cover $(Y,0) \to (X,0)$ is Cohen–Macaulay, hence Gorenstein.

Alternatively, $(X,0)$ is the quotient of a Gorenstein singularity by a cyclic group acting freely off the singular point. Some early definitions did not require the third condition, which is of course automatic for $n = 2$.

If $(X,0)$ is $\mathbb{Q}$-Gorenstein, a 1-parameter deformation $(\mathcal{X},0) \to (\mathbb{C},0)$ is called $\mathbb{Q}$-Gorenstein if it is the quotient of a deformation of the index one cover of $(X,0)$; this is exactly the condition that $(\mathcal{X},0)$ is itself $\mathbb{Q}$-Gorenstein. These notions were introduced by Kollár and Shepherd-Barron [1], who made extensive use of the author's explicit smoothings of certain cyclic quotient surface singularities in [2] (5.9); these deformations were patently $\mathbb{Q}$-Gorenstein, and it was important to name this property.

Recently, the author and others considered rational surface singularities admitting a rational homology disk smoothing (i.e., with Milnor number 0). The 3-dimensional total space of the smoothing had a rational singularity with K $\mathbb{Q}$-Cartier, but it was not initially clear whether the smoothings were $\mathbb{Q}$-Gorenstein. (This was later established [4] by proving the stronger result that the total spaces were log-terminal.) In fact, one needs to be careful because of the example of A. Singh:

Example ([3]) There is a 3-dimensional isolated rational (hence Cohen–Macaulay) complex singularity $(X,0)$ with K_X $\mathbb{Q}$-Cartier which, however, is not $\mathbb{Q}$-Gorenstein.

The purpose of this appendix is to use F. Catanese's result to provide other examples; they are not rational, but are cones over a smooth projective variety, which could for instance be assumed to be projectively normal with ideal generated by quadrics.

Proposition 13 *Let S be a surface as in Theorem 2, with $h^1(S,\mathcal{O}_S) = 0$, L ample, $K_S = rL$ (for some $r > 1$), and $h^1(mL) \neq 0$ for some $m > 0$. Let t be greater than r and relatively prime to it. Then*

1. *The cone $R = \mathcal{R}(S,tL) := \oplus_{m\geq 0} H^0(S,\mathcal{O}_S(mtL))$ is Cohen–Macaulay.*
2. *The dualizing sheaf of R is torsion, of order t.*
3. *The index 1 cover is $\mathcal{R}(S,L) := \oplus_{m\geq 0} H^0(S,\mathcal{O}_S(mL))$, and is not Cohen–Macaulay.*

In particular, R is not $\mathbb{Q}$-Gorenstein.

Proof The Cohen–Macaulayness for R follows because $h^1(itL) = 0$, for all i, thanks to Kodaira vanishing. Let $\pi \colon V \to S$ be the geometric line bundle corresponding to $-tL$; then $H^0(V, \mathcal{O}_V) \equiv R$. Since $K_V \equiv \pi^*(K_S + tL)$, one has that $jK_R \equiv \oplus_{n\in\mathbf{Z}} H^0(S, j(K_S + tL) + nL)$; since $tK_S = r(tL)$ with r and t relatively prime, K_R has order t. Making a cyclic t-fold cover and normalizing gives that $\mathcal{R}(S, L)$ is the index 1 cover, which as Catanese noted is not Cohen–Macaulay. □

References

[1] Kollár, J. and Shepherd-Barron, N. Threefolds and deformations of surface singularities. *Invent. Math.* **91**(2) (1988), 299–338.

[2] Looijenga, E. and Wahl, J. Quadratic functions and smoothing surface singularities. *Topology* **25** (1986), 261–291.

[3] Singh, A. Cyclic covers of rings with rational singularities. *Trans. AMS* **355**(3) (2002), 1009–1024.

[4] Wahl, J. Log-terminal smoothings of graded normal surface singularities. *Michigan Math J.* **62** (2013), 475–489.

7

Threefold divisorial contractions to singularities of cE type

J. A. Chen[a]
National Taiwan University

Abstract

We survey some recent progress in the classification of three-dimensional divisorial contractions to cE points. In particular, we introduce a new structure of three-dimensional cE singularity and use this structure to explain the work of Hayakawa. We also provide some new examples.

Dedicated to Rob Lazarsfeld on the occasion of his sixtieth birthday

1 Introduction

The minimal model program has been one of the main tools in the study of birational algebraic geometry. After some recent advances in the study of the geometry of complex 3-folds, one might hope to build up an explicit classification theory for 3-folds similar to the theory of surfaces by using the minimal model program.

In the minimal model program, divisorial contractions, flips, and flops are considered to be elementary maps. Any birational map obtained from the

[a] The author was partially supported by NCTS/TPE and the National Science Council of Taiwan. He is indebted to Cascini, Hacon, Hayakawa, Kawakita, Kawamata, Kollár, and Mori for many useful discussions. Some of this work was done during visits of the author to RIMS and Imperial College London. The author would like to thank both institutes for their hospitality. He is grateful to the referee for useful comments and corrections.

minimal model program consists of a combination of the above-mentioned maps. Let us briefly recall some known results about three-dimensional birational maps. First of all, Mori and then Cutkosky classified birational maps from nonsingular 3-folds and Gorenstein 3-folds respectively [4, 20]. Tziolas has produced a series of work on divisorial contractions to curves passing through Gorenstein singularities (cf. [23–25]). The recent project of Mori and Prokhorov (cf. [21, 22]) on extremal contractions provides a treatment which is valid for divisorial contractions to curves and for conic bundles. They classified completely divisorial contractions to curves of type IA, IC, and IIB. Flops are studied in Kollár's article [16]. Flips are still quite mysterious, except for some examples in [2, 18].

Divisorial contractions to points are probably the best understood, due mainly to the works of Kawamata, Hayakawa, Markushevich, and Kawakita (cf. [5–7, 10–15, 19]). Divisorial contractions to points of index > 1 are now completely classified and realized as weighted blow-ups. Therefore, it remains to consider contractions to points of index 1, i.e., terminal Gorenstein singularities. The description of contractions to index 1 points can be found in [13]. In fact, contractions to cA points are classified completely in [13]. Recently, Hayakawa started a project to classify contractions to cD and cE points [8, 9]. The project is not yet complete. Especially, the existence of divisorial contractions with discrepancy > 1 listed in [13, Table 3, e2, e3, e7] is unknown.

The purpose of this paper is to analyze the known examples in [9] and give the structure of various weighted blow-ups. We introduce a new structure of three-dimensional cE singularities and use this structure to explain the work of Hayakawa. We also provide some new examples in the last section.

2 Normal form of cE singularities

For any $F \in \mathbb{C}\{x_1, \dots, x_n\}$ the set $(F = 0)$ is a germ of a complex analytic set. For $F \in \mathbb{C}[[x_1, \dots x_n]]$, by the singularity $(F = 0)$, we mean the scheme $\mathrm{Spec}\mathbb{C}[[x_1, \dots, x_n]]/(F)$.

$F, G \in \mathbb{C}[[x_1, \dots, x_n]]$ (resp. $\mathbb{C}\{x_1, \dots, x_n\}$) are called equivalent if there is an automorphism of $\mathbb{C}[[x_1, \dots, x_n]]$ (resp. $\mathbb{C}\{x_1, \dots, x_n\}$) given by $x_i \mapsto \phi_i(x_1, \dots, x_n)$ and a unit $u(x_1, \dots, x_n)$ such that

$$u(x_1, \dots, x_n)G(x_1, \dots, x_n) = F(\phi_1, \dots, \phi_n).$$

Note that if $F, G \in \mathbb{C}\{x_1, \dots, x_n\}$ have isolated singularities at the origin, then F and G are equivalent in $\mathbb{C}\{x_1, \dots, x_n\}$ if and only if they are equivalent in $\mathbb{C}[[x_1, \dots, x_n]]$ (see, e.g., [1, 17]).

For a power series F, F_d denotes the degree-d homogeneous part and $F_{\geq d}$ (resp. $F_{>d}$) denotes the part of degree $\geq d$ (resp. $> d$).

Theorem 2.1 ([17]) *Assume that $F(x, y, z, u)$ defines a terminal singularity of type cE. Then F is equivalent to one of the following:*

- cE_6 : $x^2 + y^3 + yg_{\geq 3}(z, u) + h_{\geq 4}(z, u)$, *where* $h_4 \neq 0$.
- cE_7 : $x^2 + y^3 + yg_{\geq 3}(z, u) + h_{\geq 5}(z, u)$, *where* $g_3 \neq 0$.
- cE_8 : $x^2 + y^3 + yg_{\geq 4}(z, u) + h_{\geq 5}(z, u)$, *where* $h_5 \neq 0$.

We call such a form a normal form of F.

The following consequence is immediate:

Lemma 2.2 *Normal forms are equivalent if and only if there exists an automorphism of the form*

$$\begin{cases} x \mapsto x; \\ y \mapsto y; \\ z \mapsto \phi_3(z, u); \\ u \mapsto \phi_4(z, u). \end{cases}$$

Therefore, changing z, u by a linear transformation and up to a constant, we may and do assume that in the case of cE_6,

$$h_4 \in \{z^4, z^3(z + u), (z^2 + zu)^2, z^2(z^2 + u^2), z(z^3 + u^3)\}.$$

In particular, $z^4 \in h_4$.

In the case of cE_7, $g_3 \neq 0$, we may and do assume that

$$g_3 \in \{z^3, z^2(z + u), z(z^2 + u^2)\}.$$

In particular, $z^3 \in g_3$.

In the case of cE_8, $h_5 \neq 0$, we may and do assume that

$$h_5 \in \left\{ \begin{array}{l} z^5, z^4(z + u), z^3(z + u)^2, z^3(z^2 + u^2), \\ z^2(z + u)^2(z - u), z^2(z^3 + u^3), z(z^4 + u^4) \end{array} \right\}.$$

In particular, $z^5 \in h_5$.

Remark 2.3 Consider

$$F = x^2 + y^3 + yg(z, u) + h(z, u),$$

which possibly contains lower-degree terms in g or h. An isolated singularity with the above description is called a cE-like singularity.

An isolated cE-like singularity is at worst of type cD (resp. cE_6, cE_7, cE_8) if $g_m \neq 0$ for some $m \leq 2$ or $h_m \neq 0$ for some $m \leq 3$ (resp. $h_m \neq 0$ for some $m \leq 4$, $g_m \neq 0$ for some $m \leq 3$, $h_m \neq 0$ for some $m \leq 5$).

3 Admissible weights and canonical form

Given a three-dimensional terminal Gorenstein singularity $(P \in X)$, it is known that there exists a divisorial contraction to $(P \in X)$ with discrepancy 1, which is realized as a weighted blow-up (cf. [19]). In this section we consider weights that might be admissible for a weighted blow-up with discrepancy 1.

Given a terminal Gorenstein singularity $(P \in X)$, we always identify it with $(F = 0) \subset \mathbb{C}^4$ for some F of normal form. We consider a weighted blow-up $wBl_w\colon \mathcal{Y} \to \mathcal{X} = \mathbb{C}^4$ with weight $w = (a, b, c, d)$. Let $\mathcal{E}$ be the exceptional divisor and write $\mathcal{Y} = \cup_{i=1}^4 U_i$. Let Q_i denote the origin of U_i. There is an induced map $wBl_w\colon Y \to X$, where Y is realized as the proper transform of X in the weighted blow-up of $\mathbb{C}^4$. Let $E := \mathcal{E} \cap Y$.

In considering weighted blow-ups $wBl_w\colon Y \to X$, we have the following questions:

1. Is E irreducible and reduced? If so, then E gives rise to a valuation with discrepancy $a + b + c + d - wt_w(F) - 1$. In other words, $K_Y = wBl_w^* K_X + \alpha E$ with $\alpha = a + b + c + d - wt_w(F) - 1$, whenever it makes sense. We thus call wBl_w a *divisorial blow-up* in this situation.
2. Does Y have only isolated singularities? If not, then clearly Y is not terminal.
3. Does Y have terminal singularities? If so, then $wBl_w\colon Y \to X$ is indeed a divisorial contraction.

Since divisorial contractions with minimal discrepancy 1 play a pivotal role in the study of geometry over terminal Gorenstein points, we would like to first consider all possible weights such that $\alpha = a+b+c+d-wt_w(F)-1 = 1$, which we call *admissible weights* of F. However, we exclude the weights $(2, 1, 1, 1)$ and $(1, 1, 1, 1)$ because $\mathcal{E} \cap Y$ is never reduced, even though $\alpha = 1$ is satisfied.

Proposition 3.1 *The admissible weights for cE_6 are*

$$(6, 4, 3, 1), (4, 3, 2, 1), (3, 2, 2, 1), (2, 2, 1, 1).$$

The admissible weights for cE_7 are

$$(9, 6, 4, 1), (7, 5, 3, 1), (6, 4, 3, 1), (5, 4, 2, 1), (5, 3, 2, 1),$$

$$(4, 3, 2, 1), (3, 2, 2, 1), (3, 3, 1, 1), (3, 2, 1, 1), (2, 2, 1, 1).$$

The admissible weights for cE_8 are

$$(15, 10, 6, 1), (12, 8, 5, 1), (10, 7, 4, 1), (9, 6, 4, 1), (8, 5, 3, 1), (7, 5, 3, 1),$$

$$(6, 4, 3, 1), (5, 4, 2, 1), (5, 3, 2, 1), (4, 3, 2, 1), (3, 2, 2, 1), (3, 2, 1, 1), (2, 2, 1, 1).$$

Proof We first consider the cE_6 case. We have

$$\begin{cases} wt(x^2), wt(y^3), wt(z^4) \geq wt(F), \\ wt(xyzu) - wt(F) - 2 = 0. \end{cases}$$

It follows that

$$wt(xyz) \geq \left(\frac{1}{2} + \frac{1}{3} + \frac{1}{4}\right) wt(F)$$

and therefore

$$2 \geq \frac{1}{12} wt(F) + wt(u).$$

It is then straightforward to solve for the admissible weights.

We next consider the cE_7 case. We have

$$\begin{cases} wt(x^2), wt(y^3), wt(yz^3) \geq wt(F), \\ wt(xyzu) - wt(F) - 2 = 0. \end{cases}$$

It follows that

$$wt(x^3y^3z^3) = wt(x^3 \cdot y^2 \cdot yz^3) \geq \left(\frac{3}{2} + \frac{2}{3} + 1\right) wt(F)$$

and therefore

$$2 \geq \frac{1}{18} wt(F) + wt(u).$$

It is then straightforward to solve for the admissible weights.

The case of cE_8 is similar to the case of cE_6. We have

$$2 \geq \frac{1}{30} wt(F) + wt(u).$$

□

Definition 3.2 Given an admissible weight w of the form $(a, b, k, 1)$ with $a \geq b \geq k \geq 1$, we define the following notions:

- The level of w, denoted $lev(w)$, is k.
- $\sigma(w) := a + b + k - 1$.
- $\sigma_y(w) := a + k - 1$.
- A weight denoted by $w_{\sigma(w)}$, e.g., $w_{12} = (6, 4, 3, 1)$. There are two weights with $\sigma = 6$: $w_6 = (3, 2, 2, 1)$ and $w'_6 = (3, 3, 1, 1)$.
- w is in the main series, or of stage 0, if $w = (3k - 3, 2k - 2, k, 1)$. This consists of $wt_{30} = (15, 10, 6, 1), w_{24} = (12, 8, 5, 1), w_{18} = (9, 6, 4, 1), w_{12} = (6, 4, 3, 1),\ w_6 = (3, 2, 2, 1)$.

  Note that $\sigma = 6k - 6, \sigma_y = 4k - 4$ in this situation.

- w is in series I, or of stage 1, if $w = (3k - 2, 2k - 1, k, 1)$.
  This consists of $w_{20} = (10, 7, 4, 1), w_{14} = (7, 5, 3, 1), w_8 = (4, 3, 2, 1)$.
  Note that $\sigma = 6k - 4, \sigma_y = 4k - 3$ in this situation.
- w is in series II, or of stage 2, if $w = (3k - 1, 2k - 1, k, 1)$.
  This consists of $w_{15} = (8, 5, 3, 1), w_9 = (5, 3, 2, 1)$.
  Note that $\sigma = 6k - 3, \sigma_y = 4k - 2$ in this situation.
- w is in series III, or of stage 3, if $w = (3k - 1, 2k, k, 1)$.
  This consists of $w_{10} = (5, 4, 2, 1), w_4 = (2, 2, 1, 1)$.
  Note that $\sigma = 6k - 2, \sigma_y = 4k - 2$ in this situation.
- w is in series IV, or of stage 4, if $w = (3k, 2k, k, 1)$.
  This consists of $w_5 = (3, 2, 1, 1)$.
  Note that $\sigma = 6k - 1, \sigma_y = 4k - 1$ in this situation.
- w is in series V, or of stage 5, if $w = (3k, 2k + 1, k, 1)$.
  This consists of $w'_6 = (3, 3, 1, 1)$.
  Note that $\sigma = 6k, \sigma_y = 4k - 1$ in this situation and this only happens in the case of cE_7.

Definition 3.3 Given a normal form F, we say that F is of level $\geq k$ if $wt_w(F) \geq \sigma(w)$ for weight w of level $k \geq 2$ in the main series. Otherwise, we say that F is of level 1.

Lemma 3.4 *If F is of level $\geq k_1$, then F is of level $\geq k_2$ for any $k_1 \geq k_2 \geq 1$.*

Proof We write $g = \sum a_{ij}z^i u^j$, $h = \sum b_{ij}z^i u^j$ and let $I_g := \{(i, j)|a_{ij} \neq 0\}$, $I_h := \{(i, j)|b_{ij} \neq 0\}$.

If F is of level $\geq k_1$, then for $(i, j) \in I_h$ (resp. I_g), $k_1 i + j \geq 6(k_1 - 1)$ (resp. $4(k_1 - 1)$). It follows that for $(i, j) \in I_h$

$$k_2 i + j \geq \frac{k_2}{k_1}(k_1 i + j) \geq \frac{k_2}{k_1} 6(k_1 - 1) \geq 6(k_2 - 1).$$

Similarly, for $(i, j) \in I_g$, one has $k_2 i + j \geq 4(k_2 - 1)$. Therefore, F is of level $\geq k_2$. □

Definition 3.5 Fix a normal form F of level k, we say that F is of stage $\geq m$ if $wt_w(F) \geq \sigma(w)$ for the admissible weight w of level k of stage $m \geq 1$. Otherwise, we say that F is of level k of stage 0.

Lemma 3.6 *If F is of stage $\geq m_1$, then F is of stage $\geq m_2$ for any $m_1 \geq m_2$.*

Proof Notice that weights of level k and stage 0 consist of $v_0 = (3k - 3, 2k - 2, k, 1)$ with $\sigma = 6k - 6$. Hence F is of level k stage 0 if and only if $wt_{v_0}(g) \geq 4k - 4$ and $wt_{v_0}(h) \geq 6k - 6$.

Weights of stage 1 consist of $v_1 = (3k-2, 2k-1, k, 1)$ with $\sigma = 6k-4$. Hence F is of level k stage ≥ 1 if and only if $wt_{v_1}(g) \geq 4k-3$ and $wt_{v_1}(h) \geq 6k-4$. Since $wt_{v_0}(g) = wt_{v_1}(g)$ and $wt_{v_0}(h) = wt_{v_1}(h)$, it follows immediately that if F is of stage ≥ 1 then F is of stage 0.

Similarly, weights of stage 2 consist of $v_2 = (3k-1, 2k-1, k, 1)$ with $\sigma = 6k-3$. Hence F is of level k stage ≥ 2 if and only if $wt_{v_2}(g) \geq 4k-2$ and $wt_{v_2}(h) \geq 6k-3$. It follows immediately that if F is of stage ≥ 2 then F is of stage ≥ 1.

The comparisons of other stages are similar. □

Definition 3.7 Given F, we say that F is of level k if it is of level $\geq k$ but not of level $\geq k+1$. We say that F is of stage 0 if it is not of stage ≥ 1 and of stage m if it is of stage $\geq m$ but not of stage $\geq m+1$.

A normal form F is said to be a *canonical form* if it admits the highest level and then the highest stage among all equivalent normal forms.

Given a cE singularity $P \in X$, we associate a canonical form F so that $(P \in X) \cong o \in (F = 0) \subset \mathbb{C}^4$. We define the level and stage of $P \in X$ to be the level and stage of its canonical form F.

Therefore, we may classify isolated cE points $P \in X \cong o \in (F = 0) \subset \mathbb{C}^4$ according to their level and stage.

4 Admissible weighted blow-ups

In this section we study weighted blow-ups of cE singularities by admissible weights. Let us first fix some notation. Fix an admissible weight $w = (a, b, k, 1)$ such that $wt_w(F) \geq \sigma(w)$. We may write

$$F = F^w_\sigma + F^w_{>\sigma} \quad \text{or simply } F_\sigma + F_{>\sigma},$$

where F_σ (resp. $F_{>\sigma}$) denotes the homogeneous part of weighted degree $= \sigma(w)$ (resp. the part of weighted degree $> \sigma(w)$). More explicitly, we may also similarly write

$$F = x^2 + y^3 + yg_{\sigma_y} + yg_{>\sigma_y} + h_\sigma + h_{>\sigma}$$

or

$$F = x^2 + y^3 + y \sum_{ki+j\geq\sigma_y} a_{ij}z^iu^j + \sum_{ki+j\geq\sigma} b_{ij}z^iu^j.$$

We set $I_g := \{(i, j)|a_{ij} \neq 0\}$ and $I_h := \{(i, j)|b_{ij} \neq 0\}$.

Lemma 4.1 *Let F be of level k_1. Let $wBl_w : Y \to X$ be a weighted blow-up of weight w of level k_2 such that $k_1 > k_2 \geq 2$ and $wt_w(F) \geq \sigma(w)$.*

Suppose that

- *w is of stage 0, 1, 2, or*
- *w is of stage 3 (hence $k_2 = 2$ or 1) and $k_1 \geq k_2 + 2$, or*
- *w is of stage 4 (hence $k_2 = 1$) and $k_1 \geq 3$.*

Then Y is not terminal.

Proof We shall prove that $Q_4 \in Y \cap U_4$ is not terminal. Since F is of level k_1, we have $k_1 i + j \geq 6k_1 - 6$ (resp. $4k_1 - 4$) for $(i, j) \in I_h$ (resp. I_g).

Let w be a weight of level k_2, then $Y \cap U_4$ is given by

$$\tilde{F} = x^2 u^{wt_w(x^2)-\sigma(w)} + y^3 u^{wt_w(y^3)-\sigma(w)} + y \sum a_{ij} z^i u^{j'} + \sum b_{ij} z^i u^{j''},$$

where

$$\begin{cases} j' = k_2 i + j - \sigma_y(w); \\ j'' = k_2 i + j - \sigma(w). \end{cases}$$

Therefore, if $(i, j) \in I_g$, then

$$i + j' + \sigma_y(w) = (k_2 + 1)i + j \geq \frac{k_2 + 1}{k_1}(k_1 i + j) \geq \frac{k_2 + 1}{k_1} 4(k_1 - 1) \geq 4k_2. \quad (\dagger_g)$$

Similarly, if $(i, j) \in I_h$, then

$$i + j'' + \sigma(w) \geq 6k_2. \quad (\dagger_h)$$

Case 1 w is of stage 0.
Note that we have $\sigma(w) = 6k_2 - 6, \sigma_y(w) = 4k_2 - 4$. By $\dagger_g, \dagger_h$, $\tilde{F}$ is of the form

$$x^2 + y^3 + y \sum a_{ij} z^i u^{j'} + \sum b_{ij} z^i u^{j''},$$

with $i + j' \geq 4$ and $i + j'' \geq 6$. Hence Q_4 is not terminal.

Case 2 w is of stage 1.
Now we have $\sigma(w) = 6k_2 - 4, \sigma_y(w) = 4k_2 - 3$.

By $\dagger_g, \dagger_h$, $Y \cap U_4$ is given by

$$\tilde{F} = x^2 + y^3 u + y \sum a_{ij} z^i u^{j'} + \sum b_{ij} z^i u^{j''},$$

with $i + j' \geq 3$ and $i + j'' \geq 4$. Hence Q_4 is not terminal.

Case 3 w is of stage 2.
Now we have $\sigma(w) = 6k_2 - 3, \sigma_y(w) = 4k_2 - 2$. Note that $Y \cap U_4$ is given by

$$\tilde{F} = x^2 u + y^3 + y \sum a_{ij} z^i u^{j'} + \sum b_{ij} z^i u^{j''}.$$

By $\dagger_g, \dagger_h$, $i + j' \geq 2$ and $i + j'' \geq 3$. Hence Q_4 is not terminal.

Case 4 w is of stage 3 and $k_2 = 2$.
Now $w = (5, 4, 2, 1)$ with $\sigma = 10, \sigma_y = 6$ and

$$\tilde{F} = x^2 + y^3u^2 + y \sum a_{ij}z^i u^{j'} + \sum b_{ij}z^i u^{j''}.$$

Note that $\dagger_g$ shows that $i + j' + \sigma_y(w) \geq 12\frac{k_1-1}{k_1}$ and hence $i + j' \geq 3$ if $k_1 \geq 4$. Similarly, $\dagger_h$ shows that $i + j'' + \sigma(w) \geq 18\frac{k_1-1}{k_1}$ and hence $i + j'' \geq 4$ if $k_1 \geq 4$. Hence Q_4 is not terminal if $k_1 \geq 4$.

Case 5 w is of stage 3 and $k_2 = 1$.
Now $w = (2, 2, 1, 1)$ with $\sigma = 4, \sigma_y = 2$ and

$$\tilde{F} = x^2 + y^3u^2 + y \sum a_{ij}z^i u^{j'} + \sum b_{ij}z^i u^{j''}.$$

Note that $\dagger_g$ shows that $i + j' + \sigma_y(w) \geq 8\frac{k_1-1}{k_1}$ and hence $i + j' \geq 3$ if $k_1 \geq 3$. Similarly, $\dagger_h$ shows that $i + j'' + \sigma(w) \geq 12\frac{k_1-1}{k_1}$ and hence $i + j'' \geq 4$ if $k_1 \geq 3$. Hence Q_4 is not terminal if $k_1 \geq 3$.

Case 6 w is of stage 4 and $k_2 = 1$.
Now $w = (3, 2, 1, 1)$ with $\sigma = 5, \sigma_y = 3$ and

$$\tilde{F} = x^2u + y^3u + y \sum a_{ij}z^i u^{j'} + \sum b_{ij}z^i u^{j''}.$$

Note that $\dagger_g$ shows that $i + j' + \sigma_y(w) \geq 8\frac{k_1-1}{k_1}$ and hence $i + j' \geq 2$ if $k_1 \geq 3$. Similarly, $\dagger_h$ shows that $i + j'' + \sigma(w) \geq 12\frac{k_1-1}{k_1}$ and hence $i + j'' \geq 3$ if $k_1 \geq 3$. Hence Q_4 is not terminal if $k_1 \geq 3$. □

We will need the following useful criterion, due to Hayakawa (cf. [7]), to determine whether a weighted blow-up is a divisorial contraction or not. For the reader's convenience, we reproduce the proof.

Theorem 4.2 *Given $P \in X \cong o \in (F = 0) \subset \mathbb{C}^4$ the germ of a terminal singularity of cE type, where F is a normal form, let $f = wBl_v : Y \to X$ be the weighted blow-up with weight $v = (a, b, k, 1)$ and exceptional divisor E. Suppose that*

- *E is irreducible;*
- *$a + b + k - wt_v(F) = 1$;*
- *$Y \cap U_4$ is terminal.*

Then $Y \to X$ is a divisorial contraction.

Proof Suppose that E is irreducible, then $K_Y = f^*K_X + a(E,X)E$ with $a(E,X) = a + b + k - wt_v(F) - 1$. Let $D = (u = 0) \subset \mathrm{Div}(X)$ and D_Y be its proper transform in Y. One has $f^*D = D_Y + E$. Hence

$$f^*(K_X + D) = K_Y + D_Y$$

and $D_Y \sim_X -K_Y$.

Let $g : Z \to Y$ be a resolution of Y. For any exceptional divisor F in Z such that $g(Z) \subset D_Y$, one has $g^*D_Y = D_Z + mF + \ldots$ for some $m > 0$. It follows that $a(F, Y) = m > 0$. Therefore, it remains to consider exceptional divisors whose center in Y is not contained in D_Y. Hence Y is terminal if $Y - D_Y = Y \cap U_4$ is terminal. Once Y is terminal, it is then easy to see that $Y \to X$ is a divisorial contraction. □

Lemma 4.3 *Let $Y \to X \ni P$ be a divisorial contraction with discrepancy* 1 *to a terminal Gorenstein singularity $P \in X$. Let E be its exceptional divisor. Then for any divisor F with discrepancy* 1 *over P, its center in Y is a singular point of index* > 1 *and contained in E.*

Therefore, to search for exceptional divisors over P with discrepancy 1, it suffices to search for exceptional divisors over singular points of index > 1 in E. We denote by $\mathrm{Sing}(Y)_{>1}$ the set of singular points of index > 1 on Y (which is contained in E).

4.1 Weights of stage 0

Proposition 4.4 *Let $(F = 0)$ be a cE singularity of level $\geq k$. We consider a weighted blow-up with weight $w = (3k-3, 2k-2, k, 1)$ of level k of stage* 0. *Then the exceptional divisor is irreducible.*

Proof This is clear since the exceptional divisor is defined by $(F^w_\sigma = 0)$ and F^w_σ contains x^2, y^3. □

Now suppose that F is of level k. Fix a weight $w = (3k-3, 2k-2, k, 1)$. We may write

$$F = x^2 + y^3 + yg_{\sigma_y} + yg_{\sigma_y+1} + yg_{>\sigma_y+1} + h_\sigma + h_{\sigma+1} + h_{>\sigma+1}.$$

Lemma 4.5 *Suppose that F is of level k. Consider $Y \to X$ the weighted blow-up with weight $w = (3k-3, 2k-2, k, 1)$. Then $\mathrm{Sing}(Y) \cap U_4$ is isolated unless:*

$$\exists\ s(z,u) \text{ s.t.} \begin{cases} g_{\sigma_y} = -3s(z,u)^2, \\ h_\sigma = 2s(z,u)^3, \\ h_{\sigma+1} = -s(z,u)g_{\sigma_y+1}. \end{cases} \qquad \natural$$

Proof It is clear that $\mathrm{Sing}(Y) \cap U_1 = \emptyset$. Hence we have $\mathrm{Sing}(Y) \cap U_4 \subset (x = 0)$. Moreover, $\mathrm{Sing}(Y) \subset E$. Hence we have $\mathrm{Sing}(Y) \cap U_4 \subset (u = 0)$.

Therefore, let $\tilde{F}$ be the defining equation of $Y \cap U_4$. We have

$$\begin{aligned}\mathrm{Sing}(Y) \cap U_4 &\subset (x = u = 0) \cap (\tilde{F} = \tilde{F}_y = \tilde{F}_u = 0) \\ &\subset (x = u = 0) \cap \Sigma,\end{aligned}$$

where Σ is defined as

$$\begin{cases} y^3 + yg_{\sigma_y} + h_\sigma = 0, \\ 3y^2 + g_{\sigma_y} = 0, \\ yg_{\sigma_y+1} + h_{\sigma+1} = 0. \end{cases}$$

If g_{σ_y} is not a perfect square, then $3y^2 + g_{\sigma_y}$ is irreducible and hence Σ is finite. If g_{σ_y} is a perfect square, then we write it as $g_{\sigma_y} = -3s^2$. One sees that Σ is finite unless $y - s$ or $y + s$ divides the above three polynomials. The statement now follows. □

Once we know that $\mathrm{Sing}(Y) \cap U_4$ is isolated, then it is easy to check whether the singularities in $Y \cap U_4$ are terminal or not. Notice that $\tilde{F}$ is of the form $x^2 + y^3 + y\tilde{g} + \tilde{h}$. It is straightforward to see that $\tilde{F}$ is at worst of cE type at $Q_4 \in U_4$ if and only if F is not of level $k + 1$.

Next we study $(0, 0, \gamma, 0) \in U_4$. We rewrite

$$F = x^2 + y^3 + y \sum a'_{ij}(z - \gamma u^k)^i u^j + \sum b'_{ij}(z - \gamma u^k)^i u^j$$

and correspondingly

$$\tilde{F} = x^2 + y^3 + y \sum a'_{ij}(z - \gamma)^i u^{j'} + \sum b'_{ij}(z - \gamma)^i u^{j''},$$

with $j' = ki + j - \sigma_y$ and $j'' = ki + j - \sigma$.

If $\tilde{F}$ has a non-terminal singularity at $(0, 0, \gamma, 0) \in U_4$, then $i + j' \geq 4$ and $i + j'' \geq 6$. It follows that F has level $\geq k + 1$, a contradiction. Notice also that singularity the at $(\alpha, \beta, \gamma, \delta) \in U_4$ with $(\alpha, \beta) \neq (0, 0)$ is at worst of cA type. We thus conclude the following:

Theorem 4.6 *Given a canonical form F, suppose that F is of level k stage 0. Then the weighted blow-up with weight $w = (3k - 3, 2k - 2, k, 1)$ is a divisorial contraction if ♮ does not hold.*

Suppose that ♮ holds. Then by considering a change of coordinate $\bar{y} = y - s(z, u)$, one sees that F is equivalent to

$$G := x^2 + y^3 + s(z, u)y^2 + yg_{>\sigma_y+1} + h_{>\sigma+1}.$$

Note that $wt_w(s(z,u)) = 2k-2$. Hence consider $w_1 = (3k-2, 2k-1, k, 1)$ the weight of level k of stage 1. One sees that $wt_{w_1}(G) \geq 6k-4 = \sigma(w_1)$, that is, G is of level k of stage ≥ 1. Therefore, we may say that G is a *stage lifting* of F.

5 Divisorial contractions to cE points with discrepancy 1

We classify divisorial contractions to cE points with discrepancy 1 of higher level in this section. Moreover, we determine the number of exceptional divisors with discrepancy 1 over cE_6 points.

5.1 cE Singularity of level 6

Proposition 5.1 *Suppose that $(P \in X)$ is of level 6, then there are eight different valuations with discrepancy 1 corresponding to weighted blow-ups of different weights. Among them, there is only one divisorial contraction with discrepancy 1 over $(P \in X)$, which is given by $wBl_{w_{30}}$.*

Proof Let F be a canonical form of $(P \in X)$. If F is of level 6, then $wt_{w_{30}}(F) \geq 30$. This only happens in cE_8 singularity and $h_5 = z^5$.

1. Let $f = wBl_{w_{30}} : Y \to X$, where $w_{30} = (15, 10, 6, 1)$. It follows from Theorem 4.6 that Y is terminal.
2. $\mathrm{Sing}(Y)_{>1} = \{R_{23}, R_{13}, R_{12}\}$, where R_{ij} is a quotient singularity in $(x_i = x_j = 0) \subset E$ of index $gcd(a_i, a_j)$.
3. Take an economic resolution $Z \to Y$. We have exceptional divisors F_1, G_1, G_2, $H_1, \ldots, H_4$ with discrepancy 1, where F_i (resp. G_i, H_i) denotes the exceptional divisor over R_{23} (resp. R_{13}, R_{12}). Computation shows that all these divisors are of discrepancy 1 over X. Together with E, there are eight divisors with discrepancy 1.

Take F_1 for example, obtained by a Kawamata blow-up of weights $\frac{1}{2}(1,1,1)$ over R_{23}. In fact, this can be realized as a weighted blow-up with vector w_{15}. More precisely, let $\mathfrak{X}_1 \to \mathfrak{X}_0$ be the weighted blow-up with vector w_{30} and let $g : \mathfrak{X}_2 \to \mathfrak{X}_1$ be the weighted blow-up with vector w_{15}. Since $w_{15} = \frac{1}{2}w_{30} + \frac{1}{2}e_1 + \frac{1}{2}e_4$, the map g is obtained by subdivision of $\sigma_2 = \langle e_1, w_{30}, e_3, e_4 \rangle$ and $\sigma_3 = \langle e_1, e_2, w_{30}, e_4 \rangle$ along w_{15}. One sees that g is the weighted blow-up along the curve $(y = z = 0) \subset \mathcal{E}$, which is singular of type $\frac{1}{2}(1,1,1) \times \mathbb{P}^1$. The proper transform of Y in $\mathfrak{X}_2$, denoted Z, and its induced map $g : Z \to Y$ then gives the Kawamata blow-up.

One can consider the exceptional divisors $G_1, \ldots, H_4$ similarly. The computation can be summarized as follows:

Div	$center_Y$	$a(Z/Y)$	wt_Y	$a(Z/X)$	wt_X
F_1	R_{23}	1/2	$\frac{1}{2}(1,1,1)$	1	$(8,5,3,1) = w_{15}$
G_1	R_{13}	1/3	$\frac{1}{3}(1,1,2)$	1	$(10,7,4,1) = w_{20}$
G_2	R_{13}	2/3	$\frac{1}{3}(2,2,1)$	1	$(5,4,2,1) = w_{10}$
H_1	R_{12}	1/5	$\frac{1}{5}(1,1,4)$	1	$(12,8,5,1) = w_{24}$
H_2	R_{12}	2/5	$\frac{1}{5}(2,2,3)$	1	$(9,6,4,1) = w_{18}$
H_3	R_{12}	3/5	$\frac{1}{5}(3,3,2)$	1	$(6,4,3,1) = w_{12}$
H_4	R_{12}	4/5	$\frac{1}{5}(4,4,1)$	1	$(3,2,2,1) = w_6$

where $center_Y$ denotes the center in Y, $a(Z/Y)$ (resp. $a(Z/X)$) denotes the discrepancy over Y (resp. over X), and wt_Y (resp. wt_X) denotes the weights over Y (resp. over X).

4. $wBl_v(X)$ is not terminal if $v = w_{24}, w_{18}, w_{12}, w_6, w_{20}, w_{10}, w_{15}$. By Lemma 4.1, it is straightforward to see that any of these blow-ups is a divisorial blow-up, i.e., its exceptional divisor is irreducible.

5. $wBl_{w_{30}}$ is the unique divisorial contraction with discrepancy 1. Suppose that $f' : Y' \to X$ is another divisorial contraction to P with discrepancy 1. Let E' be its exceptional divisor. Then $E' \in \{F_1, G_1, G_2, H_1, \dots, H_4\}$ as valuations. If $E' = F_1$, then we consider $f'' = wBl_{w_{24}} : Y'' \to X$ for the corresponding vector $w_{24} = (12, 8, 5, 1)$. One sees that E'' is irreducible and hence by [10, Lemma 4.3], $f'' \cong f'$ and $Y'' \cong Y'$. However, this is absurd since Y'' is not terminal.

As we have seen, all exceptional divisors with discrepancy 1 (other than E) correspond to a weighted blow-up with the weights given above and none of these weighted blow-ups is a divisorial contraction. We thus conclude that there is no divisorial contraction other than $wBl_{w_{30}}$. □

5.2 cE Singularity of level 5

Proposition 5.2 *Suppose that $P \in X$ is of level 5, then there are seven different valuations with discrepancy 1 corresponding to weighted blow-ups of different weights. Among them, there is only one divisorial contraction over $P \in X$ which is given by $wBl_{w_{24}}$.*

Proof Let F be the canonical form. If F is of level 5, then it can only happen for a cE_8 singularity and $h_5 = z^5$.

1. Let $f = wBl_{w_{24}} : Y \to X$, where $w_{24} = (12, 8, 5, 1)$. Since $z^5 \in h$, one sees that $\natural$ does not hold. It follows from Theorem 4.6 that $Y \to X$ is a divisorial contraction.

2. $\text{Sing}(Y)_{>1} = \{R_{12}, Q_3\}$, where R_{12} is a point in $(x = y = 0) \subset E$ of index 4 and Q_3 is quotient of index 5.
3. Take an economic resolution $Z \to Y$. We have exceptional divisors $\{F_i\}_{i\leq 3}$ and $\{G_j\}_{i\leq 4}$ over R_{12} and Q_3 respectively. Computation yields the following:

Div	$center_Y$	$a(Z/Y)$	wt_Y	$a(Z/X)$	wt_X
F_1	R_{12}	1/4	$\frac{1}{4}(3,1,1)$	1	$(9,6,4,1) = w_{18}$
F_2	R_{12}	2/4	$\frac{1}{4}(2,2,2)$	1	$(6,4,3,1) = w_{12}$
F_3	R_{12}	3/4	$\frac{1}{4}(1,3,3)$	1	$(3,2,2,1) = w_6$
G_1	Q_3	1/5	$\frac{1}{5}(2,3,4,1)$	1	$(10,7,4,1) = w_{20}$
G_2	Q_3	2/5	$\frac{1}{5}(4,1,3,2)$	1	$(8,5,3,1) = w_{15}$
G_3	Q_3	3/5	$\frac{1}{5}(1,4,2,3)$	1	$(5,4,2,1) = w_{10}$
G_4	Q_3	4/5	$\frac{1}{5}(3,2,6,4)$	2	

4. $wBl_v(X)$ is a divisorial blow-up but not terminal if v is one of $w_{18}, w_{12}, w_6, w_{20}, w_{15}, w_{10}$ by Lemma 4.1.

We conclude that $wBl_{w_{24}}$ is the unique divisorial contraction with discrepancy 1. □

5.3 *cE* Singularity of level 4

We need to consider different stages.

Stage 1 Weights $w_{20} = (10, 7, 4, 1)$.

Proposition 5.3 *Suppose that F is of level 4 of stage 1, then there are six different valuations with discrepancy 1 corresponding to weighted blow-ups of different weights. Among them, there is only one divisorial contraction over $P \in X$ which is given by $wBl_{w_{20}}$.*

Proof This only happens in cE_8 singularity.
1. Let $f = wBl_{w_{20}} : Y \to X$, where $w_{20} = (10, 7, 4, 1)$. Since $z^5 \in h$, one sees that the exceptional divisor of f is irreducible. Since F is not of level 5, it is straightforward to check that $Y \to X$ is a divisorial contraction. $\text{Sing}(Y)_{>1} = \{R_{13}, Q_2\}$, where R_{13} is a point of index 2 and Q_2 is a quotient of index 7.
2. Take an economic resolution $Z \to Y$. We have exceptional divisors F_1 and $\{G_j\}_{i\leq 6}$ over R_{13} and Q_2 respectively. Computation shows that F_1, G_1, G_2, G_3, G_5 are with discrepancy 1 and G_4, G_6 are with discrepancy 2. Hence there are six divisors with discrepancy 1.

3. Similarly, the exceptional divisors F_1, G_1, G_2, G_3, G_5 correspond to vectors $w_{10}, w_{18}, w_{15}, w_{12}, w_6$ respectively. Clearly, a weighted blow-up with any of these weights has irreducible exceptional divisor. Moreover, $wBl_v(X)$ is not terminal if v is one of $w_{10}, w_{15}, w_{12}, w_6$ by Lemma 4.1.

4. We consider $wBl_{w_{18}}$. Since $\natural$ holds, one sees that $wBl_{w_{18}}(X)$ is not terminal for it contains non-isolated singularities.

We conclude that $wBl_{w_{20}}$ is the unique divisorial contraction with discrepancy 1 and there are six divisors with discrepancy 1, realized by weighted blow-ups. □

Stage 0 $w_{18} = (9, 6, 4, 1)$.

Case 1 F is of type cE_7.
Since $z^3 \in g$, $\natural$ does not hold. By Theorem 4.6, the weighted blow-up with weight w_{18} is a divisorial contraction. There are singularities R_{12}, R_{23}, Q_3 of quotient type of index $3, 2, 4$ respectively. An economic resolution produces exceptional divisors $F_1, F_2, G_1, H_1, H_2, H_3$. All of these are of discrepancy 1 over X. They correspond to weights $w_{12}, w_6, w_9, w_{14}, w_{10}, w'_6$ respectively. Clearly, $w_{12}, w_6, w_{14}, w_{10}$ give rise to divisorial blow-ups which are not terminal by Lemma 4.1.

It is clear that $wBl_{w'_6}$ has an irreducible exceptional divisor. Note that $Y \cap U_4$ is given by

$$\tilde{F} = x^2 + y^3u^3 + y\sum a_{ij}z^iu^{j'} + \sum b_{ij}z^iu^{j''}.$$

One has

$$\begin{aligned} i + j'' &= 2i + j - 6 \geq \tfrac{1}{2}(4i + j) - 6 \geq 3, \quad \text{if } (i, j) \in I_h; \\ i + j' &= 2i + j - 3 \geq \tfrac{1}{2}(4i + j) - 3 \geq 3, \quad \text{if } (i, j) \in I_g. \end{aligned}$$

In fact $4i + 2j > 18$ if $4i + j = 18$. Hence $i + j'' > 3$ if $(i, j) \in I_h$ and therefore Q_4 is not terminal.

We now consider $w_9 = (5, 3, 2, 1)$. The weighted blow-up with weight w_9 has irreducible exceptional divisor if there is some $z^iu^j \in h$ with $2i + j = 9$. However, this can not happen since $4i + j \geq 18$. Hence the exceptional divisor defined by $y^3 + yz^3$ is reducible. We need to consider an isomorphism $P' \in X' \subset \mathbb{C}^5$ such that

$$P' \in X' = \left(\begin{array}{l} x^2 + yt + yg_{>\sigma_y(w_9)} + h_{\geq\sigma(w_9)} = 0, \\ t = y^2 + z^3 \end{array} \right) \subset \mathbb{C}^5.$$

We take $wBl_v : Y' \to X$ with $v = (5, 3, 2, 1, 7)$. One sees that wBl_v has irreducible exceptional divisor. Note that Y' is non-terminal, since $Y' \cap U_4$ is not a hypersurface singularity.

Hence there are seven divisors with discrepancy 1 realized by weighted blow-ups and there is exactly one divisorial contraction with discrepancy 1 among these weighted blow-ups.

Case 2 F is of type cE_8 and ♮ holds.
Then we need to consider its stage lifting G. Now G is of level 4 of stage 1. The same argument in stage 1 holds for G. We conclude that there is a unique divisorial contraction and six divisors with discrepancy 1. All of them are realized by weighted blow-ups.

Case 3 F is of type cE_8 and ♮ does not hold.
Then one sees that the weighted blow-up of weight w_{18} is a divisorial contraction. There are singularities R_{12}, Q_3, where R_{12} is of quotient type of index 3 and Q_3 is of type $cAx/4$ with axial weight 2. By resolving R_{12} as in Case 1 of cE_7, there are two exceptional divisors with discrepancy 1 corresponding to w_{12} and w_6 respectively.

It remains to consider Q_3, whose nature varies depending on the appearance of z^4u^2.

If $z^4u^2 \notin F$, then $Y \cap U_3$ is given by

$$(\tilde{F} = x^2 + y^3 + z^2 + \ldots = 0) \subset \mathbb{C}^4/\frac{1}{4}(1,2,3,1).$$

We thus have a resolution over Q_3 given by the following vectors:

$a(Z/Y)$	wt_Y	$a(Z/X)$	wt_X
1/4	$\frac{1}{4}(5,2,3,1)$	1	$(8,5,3,1) = w_{15}$
1/2	$\frac{1}{2}(3,2,3,1)$	2	
1/2	$\frac{1}{2}(1,2,1,1)$	1	$(5,4,2,1) = w_{10}$
3/4	$\frac{1}{4}(3,2,5,3)$	2	
1	$(1,1,1,1)$	2	

Clearly, w_{15}, w_{10} give rise to divisorial blow-ups which are not terminal by Lemma 4.1.

If $z^4u^2 \in F$, then we change coordinates by $\bar{x} = x + z^2u$ to get

$$G = \bar{x}^2 - 2z^2u\bar{x} + y^3 + \ldots$$

Then $Y \cap U_3$ is given by

$$(\tilde{G} = \bar{x}^2 - 2u\bar{x} + y^3 + \ldots = 0) \subset \mathbb{C}^4/\frac{1}{4}(1,2,3,1).$$

We thus have resolution over Q_3 given by the following vectors:

$a(Z/Y)$	wt_Y	$a(Z/X)$	wt_X
1/4	$\frac{1}{4}(5,2,3,1)$	1	$(8,5,3,1) = w_{15}$
1/4	$\frac{1}{4}(1,2,3,1)$	1	$(7,5,3,1) = w_{14}$
1/2	$\frac{1}{2}(1,2,1,1)$	1	$(5,4,2,1) = w_{10}$
3/4	$\frac{1}{4}(3,2,5,3)$	2	
1	$(1,1,1,1)$	2	

Clearly, w_{15}, w_{10} give rise to divisorial blow-ups which are not terminal by Lemma 4.1. Indeed,

$$G = \bar{x}^2 - 2z^2u\bar{x} + y^3 + yg_{>\sigma_y(w_{14})} + h_{>\sigma(w_{14})}.$$

We need to consider an isomorphism $P' \in X' \subset \mathbb{C}^5$:

$$P' \in X' = \begin{pmatrix} \bar{x}t + y^3 + yg_{>\sigma_y(w_{14})} + h_{>\sigma(w_{14})} = 0, \\ t = \bar{x} - 2z^2u \end{pmatrix} \subset \mathbb{C}^5.$$

The weighted blow-up with weight $(7,5,3,1,8)$, which is clearly a divisorial blow-up but not a divisorial contraction, realizes the exceptional divisor corresponding to w_{14}.

We thus conclude that there are six (resp. five) exceptional divisors with discrepancy 1 if $z^4u^2 \in F$ (resp. $z^4u^2 \notin F$). In any event, there is exactly one divisorial contraction with discrepancy 1.

5.4 cE_6 Singularities

We consider cE_6 singularities in this subsection. Instead of providing a detailed classification, we are interested in determining the number of divisorial contractions and exceptional divisors with discrepancy 1 over a given singularity.

Level 3 It is clear that $w_{12} = (6,4,3,1)$ is the only admissible weight of level 3. Since $h_4 = z^4$, ♮ does not hold and therefore the weighted blow-up $wBl_{w_{12}}: Y \to X$ with weight w_{12} is a divisorial contraction. There are two singularities of type $\frac{1}{3}(1,2,1)$ and one singularity of type $\frac{1}{2}(1,1,1)$. Let $Z \to Y$ be the economic resolution over these singular points. One sees that there are six exceptional divisors with discrepancy 1, say $F_{11}, F_{12}, F_{21}, F_{22}, G_1$, and E.

Then G_1 corresponds to the divisorial weighted blow-up of weight $(3,2,2,1)$, which is not a divisorial contraction by Lemma 4.1. Next, we consider the coordinate change $\bar{x} = x \pm z^2$ to get

$$G = \bar{x}^2 \mp 2\bar{x}z^2 + y^3 + yg_{\geq 3} + h_{\geq 5}.$$

Take weighted blow-ups with weight $(5, 3, 2, 1)$ so that the exceptional set is irreducible. This realizes two more divisors.

Since $wt_{(3,1)} g_3 \geq 8$ (resp. $wt_{(3,1)} h_5 \geq 12$), it is clear that $g_3 = z^2 g'$ (resp. $h_5 = z^2 h'$) for some g' (resp. h'). The singularity is also isomorphic to

$$\left(\begin{array}{l} \bar{x}^2 + y^3 + z^2 t + y g_{\geq 4} + h_{\geq 6} = 0 \\ t = \mp 2\bar{x} + y g' + h' \end{array} \right).$$

Take weighted blow-ups with weight $(3, 2, 1, 1, 4)$ so that the exceptional set is irreducible. Then one realizes two more divisors with discrepancy 1. These five additional weighted blow-ups do not produce divisorial contractions.

Level 2 and Stage 1 Now we consider $w_8 = (4, 3, 2, 1)$. Since F is not of level 3, one sees that $wBl_{w_8} : Y \to X$ is a divisorial contraction if its exceptional divisor is irreducible. There are two singularities of type $\frac{1}{2}(1, 1, 1)$ and one singularity of type $\frac{1}{3}(1, 2, 1)$. Let $Z \to Y$ be the economic resolution over these singular points. One sees that there are four exceptional divisors with discrepancy 1, say F_1, F_2, G_1, and E. G_1 corresponds to the weighted blow-up of weight $(3, 2, 2, 1)$, which is clearly not a divisorial contraction by Lemma 4.1. To realize the other two divisors, a similar argument as in Level 3 shows that weighted blow-ups with weight $(3, 2, 1, 1, 4)$ realize the other two divisors.

Level 2 and Stage 0 Suppose that $\natural$ holds for F, then we can consider its stage lifting G. The situation is then exactly the same as in the above level 2 and stage 1 case. That is, there are four exceptional divisors with discrepancy 1.

Suppose that $\natural$ doesn't hold. Then wBl_{w_6} is a divisorial contraction, where Q_3 is the only singularity of index > 1.

If h_4 is not a square, then Q_3 is of type $cA/2$ with axial weight 3 and τ-$wt = 1$ (cf. [7, Section 8]), hence there is only one exceptional divisor F_1 with discrepancy $\frac{1}{2}$ over Y by weighted blow-up with weight $\frac{1}{2}(1, 2, 1, 1)$. This is a divisor with discrepancy 1 over X corresponding to a weighted blow-up with weight $(2, 2, 1, 1)$.

If $h_4 = -q(z, u)^2$ is a square, then we change coordinates by $\bar{x} = x - q(z, u)$. Now F is equivalent to

$$G = \bar{x}^2 \mp 2\bar{x}q(z, u) + y^3 + y g_{\geq 3} + h_{\geq 5}.$$

By considering $wBl_{w_6} : Y \to X$, one sees that Q_3 is of type $cA/2$ with axial weight 3 and τ-$wt = 2$ or 3. Computation shows that there are two or three exceptional divisors with discrepancy $\frac{1}{2}$ over Q_3, however there is only one divisor with discrepancy 1 over $P \in X$. This divisor corresponds to divisorial weighted blow-up of G with weights $(3, 2, 1, 1)$ and $(4, 2, 1, 1)$ respectively.

In any case, there are two divisors with discrepancy 1.

Level 1 If $h_4 = -q(z,u)^2$ is a square, then we consider another stage lifting by $\bar{x} = x \pm q(z,u)$,

$$G = \bar{x}^2 \mp 2\bar{x}q(z,u) + y^3 + yg_{\geq 3} + h_{\geq 5}.$$

Now G has level 1 of stage 4, and hence we consider the admissible weighted blow-up $wBl_{w_5}\colon Y \to X$ with weight $w_5 = (3,2,1,1)$. Since F is of level 1, there exists yu^3 or u^5 in F. One can thus check that Y is terminal by considering $Y \cap U_4$. There are singularities Q_1, Q_2 of type $\frac{1}{3}(2,1,1)$ and $\frac{1}{2}(1,1,1)$ respectively. Let $Z \to Y$ be the economic resolution. We have exceptional divisors F_1, F_2, G_1. Computation shows that $a(F_1, X) = 1$ and $a(F_2, X) = a(G_1, X) = 2$.

Indeed, F_1 corresponds to the weight $(2,2,1,1)$. To realize this valuation as an exceptional divisor of a weighted blow-up, we proceed as in the level 3 case. The singularity is isomorphic to

$$\begin{pmatrix} \bar{x}t + y^3 + yg_{\geq 3} + h_{\geq 5} \\ t = \bar{x} \mp 2q(z,u) \end{pmatrix}.$$

Take weighted blow-ups with weight $(2,2,1,1,3)$. We thus conclude that there are exactly two divisors with discrepancy 1 in this case.

If h_4 is not a square, then we consider $w_4 = (2,2,1,1)$. One sees that $wBl_{w_4}\colon Y \to X$ is a divisorial contraction. The higher-index singular point is Q_2 of type $cAx/2$, given by

$$x^2 + y^2 + h_4 + \text{ other terms} = 0 \subset \mathbb{C}^4/\frac{1}{2}(2,1,1,1).$$

By [5, Theorem 8.4], the only exceptional divisor of discrepancy $\frac{1}{2}$ is given by a weighted blow-up of weight $\frac{1}{2}(2,3,1,1)$. Hence its discrepancy over X is 2. We thus conclude that there is a unique divisorial contraction and unique exceptional divisor of discrepancy 1 in this case.

We thus have the following observation:

Corollary 5.4 *Let F be a canonical form of cE_6. Suppose that F has exactly one exceptional divisor with discrepancy* 1*. Then F is of level* 1 *and h_4 is not a square.*

6 Divisorial contractions with higher discrepancies

By the classification of Kawakita [13, Theorem 1.2.ii], divisorial contractions to a cE point with discrepancy > 1 are not of ordinary type and a brief description was given in [13, Table 3]. The purpose of this section is to give some more examples which were not previously known and also to provide some characterization of cE singularities admitting divisorial contractions with higher discrepancy.

The only previously known example with higher discrepancy is the following:

Example 6.1 ([13, Example 8.9]) Let $P \in X$ be the germ

$$o \in (x^2 + y^3 + yz^3 + u^7 = 0) \subset \mathbb{C}^4.$$

P is of type cE_7. Take the weighted blow-up with weight $(7, 5, 3, 2)$. Then it is a divisorial contraction with discrepancy 2.

Let $P \in X$ be the germ

$$o \in (x^2 + y^3 + z^5 + u^7 = 0) \subset \mathbb{C}^4.$$

P is of type cE_8. Take the weighted blow-up with weight $(7, 5, 3, 2)$. Then it is a divisorial contraction with discrepancy 2.

We provide another example:

Example 6.2 Let $P \in X$ be the germ

$$o \in \left(\begin{array}{c} x^2 + yt + u^5 = 0 \\ t = y^2 + z^3 \end{array} \right) \subset \mathbb{C}^5.$$

P is of type cE_7. Take the weighted blow-up with weight $(5, 3, 2, 2, 7)$. Then it is a divisorial contraction with discrepancy 2.

In fact, by the studies in [3, Case IIc], it is known that there is only one exceptional divisor with discrepancy 1 over $P \in X$. The unique divisorial contraction with discrepancy 1 is given by a weighted blow-up with weight $(3, 2, 1, 1, 3)$.

Example 6.3 If $P \in X$ is a germ of cE_6 such that it admits a divisorial contraction $Y \to X \ni P$ of discrepancy 3, then a singularity of higher index in Y is a $cAx/4$ point of axial weight 2. By [3, Case IId], there is exactly one exceptional divisor with discrepancy 1 over $P \in X$. By Corollary 5.4, one see that its canonical form must be of level 1 and h_4 is not a square. However, we do not know if such an example exists or not.

References

[1] Arnold, V. I. Gusein-Zade, S. M., and Varchenko, A. N. *Singularities of Differentiable Maps I, II.* Berlin: Birkhäuser, 1985, 1988.

[2] Brown, G. Pluricanonical cohomology across flips. *Bull. London Math. Soc.* **31**(5) (1999) 513–522.

[3] Chen, J. A. Birational maps of 3-folds. arXiv 1304.5949. *Algebraic Geometry in East Asia, Taipei, Advanced Studies in Pure Mathematics*, to appear.

[4] Cutkosky, S. D. Elementary contractions of Gorenstein threefolds. *Math. Ann.* **280**(3) (1988) 521–525.
[5] Hayakawa, Blowing ups of 3-dimensional terminal singularities. *Publ. Res. Inst. Math. Sci.* **35**(3) (1999) 515–570.
[6] Hayakawa, T. Blowing ups of 3-dimensional terminal singularities. II. *Publ. Res. Inst. Math. Sci.* **36**(3) (2000) 423–456.
[7] Hayakawa, T. Divisorial contractions to 3-dimensional terminal singularities with discrepancy one. *J. Math. Soc. Japan* **57**(3) (2005) 651–668.
[8] Hayakawa, T. Divisorial contractions to cD points. Forthcoming.
[9] Hayakawa, T. Divisorial contractions to cE points. Forthcoming.
[10] Kawakita, M. Divisorial contractions in dimension three which contract divisors to smooth points. *Invent. Math.* **145** (2001) 105–119.
[11] Kawakita, M. Divisorial contractions in dimension three which contract divisors to compound A_1 points. *Compos. Math.* **133** (2002) 95–116.
[12] Kawakita, M. General elephants of three-fold divisorial contractions. *J. Amer. Math. Soc.* **16**(2) (2002) 331–362.
[13] Kawakita, M. Three-fold divisorial contractions to singularities of higher indices. *Duke Math. J.* **130** (2005) 57–126.
[14] Kawakita, M. Supplement to classification of three-fold divisorial contractions. *Nagoya Math. J.* **206** (2012) 67–73.
[15] Kawamata, Y. Divisorial contractions to 3-dimensional terminal quotient singularities. In *Higher-dimensional Complex Varieties (Trento, 1994)*. Berlin: de Gruyter, pp. 241–246, 1996.
[16] Kollár, J. Flops. *Nagoya Math. J.* **113** (1989) 15–36.
[17] Kollár, J. Real algebraic threefolds. I. Terminal singularities. *Collect. Math.* **49**(2&3) (1998) 335–360. Dedicated to the memory of Fernando Serrano.
[18] Kollár, J. and Mori, S. Classification of three-dimensional flips. *J. Amer. Math. Soc.* **5**(3) (1992) 533–703.
[19] Markushevich, D. Minimal discrepancy for a terminal cDV singularity is 1. *J. Math. Sci. Univ. Tokyo* **3**(2) (1996) 445–456.
[20] Mori, S. Threefolds whose canonical bundles are not numerically effective. *Ann. Math.* **116** (1982) 133–176.
[21] Mori, S. and Prokhorov, Y. Threefold extremal contractions of type (IA). *Kyoto J. Math.*, **51**(2) (2011) 393–438.
[22] Mori, S. and Prokhorov, Y. Threefold extremal contractions of types (IC) and (IIB). *Proceedings of the Edinburgh Mathematical Society*, to appear.
[23] Tziolas, N. Terminal 3-fold divisorial contractions of a surface to a curve. I. *Compos. Math.* **139**(3) (2003) 239–261.
[24] Tziolas, N. Three dimensional divisorial extremal neighborhoods. *Math. Ann.* **333**(2) (2005) 315–354.
[25] Tziolas, N. $\mathbb{Q}$-Gorenstein deformations of nonnormal surfaces. *Amer. J. Math.* **131** (2009) 171–193.

8

Special prime Fano fourfolds of degree 10 and index 2

O. Debarre
École Normale Supérieure et CNRS

A. Iliev
Seoul National University

L. Manivel
Université de Grenoble I et CNRS

Abstract

We analyze (complex) prime Fano fourfolds of degree 10 and index 2. Mukai gave in [M1] a complete geometric description; in particular, most of them are contained in a Grassmannian $G(2,5)$. As in the case of cubic fourfolds, they are unirational and some are rational, as already remarked by Roth in 1949.

We show that their middle cohomology is of K3 type and that their period map is dominant, with smooth 4-dimensional fibers, onto a 20-dimensional bounded symmetric period domain of type IV. Following Hassett, we say that such a fourfold is *special* if it contains a surface whose cohomology class does not come from the Grassmannian $G(2,5)$. Special fourfolds correspond to a countable union of hypersurfaces (the Noether–Lefschetz locus) in the period domain, labelled by a positive integer d. We describe special fourfolds for some low values of d. We also characterize those integers d for which special fourfolds do exist.

Dedicated to Robert Lazarsfeld on the occasion of his sixtieth birthday

1 Introduction

One of the most vexing classical questions in complex algebraic geometry is whether there exist irrational smooth cubic hypersurfaces in $\mathbf{P}^5$. They are all

From *Recent Advances in Algebraic Geometry*, edited by Christopher Hacon, Mircea Mustaţă and Mihnea Popa © 2014 Cambridge University Press.

unirational, and rational examples are easy to construct (such as Pfaffian cubic fourfolds) but no smooth cubic fourfold has yet been proven to be irrational. The general feeling seems to be that the question should have an affirmative answer but, despite numerous attempts, it is still open.

In a couple of very interesting articles on cubic fourfolds ([H1], [H2]), Hassett adopted a Hodge-theoretic approach and, using the period map (proven to be injective by Voisin in [V]) and the geometry of the period domain, a 20-dimensional bounded symmetric domain of type IV, he related geometric properties of a cubic fourfold to arithmetical properties of its period point.

We do not solve the rationality question in this paper, but investigate instead similar questions for another family of Fano fourfolds (see Section 2 for their definition). Again, they are all unirational (see Section 3) and rational examples were found by Roth (see [R]; also [P] and Section 7), but no irrational examples are known.

We prove in Section 4 that the moduli stack $\mathscr{X}_{10}$ associated with these fourfolds is smooth of dimension 24 (Proposition 4.1) and that the period map is smooth and dominant onto, again, a 20-dimensional bounded symmetric domain of type IV (Theorem 4.3). We identify the underlying lattice in Section 5. Then, following [H1], we define in Section 6.1 hypersurfaces in the period domain which parametrize "special" fourfolds X, whose period point satisfies a nontrivial arithmetical property depending on a positive integer d, the *discriminant.* As in [H1], we characterize in Proposition 6.6 those integers d for which the nonspecial cohomology of a special X of discriminant d is essentially the primitive cohomology of a K3 surface; we say that this K3 surface is *associated* with X. Similarly, we characterize in Proposition 6.7 those d for which the nonspecial cohomology of a special X of discriminant d is the nonspecial cohomology of a cubic fourfold in the sense of [H1].

In Section 7, we give geometric constructions of special fourfolds X for $d \in \{8, 10, 12\}$; in particular, we discuss some rational examples (already present in [R] and [P]). When $d = 10$, the associated K3 surface (in the sense of Proposition 6.6) does appear in the construction of X; when $d = 12$, so does the associated cubic fourfold (in the sense of Proposition 6.7) and it is birationally isomorphic to X.

In Section 8, we characterize the positive integers d for which there exist (smooth) special fourfolds of discriminant d. As in [H1], our construction relies on the surjectivity of the period map for K3 surfaces. Finally, in Section 9, we ask a question about the image of the period map.

So in some sense, the picture is very similar to what we have for cubic fourfolds, with one big difference: the Torelli theorem does not hold. In a forthcoming article, building on the link between our fourfolds and EPW sextics discovered in [IM], we will analyze the (4-dimensional) fibers of the period map.

2 Prime Fano fourfolds of degree 10 and index 2

Let X be a (smooth) prime Fano fourfold of degree 10 (i.e., of "genus" 6) and index 2; this means that $\mathrm{Pic}(X)$ is generated by the class of an ample divisor H such that $H^4 = 10$ and $-K_X \underset{\mathrm{lin}}{\equiv} 2H$. Then H is very ample and embeds X in $\mathbf{P}^8$ as follows ([M2]; [IP], Theorem 5.2.3).

Let V_5 be a 5-dimensional vector space (our running notation is V_k for any k-dimensional vector space). Let $G(2, V_5) \subset \mathbf{P}(\wedge^2 V_5)$ be the Grassmannian in its Plücker embedding and let $CG \subset \mathbf{P}(\mathbf{C} \oplus \wedge^2 V_5) \simeq \mathbf{P}^{10}$ be the cone, with vertex $v = \mathbf{P}(\mathbf{C})$, over $G(2, V_5)$. Then

$$X = CG \cap \mathbf{P}^8 \cap Q,$$

where Q is a quadric. There are two cases:

- either $v \notin \mathbf{P}^8$, in which case X is isomorphic to the intersection of $G(2, V_5) \subset \mathbf{P}(\wedge^2 V_5)$ with a hyperplane (the projection of $\mathbf{P}^8$ to $\mathbf{P}(\wedge^2 V_5)$) and a quadric;
- or $v \in \mathbf{P}^8$, in which case $\mathbf{P}^8$ is a cone over a $\mathbf{P}^7 \subset \mathbf{P}(\wedge^2 V_5)$ and X is a double cover of $G(2, V_5) \cap \mathbf{P}^7$ branched along its intersection with a quadric.

The varieties obtained by the second construction will be called "of Gushel type" (after Gushel, who studied the 3-dimensional case in [G]). They are specializations of varieties obtained by the first construction.

Let $\mathscr{X}_{10}$ be the irreducible moduli stack for (smooth) prime Fano fourfolds of degree 10 and index 2, let $\mathscr{X}_{10}^G$ be the (irreducible closed) substack of those which are of Gushel type, and let $\mathscr{X}_{10}^0 := \mathscr{X}_{10} - \mathscr{X}_{10}^G$. We have

$$\dim(\mathscr{X}_{10}) = 24, \quad \dim(\mathscr{X}_{10}^G) = 22.$$

3 Unirationality

Let $G := G(2, V_5)$ and let $X := G \cap \mathbf{P}^8 \cap Q$ be a fourfold of type $\mathscr{X}_{10}^0$. We prove in this section that X is unirational.

The hyperplane $\mathbf{P}^8$ is defined by a nonzero skew-symmetric form ω on V_5, and the singular locus of $G^\omega := G \cap \mathbf{P}^8$ is isomorphic to $G(2, \mathrm{Ker}(\omega))$. Since X is smooth, this singular locus must be finite, hence ω must be of maximal rank and G^ω is also smooth. The variety G^ω is the unique del Pezzo fivefold of degree 5 ([IP], Theorem 3.3.1); it parameterizes isotropic 2-planes for the form ω.

If $V_1^\omega \subset V_5$ is the kernel of ω, the 3-plane $\mathbf{P}_0^3 := \mathbf{P}(V_1^\omega \wedge V_5)$ of lines passing through $[V_1^\omega] \in \mathbf{P}(V_5)$ is contained in G^ω, hence X contains $\Sigma_0 := \mathbf{P}_0^3 \cap Q$, a "$\sigma$-quadric" surface,[1] possibly reducible. Any irreducible σ-quadric contained in X is Σ_0.

Proposition 3.1 *Any fourfold X of type $\mathscr{X}_{10}^0$ is unirational. More precisely, there is a rational double cover $\mathbf{P}^4 \dashrightarrow X$.*

Proof If $p \in \Sigma_0$, the associated $V_{2,p} \subset V_5$ contains V_1^ω, hence its ω-orthogonal complement is a hyperplane $V_{2,p}^\perp \subset V_5$ containing $V_{2,p}$. Then

$$Y := \bigcup_{p \in \Sigma_0} \mathbf{P}(V_{2,p}) \times \mathbf{P}(V_{2,p}^\perp / V_{2,p})$$

is the fiber product of the projectivizations of two vector bundles over Σ_0, hence is rational.

A general point of Y defines a flag $V_1^\omega \subset V_{2,p} \subset V_3 \subset V_{2,p}^\perp \subset V_5$, hence a line in $G(2, V_5)$ passing through p and contained in $\mathbf{P}^8$. This line meets $X - \Sigma_0$ at a unique point, and this defines a rational map $Y \dashrightarrow X$.

This map has degree 2: if x is general in X, lines in $G(2, V_5)$ through x meet $\mathbf{P}(V_1^\omega \wedge V_5)$ in points p such that $V_{2,p} = V_1^\omega \oplus V_1$, with $V_1 \subset V_{2,x}$, hence the intersection is $\mathbf{P}(V_1^\omega \wedge V_{2,x})$. This is a line, therefore it meets Σ_0 in two points. □

4 Cohomology and the local period map

This section contains more or less standard computations of various cohomology groups of fourfolds of type $\mathscr{X}_{10}$.

As in Section 3, we set $G := G(2, V_5)$ and let $G^\omega := G \cap \mathbf{P}^8$ be a smooth hyperplane section of G.

4.1 The Hodge diamond of X

The inclusion $G^\omega \subset G$ induces isomorphisms

$$H^k(G, \mathbf{Z}) \xrightarrow{\sim} H^k(G^\omega, \mathbf{Z}) \qquad \text{for all } k \in \{0, \ldots, 5\}. \tag{1}$$

[1] This means that the lines in $\mathbf{P}(V_5)$ parameterized by Σ_0 all pass through a fixed point. Since X is smooth, it contains no 3-planes by the Lefschetz theorem, hence Σ_0 is indeed a surface.

The Hodge diamond for a fourfold $X := G^\omega \cap Q$ of type $\mathscr{X}_{10}^0$ was computed in [IM], Lemma 4.1; its upper half is as follows:

$$\begin{array}{ccccccccc} & & & & 1 & & & & \\ & & & 0 & & 0 & & & \\ & & 0 & & 1 & & 0 & & \\ & 0 & & 0 & & 0 & & 0 & \\ 0 & & 1 & & 22 & & 1 & & 0 \end{array} \tag{2}$$

When X is of Gushel type, the Hodge diamond remains the same. In all cases, the rank-2 lattice $H^4(G, \mathbf{Z})$ embeds into $H^4(X, \mathbf{Z})$ and *we define the vanishing cohomology* $H^4(X, \mathbf{Z})_{\mathrm{van}}$ *as the orthogonal complement* (for the intersection form) *of the image of* $H^4(G, \mathbf{Z})$ *in* $H^4(X, \mathbf{Z})$. It is a lattice of rank 22.

4.2 The local deformation space

We compute the cohomology groups of the tangent sheaf T_X of a fourfold X of type $\mathscr{X}_{10}$.

Proposition 4.1 *For any fourfold* X *of type* $\mathscr{X}_{10}$*, we have*

$$H^p(X, T_X) = 0 \qquad \text{for} \quad p \neq 1$$

and $h^1(X, T_X) = 24$. *In particular, the group of automorphisms of* X *is finite and the local deformation space* $\mathrm{Def}(X)$ *is smooth of dimension 24.*

Proof For $p \geq 2$, the conclusion follows from the Kodaira–Akizuki–Nakano theorem since $T_X \simeq \Omega_X^3(2)$. We assume that $X = G^\omega \cap Q$ is not of Gushel type (the proof in the case where X is of Gushel type is similar, and left to the reader).

Let us prove $H^0(X, T_X) = 0$. We have inclusions $X \subset G^\omega \subset G$. The conormal exact sequence $0 \to \mathscr{O}_X(-2) \to \Omega^1_{G^\omega}|_X \to \Omega^1_X \to 0$ induces an exact sequence

$$0 \to \Omega_X^2 \to \Omega^3_{G^\omega}(2)|_X \to T_X \to 0.$$

Since $H^1(X, \Omega_X^2)$ vanishes, it is enough to show $H^0(X, \Omega^3_{G^\omega}(2)|_X) = 0$. Since $H^1(G^\omega, \Omega^3_{G^\omega}) = 0$, it is enough to show that $H^0(G^\omega, \Omega^3_{G^\omega}(2))$, or equivalently its Serre dual $H^5(G^\omega, \Omega^2_{G^\omega}(-2))$, vanishes.

The conormal exact sequence of G^ω in G induces an exact sequence

$$0 \to \Omega^1_{G^\omega}(-3) \to \Omega^2_G(-2)|_{G^\omega} \to \Omega^2_{G^\omega}(-2) \to 0.$$

The desired vanishing follows since $H^5(G, \Omega^2_G(-2)) = H^6(G, \Omega^2_G(-3)) = 0$ by Bott's theorem. Since X is (anti)canonically polarized, this vanishing implies that its group of automorphisms is a discrete subgroup of PGL(9, $\mathbf{C}$), hence is finite.

We also leave the computation of $h^1(X, T_X) = -\chi(X, T_X) = -\chi(X, \Omega^1_X(-2))$ to the reader. □

4.3 The local period map

Let X be a fourfold of type $\mathscr{X}_{10}$ and let Λ be a fixed lattice isomorphic to $H^4(X, \mathbf{Z})_{\mathrm{van}}$. By Proposition 4.1, X has a smooth (simply connected) local deformation space Def(X) of dimension 24. By (2), the Hodge structure of $H^4(X)_{\mathrm{van}}$ is of K3 type, hence we can define a morphism

$$\mathrm{Def}(X) \to \mathbf{P}(\Lambda \otimes \mathbf{C})$$

with values in the smooth 20-dimensional quadric

$$\mathscr{Q} := \{\omega \in \mathbf{P}(\Lambda \otimes \mathbf{C}) \mid (\omega \cdot \omega) = 0\}.$$

We show below (Theorem 4.3) that the restriction $p\colon \mathrm{Def}(X) \to \mathscr{Q}$, the *local period map,* is a submersion.

Recall from Section 3 that the hyperplane $\mathbf{P}^8$ is defined by a skew-symmetric form on V_5 whose kernel is a 1-dimensional subspace V_1^ω of V_5.

Lemma 4.2 *There is an isomorphism* $H^1(G^\omega, \Omega^3_{G^\omega}(2)) \simeq V_5/V_1^\omega$.

Proof From the normal exact sequence of the embedding $G^\omega \subset G$, we deduce the exact sequences

$$0 \to \Omega^1_{G^\omega} \to \Omega^2_G(1)|_{G^\omega} \to \Omega^2_{G^\omega}(1) \to 0, \qquad (3)$$

$$0 \to \Omega^2_{G^\omega}(1) \to \Omega^3_G(2)|_{G^\omega} \to \Omega^3_{G^\omega}(2) \to 0. \qquad (4)$$

By Bott's theorem, $\Omega^2_G(1)$ is acyclic, so we have

$$H^q(G^\omega, \Omega^2_G(1)|_{G^\omega}) \simeq H^{q+1}(G, \Omega^2_G) \simeq \delta_{q,1}\mathbf{C}^2.$$

On the other hand, by (1), we have $H^q(G^\omega, \Omega^1_{G^\omega}) \simeq \delta_{q,1}\mathbf{C}$. Therefore, we also get, by (3), $H^q(G^\omega, \Omega^2_{G^\omega}(1)) \simeq \delta_{q,1}V_1^\omega$.

By Bott's theorem again, $\Omega^3_G(1)$ is acyclic, hence using (4) we obtain

$$H^q(G^\omega, \Omega^3_G(2)|_{G^\omega}) \simeq H^q(G, \Omega^3_G(2)) \simeq \delta_{q,1}V_5.$$

This finishes the proof of the lemma. □

Theorem 4.3 *For any fourfold X of type $\mathscr{X}_{10}$, the local period map $p\colon \mathrm{Def}(X) \to \mathscr{Q}$ is a submersion.*

Proof The tangent map to p at the point $[X]$ defined by X has the same kernel as the morphism

$$T\colon H^1(X,T_X) \to \mathrm{Hom}(H^{3,1}(X), H^{3,1}(X)^{\perp}/H^{3,1}(X))$$
$$\simeq \mathrm{Hom}(H^1(X,\Omega_X^3), H^2(X,\Omega_X^2))$$

defined by the natural pairing $H^1(X,T_X)\otimes H^1(X,\Omega_X^3) \to H^2(X,\Omega_X^2)$ (by (2), $H^1(X,\Omega_X^3)$ is 1-dimensional).

Again, we will only explain the proof when X is not of Gushel type, i.e. , when it is a smooth quadratic section of G^ω, leaving the Gushel case to the reader. Recall the isomorphism $T_X \simeq \Omega_X^3(2)$. The normal exact sequence of the embedding $X \subset G^\omega$ yields the exact sequence $0 \to \Omega_X^2 \to \Omega_{G^\omega}^3(2)|_X \to T_X \to 0$.

Moreover, the induced coboundary map

$$H^1(X,T_X) \to H^2(X,\Omega_X^2)$$

coincides with the cup product by a generator of $H^1(X,\Omega_X^3) \simeq \mathbf{C}$, hence is the morphism T. Since $H^{2,1}(X) = 0$ (see (2)), its kernel K is isomorphic to $H^1(X,\Omega_{G^\omega}^3(2)|_X)$.

In order to compute this cohomology group, we consider the exact sequence $0 \to \Omega_{G^\omega}^3 \to \Omega_{G^\omega}^3(2) \to \Omega_{G^\omega}^3(2)|_X \to 0$. Since, by (1), we have $H^1(G^\omega,\Omega_{G^\omega}^3) = H^2(G^\omega,\Omega_{G^\omega}^3) = 0$, we get

$$K \simeq H^1(X,\Omega_{G^\omega}^3(2)|_X) \simeq H^1(G^\omega,\Omega_{G^\omega}^3(2)) \simeq V_5/V_1^\omega$$

by Lemma 4.2. Since Def(X) is smooth of dimension 24 and Q is smooth of dimension 20, this concludes the proof of the theorem in this case. □

The fact that the period map is dominant implies a Noether–Lefschetz-type result.

Corollary 4.4 *If X is a very general fourfold of type $\mathscr{X}_{10}$, we have $H^{2,2}(X)\cap H^4(X,\mathbf{Q}) = H^4(G,\mathbf{Q})$ and the Hodge structure $H^4(X,\mathbf{Q})_{\mathrm{van}}$ is simple.*

Proof For $H^{2,2}(X)\cap H^4(X,\mathbf{Q})_{\mathrm{van}}$ to be nonzero, the corresponding period must be in one of the (countably many) hypersurfaces $\alpha^\perp \cap \mathscr{Q}$, for some $\alpha \in \mathbf{P}(\Lambda\otimes\mathbf{Q})$. Since the local period map is dominant, this does not happen for X very general.

For any X, a standard argument (see, e.g., [Z], Theorem 1.4.1) shows that the transcendental lattice $(H^4(X,\mathbf{Z})_{\mathrm{van}}\cap H^{2,2}(X))^\perp$ inherits a simple rational Hodge structure. For X very general, the transcendental lattice is $H^4(X,\mathbf{Z})_{\mathrm{van}}$. □

Remark 4.5 If X is of Gushel type, we may consider, inside Def(X), the locus Def$^G(X)$ where the deformation of X remains of Gushel type and the restriction $p^G\colon \mathrm{Def}^G(X) \to \mathscr{Q}$ of the local period map. One can show that the kernel of $T_{p^G,[X]}$ is 2-dimensional. In particular, p^G is a submersion at $[X]$.

Also, the conclusion of Corollary 4.4 remains valid for very general fourfolds of Gushel type.

5 The period domain and the period map

5.1 The vanishing cohomology lattice

Let $(L,\cdot)$ be a lattice, we denote by $L^\vee$ its dual $\mathrm{Hom}_{\mathbf{Z}}(L,\mathbf{Z})$. The symmetric bilinear form on L defines an embedding $L \subset L^\vee$. The *discriminant group* is the finite abelian group $D(L) := L^\vee/L$; it is endowed with the symmetric bilinear form $b_L\colon D(L) \times D(L) \to \mathbf{Q}/\mathbf{Z}$ defined by $b_L([w],[w']) := w \cdot_{\mathbf{Q}} w' \pmod{\mathbf{Z}}$ ([N], Section 1, 3°). We define the *divisibility* $\mathrm{div}(w)$ of a nonzero element w of L as the positive generator of the ideal $w \cdot L \subset \mathbf{Z}$, so that $w/\mathrm{div}(w)$ is primitive in $L^\vee$. We set $w_* := [w/\mathrm{div}(w)] \in D(L)$. If w is primitive, $\mathrm{div}(w)$ is the order of w_* in $D(L)$.

Proposition 5.1 *Let X be a fourfold of type $\mathscr{X}_{10}$. The* vanishing cohomology lattice $H^4(X,\mathbf{Z})_{\mathrm{van}}$ *is even and has signature* $(20,2)$ *and discriminant group* $(\mathbf{Z}/2\mathbf{Z})^2$. *It is isometric to*

$$\Lambda := 2E_8 \oplus 2U \oplus 2A_1. \tag{5}$$

Proof By (2), the Hodge structure on $H^4(X)$ has weight 2 and the unimodular lattice $\Lambda_X := H^4(X,\mathbf{Z})$, endowed with the intersection form, has signature $(22,2)$. Since $22-2$ is not divisible by 8, this lattice must be odd, hence of type $22\langle 1\rangle \oplus 2\langle -1\rangle$, often denoted by $I_{22,2}$ ([S], Chapitre V, Section 2, Corollaire 1 of Théorème 2 and Théorème 4).

The intersection form on the lattice $\Lambda_G := H^4(G(2,V_5),\mathbf{Z})|_X$ has matrix $\begin{pmatrix} 2 & 2 \\ 2 & 4 \end{pmatrix}$ in the basis $(\sigma_{1,1}|_X, \sigma_2|_X)$. It is of type $2\langle 1\rangle$ and embeds as a primitive sublattice in $H^4(X,\mathbf{Z})$. The vanishing cohomology lattice $\Lambda_X^0 := H^4(X,\mathbf{Z})_{\mathrm{van}} = \Lambda_G^\perp$ therefore has signature $(20,2)$ and $D(\Lambda_X^0) \simeq D(\Lambda_G) \simeq (\mathbf{Z}/2\mathbf{Z})^2$ ([N], Proposition 1.6.1).

An element x of $I_{22,2}$ is *characteristic* if

$$\forall y \in I_{22,2} \quad x \cdot y \equiv y^2 \pmod 2.$$

The lattice $x^\perp$ is then even. One has from [BH], Section 16.2,

$$\begin{aligned} c_1(T_X) &= 2\sigma_1|_X, \\ c_2(T_X) &= 4\sigma_1^2|_X - \sigma_2|_X. \end{aligned} \tag{6}$$

Wu's formula (see [W]) then gives

$$\forall y \in \Lambda_X \quad y^2 \equiv y \cdot (c_1^2 + c_2) \equiv y \cdot \sigma_2|_X \pmod 2. \tag{7}$$

In other words, $\sigma_2|_X$ is characteristic, hence Λ_X^0 is an even lattice. As one can see from Table 15.4 in [CS], there is only one genus of even lattices with signature $(20, 2)$ and discriminant group $(\mathbf{Z}/2\mathbf{Z})^2$ (it is denoted by $II_{20,2}(2_I^2)$ in that table); moreover, there is only one isometry class in that genus ([CS], Theorem 21). In other words, any lattice with these characteristics, such as the one defined in (5), is isometric to Λ_X^0. □

One can also check that Λ is the orthogonal complement in $I_{22,2}$ of the lattice generated by the vectors

$$u := e_1 + e_2 \quad \text{and} \quad v' := e_1 + \cdots + e_{22} - 3f_1 - 3f_2$$

in the canonical basis $(e_1, \ldots, e_{22}, f_1, f_2)$ for $I_{22,2}$. Putting everything together, we see that there is an isometry $\gamma \colon \Lambda_X \xrightarrow{\sim} I_{22,2}$ such that

$$\gamma(\sigma_{1,1}|_X) = u, \quad \gamma(\sigma_2|_X) = v', \quad \gamma(\Lambda_X^0) \simeq \Lambda. \tag{8}$$

We let $\Lambda_2 \subset I_{22,2}$ be the rank-2 sublattice $\langle u, v' \rangle = \langle u, v \rangle$, where $v := v' - u$. Then u and v both have divisibility 2, $D(\Lambda_2) = \langle u_*, v_* \rangle$, and the matrix of b_{Λ_2} associated with these generators is $\begin{pmatrix} 1/2 & 0 \\ 0 & 1/2 \end{pmatrix}$.

5.2 Lattice automorphisms

One can construct $I_{20,2}$ as an overlattice of Λ as follows. Let e and f be respective generators for the last two A_1-factors of Λ (see (5)). They both have divisibility 2 and $D(\Lambda) \simeq (\mathbf{Z}/2\mathbf{Z})^2$, with generators e_* and f_*; the form b_Λ has matrix $\begin{pmatrix} 1/2 & 0 \\ 0 & 1/2 \end{pmatrix}$. In particular, $e_* + f_*$ is the only isotropic nonzero element in $D(\Lambda)$. By [N], Proposition 1.4.1, this implies that there is a unique unimodular overlattice of Λ. Since there is just one isometry class of unimodular lattices of signature $(20, 2)$, this is $I_{20,2}$.

Note that Λ is an even sublattice of index 2 of $I_{20,2}$, so it is the maximal even sublattice $\{x \in I_{20,2} \mid x^2 \text{ even}\}$ (it is contained in that sublattice, and has the same index in $I_{20,2}$).

Every automorphism of $I_{20,2}$ will preserve the maximal even sublattice, so $O(I_{20,2})$ is a subgroup of $O(\Lambda)$. On the other hand, the group $O(D(\Lambda))$ has order 2 and fixes $e_* + f_*$. It follows that every automorphism of Λ fixes $I_{20,2}$, and we obtain $O(I_{20,2}) \simeq O(\Lambda)$.

Now let us try to extend to $I_{22,2}$ an automorphism $\mathrm{Id}\oplus h$ of $\Lambda_2\oplus\Lambda$. Again, this automorphism permutes the overlattices of $\Lambda_2\oplus\Lambda$, such as $I_{22,2}$, according to its action on $D(\Lambda_2)\oplus D(\Lambda)$. By [N], overlattices correspond to isotropic subgroups of $D(\Lambda_2)\oplus D(\Lambda)$ that map injectively to both factors. Among them is $I_{22,2}$; after perhaps permuting e and f, it corresponds to the (maximal isotropic) subgroup

$$\{0, u_* + e_*, v_* + f_*, u_* + v_* + e_* + f_*\}.$$

Any automorphism of Λ leaves $e_* + f_*$ fixed. So either h acts trivially on $D(\Lambda)$, in which case $\mathrm{Id}\oplus h$ leaves $I_{22,2}$ fixed, hence extends to an automorphism of $I_{22,2}$; or h switches the other two nonzero elements, in which case $\mathrm{Id}\oplus h$ does not extend to $I_{22,2}$.

In other words, the image of the restriction map

$$\{g \in O(I_{22,2}) \mid g|_{\Lambda_2} = \mathrm{Id}\} \hookrightarrow O(\Lambda)$$

is the *stable orthogonal group*

$$\widetilde{O}(\Lambda) := \mathrm{Ker}(O(\Lambda) \to O(D(\Lambda))). \tag{9}$$

It has index 2 in $O(\Lambda)$ and a generator for the quotient is the involution $r \in O(\Lambda)$ that exchanges e and f and is the identity on $\langle e, f\rangle^\perp$. Let r_2 be the involution of Λ_2 that exchanges u and v. It follows from the discussion above that the involution $r_2 \oplus r$ of $\Lambda_2 \oplus \Lambda$ extends to an involution r_I of $I_{22,2}$.

5.3 The period domain and the period map

Fix a lattice Λ as in (5); it has signature $(20, 2)$. The manifold

$$\Omega := \{\omega \in \mathbf{P}(\Lambda \otimes \mathbf{C}) \mid (\omega \cdot \omega) = 0\,, (\omega \cdot \bar{\omega}) < 0\}$$

is a homogeneous space for the real Lie group $\mathrm{SO}(\Lambda \otimes \mathbf{R}) \simeq \mathrm{SO}(20, 2)$. This group has two components, and one of them reverses the orientation on the negative definite part of $\Lambda \otimes \mathbf{R}$. It follows that Ω has two components, Ω^+ and Ω^-, both isomorphic to the 20-dimensional open complex manifold $\mathrm{SO}_0(20, 2)/\,\mathrm{SO}(20) \times \mathrm{SO}(2)$, a bounded symmetric domain of type IV.

Let $\mathscr{U}$ be a smooth (irreducible) quasi-projective variety parameterizing all fourfolds of type $\mathscr{X}_{10}$. Let u be a general point of $\mathscr{U}$ and let X be the corresponding fourfold. The group $\pi_1(\mathscr{U}, u)$ acts on the lattice $\Lambda_X := H^4(X, \mathbf{Z})$ by isometries and the image Γ_X of the morphism $\pi_1(\mathscr{U}, u) \to O(\Lambda_X)$ is called the monodromy group. The group Γ_X is contained in the subgroup (see (9))

$$\widetilde{O}(\Lambda_X) := \{g \in O(\Lambda_X) \mid g|_{\Lambda_G} = \mathrm{Id}\}.$$

Choose an isometry $\gamma\colon \Lambda_X \xrightarrow{\sim} I_{22,2}$ satisfying (8). It induces an isomorphism $\widetilde{O}(\Lambda_X) \simeq \widetilde{O}(\Lambda)$. The group $\widetilde{O}(\Lambda)$ acts on the manifold Ω defined above and, by a theorem of Baily and Borel, the quotient $\mathscr{D} := \widetilde{O}(\Lambda)\backslash\Omega$ has the structure of an irreducible quasi-projective variety. One defines as usual a period map $\mathscr{U} \to \mathscr{D}$ by sending a variety to its period; it is an algebraic morphism. It descends to "the" period map

$$\wp\colon \mathscr{X}_{10} \to \mathscr{D}.$$

By Theorem 4.3 (and Remark 4.5), $\wp$ is dominant with 4-dimensional smooth fibers *as a map of stacks.*

Remark 5.2 As in the 3-dimensional case ([DIM1]), we do not know whether our fourfolds have a coarse moduli space, even in the category of algebraic spaces (the main unresolved issue is whether the corresponding moduli functor is separated). If such a space $\mathbf{X}_{10}$ exists, note however that it is *singular along the Gushel locus:* any fourfold X of Gushel type has a canonical involution; if X has no other nontrivial automorphisms, $\mathbf{X}_{10}$ is then locally around $[X]$ the product of a smooth 22-dimensional germ and the germ of a surface node. The fiber of the period map $\mathbf{X}_{10} \to \mathscr{D}$ then has multiplicity 2 along the surface corresponding to Gushel fourfolds (see Remark 4.5).

6 Special fourfolds

Following [H1], Section 3, we say that a fourfold X of type $\mathscr{X}_{10}$ is *special* if it contains a surface whose cohomology class "does not come" from $G(2, V_5)$. Since the Hodge conjecture is true (over $\mathbf{Q}$) for Fano fourfolds (more generally, by [CM], for all uniruled fourfolds), this is equivalent to saying that the rank of the (positive definite) lattice $H^{2,2}(X) \cap H^4(X, \mathbf{Z})$ is at least 3, hence by Corollary 4.4, a very general X is not special. The set of special fourfolds is sometimes called the Noether–Lefschetz locus.

6.1 Special loci

For each primitive, positive definite, rank-3 sublattice $K \subset I_{22,2}$ containing the lattice Λ_2 defined at the end of Section 5.1, we define an irreducible hypersurface of Ω^+ by setting

$$\Omega_K := \{\omega \in \Omega^+ \mid K \subset \omega^\perp\}.$$

A fourfold X is *special* if and only if its period is in one of these (countably many) hypersurfaces. We now investigate these lattices K.

Lemma 6.1 *The discriminant d of K is positive and $d \equiv 0, 2$, or $4 \pmod 8$.*

Proof Since K is positive definite, d must be positive. Completing the basis (u, v) of Λ_2 from Section 5.1 to a basis of K, we see that the matrix of the intersection form in that basis is $\begin{pmatrix} 2 & 0 & a \\ 0 & 2 & b \\ a & b & c \end{pmatrix}$, whose determinant is $d = 4c - 2(a^2 + b^2)$. By Wu's formula (7) (or equivalently, since v is characteristic), we have $c \equiv a + b \pmod 2$, hence $d \equiv 2(a^2 + b^2) \pmod 8$. This proves the lemma. □

We keep the notation of Section 5.

Proposition 6.2 *Let d be a positive integer such that $d \equiv 0, 2$, or $4 \pmod 8$ and let $\mathcal{O}_d$ be the set of orbits for the action of the group*

$$\widetilde{O}(\Lambda) = \{g \in O(I_{22,2}) \mid g|_{\Lambda_2} = \mathrm{Id}\} \subset O(\Lambda)$$

on the set of primitive, positive definite, rank-3, discriminant-d, sublattices $K \subset I_{22,2}$ containing Λ_2. Then:

(a) if $d \equiv 0 \pmod 8$, $\mathcal{O}_d$ has one element and $K \simeq \begin{pmatrix} 2 & 0 & 0 \\ 0 & 2 & 0 \\ 0 & 0 & d/4 \end{pmatrix}$;

(b) if $d \equiv 2 \pmod 8$, $\mathcal{O}_d$ has two elements, which are interchanged by the involution r_I of $I_{22,2}$, and $K \simeq \begin{pmatrix} 2 & 0 & 0 \\ 0 & 2 & 1 \\ 0 & 1 & (d+2)/4 \end{pmatrix}$;

(c) if $d \equiv 4 \pmod 8$, $\mathcal{O}_d$ has one element and $K \simeq \begin{pmatrix} 2 & 0 & 1 \\ 0 & 2 & 1 \\ 1 & 1 & (d+4)/4 \end{pmatrix}$.

In case (b), one orbit is characterized by the properties $K \cdot u = \mathbf{Z}$ and $K \cdot v = 2\mathbf{Z}$, and the other by $K \cdot u = 2\mathbf{Z}$ and $K \cdot v = \mathbf{Z}$.

Proof By a theorem of Eichler (see, e.g., [GHS], Lemma 3.5), the $\widetilde{O}(\Lambda)$-orbit of a primitive vector w in the even lattice Λ is determined by its length w^2 and its class $w_* \in D(\Lambda)$.

If $\mathrm{div}(w) = 1$, we have $w_* = 0$ and the orbit is determined by w^2. The lattice $\Lambda_2 \oplus \mathbf{Z}w$ is primitive: if $\alpha u + \beta v + \gamma w = mw'$, and if $w \cdot w'' = 1$, we obtain $\gamma = mw' \cdot w''$, hence $\alpha u + \beta v = m((w' \cdot w'')w - w')$ and m divides α, β, and γ. Its discriminant is $4w^2 \equiv 0 \pmod 8$.

If $\mathrm{div}(w) = 2$, we have $w_* \in \{e_*, f_*, e_* + f_*\}$. Recall from Section 5.2 that $\frac{1}{2}(u + e)$, $\frac{1}{2}(v + f)$, and $\frac{1}{2}(u + v + e + f)$ are all in $I_{22,2}$. It follows that exactly one of $\frac{1}{2}(u+w)$, $\frac{1}{2}(v+w)$, and $\frac{1}{2}(u+v+w)$ is in $I_{22,2}$, and $\Lambda_2 \oplus \mathbf{Z}w$ has index 2 in its saturation K in $I_{22,2}$. In particular, K has discriminant w^2. If $w_* \in \{e_*, f_*\}$, this is $\equiv 2 \pmod 8$; if $w_* = e_* + f_*$, this is $\equiv 4 \pmod 8$.

Now if K is a lattice as in the statement of the proposition, we let $K^\perp$ be its orthogonal complement in $I_{22,2}$, so that the rank-1 lattice $K^0 := K \cap \Lambda$ is the orthogonal complement of $K^\perp$ in Λ. From $K^0 \subset \Lambda$, we can therefore recover $K^\perp$, then $K \supset \Lambda_2$. The preceding discussion applied to a generator w of K^0 gives the statement, except that we still have to prove that there are indeed elements w of the various types for all d, i.e., we need to construct elements in each orbit to show they are not empty.

Let u_1 and u_2 be standard generators for a hyperbolic factor U of Λ. For any integer m, set $w_m := u_1 + mu_2$. We have $w_m^2 = 2m$ and $\operatorname{div}(w_m) = 1$. The lattice $\Lambda_2 \oplus \mathbf{Z}w_m$ is saturated with discriminant $8m$.

We have $(e + 2w_m)^2 = 8m + 2$ and $\operatorname{div}(e + 2w_m) = 2$. The saturation of the lattice $\Lambda_2 \oplus \mathbf{Z}(e + 2w_m)$ has discriminant $d = 8m + 2$, and similarly upon replacing e with f (same d) or $e + f$ ($d = 8m + 4$). □

Let K be a lattice as above. The image in $\mathscr{D} = \widetilde{O}(\Lambda)\backslash\Omega$ of the hypersurface $\Omega_K \subset \Omega^+$ depends only on the $\widetilde{O}(\Lambda)$-orbit of K. Also, the involution $r \in O(\Lambda)$ induces a nontrivial involution $r_{\mathscr{D}}$ of $\mathscr{D}$.

Corollary 6.3 *The periods of the special fourfolds of discriminant d are contained in*

(a) if $d \equiv 0 \pmod 4$, an irreducible hypersurface $\mathscr{D}_d \subset \mathscr{D}$;
(b) if $d \equiv 2 \pmod 8$, the union of two irreducible hypersurfaces $\mathscr{D}'_d$ and $\mathscr{D}''_d$, which are interchanged by the involution $r_{\mathscr{D}}$.

Assume $d \equiv 2 \pmod 8$ (case (b)). Then, $\mathscr{D}'_d$ (resp. $\mathscr{D}''_d$) corresponds to lattices K with $K \cdot u = \mathbf{Z}$ (resp. $K \cdot v = \mathbf{Z}$). In other words, given a fourfold X of type $\mathscr{X}_{10}$ whose period point is in $\mathscr{D}_d = \mathscr{D}'_d \cup \mathscr{D}''_d$, it is in $\mathscr{D}'_d$ if $K \cdot \sigma_1^2 \subset 2\mathbf{Z}$, and it is in $\mathscr{D}''_d$ if $K \cdot \sigma_{1,1} \subset 2\mathbf{Z}$.

Remark 6.4 The divisors $\mathscr{D}_d$ appear in the theory of modular forms under the name of *Heegner divisors*. In the notation of [B]:

- when $d \equiv 0 \pmod 8$, we have $\mathscr{D}_d = h_{d/8,0}$;
- when $d \equiv 2 \pmod 8$, we have $\mathscr{D}'_d = h_{d/2,e_*}$ and $\mathscr{D}''_d = h_{d/2,f_*}$;
- when $d \equiv 4 \pmod 8$, we have $\mathscr{D}_d = h_{d/2,e_*+f_*}$.

Remark 6.5 Zarhin's argument, already used in the proof of Corollary 4.4, proves that if X is a fourfold whose period is very general in any given $\mathscr{D}_d$, the lattice $K = H^4(X, \mathbf{Z}) \cap H^{2,2}(X)$ has rank exactly 3 and the rational Hodge structure $K^\perp \otimes \mathbf{Q}$ is simple.

6.2 Associated K3 surface

As we will see in the next section, K3 surfaces sometimes occur in the geometric description of special fourfolds X of type $\mathscr{X}_{10}$. This is related to the fact that, for some values of d, the nonspecial cohomology of X looks like the primitive cohomology of a K3 surface.

Following [H1], we determine, in each case of Proposition 6.2, the discriminant group of the *nonspecial lattice* $K^\perp$ and the symmetric form $b_{K^\perp} = -b_K$. We then find all cases when the nonspecial lattice of X is isomorphic (with a change of sign) to the primitive cohomology lattice of a (pseudo-polarized, degree-d) K3 surface. Although this property is only lattice-theoretic, the surjectivity of the period map for K3 surfaces then produces an actual K3 surface, which is said to be "associated with X." For $d = 10$, we will see in Sections 7.1 and 7.3 geometric constructions of the associated K3 surface.

Finally, there are other cases where geometry provides an "associated" K3 surface S (see Section 7.6), but not in the sense considered here: the Hodge structure of S is only isogeneous to that of the fourfold. So there might be integers d not in the list provided by the proposition below, for which special fourfolds of discriminant d are still related in some way to K3 surfaces (of degree different from d).

Proposition 6.6 *Let d be a positive integer such that $d \equiv 0, 2,$ or $4 \pmod 8$ and let (X, K) be a special fourfold of type $\mathscr{X}_{10}$ with discriminant d. Then:*

(a) if $d \equiv 0 \pmod 8$, we have $D(K^\perp) \simeq (\mathbf{Z}/2\mathbf{Z})^2 \times (\mathbf{Z}/(d/4)\mathbf{Z})$;
(b) if $d \equiv 2 \pmod 8$, we have $D(K^\perp) \simeq \mathbf{Z}/d\mathbf{Z}$ and we may choose this isomorphism so that $b_{K^\perp}(1,1) = -\frac{d+8}{2d} \pmod{\mathbf{Z}}$;
(c) if $d \equiv 4 \pmod 8$, we have $D(K^\perp) \simeq \mathbf{Z}/d\mathbf{Z}$ and we may choose this isomorphism so that $b_{K^\perp}(1,1) = -\frac{d+2}{2d} \pmod{\mathbf{Z}}$.

The lattice $K^\perp$ is isomorphic to the opposite of the primitive cohomology lattice of a pseudo-polarized K3 surface (necessarily of degree d) if and only if we are in case (b) or (c) and the only odd primes that divide d are $\equiv 1 \pmod 4$.

In these cases, there exists a pseudo-polarized, degree-d, K3 surface S such that the Hodge structure $H^2(S, \mathbf{Z})^0(-1)$ is isomorphic to $K^\perp$. Moreover, if the period point of X is not in $\mathscr{D}_8$, the pseudo-polarization is a polarization.

The first values of d that satisfy the conditions for the existence of an associated K3 surface are: 2, 4, 10, 20, 26, 34, 50, 52, 58, 68, 74, 82, 100, ...

Proof Since $I_{22,2}$ is unimodular, we have $(D(K^\perp), b_{K^\perp}) \simeq (D(K), -b_K)$ ([N], Proposition 1.6.1). Case (a) follows from Proposition 6.2.

Let e, f, and g be the generators of K corresponding to the matrix given in Proposition 6.2. The matrix of $b_{K^\perp}$ in the dual basis $(e^\vee, f^\vee, g^\vee)$ of $K^\perp$ is the inverse of that matrix.

In case (b), one checks that $e^\vee + g^\vee$ generates $D(K)$, which is isomorphic to $\mathbf{Z}/d\mathbf{Z}$. Its square is $\frac{1}{2} + \frac{4}{d} = \frac{d+8}{2d}$.

In case (c), one checks that $e^\vee$ generates $D(K)$, which is isomorphic to $\mathbf{Z}/d\mathbf{Z}$. Its square is $\frac{d+2}{2d}$.

The opposite of the primitive cohomology lattice of a pseudo-polarized K3 surface of degree d has discriminant group $\mathbf{Z}/d\mathbf{Z}$ and the square of a generator is $\frac{1}{d}$. So case (a) is impossible.

In case (b), the forms are conjugate if and only if $-\frac{d+8}{2d} \equiv \frac{n^2}{d} \pmod{\mathbf{Z}}$ for some integer n prime to d, or $-\frac{d+8}{2} \equiv n^2 \pmod{d}$. Set $d = 2d'$ (so that $d' \equiv 1 \pmod 4$); then this is equivalent to saying that $d' - 4$ is a square in the ring $\mathbf{Z}/d\mathbf{Z}$. Since d' is odd, this ring is isomorphic to $\mathbf{Z}/2\mathbf{Z} \times \mathbf{Z}/d'\mathbf{Z}$, hence this is equivalent to asking that -4, or equivalently -1, is a square in $\mathbf{Z}/d'\mathbf{Z}$. This happens if and only if the only odd primes that divide d' (or d) are $\equiv 1 \pmod 4$.

In case (c), the reasoning is similar: we need $-\frac{d+2}{2d} \equiv \frac{n^2}{d} \pmod{\mathbf{Z}}$ for some integer n prime to d. Set $d = 4d'$, with d' odd. This is equivalent to $-2 \equiv 2n^2 \pmod{d'}$, and we conclude as above.

As already explained, the existence of the polarized K3 surface (S, f) follows from the surjectivity of the period map for K3 surfaces. Finally, if $\wp([X])$ is not in $\mathscr{D}_8$, there are no classes of type $(2, 2)$ with square 2 in $H^4(X, Z)_{\mathrm{van}}$, hence no (-2)-curves on S orthogonal to f, so f is a polarization. □

6.3 Associated cubic fourfold

Cubic fourfolds also sometimes occur in the geometric description of special fourfolds X of type $\mathscr{X}_{10}$ (see Section 7.2). We determine for which values of d the nonspecial cohomology of X is isomorphic to the nonspecial cohomology of a special cubic fourfold. Again, this is only a lattice-theoretic association, but the surjectivity of the period map for cubic fourfolds then produces a (possibly singular) actual cubic. We will see in Section 7.2 that some special fourfolds X of discriminant 12 are actually birationally isomorphic to their associated special cubic fourfold.

Proposition 6.7 *Let d be a positive integer such that $d \equiv 0, 2,$ or $4 \pmod 8$ and let (X, K) be a special fourfold of type $\mathscr{X}_{10}$ with discriminant d. The lattice*

$K^\perp$ is isomorphic to the nonspecial cohomology lattice of a (possibly singular) special cubic fourfold (necessarily of discriminant d) if and only if:

(a) either $d \equiv 2$ or $20 \pmod{24}$, and the only odd primes that divide d are $\equiv \pm 1 \pmod{12}$;

(b) or $d \equiv 12$ or $66 \pmod{72}$, and the only primes ≥ 5 that divide d are $\equiv \pm 1 \pmod{12}$.

In these cases, if moreover the period point of X is general in $\mathscr{D}_d$ and $d \neq 2$, there exists a smooth *special cubic fourfold whose nonspecial Hodge structure is isomorphic to $K^\perp$.*

The first values of d that satisfy the conditions for the existence of an associated cubic fourfold are: 2, 12, 26, 44, 66, 74, 92, 122, 138, 146, 156, 194, . . .

Proof Recall from [H1], Section 4.3, that (possibly singular) special cubic fourfolds of positive discriminant d exist for $d \equiv 0$ or $2 \pmod 6$ (for $d = 2$, the associated cubic fourfold is the (singular) determinantal cubic; for $d = 6$, it is nodal). Combining that condition with that of Lemma 6.1, we obtain the necessary condition $d \equiv 0, 2, 8, 12, 18, 20 \pmod{24}$. Write $d = 24d' + e$, with $e \in \{0, 2, 8, 12, 18, 20\}$.

Then, one needs to check whether the discriminant forms are isomorphic. Recall from [H1], Proposition 3.2.5, that the discriminant group of the nonspecial lattice of a special cubic fourfold of discriminant d is isomorphic to $(\mathbf{Z}/3\mathbf{Z}) \times (\mathbf{Z}/(d/3)\mathbf{Z})$ if $d \equiv 0 \pmod 6$, and to $\mathbf{Z}/d\mathbf{Z}$ if $d \equiv 2 \pmod 6$. This excludes $e = 0$ or 8; for $e = 12$, we need $d' \not\equiv 1 \pmod 3$ and for $e = 18$, we need $d' \not\equiv 0 \pmod 3$. In all these cases, the discriminant group is cyclic.

When $e = 2$, the discriminant forms are conjugate if and only if $-\frac{d+8}{2d} \equiv n^2 \frac{2d-1}{3d} \pmod{\mathbf{Z}}$ for some integer n prime to d (Proposition 6.6 and [H1], Proposition 3.2.5), or equivalently, since 3 is invertible modulo d, if and only if $\frac{d}{2} + 12 \equiv 3\frac{d+8}{2} \equiv n^2 \pmod d$. This is equivalent to saying that $12d' + 13$ is a square in $\mathbf{Z}/d\mathbf{Z} \simeq (\mathbf{Z}/(12d' + 1)\mathbf{Z}) \times (\mathbf{Z}/2\mathbf{Z})$, or that 3 is a square in $\mathbf{Z}/(12d' + 1)\mathbf{Z}$. Using quadratic reciprocity, we see that this is equivalent to saying that the only odd primes that divide d are $\equiv \pm 1 \pmod{12}$.

When $e = 20$, we need $-\frac{d+2}{2d} \equiv n^2 \frac{2d-1}{3d} \pmod{\mathbf{Z}}$ for some integer n prime to d, or equivalently, $\frac{d}{2} + 3 \equiv n^2 \pmod d$. Again, we get the same condition.

When $e = 12$, we need $9 \nmid d$ and $-\frac{d+2}{2d} \equiv n^2\left(\frac{2}{3} - \frac{3}{d}\right) \pmod{\mathbf{Z}}$ for some integer n prime to d, or equivalently, $-12d' - 7 \equiv n^2(16d' + 5) \pmod d$. Modulo 3, we get that $1 - d'$ must be a nonzero square, hence $3 \mid d'$. Modulo 4, there are no conditions. Then we need $1 \equiv 3n^2 \pmod{2d' + 1}$ and we conclude as above.

Finally, when $e = 18$, we need $9 \nmid d$ and $-\frac{d+8}{2d} \equiv n^2\left(\frac{2}{3} - \frac{3}{d}\right) \pmod{\mathbf{Z}}$ for some n prime to d, or equivalently, $-12d' - 13 \equiv n^2(16d' + 9) \pmod{d}$. Modulo 3, we get $d' \equiv 2 \pmod 3$, and then $4 \equiv 3n^2 \pmod{4d' + 3}$ and we conclude as above.

At this point, we have a Hodge structure on $K^\perp$ which is, as a lattice, isomorphic to the nonspecial cohomology of a special cubic fourfold. It corresponds to a point in the period domain $\mathscr{C}$ of cubic fourfolds. To make sure that it corresponds to a (then unique) smooth cubic fourfold, we need to check that it is not in the special loci $\mathscr{C}_2 \cup \mathscr{C}_6$ ([La], Theorem 1.1). If the period point of X is general in $\mathscr{D}_d$, the period point in $\mathscr{C}$ is general in $\mathscr{C}_d$, hence is not in $\mathscr{C}_2 \cup \mathscr{C}_6$ if $d \notin \{2, 6\}$. □

Remark 6.8 One can be more precise and figure out explicit conditions on $\wp([X])$ for the associated cubic fourfold to be smooth (but calculations are complicated). For example, when $d = 12$, we find that it is enough to assume $\wp([X]) \notin \mathscr{D}_2 \cup \mathscr{D}_4 \cup \mathscr{D}_8 \cup \mathscr{D}_{16} \cup \mathscr{D}_{28} \cup \mathscr{D}_{60} \cup \mathscr{D}_{112} \cup \mathscr{D}_{240}$.

7 Examples of special fourfolds

Assume that a fourfold X of type $\mathscr{X}_{10}$ contains a smooth surface S. Then, by (6),

$$c(T_X)|_S = 1 + 2\sigma_1|_S + (4\sigma_1^2|_S - \sigma_2|_S) = c(T_S)c(N_{S/X}).$$

This implies $c_1(T_S) + c_1(N_{S/X}) = 2\sigma_1|_S$ and

$$4\sigma_1^2|_S - \sigma_2|_S = c_1(T_S)c_1(N_{S/X}) + c_2(T_S) + c_2(N_{S/X}).$$

We obtain

$$(S)_X^2 = c_2(N_{S/X}) = 4\sigma_1^2|_S - \sigma_2|_S - c_1(T_S)(2\sigma_1|_S - c_1(T_S)) - c_2(T_S).$$

Write $[S] = a\sigma_{3,1} + b\sigma_{2,2}$ in $G(2, V_5)$. Using Noether's formula, we obtain

$$(S)_X^2 = 3a + 4b + 2K_S \cdot \sigma_1|_S + 2K_S^2 - 12\chi(\mathscr{O}_S). \tag{10}$$

The determinant of the intersection matrix in the basis $(\sigma_{1,1}|_X, \sigma_2|_X - \sigma_{1,1}|_X, [S])$ is then

$$d = 4(S)_X^2 - 2(b^2 + (a - b)^2). \tag{11}$$

We remark that $\sigma_2|_X - \sigma_{1,1}|_X$ is the class of the unique σ-quadric surface Σ_0 contained in X (see Section 3).

The results of this section are summarized in Section 7.7.

7.1 Fourfolds containing a σ-plane (divisor $\mathscr{D}''_{10}$)

A σ-plane is a 2-plane in $G(2, V_5)$ of the form $\mathbf{P}(V_1 \wedge V_4)$; its class in $G(2, V_5)$ is $\sigma_{3,1}$. Fourfolds of type $\mathscr{X}^0_{10}$ containing such a 2-plane were already studied by Roth ([R], Section 4) and Prokhorov ([P], Section 3).

Proposition 7.1 *Inside $\mathscr{X}_{10}$, the family $\mathscr{X}_{\sigma\text{-plane}}$ of fourfolds containing a σ-plane is irreducible of codimension 2. The period map induces a dominant map $\mathscr{X}_{\sigma\text{-plane}} \to \mathscr{D}''_{10}$ whose general fiber has dimension 3 and is rationally dominated by a $\mathbf{P}^1$-bundle over a degree-10 K3 surface.*

A general member of $\mathscr{X}_{\sigma\text{-plane}}$ is rational.

During the proof, we present an explicit geometric construction of a general member X of $\mathscr{X}_{\sigma\text{-plane}}$, starting from a general degree-10 K3 surface $S \subset \mathbf{P}^6$, a general point p on S, and a smooth quadric Y containing the projection $\widetilde{S} \subset \mathbf{P}^5$ from p. The birational isomorphism $Y \dashrightarrow X$ is given by the linear system of cubics containing $\widetilde{S}$.

Proof A parameter count ([IM], Lemma 3.6) shows that $\mathscr{X}_{\sigma\text{-plane}}$ is irreducible of codimension 2 in $\mathscr{X}_{10}$. Let $P \subset X$ be a σ-plane. From (10), we obtain $(P)^2_X = 3$ and from (11), $d = 10$. Since $\sigma_1^2 \cdot P$ is odd, we are in $\mathscr{D}''_{10}$.

For X general in $\mathscr{X}_{\sigma\text{-plane}}$ (see [P], Section 3, for the precise condition), the image of the projection $\pi_P \colon X \dashrightarrow \mathbf{P}^5$ from P is a smooth quadric $Y \subset \mathbf{P}^5$ and, if $\widetilde{X} \to X$ is the blow-up of P, the projection π_P induces a birational morphism $\widetilde{X} \to Y$ which is the blow-up of a smooth degree-9 surface $\widetilde{S}$, itself the blow-up of a smooth degree-10 K3 surface S at one point ([P], Proposition 2).

Conversely, let $S \subset \mathbf{P}^6$ be a degree-10 K3 surface. When S is general, the projection from a general point p on S induces an embedding $\widetilde{S} \subset \mathbf{P}^5$ of the blow-up of S at p. Given any *smooth* quadric Y containing $\widetilde{S}$, one can reverse the construction above and produce a fourfold X containing a σ-plane (we will give more details about this construction and explicit genericity assumptions during the proof of Theorem 8.1).

There are isomorphisms of polarized integral Hodge structures

$$\begin{aligned} H^4(\widetilde{X}, \mathbf{Z}) &\simeq H^4(X, \mathbf{Z}) \oplus H^2(P, \mathbf{Z})(-1) \\ &\simeq H^4(Y, \mathbf{Z}) \oplus H^2(\widetilde{S}, \mathbf{Z})(-1) \\ &\simeq H^4(Y, \mathbf{Z}) \oplus H^2(S, \mathbf{Z})(-1) \oplus \mathbf{Z}(-2). \end{aligned}$$

For S very general, the Hodge structure $H^2(S, \mathbf{Q})_0$ is simple, hence it is isomorphic to the nonspecial cohomology $K^\perp \otimes \mathbf{Q}$ (where K is the lattice spanned by $H^4(G(2, V_5), \mathbf{Z})$ and $[P]$ in $H^4(X, \mathbf{Z})$). Moreover, the lattice $H^2(S, \mathbf{Z})_0(-1)$

embeds isometrically into $K^\perp$. Since they both have rank 21 and discriminant 10, they are isomorphic. The surface S is thus the (polarized) K3 surface associated with X as in Proposition 6.6.

Since the period map for polarized degree-10 K3 surfaces is dominant onto their period domain, the period map for $\mathscr{X}_{\sigma\text{-plane}}$ is dominant onto $\mathscr{D}''_{10}$ as well. Since the Torelli theorem for K3 surfaces holds, S is determined by the period point of X, hence the fiber $\wp^{-1}([X])$ is rationally dominated by the family of pairs (p, Y), where $p \in S$ and Y belongs to the pencil of quadrics in $\mathbf{P}^5$ containing $\widetilde{S}$. □

With the notation above, the inverse image of the quadric $Y \subset \mathbf{P}^5$ by the projection $\mathbf{P}^8 \dashrightarrow \mathbf{P}^5$ from P is a rank-6 non-Plücker quadric in $\mathbf{P}^8$ containing X, with vertex P. We show in Section 7.5 that $\mathscr{X}_{\sigma\text{-plane}}$ is contained in the irreducible hypersurface of $\mathscr{X}_{10}$ parameterizing the fourfolds X contained in such a quadric.

7.2 Fourfolds containing a ρ-plane (divisor $\mathscr{D}_{12}$)

A ρ-plane is a 2-plane in $G(2, V_5)$ of the form $\mathbf{P}(\wedge^2 V_3)$; its class in $G(2, V_5)$ is $\sigma_{2,2}$. Fourfolds of type $\mathscr{X}_{10}$ containing such a 2-plane were already studied by Roth ([R], Section 4).

Proposition 7.2 *Inside $\mathscr{X}_{10}$, the family $\mathscr{X}_{\rho\text{-plane}}$ of fourfolds containing a ρ-plane is irreducible of codimension 3. The period map induces a dominant map $\mathscr{X}_{\rho\text{-plane}} \to \mathscr{D}_{12}$ whose general fiber is the union of two rational surfaces.*

A general member of $\mathscr{X}_{\rho\text{-plane}}$ is birationally isomorphic to a cubic fourfold containing a smooth cubic surface scroll.

The proof presents a geometric construction of a general member of $\mathscr{X}_{\rho\text{-plane}}$, starting from any smooth cubic fourfold $Y \subset \mathbf{P}^5$ containing a smooth cubic surface scroll T. The birational isomorphism $Y \dashrightarrow X$ is given by the linear system of quadrics containing T.

Proof A parameter count ([IM], Lemma 3.6) shows that $\mathscr{X}_{\rho\text{-plane}}$ is irreducible of codimension 3 in $\mathscr{X}_{10}$. Let $P = \mathbf{P}(\wedge^2 V_3) \subset X$ be a ρ-plane. From (10), we obtain $(P)^2_X = 4$. From (11), we obtain $d = 12$ and we are in $\mathscr{D}_{12}$.

As shown in [R], Section 4, the image of the projection $\pi_P \colon X \dashrightarrow \mathbf{P}^5$ from P is a cubic hypersurface Y and the image of the intersection of X with the Schubert hypersurface

$$\Sigma_P = \{V_2 \subset V_5 \mid V_2 \cap V_3 \neq 0\} \subset G(2, V_5)$$

is a cubic surface scroll T (contained in Y). If $\widetilde{X} \to X$ is the blow-up of P, with exceptional divisor E_P, the projection π_P induces a birational morphism $\widetilde{\pi}_P \colon \widetilde{X} \to Y$. One checks (with the same arguments as in [P], Section 3) that all fibers have dimension ≤ 1 and hence that $\widetilde{\pi}_P$ is the blow-up of the smooth surface T. The image $\widetilde{\pi}_P(E_P)$ is the (singular) hyperplane section $Y_0 := Y \cap \langle T \rangle$.

Conversely, a general cubic fourfold Y containing a smooth cubic scroll contains two families (each parameterized by $\mathbf{P}^2$) of such surfaces (see [HT1] and [HT2], Example 7.12). For each such smooth cubic scroll, one can reverse the construction above and produce a smooth fourfold X containing a ρ-plane.

As in Section 7.1, there are isomorphisms of polarized integral Hodge structures

$$H^4(\widetilde{X}, \mathbf{Z}) \simeq H^4(X, \mathbf{Z}) \oplus H^2(P, \mathbf{Z})(-1) \simeq H^4(Y, \mathbf{Z}) \oplus H^2(T, \mathbf{Z})(-1).$$

Let K be the lattice spanned by $H^4(G(2, V_5), \mathbf{Z})$ and $[P]$ in $H^4(X, \mathbf{Z})$. For X very general in $\mathscr{X}_{\rho\text{-plane}}$, the Hodge structure $K^\perp \otimes \mathbf{Q}$ is simple (Remark 6.5), hence it is isomorphic to the Hodge structure $\langle h^2, [T] \rangle^\perp \subset H^4(Y, \mathbf{Q})$. Moreover, the lattices $K^\perp$ and $\langle h^2, [T] \rangle^\perp \subset H^4(Y, \mathbf{Z})$, which both have rank 21 and discriminant 12 (see [H1], Section 4.1.1), are isomorphic. This case fits into the setting of Proposition 6.7: the special cubic fourfold Y is associated with X.

Finally, since the period map for cubic fourfolds containing a cubic scroll surface is dominant onto the corresponding hypersurface in their period domain, the period map for $\mathscr{X}_{\rho\text{-plane}}$ is dominant onto $\mathscr{D}_{12}$ as well. Since the Torelli theorem holds for cubic fourfolds ([V]), Y is determined by the period point of X, hence the fiber $\wp^{-1}([X])$ is rationally dominated by the family of smooth cubic scrolls contained in Y. It is therefore the union of two rational surfaces. □

With the notation above, let $V_4 \subset V_5$ be a general hyperplane containing V_3. Then $G(2, V_4) \cap X$ is the union of P and a cubic scroll surface.

7.3 Fourfolds containing a τ-quadric surface (divisor $\mathscr{D}'_{10}$)

A τ-quadric surface in $G(2, V_5)$ is a linear section of $G(2, V_4)$; its class in $G(2, V_5)$ is $\sigma_1^2 \cdot \sigma_{1,1} = \sigma_{3,1} + \sigma_{2,2}$.

Proposition 7.3 *The closure $\overline{\mathscr{X}}_{\tau\text{-quadric}} \subset \mathscr{X}_{10}$ of the family of fourfolds containing a τ-quadric surface is an irreducible component of $\wp^{-1}(\mathscr{D}'_{10})$. The period map induces a dominant map $\mathscr{X}_{\tau\text{-quadric}} \to \mathscr{D}'_{10}$ whose general fiber* *is birationally isomorphic to the quotient by an involution of the symmetric square of a degree-10 K3 surface.*

A general member of $\mathscr{X}_{\tau\text{-quadric}}$ is rational.

During the proof, we present a geometric construction of a general member of $\mathscr{X}_{\tau\text{-quadric}}$, starting from a general degree-10 K3 surface $S \subset \mathbf{P}^6$ and two general points on S: if $S_0 \subset \mathbf{P}^4$ is the (singular) projection of S from these two points, the birational isomorphism $\mathbf{P}^4 \dashrightarrow X$ is given by the linear system of quartics containing S_0.

Proof A parameter count shows that $\mathscr{X}_{\tau\text{-quadric}}$ is irreducible of codimension 1 in $\mathscr{X}_{10}$ (one can also use the parameter count at the end of the proof). Let $\Sigma \subset X$ be a smooth τ-quadric surface. From (10), we obtain $(\Sigma)^2_X = 3$ and from (11), $d = 10$. Since $\sigma_1^2 \cdot \Sigma$ is even, we are in $\mathscr{D}'_{10}$. The family $\mathscr{X}_{\tau\text{-quadric}}$ is therefore a component of the divisor $\wp^{-1}(\mathscr{D}'_{10})$.

The projection from the 3-plane $\langle\Sigma\rangle$ induces a birational map $X \dashrightarrow \mathbf{P}^4$ (in particular, X is rational!). If $\varepsilon\colon \widetilde{X} \to X$ is the blow-up of Σ, one checks that it induces a birational *morphism* $\pi\colon \widetilde{X} \to \mathbf{P}^4$ which is more complicated than just the blow-up of a smooth surface (compare with Section 7.1).

In the first part of the proof, we analyze the birational structure of π by factorizing it into a composition of blow-ups with smooth centers and their inverses (see diagram (12)). This gives an explicit construction of X, and in the second part of the proof we prove that any such construction does give an X containing a τ-quadric surface.

Since Σ is contained in a $G(2, V_4)$, the quartic surface $X \cap G(2, V_4)$ is the union of Σ and another τ-quadric surface $\Sigma^\star$. The two 3-planes $\langle\Sigma\rangle$ and $\langle\Sigma^\star\rangle$ meet along a 2-plane, hence (the strict transform of) $\Sigma^\star$ is contracted by π to a point. Generically, the only quadric surfaces contained in X are the σ-quadric surface Σ_0 (defined in Section 3) and the τ-quadric surfaces Σ and $\Sigma^\star$. Using the fact that X is an intersection of quadrics, one checks that $\Sigma^\star$ is the only surface contracted (to a point) by π.

Let $\ell' \subset \widetilde{X}$ be a line contracted by ε. If $\ell \subset \widetilde{X}$ is (the strict transform of) a line contained in $\Sigma^\star$, it meets Σ and is contracted by π. Since $\widetilde{X}$ has Picard number 2, the rays $\mathbf{R}^+[\ell]$ and $\mathbf{R}^+[\ell']$ are extremal, hence span the cone of curves of $\widetilde{X}$. These two classes have $(-K_{\widetilde{X}})$-degree 1, hence $\widetilde{X}$ is a Fano fourfold. Extremal contractions on smooth fourfolds have been classified ([AM], Theorem 4.1.3). In our case, we have:

- π is a divisorial contraction, its (irreducible) exceptional divisor D contains $\Sigma^\star$, and $D \underset{\text{lin}}{\equiv} 3H - 4E$;

- $S_0 := \pi(D)$ is a surface with a single singular point $s := \pi(\Sigma^\star)$, where it is locally the union of two smooth 2-dimensional germs meeting transversely;
- outside of s, the map π is the blow-up of S_0 in $\mathbf{P}^4$.

Let $\widehat{X} \to \widetilde{X}$ be the blow-up of $\Sigma^\star$, with exceptional divisor $\widehat{E}$, and let $\widehat{\mathbf{P}^4} \to \mathbf{P}^4$ be the blow-up of s, with exceptional divisor $\mathbf{P}^3_s$. The strict transform $\widehat{S}_0 \subset \widehat{\mathbf{P}}^4$ of S_0 is the blow-up of its (smooth) normalization S'_0 at the two points lying over s and meets $\mathbf{P}^3_s$ along the disjoint union of the two exceptional curves L_1 and L_2. There is an induced morphism $\widehat{X} \to \widehat{\mathbf{P}}^4$ which is an extremal contraction ([AM], Theorem 4.1.3), hence is the blow-up of the smooth surface $\widehat{S}_0$, with exceptional divisor the strict transform $\widehat{D} \subset \widehat{X}$ of D; it induces by restriction a morphism $\widehat{E} \to \mathbf{P}^3_s$ which is the blow-up of $L_1 \sqcup L_2$.

It follows that we have isomorphisms of polarized Hodge structures

$$\begin{aligned} H^4(\widehat{X}, \mathbf{Z}) &\simeq H^4(X, \mathbf{Z}) \oplus H^2(\Sigma, \mathbf{Z})(-1) \oplus H^2(\Sigma^\star, \mathbf{Z})(-1) \\ &\simeq H^4(\mathbf{P}^4, \mathbf{Z}) \oplus H^2(\mathbf{P}^3_s, \mathbf{Z})(-1) \oplus \mathbf{Z}[L_1] \oplus \mathbf{Z}[L_2] \oplus H^2(S'_0, \mathbf{Z})(-1). \end{aligned}$$

In particular, we have $b_2(S'_0) = 24 + 2 + 2 - 1 - 1 - 1 - 1 = 24$ and $h^{2,0}(S'_0) = h^{3,1}(\widehat{X}) = 1$; moreover, the Picard number of S'_0 is 3 for X general. The situation is as follows:

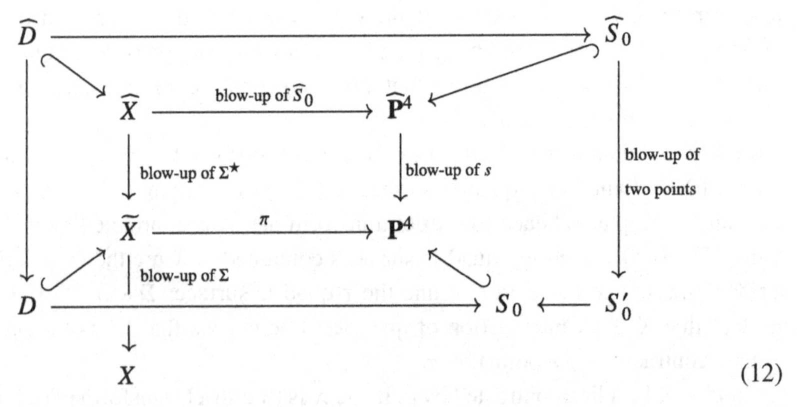

(12)

To compute the degree d of S_0, we consider the (smooth) inverse image $P \subset \widetilde{X}$ of a 2-plane in $\mathbf{P}^4$. It is isomorphic to the blow-up of $\mathbf{P}^2$ at d points, hence $K_P^2 = 9 - d$. On the other hand, we have by adjunction

$$K_P \underset{\text{lin}}{\equiv} (K_{\widetilde{X}} + 2(H - E))|_P \underset{\text{lin}}{\equiv} (-2H + E + 2(H - E))|_P = -E|_P,$$

hence $K_P^2 = E^2 \cdot (H - E)^2 = 1$ and $d = 8$.

Consider now a general hyperplane $h \subset \mathbf{P}^4$. Its intersection with S_0 is a smooth connected curve C of degree 8, and its inverse image in $\widetilde{X}$ is the blow-up of h along C, with exceptional divisor its intersection with D. From [IP], Lemma 2.2.14, we obtain

$$D^3 \cdot (H - E) = -2g(C) + 2 + K_h \cdot C = -2g(C) + 2 - 4\deg(C) = -2g(C) - 30,$$

from which we get $g(C) = 6$. In particular, $c_1(S'_0) \cdot h = 2$. On the other hand, using a variant of the formula for smooth surfaces in $\mathbf{P}^4$, we obtain

$$d^2 - 2 = 10d + c_1^2(S'_0) - c_2(S'_0) + 5c_1(S'_0) \cdot h,$$

hence $c_1^2(S'_0) - c_2(S'_0) = -28$. We then use a formula from [P], Lemma 2:

$$\begin{aligned}
\widehat{D}^4 &= (c_2(\widetilde{\mathbf{P}}^4) - c_1^2(\widetilde{\mathbf{P}}^4)) \cdot \widehat{S}_0 + c_1(\widetilde{\mathbf{P}}^4)|_{\widehat{S}_0} \cdot c_1(\widehat{S}_0) - c_2(\widehat{S}_0) \\
&= (-15h^2 - 7[\mathbf{P}^3_s]^2) \cdot \widehat{S}_0 + (-5h^2 + 3[\mathbf{P}^3_s])|_{\widehat{S}_0} \cdot c_1(\widehat{S}_0) - c_2(\widehat{S}_0) \\
&= (-15h^2 - 7[\mathbf{P}^3_s]^2) \cdot \widehat{S}_0 + (-5h^2 + 3[\mathbf{P}^3_s])|_{\widehat{S}_0} \cdot c_1(\widehat{S}_0) - c_2(\widehat{S}_0) \\
&= -120 + 14 - 10 - 6 - c_2(\widehat{S}_0).
\end{aligned}$$

Since $\widehat{D}^4 = D^4 = (3H - 4E)^4 = -150$, we obtain $c_2(\widehat{S}_0) = 28$, hence $c_2(S'_0) = 26$ and $c_1^2(S'_0) = -2$. Noether's formula implies $\chi(S'_0, \mathcal{O}_{S'_0}) = 2$, hence $h^1(S'_0, \mathcal{O}_{S'_0}) = 0$. The classification of surfaces implies that S'_0 is the blow-up at two points of a K3 surface S of degree 10. By the simplicity argument used before, the integral polarized Hodge structures $H^2(S, \mathbf{Z})_0(-1)$ and $K^\perp$ are isomorphic: S is the (polarized) K3 surface associated with X via Proposition 6.6.

What happens if we start from the τ-quadric $\Sigma^\star$ instead of Σ? Blowing up Σ and then the strict transform of $\Sigma^\star$ is not the same as doing it in the reverse order, but the end products have a common open subset $\widetilde{X}^0$ (whose complements have codimension 2). The morphisms $\widetilde{X}^0 \to \widetilde{\mathbf{P}}^4 \to \mathbf{P}^3$ (where the second morphism is induced by projection from s) are then the same, because they are induced by the projection of X from the 4-plane $\langle \Sigma, \Sigma^\star \rangle$, and the locus where they are not smooth is the common projection S_1 in $\mathbf{P}^3$ of the surfaces $S_0 \subset \mathbf{P}^4$ and $S_0^\star \subset \mathbf{P}^4$ from their singular points.

This surface S_1 is also the projection of the K3 surface $S \subset \mathbf{P}^6$ from the 2-plane spanned by p, p', q, q'. The end result is therefore the same K3 surface S (as it should be, because its period is determined by that of X), but the pair of points is now q, q'. We let ι_S denote the birational involution on $S^{[2]}$ defined by $p + p' \mapsto q + q'$ (in [O], Proposition 5.20, O'Grady proves that for S general, the involution ι_S is biregular on the complement of a 2-plane).

Conversely, let $S = G(2, V_5) \cap Q' \cap \mathbf{P}^6$ be a general K3 surface of degree 10 and let p (corresponding to $V_2 \subset V_5$) and p' (corresponding to $V'_2 \subset V_5$)

be two general points on S. If $V_4 := V_2 \oplus V_2'$, the intersection $S \cap G(2, V_4)$ is a set of four points p, p', q, q' in the 2-plane $\mathbf{P}(\wedge^2 V_4) \cap \mathbf{P}^6$. Projecting S from the line pp' gives a nonnormal degree-8 surface $S_0 := S_{pp'} \subset \mathbf{P}^4$, where q and q' have been identified. Its normalization S_0' is the blow-up of S at p and p'. Now let $\widehat{\mathbf{P}}^4 \to \mathbf{P}^4$ be the blow-up of the singular point of S_0, and let $\widehat{X} \to \widehat{\mathbf{P}}^4$ be the blow-up of the strict transform of S_0 in $\widehat{\mathbf{P}}^4$. The strict transform in $\widehat{X}$ of the exceptional divisor $\mathbf{P}^3_s \subset \widehat{\mathbf{P}}^4$ can be blown down by $\widehat{X} \to \widetilde{X}$.

The resulting smooth fourfold $\widetilde{X}$ is a Fano variety with Picard number 2. One extremal contraction is $\pi \colon \widetilde{X} \to \mathbf{P}^4$. The other extremal contraction gives the desired X. This construction depends on 23 parameters (19 for the surface S and 4 for $p, p' \in S$).

All this implies (as in the proofs of Propositions 7.1 and 7.2) that the period map for $\mathscr{X}_{\rho\text{-plane}}$ is dominant onto $\mathscr{D}_{10}'$, with fiber birationally isomorphic to $S^{[2]}/\iota_S$. □

7.4 Fourfolds containing a cubic scroll (divisor $\mathscr{D}_{12}$)

We consider rational cubic scroll surfaces obtained as smooth hyperplane sections of the image of a morphism $\mathbf{P}(V_2) \times \mathbf{P}(V_3) \to G(2, V_5)$, where $V_5 = V_2 \oplus V_3$; their class in $G(2, V_5)$ is $\sigma_1^2 \cdot \sigma_2 = 2\sigma_{3,1} + \sigma_{2,2}$.

Proposition 7.4 *The closure $\overline{\mathscr{X}}_{\text{cubic scroll}} \subset \mathscr{X}_{10}$ of the family of fourfolds containing a cubic scroll surface is the irreducible component of $\wp^{-1}(\mathscr{D}_{12})$ that contains the family $\mathscr{X}_{\rho\text{-plane}}$.*

Proof Let us count parameters. We have $6 + 6 = 12$ parameters for the choice of V_2 and V_3, hence a priori 12 parameters for cubic scroll surfaces in the isotropic Grassmannian $G_\omega(2, V_5)$. However, one checks that there is a 1-dimensional family of V_3 which all give the same cubic scroll, so there are actually only 11 parameters. Then, for X to contain a given cubic scroll F represents $h^0(F, \mathscr{O}_F(2, 2)) = 12$ conditions. It follows that $\mathscr{X}_{\text{cubic scroll}}$ is irreducible of codimension $12 - 11 = 1$ in $\mathscr{X}_{10}$.

Let $F \subset X$ be a cubic scroll. Since K_F has type $(-1, -2)$, we obtain $(F)^2_X = 4$ from (10). From (11), we obtain $d = 12$ and we are in $\mathscr{D}_{12}$. The family $\overline{\mathscr{X}}_{\text{cubic scroll}}$ is therefore a component of the hypersurface $\wp^{-1}(\mathscr{D}_{12})$.

In the degenerate situation where $V_4 = V_2 + V_3$ is a hyperplane, the associated rational cubic scroll is contained in $G(2, V_4)$ and is a cubic scroll surface as in the comment at the end of Section 7.2. It follows that $\mathscr{X}_{\rho\text{-plane}}$ is contained in $\overline{\mathscr{X}}_{\text{cubic scroll}}$. □

7.5 Fourfolds containing a quintic del Pezzo surface (divisor $\mathscr{D}''_{10}$)

We consider quintic del Pezzo surfaces obtained as the intersection of $G(2, V_5)$ with a $\mathbf{P}^5$; their class is $\sigma_1^4 = 3\sigma_{3,1} + 2\sigma_{2,2}$ in $G(2, V_5)$. Fourfolds of type $\mathscr{X}_{10}$ containing such a surface were already studied by Roth ([R], Section 4).

Proposition 7.5 *The closure $\overline{\mathscr{X}}_{\text{quintic}} \subset \mathscr{X}_{10}$ of the family of fourfolds containing a quintic del Pezzo surface is the irreducible component of $\wp^{-1}(\mathscr{D}''_{10})$ that contains $\mathscr{X}_{\sigma\text{-plane}}$.*

A general member of $\mathscr{X}_{\text{quintic}}$ is rational.

Proof Let us count parameters. We have $\dim G(5, \mathbf{P}^8) = 18$ parameters for the choice of the $\mathbf{P}^5$ that defines a del Pezzo surface T. Then, for X to contain a given quintic del Pezzo surface T represents $h^0(\mathbf{P}^5, \mathscr{O}(2)) - h^0(\mathbf{P}^5, \mathscr{I}_T(2)) = 21 - 5 = 16$ conditions.

Since $h^0(\mathbf{P}^8, \mathscr{I}_X(2)) = 6 = h^0(\mathbf{P}^5, \mathscr{I}_T(2)) + 1$, there exists a unique (non-Plücker) quadric $Q \subset \mathbf{P}^8$ containing X and $\mathbf{P}^5$. This quadric has rank ≤ 6, hence it is a cone with vertex a 2-plane over a (in general) smooth quadric in $\mathbf{P}^5$. Such a quadric contains two 3-dimensional families of 5-planes. The intersection of such a 5-plane with X is, in general, a quintic del Pezzo surface, hence X contains (two) 3-dimensional families of quintic del Pezzo surfaces. It follows that $\mathscr{X}_{\text{quintic}}$ has codimension $16 - 18 + 3 = 1$ in $\mathscr{X}_{10}$.

Let $T \subset X$ be a quintic del Pezzo surface. From (10), we obtain $(T)^2_X = 5$ and from (11), $d = 10$. Since $\sigma_{1,1} \cdot T$ is odd, we are in $\mathscr{D}''_{10}$. The family $\overline{\mathscr{X}}_{\text{quintic}}$ is therefore a component of the divisor $\wp^{-1}(\mathscr{D}''_{10})$.

The lattice spanned by $H^4(G(2, V_5), \mathbf{Z})$ and $[T]$ in $H^4(X, \mathbf{Z})$ is the same as for fourfolds containing a σ-plane P, and $[T] = \sigma_2|_X - [P]$. We will now explain this fact geometrically.

If X contains a quintic del Pezzo surface, we saw that X is contained in a (non-Plücker) quadric $Q \subset \mathbf{P}^8$ of rank ≤ 6. Conversely, if X is contained in such a quadric, this quadric contains 5-planes and the intersection of such a 5-plane with X is, in general, a quintic del Pezzo surface.

If follows that $\mathscr{X}_{\text{quintic}}$ has the same closure in $\mathscr{X}_{10}$ as the set of X contained in a non-Plücker rank-6 quadric Q. When the vertex of Q is contained in X, it is a σ-plane, hence $\overline{\mathscr{X}}_{\text{quintic}}$ contains $\mathscr{X}_{\sigma\text{-plane}}$.

Finally, note after [R], Section 5.(5), that the general fibers of the projection $X \dashrightarrow \mathbf{P}^2$ from $\langle T \rangle$ are again degree-5 del Pezzo surfaces (they are residual surfaces to T in the intersection of X with a 6-plane $\langle T, x \rangle$, and this intersection is contained in $\langle T, x \rangle \cap Q$, which is the union of two hyperplanes). It follows from a theorem of Enriques that X is rational ([E], [SB]). □

7.6 Nodal fourfolds (divisor $\mathscr{D}_8$)

Let X be a general prime *nodal* Fano fourfold of index 2 and degree 10. As in the 3-dimensional case ([DIM2], Lemma 4.1), X is the intersection of a smooth $G^\omega := G(2,V_5)\cap\mathbf{P}^8$ with a nodal quadric Q, singular at a point O general in G^ω.

One checks that, as in the case of cubic fourfolds (see [V], Section 4; [H1], Proposition 4.2.1), the limiting Hodge structure is pure, and the period map extends to the moduli stack $\overline{\mathscr{X}_{10}}$ of our fourfolds with at most one node as

$$\overline{\wp}\colon \overline{\mathscr{X}_{10}} \to \mathscr{D}.$$

Proposition 7.6 *The closure $\overline{\mathscr{X}_{\text{nodal}}} \subset \overline{\mathscr{X}_{10}}$ of the family of nodal fourfolds is an irreducible component of $\overline{\wp}^{-1}(\mathscr{D}_8)$.*

Proof Let X be a prime nodal Fano fourfold with a node at O, obtained as above. If $\widetilde{X} \to X$ is the blow-up of O, the (pure) limiting Hodge structure is the direct sum of $\langle\delta\rangle$, where δ is the vanishing cycle, with self-intersection 2, and $H^4(\widetilde{X},\mathbf{Z})$. In the basis $(\sigma_{1,1}|_X, \sigma_2|_X - \sigma_{1,1}|_X, \delta)$, the corresponding lattice K has intersection matrix $\begin{pmatrix} 2 & 0 & 0\\ 0 & 2 & 0\\ 0 & 0 & 2\end{pmatrix}$, hence we are in $\mathscr{D}_8$.

The point O defines a pencil of Plücker quadrics, singular at O, and the image G^ω_O of G^ω by the projection $p_O\colon \mathbf{P}^8 \dashrightarrow \mathbf{P}^7_O$ is the base-locus of a pencil of rank-6 quadrics (see [DIM2], Section 3). One checks that G^ω_O contains the 4-plane $\mathbf{P}^4_O := p_O(\mathbf{T}_{G^\omega,O})$ and that G^ω_O is singular along a cubic surface contained in $\mathbf{P}^4_O$. If $\widetilde{\mathbf{P}}^7_O \to \mathbf{P}^7_O$ is the blow-up of $\mathbf{P}^4_O$, the strict transform $\widetilde{G}^\omega_O \subset \widetilde{\mathbf{P}}^7_O \subset \mathbf{P}^7_O{\times}\mathbf{P}^2$ of G^ω_O is smooth and the projection $\widetilde{G}^\omega_O \to \mathbf{P}^2$ is a $\mathbf{P}^3$-bundle (this can be checked by explicit computations as in [DIM2], Section 9.2).

The image $X_O := p_O(X)$ is thus the base locus in $\mathbf{P}^7_O$ of a net of quadrics $\mathbf{P}$, containing a special line of rank-6 Plücker quadrics. The strict transform $\widetilde{X}_O \subset \widetilde{G}^\omega_O$ of X_O is smooth. The induced projection $\widetilde{X}_O \to \mathbf{P}^2$ is a quadric bundle, with discriminant a smooth sextic curve $\Gamma^\star_6 \subset \mathbf{P}^2$ (compare with [DIM2], Proposition 4.2) and associated double cover $S \to \mathbf{P}^2$ ramified along $\Gamma^\star_6$. It follows that S is a (smooth) K3 surface with a degree-2 polarization. By [L], Theorem II.3.1, there is an exact sequence

$$0 \longrightarrow H^4(\widetilde{X}_O,\mathbf{Z})_0 \xrightarrow{\Phi} H^2(S,\mathbf{Z})_0(-1) \longrightarrow \mathbf{Z}/2\mathbf{Z} \longrightarrow 0.$$

Both desingularizations $\widetilde{X} \to X_O$ and $\widetilde{X}_O \to X_O$ are small and their fibers all have dimension ≤ 1; by [FW], Proposition 3.1, the graph of the rational map $\widetilde{X} \dashrightarrow \widetilde{X}_O$ induces an isomorphism $H^4(\widetilde{X}_O,\mathbf{Z}) \xrightarrow{\sim} H^4(\widetilde{X},\mathbf{Z})$ of polarized Hodge structures. The usual simplicity argument implies that this isomorphism sends

$H^4(\widetilde{X}_O,\mathbf{Z})_0$ onto the nonspecial cohomology $K^\perp$, which therefore has index 2 in $H^2(S,\mathbf{Z})_0(-1)$.

When X is general, so is S among degree-2 K3 surfaces, hence the image $\overline{\wp}(\mathscr{X}_{\text{nodal}})$ has dimension 19. It follows that $\mathscr{X}_{\text{nodal}}$ is an irreducible component of $\overline{\wp}^{-1}(\mathscr{D}_8)$. □

7.7 Summary of results

We summarize the results of this section in the table below. Please refer to the corresponding subsections for exact statements.

X contains a	Dimension of family	Image in period domain	Fiber of period map	General X birational to
σ-plane	22	$\mathscr{D}''_{10}$	$\mathbf{P}^1$-bundle over K3	$\mathbf{P}^4$
quintic del Pezzo	23	$\mathscr{D}''_{10}$	?	$\mathbf{P}^4$
τ-quadric	23	$\mathscr{D}'_{10}$	$K3^{[2]}/\text{inv.}$	$\mathbf{P}^4$
ρ-plane	21	$\mathscr{D}_{12}$	2 rational surfaces	cubic
cubic scroll	23	$\mathscr{D}_{12}$	?	
node	23	$\mathscr{D}_8$	?	int. of 3 quadrics

8 Construction of special fourfolds

Again following Hassett (particularly [H1], Section 4.3), we construct special fourfolds with given discriminant. Hassett's idea was to construct, using the surjectivity of the (extended) period map for K3 surfaces, nodal cubic fourfolds whose Picard group also contains a rank-2 lattice with discriminant d and to smooth them using the fact that the period map remains a submersion on the nodal locus ([V], p. 597). This method should work in our case, but would require first making the construction of Section 7.6 of a nodal fourfold X of type $\mathscr{X}_{10}$ from a given degree-2 K3 surface more explicit, and second proving that the extended period map remains submersive at any point of the nodal locus.

We prefer here to use the simpler construction of Section 7.1 to prove the following:

Theorem 8.1 *The image of the period map $\wp\colon \mathscr{X}^0_{10} \to \mathscr{D}$ meets all divisors $\mathscr{D}_d$, for $d \equiv 0 \pmod 4$ and $d \geq 12$, and all divisors $\mathscr{D}'_d$ and $\mathscr{D}''_d$, for $d \equiv 2 \pmod 8$ and $d \geq 10$, except possibly $\mathscr{D}''_{18}$.*

Actually, the divisor $\mathscr{D}''_{18}$ also meets the image of the period map: in a forthcoming article, we construct birational transformations that take elements of $\wp^{-1}(\mathscr{D}'_d)$ to elements of $\wp^{-1}(\mathscr{D}''_d)$.

Proof Our starting point is Lemma 4.3.3 of [H1]: let Γ be a rank-2 indefinite even lattice containing a primitive element h with $h^2 = 10$, and assume there is no $c \in \Gamma$ with

- either $c^2 = -2$ and $c \cdot h = 0$;
- or $c^2 = 0$ and $c \cdot h = 1$;
- or $c^2 = 0$ and $c \cdot h = 2$.

Then there exists a K3 surface S with $\operatorname{Pic}(S) = \Gamma$ and h is very ample on S, hence embeds it in $\mathbf{P}^6$. Assuming moreover that S is not trigonal, e.g., that there are no classes $c \in \Gamma$ with $c^2 = 0$ and $c \cdot h = 3$, it has Clifford index 2 and is therefore obtained as the intersection of a Fano threefold $Z := G(2, V_5) \cap \mathbf{P}^6$ with a quadric ([M3], (3.9); [JK], Theorem 10.3 and Proposition 10.5).

In particular, S is an intersection of quadrics, and since a general point p of S is not on a line contained in S, the projection from p of S is a (degree-9) smooth surface $\widetilde{S}_p \subset \mathbf{P}^5$.

On the other hand, if $\Pi \subset \mathbf{P}(\wedge^2 V_5^\vee)$ is the 2-plane of hyperplanes that cut out $\mathbf{P}^6$ in $\mathbf{P}(\wedge^2 V_5)$, one has ([PV], Corollary 1.6)

$$\operatorname{Sing}(Z) = \Pi^\perp \cap \bigcup_{[\omega]\in\Pi} G(2, \operatorname{Ker}(\omega)).$$

Since S is smooth, $\operatorname{Sing}(Z)$ is finite. If $\operatorname{Sing}(Z) \neq \varnothing$, some $[\omega_0] \in \Pi$ must have rank 2 and one checks that there exists $V_2 \subset \operatorname{Ker}(\omega_0)$ such that Z contains a family of lines through $[V_2]$, parameterized by a rational cubic curve. The intersection of the cone swept out by these lines and the quadric that defines S in Z is a sextic curve of genus 2 in S. Its class c' thus satisfies $c'^2 = 2$ and $c' \cdot h = 6$, hence $(h - c')^2 = 0$ and $(h - c') \cdot h = 4$.

So if we assume finally that there are no classes $c \in \Gamma$ with $c^2 = 0$ and $c \cdot h = 4$, the threefold Z is smooth. It is then known ([PV], Theorem 7.5 and Proposition 7.6) that $\operatorname{Aut}(Z)$ is isomorphic to $PGL(2, \mathbf{C})$ and acts on Z with three (rational) orbits of respective dimensions 3, 2, and 1. Since S is not rational, it must meet the open orbit. Take $p \in S$ in that orbit.

It is then classical ([PV], Section 7) that Z contains three lines passing through p, so that the projection $\widetilde{Z}_p \subset \mathbf{P}^5$ from p of Z has exactly three singular points, which are also on the smooth surface $\widetilde{S}_p$. One then checks on explicit equations of $\widetilde{Z}_p$ ([I], Section 3.1) that $\widetilde{Z}_p$ is contained in a smooth quadric $Y \subset \mathbf{P}^5$. Consider the blow-up $\widetilde{Y} \to Y$ of $\widetilde{S}_p$. The inverse image

$E \subset \widetilde{Y}$ of $\widetilde{Z}_p$ is then a small resolution, which is isomorphic to the blow-up of p in Z.

The morphism $E \to \widetilde{Z}_p$ is in particular independent of the choice of S, p, and Y and it follows from the description of the general case in the proof of Proposition 7.1 that E is a $\mathbf{P}^1$-bundle over $\mathbf{P}^2$. More precisely, the linear system $|H|$ on $\widetilde{Y}$ given by cubics containing $\widetilde{S}_p$ induces on E a morphism $E \to \mathbf{P}^2$ with $\mathbf{P}^1$-fibers (in the notation of that proof, E is the exceptional divisor of the blow-up $\widetilde{X} \to X$ of the plane P).

The linear system $|H|$ is base-point-free and injective outside of E (because $\widetilde{Z}_p$ is a quadratic section of Y which contains $\widetilde{S}_p$) and base-point-free on E as we just saw. Since $H^4 = 10$ and $h^0(\widetilde{Y}, H) = 9$, it defines a birational morphism $\widetilde{Y} \twoheadrightarrow X \subset \mathbf{P}^8$ which maps E onto a 2-plane $P \subset X$. This morphism is one of the two $K_{\widetilde{Y}}$-negative extremal contractions of $\widetilde{Y}$ (the other one being the blow-up $\widetilde{Y} \to Y$); its fibers all have dimension ≤ 1, hence X is *smooth* and the contraction is the blow-up of P ([AM], Theorem 4.1.3).

It is then easy to check that X is a (special) Fano fourfold of type $\mathscr{X}_{10}^0$ containing P as a σ-plane. As explained in the proof of Proposition 7.1, its nonspecial cohomology is isomorphic to the primitive cohomology of S.

We will now apply this construction with various lattices Γ to produce examples of smooth fourfolds X which will all be in $\mathscr{X}_{\sigma\text{-plane}}$, hence with period point in $\mathscr{D}''_{10}$, but whose lattice $H^{2,2}(X) \cap H^4(X, \mathbf{Z})$ will contain other sublattices of rank 3 with various discriminants.

Apply first Hassett's lemma with the rank-2 lattice Γ with matrix $\begin{pmatrix} 10 & 0 \\ 0 & -2e \end{pmatrix}$ in a basis (h, w). When $e > 1$, the conditions we need on Γ are satisfied and we obtain a K3 surface S and a smooth fourfold X containing a σ-plane P, such that $H^4(X, \mathbf{Z}) \cap H^{2,2}(X)$ contains a lattice $K_{10} = \langle u, v, w''_{10} \rangle$ with matrix $\begin{pmatrix} 2 & 0 & 0 \\ 0 & 2 & 1 \\ 0 & 1 & 3 \end{pmatrix}$ and discriminant 10 (here $w''_{10} = [P]$; see proof of Proposition 7.1). Moreover, $H^2(S, \mathbf{Z})_0(-1) \simeq K_{10}^{\perp}$ as polarized integral Hodge structures. The element $w \in \Gamma \cap H^2(S, \mathbf{Z})_0$ corresponds to $w_X \in K_{10}^{\perp} \cap H^{2,2}(X)$, and $w_X^2 = -w^2 = 2e$. Therefore, $H^4(X, \mathbf{Z}) \cap H^{2,2}(X)$ is the lattice $\langle u, v, w''_{10}, w_X \rangle$, with matrix

$$\begin{pmatrix} 2 & 0 & 0 & 0 \\ 0 & 2 & 1 & 0 \\ 0 & 1 & 3 & 0 \\ 0 & 0 & 0 & 2e \end{pmatrix}.$$

It contains the lattice $\langle u, v, w_X \rangle$, with matrix $\begin{pmatrix} 2 & 0 & 0 \\ 0 & 2 & 0 \\ 0 & 0 & 2e \end{pmatrix}$. Therefore, the period point of X belongs to $\mathscr{D}_{8e}$, and this proves the theorem when $d \equiv 0 \pmod 8$.

It also contains the lattice $\langle u, v, w''_{10} + w_X \rangle$, with matrix $\begin{pmatrix} 2 & 0 & 0 \\ 0 & 2 & 1 \\ 0 & 1 & 2e+3 \end{pmatrix}$ and discriminant $8e + 10$, hence we are also in $\mathscr{D}''_{8e+10}$.

Now let $e \geq 0$ and apply Hassett's lemma with the lattice Γ with matrix $\begin{pmatrix} 10 & 5 \\ 5 & -2e \end{pmatrix}$ in a basis (h, g). The orthogonal complement of h is spanned by $w := h - 2g$. One checks that primitive classes $c \in \Gamma$ such that $c^2 = 0$ satisfy $c \cdot h \equiv 0 \pmod 5$. All the conditions we need are thus satisfied and we obtain a K3 surface S and a smooth fourfold X such that $H^4(X, \mathbf{Z}) \cap H^{2,2}(X)$ contains a lattice K_{10} of discriminant 10 and $H^2(S, \mathbf{Z})_0(-1) \simeq K_{10}^{\perp}$ as polarized Hodge structures. Again, w corresponds to $w_X \in K_{10}^{\perp} \cap H^{2,2}(X)$ with $w_X^2 = -w^2 = 8e + 10$. Set

$$K := (\Lambda_2 \oplus \mathbf{Z} w_X)^{\text{sat}}.$$

To compute the discriminant of K, we need to know the ideal $w_X \cdot \Lambda$. As in the proof of Proposition 6.2, let w_{10} be a generator of $K_{10} \cap \Lambda$; it satisfies $w_{10}^2 = 10$. Then $K_{10}^{\perp} \oplus \mathbf{Z} w_{10}$ is a sublattice of Λ and, taking discriminants, we find that the index is 5. Let u be an element of Λ whose class generates the quotient. We have

$$w_X \cdot \Lambda = \mathbf{Z} w_X \cdot u + w_X \cdot (K_{10}^{\perp} \oplus \mathbf{Z} w_{10}) = \mathbf{Z} w_X \cdot u + w_X \cdot K_{10}^{\perp} = \mathbf{Z} w_X \cdot u + w \cdot H^2(S, \mathbf{Z})_0.$$

One checks directly on the K3 lattice that $w \cdot H^2(S, \mathbf{Z})_0 = 2\mathbf{Z}$. Since $5u \in K_{10}^{\perp} \oplus \mathbf{Z} w_{10}$, we have $5 w_X \cdot u \in 2\mathbf{Z}$, hence $w_X \cdot u \in 2\mathbf{Z}$. All in all, we have proved $w_X \cdot \Lambda = 2\mathbf{Z}$, hence the proof of Proposition 6.2 implies that the discriminant of K is $w_X^2 = 8e + 10$. Therefore, the period point of X belongs to $\mathscr{D}_{8e+10}$.

Since the period point of X is in $\mathscr{D}''_{10}$, we saw in the proof of Proposition 6.2 that $w''_{10} := \frac{1}{2}(v + w_{10})$ is in $H^4(X, \mathbf{Z})$. Similarly, either $w'_X := \frac{1}{2}(u + w_X)$ or $w''_X := \frac{1}{2}(v + w_X)$ is in K. Taking intersections with w''_{10} (and recalling $w_X \cdot w_{10} = 0$ and $v \cdot w_{10} = 1$), we see that we are in the first case, hence the period point of X is actually in $\mathscr{D}'_{8e+10}$.

More precisely, $H^4(X, \mathbf{Z}) \cap H^{2,2}(X)$ is the lattice $\langle u, v, w''_{10}, w'_X \rangle$, with matrix

$$\begin{pmatrix} 2 & 0 & 0 & 1 \\ 0 & 2 & 1 & 0 \\ 0 & 1 & 3 & 0 \\ 1 & 0 & 0 & 2e+3 \end{pmatrix}.$$

This lattice also contains the lattice $\langle u, v, w''_{10}+w'_X\rangle$, with matrix $\begin{pmatrix} 2 & 0 & 1 \\ 0 & 2 & 1 \\ 1 & 1 & 2e+6 \end{pmatrix}$ and discriminant $8e+20$, hence we are also in $\mathscr{D}_{8e+20}$.

Since we know from Section 7.2 that the period points of some smooth fourfolds X of type $\mathscr{X}^0_{10}$ lie in $\mathscr{D}_{12}$, this proves the theorem when $d \equiv 4 \pmod 8$. □

9 A question

It would be very interesting, as Laza did for cubic fourfolds ([La], Theorem 1.1), to determine the exact image in the period domain $\mathscr{D}$ of the period map for our fourfolds.

Question 9.1 Is the image of the period map equal to $\mathscr{D} - \mathscr{D}_2 - \mathscr{D}_4 - \mathscr{D}_8$?

Answering this question seems far from our present possibilities; to start with, inspired by the results of [H1], one could ask (see Theorem 8.1) whether the image of the period map is disjoint from the hypersurfaces $\mathscr{D}_2$, $\mathscr{D}_4$, and $\mathscr{D}_8$.

Acknowledgments We thank V. Gritsenko and G. Nebe for their help with lattice theory, and K. Ranestad and C.-L. Wang for drawing our attention to the references [JK] and [FW] respectively. The authors are also grateful to the referee for his suggestions.

References

[AM] Andreatta, M. and Mella, M. Morphisms of projective varieties from the viewpoint of minimal model program. *Dissertationes Math. (Rozprawy Mat.)* **413** (2003), 72 pp.

[BH] Borel, A. and Hirzebruch, F. Characteristic classes and homogeneous spaces, I. *Amer. J. Math.* **80** (1958) 458–538.

[B] Bruinier, J. H. Hilbert modular forms and their applications. In *The 1-2-3 of Modular Forms*. Berlin: Springer-Verlag, pp. 105–179, 2008.

[CM] Conte, A. and Murre, J. P. The Hodge conjecture for fourfolds admitting a covering by rational curves. *Math. Ann.* **238** (1978) 79–88.

[CS] Conway, J. H. and Sloane, N. J. A. *Sphere Packings, Lattices and Groups*, 2nd edn. With additional contributions by E. Bannai, R. E. Borcherds, J. Leech, S. P. Norton, A. M. Odlyzko, R. A. Parker, L. Queen, and B. B. Venkov. Grundlehren der Mathematischen Wissenschaften **290**, New York: Springer-Verlag, 1993.

[DIM1] Debarre, O., Iliev, A., and Manivel, L. On the period map for prime Fano threefolds of degree 10. *J. Algebraic Geom.* **21** (2012) 21–59.

[DIM2] Debarre, O., Iliev, A., and Manivel, L. On nodal prime Fano threefolds of degree 10. *Sci. China Math.* **54** (2011) 1591–1609.

[E] Enriques, F. Sulla irrazionalita da cui puo farsi dipendere la resoluzione d'un'equazione algebrica $f(x,y,z) = 0$ con funzione razionali di due parametri. *Math. Ann.* **49** (1897) 1–23.

[FW] Fu, B. and Wang, C.-L. Motivic and quantum invariance under stratified Mukai flops. *J. Differential Geom.* **80** (2008) 261–280.

[GHS] Gritsenko, V., Hulek, K., and Sankaran, G. K. Moduli spaces of irreducible symplectic manifolds. *Compos. Math.* **146** (2010) 404–434.

[G] Gushel', N. P. Fano varieties of genus 6 (in Russian). *Izv. Akad. Nauk SSSR Ser. Mat.* **46** (1982) 1159–1174, 1343. English transl.: *Math. USSR Izv.* **21** (1983) 445–459.

[H1] Hassett, B. Special cubic fourfolds. *Compos. Math.* **120** (2000) 1–23.

[H2] Hassett, B. Some rational cubic fourfolds. *J. Algebraic Geom.* **8** (1999) 103–114.

[HT1] Hassett, B. and Tschinkel, Y. Flops on holomorphic symplectic fourfolds and determinantal cubic hypersurfaces. *J. Inst. Math. Jussieu* **9** (2010) 125–153.

[HT2] Hassett, B. and Tschinkel, Y. Rational curves on holomorphic symplectic fourfolds. *Geom. Funct. Anal.* **11** (2001) 1201–1228.

[I] Iliev, A. The Fano surface of the Gushel threefold. *Compos. Math.* **94** (1994) 81–107.

[IM] Iliev, A. and Manivel, L. Fano manifolds of degree 10 and EPW sextics. *Ann. Sci. École Norm. Sup.* **44** (2011) 393–426.

[IP] Iskovskikh, V. A. and Prokhorov, Y. Fano varieties. In *Algebraic Geometry, V.* Encyclopaedia of Mathematical Sciences, Vol. 47. Berlin: Springer-Verlag, pp. 1–247, 1999.

[JK] Johnsen, T. and Knutsen, A. L. *K3 Projective Models in Scrolls.* Lecture Notes in Mathematics, Vol. 1842. Berlin: Springer-Verlag, 2004.

[L] Laszlo, Y. Théorème de Torelli générique pour les intersections complètes de trois quadriques de dimension paire. *Invent. Math.* **98** (1989) 247–264.

[La] Laza, R. The moduli space of cubic fourfolds via the period map. *Ann. Math.* **172** (2010) 673–711.

[M1] Mukai, S. Curves, $K3$ surfaces and Fano 3-folds of genus ≤ 10. In *Algebraic Geometry and Commutative Algebra,* Vol. I, pp. 357–377, Kinokuniya, Tokyo, 1988.

[M2] Mukai, S. Biregular classification of Fano 3-folds and Fano manifolds of coindex 3, *Proc. Natl. Acad. Sci. USA* **86** (1989) 3000–3002.

[M3] Mukai, S. New development of theory of Fano 3-folds: Vector bundle method and moduli problem. *Sugaku* **47** (1995) 125–144.

[N] Nikulin, V. Integral symmetric bilinear forms and some of their applications. *Izv. Akad. Nauk SSSR Ser. Mat.* **43** (1979) 111–177. English transl.: *Math. USSR Izv.* **14** (1980) 103–167.

[O] O'Grady, K. Double covers of EPW-sextics. *Michigan Math. J.* **62** (2013) 143–184.

[PV] Piontkowski, J. and Van de Ven, A. The automorphism group of linear sections of the Grassmannian $G(1,N)$. *Doc. Math.* **4** (1999) 623–664.

[P] Prokhorov, Y. Rationality constructions of some Fano fourfolds of index 2. *Moscow Univ. Math. Bull.* **48** (1993) 32–35.

[R] Roth, L. Algebraic varieties with canonical curve sections. *Ann. Mat. Pura Appl. (4)* **29** (1949) 91–97.

[S] Serre, J.-P. *Cours d'Arithmétique*. Le Mathématicien **2**. Paris: Presses Universitaires de France, 1977.

[SB] Shepherd-Barron, N. The rationality of quintic del Pezzo surfaces–a short proof. *Bull. London Math. Soc.* **24** (1992) 249–250.

[V] Voisin, C. Théorème de Torelli pour les cubiques de $\mathbf{P}^5$. *Invent. Math.* **86** (1986) 577–601 and Erratum, *Invent. Math.* **172** (2008) 455–458.

[W] Wu, W.-T. Classes caractéristiques et *i*-carrés d'une variété. *C. R. Acad. Sci. Paris* **230** (1950) 508–511.

[Z] Zarhin, Y. Hodge groups of K3 surfaces. *J. reine und angew. Math.* **341** (1983) 193–220.

9

Configuration spaces of complex and real spheres

I. Dolgachev
University of Michigan

B. Howard[a]
Institute for Defense Analysis

Abstract

We study the GIT-quotient of the Cartesian power of projective space modulo the projective orthogonal group. A classical isomorphism of this group with the inversive group of birational transformations of the projective space of one dimension less allows us to interpret these spaces as configuration spaces of complex or real spheres.

To Rob Lazarsfeld on the occasion of his 60th birthday

1 Introduction

In this paper we study the moduli space of configurations of points in complex projective space with respect to the group of projective transformations leaving invariant a nondegenerate quadric. More precisely, if $\mathbb{P}^n = \mathbb{P}(V)$ denotes the projective space of lines in a linear complex space V equipped with a nondegenerate symmetric form $\langle v, w\rangle$, we study the GIT-quotient

$$\begin{aligned}\mathbf{O}_n^m &:= \mathbb{P}(V)^m // \mathrm{PO}(V) = \mathrm{Proj}(\bigoplus_{d=0}^{\infty} H^0(\mathbb{P}(V)^m, \mathcal{O}_{\mathbb{P}(V)}(d)^{\boxtimes m})^{\mathrm{O}(V)} \\ &\cong \mathrm{Proj}(\bigoplus_{d=0}^{\infty}(S^d(V^*)^{\otimes m})^{\mathrm{O}(V)}.\end{aligned}$$

[a] The author was supported in part by NSF grant 0703674.

From *Recent Advances in Algebraic Geometry*, edited by Christopher Hacon, Mircea Mustaţă and Mihnea Popa

If $m \geq n+1 = \dim(V)$ then generic point configurations have 0-dimensional isotropy subgroups in $O(V)$, and since $\dim O(n+1) = \frac{1}{2}n(n+1)$ we expect that $\dim \mathbf{O}_n^m = mn - \frac{1}{2}n(n+1)$ when $m \geq n+1$.

Let

$$R(n;m) = \bigoplus_{d=0}^{\infty} (S^d(V^*)^{\otimes m})^{O(V)}.$$

It is a finitely generated graded algebra with graded part $R(n;m)_d$ of degree d equal to $(S^d(V^*)^{\otimes m})^{O(V)}$. After polarization, $R(n;m)_d$ becomes isomorphic to the linear space $\mathbb{C}[V^m]_{d,\ldots,d}^{O(V)}$ of $O(V)$-invariant polynomials on V^m which are homogeneous of degree d in each vector variable. The first fundamental theorem (FFT) of invariant theory for the orthogonal group [19, Chapter 2, Section 9] asserts that $\mathbb{C}[V^m]^{O(V)}$ is generated by the bracket functions $[ij]$: $(v_1, \ldots, v_m) \mapsto \langle v_i, v_j \rangle$. Using this theorem, our first result is the following:

Theorem 1.1 *Let* Sym_m *be the space of symmetric matrices of size* m *with the torus* $T^{m-1} = \{(z_1, \ldots, z_m) \in (\mathbb{C}^*)^m : z_1 \cdots z_m = 1\}$ *acting by scaling each* i*th row and* i*th column by* z_i. *Let* $\mathbb{S}_m$ *be the toric variety* $\mathbb{P}(\mathrm{Sym}_m)//T^{m-1}$. *Then* $\mathbf{O}_n^m$ *is isomorphic to a closed subvariety of* $\mathbb{S}_m$ *defined by the rank condition* $r \leq n+1$.

For example, when $m \leq n+1$, we obtain that $\mathbf{O}_n^m$ is a toric variety of dimension $\frac{1}{2}n(n+1)$.

The varieties $\mathbf{O}_1^m$ are special since the connected component of the identity of $O(2)$ is isomorphic to $SO(2) \cong \mathbb{C}^*$. This implies that $\mathbf{O}_1^m$ admits a double cover isomorphic to a toric variety $(\mathbb{P}^1)^m // SO(2)$. We compare this variety with the toric variety $X(A_{m-1})$ associated with the root system of type A_{m-1} (see [2, 15]). The variety $X(A_{m-1})$ admits a natural involution defined by the standard Cremona transformation of $\mathbb{P}^{m-1}$ and the quotient by this involution is a generalized Cayley 4-nodal cubic surface Cay_{m-1} (equal to the Cayley cubic surface if $m = 3$). We prove that $\mathbf{O}_1^m$ is isomorphic to Cay_{m-1} for odd m and equal to some blow-down of Cay_{m-1} when m is even.

The main geometric motivation for our work is the study of configuration spaces of complex and real spheres. It has been known since F. Klein and S. Lie that the inversive group[1] defining the geometry of spheres in dimension n is isomorphic to the projective orthogonal group $PO(n+1)$ (see, e.g., [8, Section 25]). Thus any problem about configurations of m spheres in $\mathbb{P}^n$ is equivalent to the same problem about configurations of m points in $\mathbb{P}^{n+1}$ with

[1] Also called the inversion group or the Laguerre group. It is a subgroup of the Cremona group of $\mathbb{P}^n$ generated by the projective affine orthogonal group $PAO(n+1)$ and the inversion transformation $[x_0, \ldots, x_n] \mapsto [x_1^2 + \ldots + x_n^2, x_0x_1, \ldots, x_0x_n]$.

respect to $\mathrm{PO}(n+1)$. The last two sections of this paper give some applications to the geometry of spheres.

2 The first fundamental theorem of invariant theory

Let V be an $(n+1)$-dimensional vector quadratic space, i.e., a vector space together with a nondegenerate symmetric bilinear form whose values we denote by $\langle v, w\rangle$. Let $G = \mathrm{O}(V)$ be the orthogonal group of V and $\mathrm{PO}(V) = \mathrm{O}(V)/\{\pm I\}$. Consider the diagonal action of G on V^m. The first fundamental theorem of invariant theory for the orthogonal group (see [14, Chapter 11, 2.1; 19, Chapter 2, Section 9]) asserts that any G-invariant polynomial function on V^m is a polynomial in the bracket functions

$$[ij] : V^m \to \mathbb{C},\ (v_1, \dots, v_m) \mapsto \langle v_i, v_j\rangle, \quad 1 \le i, j \le m.$$

The algebra of G-invariant polynomial functions $\mathbb{C}[V^m]^G$ has a natural multi-grading by $\mathbb{N}^m$ with homogeneous part $\mathbb{C}[V^m]^G_{(d_1,\dots,d_m)}$ equal to the linear space of polynomials which are homogeneous of degree d_i in each ith vector variable. This grading corresponds to the natural action of the torus $\mathbb{C}^{*m}$ by scaling the vectors in each factor. The $\mathbb{N}$-graded ring $R(n; m)$ in which we are interested is the subring $\oplus_{d=0}^{\infty}\mathbb{C}[V^m]^G_{(d,\dots,d)}$. We have

$$R(n; m) \cong \mathbb{C}[V^m]^{\mathrm{O}(V)\times T},$$

where $T = \{(z_1, \dots, z_m) \in \mathbb{C}^{*m} : z_1 \cdots z_m = 1\}$.

Let Sym_m denote the linear space of complex symmetric $m \times m$ matrices. If we view V^m as the space of linear functions $L(\mathbb{C}^m, V)$, then we can define a quadratic map

$$\Phi : V^m \to \mathrm{Sym}_m, \quad (v_1, \dots, v_m) \mapsto (\langle v_i, v_j\rangle)$$

by composing

$$\mathbb{C}^m \xrightarrow{\phi} V \xrightarrow{b} V^* \xrightarrow{{}^t\phi} (\mathbb{C}^m)^*,$$

where the middle map is defined by the symmetric bilinear form b associated with q. It is easy to see that, considering the domain and the range of Φ as affine spaces over $\mathbb{C}$, the image of ϕ is the closed subvariety $\mathrm{Sym}_m(n+1) \subset \mathrm{Sym}_m$ of symmetric matrices of rank $\le n+1$. Passing to the rings of regular functions, we get a homomorphism of rings

$$\Phi : \mathbb{C}[\mathrm{Sym}_m] :\to \mathbb{C}[V^m]. \tag{2.1}$$

The map Φ is obviously T-equivariant if we make $(z_1, \ldots, z_m)$ act by multiplying the entry x_{ij} of a symmetric matrix by $z_i z_j$. By passing to invariants, we obtain a homomorphism of graded rings

$$\Phi_T : \mathbb{C}[\mathrm{Sym}_m]^T \to \mathbb{C}[V^m]^T. \tag{2.2}$$

The FTT can be restated by saying that the image of this homomorphism is equal to the ring $R(n; m)$.

We identify $\mathbb{C}[\mathrm{Sym}_m]$ with the polynomial ring in entries X_{ij} of a general symmetric matrix $X = (X_{ij})$ of size $m \times m$. Note that the action of $(z_1, \ldots, z_m) \in (\mathbb{C}^*)^m$ on a symmetric matrix (X_{ij}) is by multiplying each entry X_{ij} by $z_i z_j$. The graded part $\mathbb{C}[\mathrm{Sym}_m]_d$ of $\mathbb{C}[\mathrm{Sym}_m]$ consists of functions which under this action are multiplied by $(z_1 \cdots z_m)^d$. They are obviously contained in $\mathbb{C}[\mathrm{Sym}_m]^T$ and define the grading of the ring $\mathbb{C}[\mathrm{Sym}_m]^T$. The homomorphism Φ^* is a homomorphism of graded rings from $\mathbb{C}[\mathrm{Sym}_m]^T$ to $R(n; m)$.

Let

$$\det X = \sum_{\sigma \in \mathfrak{S}_m} \epsilon(\sigma) X_{\sigma(1)1} \cdots X_{\sigma(N)N}$$

be the determinant of X. The monomials $d_\sigma = X_{\sigma(1)1} \cdots X_{\sigma(N)N}$ will be called the *determinantal* terms. Note that the number $k(m)$ of different determinantal terms is less than $m!$. It was known since the 19th century [16, p. 46] that the generating function for the numbers $k(m)$ is equal to

$$1 + \sum_{m=1}^{\infty} \frac{1}{m!} k(m) t^m = \frac{e^{\frac{1}{2}t + \frac{1}{4}t^2}}{\sqrt{1-t}}.$$

For example, $k(3) = 5, k(4) = 17, k(5) = 73, k(6) = 338$.

Each permutation σ decomposes into disjoint oriented cycles. Consider the directed graph on m vertices which consists of the oriented cycles in σ; i.e., we take a directed edge $i \to \sigma(i)$ for each vertex i. Suppose there is a cycle τ in σ of length ≥ 3. Write $\sigma = \tau \upsilon = \upsilon \tau$, and define $\sigma' = \tau^{-1} \upsilon$. Since our matrix is symmetric, the determinantal term $d_{\sigma'}$ corresponding to σ' has the same value as d_σ, and furthermore σ' has the same sign as σ (so there is no canceling), and so we may drop the orientation on each cycle. We may therefore envision the determinantal terms as 2-regular undirected graphs on m vertices (where 2-cycles and loops are admitted). Thus for each 2-regular graph having k cycles of length ≥ 3, there correspond 2^k determinantal terms.

Proposition 2.1 *The ring $\oplus_{d=0}^{\infty} \mathbb{C}[\mathrm{Sym}_m]^T_{2d}$ is generated by the determinantal terms.*

Proof A monomial $X_{i_1 j_1} \cdots X_{i_k j_k}$ belongs to $\mathbb{C}[\mathrm{Sym}_m]^T_d$ if and only if

$$z_{i_1} \cdots z_{i_k} z_{j_1} \cdots z_{j_k} = (z_1 \cdots z_m)^d$$

for any $z_1, \dots, z_m \in \mathbb{C}^*$. This happens if and only if each $i \in \{1, \dots, m\}$ occurs exactly d times among $i_1, \dots, i_k, j_1, \dots, j_k$. Consider the graph with set of vertices equal to $\{1, \dots, m\}$ and an edge from i to j if X_{ij} enters into the monomial. The above property is equivalent to the graph being a regular graph of valency d. The multiplication of monomials corresponds to the operation of adding graphs (in the sense that we add the sets of the edges). It remains to use the fact that any regular graph of valency $2d$ is equal to the union of regular graphs of valency 2 (this is sometimes called a "2-factorization" or Petersen's factorization theorem) [13, Section 9]. □

Corollary 2.2 *A set* $([v_1], \dots, [v_m])$ *is semi-stable for the action of* $\mathrm{O}(V)$ *on* $(\mathbb{P}^n)^m$ *if and only if there exists* $\sigma \in \mathfrak{S}_m$ *such that* $\langle v_{\sigma(1)}, v_1 \rangle \cdots \langle v_{\sigma(m)}, v_m \rangle$ *is not equal to zero.*

We can make it more explicit.

Proposition 2.3 *A point set* $([v_1], \dots, [v_m])$ *is unstable if and only if there exists* $I \subseteq J \subseteq \{1, \dots, m\}$ *such that* $|I| + |J| = m + 1$ *and* $\langle v_i, v_j \rangle = 0$ *for all* $i \in I$ *and* $j \in J$.

Proof Since $(\mathbb{C}[\mathrm{Sym}_m]^{(2)})^T$ is generated by determinantal terms, we obtain that a matrix $A = (a_{ij})$ has all determinantal terms equal to zero if and only if it represents an unstable point in $\mathbb{P}(\mathrm{Sym}_m)$ with respect to the torus action. Now the assertion becomes a simple consequence of the Hilbert–Mumford numerical criterion of stability.

It is obvious that if such subsets I and J exist then all determinantal terms vanish. So we are left with proving the existence of the subsets I and J if we have an unstable matrix.

Let $r : t \mapsto (t^{r_1}, \dots, t^{r_m})$ be a nontrivial 1-parameter subgroup of the torus T. Permuting the points, we may assume that $r_1 \leq r_2 \leq \dots \leq r_m$. We also have $r_1 + \dots + r_m = 0$. We claim that there exist i, j such that $i + j = m + 1$ and $r_i + r_j \leq 0$. If not, then each of $r_1 + r_m, r_2 + r_{m-1}, \dots, r_{\lfloor \frac{m+1}{2} \rfloor} + r_{\lceil \frac{m+1}{2} \rceil}$ is strictly positive, which contradicts $\sum_i r_i = 0$.

Since our symmetric matrix $A = (a_{ij})$ is unstable, by the Hilbert–Mumford criterion there must exist r such that $\min\{r_i + r_j : a_{ij} \neq 0\} > 0$. Permute the points if necessary so that $r_1 \leq r_2 \leq \cdots \leq r_n$. Let i_0, j_0 be such that $r_{i_0} + r_{j_0} \leq 0$ and $i_0 + j_0 = m + 1$. We may assume that $i_0 \leq j_0$ since the above condition is symmetric in i_0, j_0. Now, since the entries of r are increasing, we have that $r_i + r_j \leq 0$ for all $i \leq i_0$ and $j \leq j_0$. Hence $a_{ij} = 0$ for all $i \leq i_0$ and $j \leq j_0$. Now let $I = \{1, \dots, i_0\}$ and $J = \{1, \dots, j_0\}$. □

Similarly, we can prove the following:

Proposition 2.4 *A point set* $([v_1], \dots, [v_m])$ *is semi-stable but not stable if and only if*

$$m = \max\{|I| + |J| : I \subseteq J \subseteq \{1, \dots, m\} \textit{ and } \langle v_i, v_j \rangle = 0 \textit{ for all } i \in I, j \in J\}.$$

Proof Suppose that $A = (a_{ij} = \langle v_i, v_j \rangle)$. Let

$$m'(A) = \max\{|I| + |J| : I \subseteq J \subseteq \{1, \dots, m\} \text{ and } \langle v_i, v_j \rangle = 0 \text{ for all } i \in I, j \in J\}.$$

Suppose that A is semi-stable but not stable. Since A is not unstable, we know by the prior proposition that $m'(A) \leq m$. So we are left with showing that $m'(A) = m$.

Since A is not stable, there is a 1-parameter subgroup $r : t \mapsto (t^{r_1}, \dots, t^{r_m})$ such that $r_i + r_j \geq 0$ whenever $a_{ij} \neq 0$. We shall reorder the points so that $r_1 \leq r_2 \leq \cdots \leq r_n$. Recall also that $\sum_i r_i = 0$. Since some $r_i \neq 0$, we know that $r_1 < 0 < r_n$. We claim there is some i, j such that $i + j = m$ and $r_i + r_j < 0$. Otherwise, each of $r_1 + r_{m-1}, r_2 + r_{m-3}, \dots, r_{\lfloor \frac{m}{2} \rfloor} + r_{\lceil \frac{m}{2} \rceil}$ would be non-negative. This implies that $r_1 + \cdots + r_{m-1} \geq 0$. But since $r_n > 0$, we have that $r_1 + \cdots r_m > 0$, a contradiction. Hence, the claim is true. Now, take $i_0 \leq j_0$ such that $i_0 + j_0 = m$ and $r_{i_0} + r_{j_0} < 0$. Now, we must have that $a_{ij} = 0$ for all $i \leq i_0$ and $j \leq j_0$. Let $I = \{1, \dots, i_0\}$ and $J = \{1, \dots, j_0\}$. Then $I \subseteq J \subseteq \{1, \dots, m\}$, $|I| + |J| = m$, and $a_{ij} = 0$ for all $i \in I$, $j \in J$. Thus $m'(A) = m$.

Conversely, suppose that $m'(A) = m$. Then A is not unstable (if A were unstable, Proposition 2.3 implies that $m'(A) \geq m + 1$). Let $I \subseteq J \subseteq \{1, \dots, m\}$, such that $|I| + |J| = m$, and $a_{ij} = 0$ for all $i \in I$, $j \in J$. Reorder points if necessary so that $I = \{1, \dots, i_0\}$ and $J = \{1, \dots, j_0\}$. Let $r : t \mapsto (t^{r_1}, \dots, t^{r_m})$ be defined as follows. Let $r_i = -1$ for $i \leq i_0$, let $r_i = 1$ for $i > m - i_0$, and let $r_i = 0$ otherwise. The sum $\sum_i r_i$ is zero and not all r_i are zero, so this defines a 1-parameter subgroup of the torus T. Also, if $r_i + r_j < 0$ and $i \leq j$, then $i \leq i_0$ and $j \leq j_0$, which implies that $a_{ij} = 0$. Thus $r_i + r_j \geq 0$ whenever $a_{ij} \neq 0$. Hence A is not stable. □

The second fundamental theorem (SFT) of invariant theory for the group $\mathrm{O}(V)$ describes the kernel of the homomorphism (2.1) (see [14, p. 407; 19, Chapter 2, Section 17]).

Consider the ideal $\mathfrak{I}(n, m)$ in $\mathbb{C}[V^n]^{\mathrm{O}(V)}$ generated by the *Gram functions*

$$\gamma_{I,J} : (v_1, \dots, v_m) \mapsto \det \begin{pmatrix} \langle v_{i_1}, v_{j_1} \rangle & \dots & \langle v_{i_{n+2}}, v_{j_{n+2}} \rangle \\ \vdots & \vdots & \vdots \\ \langle v_{i_{n+2}}, v_{j_1} \rangle & \dots & \langle v_{i_{n+2}}, v_{j_{n+2}} \rangle \end{pmatrix}$$

where $I = (1 < i_1 \cdots < i_{n+2})$, $J = (1 < j_1 \cdots < j_{n+2})$ are subsets of $[1, m]$. We set $\gamma_I = \gamma_{I,I}$.

The pre-image of this ideal in $\mathbb{C}[\mathrm{Sym}^2(V^*)] \cong \mathbb{C}[\mathrm{Sym}_m]$ is the determinant ideal $\mathcal{D}_m(n+1)$ of matrices of rank $\leq n+1$. The SFT asserts that it is the kernel of the homomorphism (2.1). Then

$$\mathrm{Ker}(\Phi_T) = \mathcal{D}_m(n+1) \cap \mathbb{C}[\mathrm{Sym}_m]^T$$

and it is finitely generated by polynomials of the form $\mathbf{m}\Delta_{I,J}$, where $\mathbf{m}$ is a monomial in X_{ij} of degree $(k, k, \ldots, k) - \deg(\Delta_{I,J})$ for some $k \geq 2$.

Our naive hope was that $\mathrm{Ker}(\Phi_T)$ is generated only by polynomials of the form $\mathbf{m}\Delta_{I,J}$, for $\mathbf{m}$ having degree $(2, 2, \ldots, 2) - \deg(\Delta_{I,J})$. This is not true even if we restrict it to the open subset of semi-stable points in Sym_m with respect to the torus action. The symmetric matrix

$$A = \begin{pmatrix} 0 & 0 & 1 & 1 & 1 \\ 0 & 0 & 1 & 1 & 1 \\ 1 & 1 & 1 & 0 & 0 \\ 1 & 1 & 0 & 1 & 0 \\ 1 & 1 & 0 & 0 & 1 \end{pmatrix}$$

has rank 4, but for any (i, j) the product $a_{ij}A_{ij}$ (where A_{ij} is the complementary minor) is equal to zero. Thus our naive relations make it appear that A has rank 3. Also, $a_{31}a_{42}a_{13}a_{24}a_{55} \neq 0$, so the matrix represents a semi-stable point. It can be shown that no counterexample exists with $m < 5$.

The following T-invariant polynomial vanishes on rank 3 matrices and is nonzero when evaluated on the matrix A above:

$$a_{13}^2 a_{14}^2 a_{25}^2 \Delta_{\{2,3,4,5\},\{2,3,4,5\}}.$$

Hence we need to consider higher-degree relations. We can at least give a bound on the degree of such relations, again appealing to Petersen's factorization theorem.

Proposition 2.5 *The ideal* $\mathrm{Ker}(\Phi_T)$ *is generated by polynomials of the form* $\mathbf{m}\Delta_{I,J}$, *for* $\mathbf{m}$ *having degree at most* $(2(n+2), 2(n+2), \ldots, 2(n+2)) - \deg(\Delta_{I,J})$.

Proof It is clear that $\mathrm{Ker}(\Phi_T)$ is generated by relations of the form $\mathbf{m}\Delta_{I,J}$ where $\mathbf{m}$ is a monomial of degree $(2k, 2k, \ldots, 2k) - \deg(\Delta_{I,J})$, for arbitrary k. Suppose that $k > n + 2$. The monomial $\mathbf{m}$ corresponds to the multigraph Γ' with edges ij for each X_{ij} dividing $\mathbf{m}$, counting multiplicity. Choose any term from $\Delta_{I,J}$; similarly this term corresponds to a multigraph Γ''. The graph Γ'' has exactly $n + 2$ edges.

The union $\Gamma = \Gamma' \sqcup \Gamma''$ is a $2k$-regular graph. By Petersen's factorization theorem, we know that Γ completely factors into k disjoint 2-factors. Since

$k > n+2$, at least one of these 2-factors is disjoint from Γ''. Hence, this 2-factor must be a factor of Γ'. This means that the monomial $\mathbf{m}$ is divisible by the T-invariant monomial $\mathbf{m}_0$ corresponding to the 2-factor of Γ':

$$\mathbf{m} = \mathbf{m}_0 \cdot \mathbf{m}'.$$

Hence, the relation $\mathbf{m}\Delta_{I,J}$ is equal to $\mathbf{m}_0 \cdot (\mathbf{m}'\Delta_{I,J})$, where $\mathbf{m}'\Delta_{I,J} \in \mathrm{Ker}(\Phi_T)$ has smaller degree. □

Conjecture 2.6 A recent conjecture of Andrew Snowden (informal communication) implies that there is a bound $d_0(n)$ such that $R(n;m)$ is generated in degree $\leq d_0(n)$ for all m. Further, after choosing a minimal set of generators (each of degree $\leq d_0(n)$), his conjecture also implies that there is a bound $d_1(n)$ such that the ideal of relations is generated in degree $\leq d_1(n)$ for all m. His conjecture applies to all GIT quotients of the form $X^m//G$, where G is linearly reductive and X is a G-polarized projective variety.

One of our goals was to prove (or perhaps disprove) his conjecture for this case of $\mathbb{P}(V)^m//\mathrm{PO}(V)$. We were not able to do so. However, we have shown that the second Veronese subring is generated in lowest degree, providing small evidence of the first part of his conjecture. Furthermore, Proposition 2.5 is a small step toward proving an m-independent degree bound on the generating set of the ideal (again for the second Veronese subring only).

3 A toric variety

The variety $\mathbb{S}_m = \mathrm{Sym}_m//T = \mathrm{Proj}\ \mathbb{C}[\mathrm{Sym}_m)]^T$ is a toric variety of dimension $m(m-1)/2$. We identify the character lattice of $\mathbb{C}^{*m}$ with $\mathbb{Z}^m$. We have $\mathrm{Sym}_m = \oplus\mathbb{C}X_{ij}$, where X_{ij} is an eigenvector with the character $\mathbf{e}_i + \mathbf{e}_j$. The lattice M of characters of the torus $(\mathbb{C}^*)^{m(m-1)/2}$ acting on $\mathbb{S}_m$ is equal to the kernel of the homomorphism $\mathbb{Z}^{m(m+1)/2} \to \mathbb{Z}^m, \mathbf{e}_{ij} \mapsto \mathbf{e}_i + \mathbf{e}_j$. It is defined by the matrix A with (ij)-spot in a kth row equal:

$$a_{k,ij} = \begin{cases} 0 & \text{if } k \neq i, j; \\ 1 & \text{if } k = i \neq j, \text{or } k = j \neq i; \\ 2 & \text{if } k = i = j. \end{cases}$$

Let

$$S = \{x \in \mathbb{Z}^{m(m+1)/2}_{\geq 0} : Ax = 2d(\mathbf{e}_1 + \ldots + \mathbf{e}_m), \text{for some } d \geq 0\}$$

be the graded semigroup. Then

$$\mathbb{C}[\mathrm{Sym}_m]^T = \mathbb{C}[S].$$

In other words, the toric variety $\mathbb{S}_m$ is equal to the toric space $\mathbb{P}_\Delta$, where Δ_m is the convex polytope in $\{x \in \mathbb{R}^{m(m+1)/2} : Ax = (2, \ldots, 2)\}$ spanned by the vectors $v_\sigma, \sigma \in \mathfrak{S}_m$, such that a_{ij} is equal to the number of edges from i to j in the regular graph corresponding to the determinantal term d_σ. For example, if $m = 3$, $\sigma = (12)$ defines the v_σ with $a_{12} = 2, a_{33} = 1$ and $a_{ij} = 0$ otherwise. Thus the number of lattice points in the polytope Δ is equal to the number $k(m)$ of determinantal terms in a general symmetric matrix.

Proposition 3.1

$$\#(d\Delta_m) \cap M = \#\{\textit{regular graphs with valency } 2d\}.$$

Proof This follows easily from Proposition 2.1. □

4 Examples

Example 4.1 Let $n = 2$ and $m = 3$. We are interested in the moduli space of 3-points in $\mathbb{P}^2$ modulo the group of projective transformations leaving invariant a nonsingular conic. The group $\mathrm{PO}(3) \cong \mathrm{PSL}_2$ is a 3-dimensional group. So, we expect a 3-dimensional variety of configurations.

We have five determinantal terms given by the following graphs:

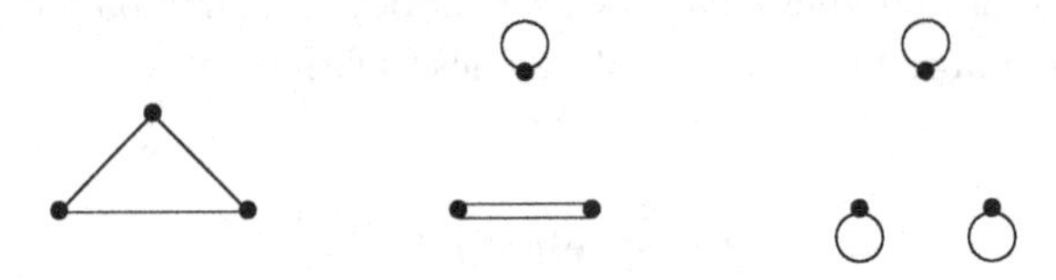

Let t_0, t_1, t_2, t_3, t_4 be generators of the ring $R(2; 3)$ corresponding, respectively, to the triangle, to the three graphs of the second type, and the one graph of the third type. We have the cubic relation

$$t_1t_2t_3 - t_0^2t_4 = 0.$$

Thus our variety is a cubic threefold in $\mathbb{P}^4$. Its singular locus consists of three lines

$$t_0 = t_1 = t_2 = 0, \; t_0 = t_1 = t_3 = 0, \; t_0 = t_2 = t_3 = 0.$$

Let H_i be the hyperplane section of the cubic by the coordinate hyperplane $t_i = 0$. Then

- H_0: point sets with two points conjugate with respect to the fundamental conic. H_0 is the union of three planes $\Lambda_i : t_0 = t_i = 0, i = 1, 2, 3$.
- H_4: one of the points lies on the fundamental conic. It is the union of three planes $\Pi_i : t_4 = t_i = 0, i = 1, 2, 3$.
- H_i is the union of two planes Λ_i and $\Pi_i, i = 1, 2, 3$.

- $\Lambda_i \cap \Lambda_j$ is a singular line on $\mathbb{S}_3$, the locus of point sets where one point is the intersection point of the polar lines of two other points.
- $\Pi_i \cap \Pi_j$: two points are on the fundamental conic.
- $\Pi_i \cap \Lambda_i$: two points are conjugate, the third point is on the conic.
- $\Pi_i \cap \Lambda_j, i \neq j$: one point is on the conic, and another point lies on the tangent to the conic at this point.
- $\Lambda_1 \cap \Lambda_2 \cap \Lambda_3$ is the point representing the orbit of ordered self-conjugate triangles.
- $\Pi_1 \cap \Pi_2 \cap \Pi_3$ is the point representing the orbit of ordered sets of points on the fundamental conic.

The singular point $\Lambda_1 \cap \Lambda_2 \cap \Lambda_3 = [0, 0, 0, 0, 1]$ represents the orbit of ordered self-polar triangles. Recall that unordered self-conjugate triangles are parameterized by the homogeneous space $\mathrm{PO}(3)/\mathfrak{S}_4$. It admits a smooth compactification isomorphic to the Fano threefold of degree 5 and index 2 [11, Theorem (2.1) and Lemma (3.3)] (see also [1, 2.1.3]).

Example 4.2 Let us look at the variety $\mathbf{O}_1^3$. It is isomorphic to the subvariety of $\mathbf{O}_2^3$ representing collinear triples of points. The equation of the determinant of the Gram matrix of three points is

$$2t_0 - t_1 - t_2 - t_3 + t_4 = 0. \tag{4.1}$$

It is a hyperplane section of $\mathbb{S}_3$ isomorphic to a cubic surface S in $\mathbb{P}^3$ with equation

$$t_1t_2t_3 + 2t_0^3 - t_0^2t_1 - t_0^2t_2 - t_0^2t_3 = 0. \tag{4.2}$$

The surface is projectively isomorphic to the 4-nodal Cayley cubic surface given by the equation

$$x_0x_1x_2 + x_0x_2x_3 + x_0x_1x_3 + x_1x_2x_3 = 0.$$

Its singular points are $[t_0, t_1, t_2, t_3] = [1, 1, 1, 1], [0, 0, 0, 1], [0, 0, 1, 0], [0, 1, 0, 0]$. Since the surface is irreducible, and all collinear sets of points satisfy (4.1), we obtain that the surface represents the locus of collinear point sets. It is also isomorphic to the variety $\mathbf{O}_1^3$ of 3-points on $\mathbb{P}^1$. The additional singular point $[1, 1, 1, 1]$ not inherited from the singular locus of $\mathbf{O}_2^3$ is the orbit of three collinear points $[v_1], [v_2], [v_3]$ such that the determinantal terms of the Gram matrix $G(v_1, v_2, v_3)$ are all equal. This is equivalent to all principal minors being equal to zero and the squares of the discriminant terms $d_{(123)}$ and $d_{(321)}$ being equal. This gives two possible points $[t_0, t_1, t_2, t_3] = [\pm 1, 1, 1, 1]$. We check that the point $[-1, 1, 1, 1]$ does not satisfy (4.2). Thus the point $[1, 1, 1, 1]$ is determined by the condition that the principal minors of the Gram matrix

$G(v_1, v_2, v_3)$ are equal to zero. This implies that $[v_1] = [v_2] = [v_3]$. It follows from the stability criterion that this point is not one of the two isotropic points.

It is immediate that $R(1;2)^{(2)}$ is freely generated by two determinantal terms and hence $\mathbf{O}_1^2 \cong \mathbb{P}^1$. The three projections $\mathbf{O}_1^3$ to $\mathbf{O}_1^2$ are a regular map. If we realize $\mathcal{S}_3$ as the image of the anticanonical system of the blow-up of six vertices of a complete quadrilateral in the plane, then the three maps are defined by the linear system of conics through three subsets of four vertices, no three lying on one side of the quadrilateral. We can show that these are the only regular maps from $\mathcal{S}_3$ to $\mathbb{P}^1$.

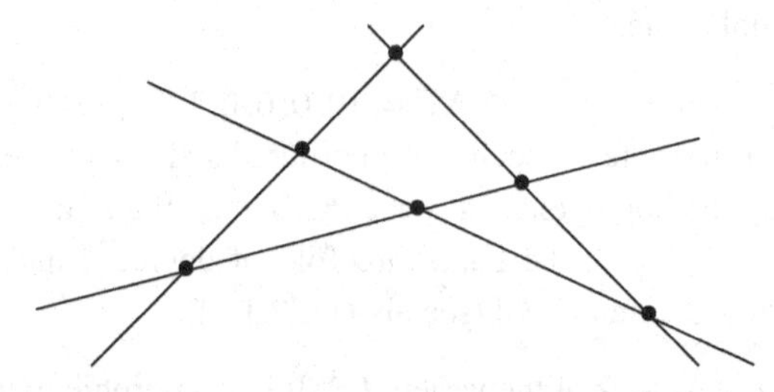

Finally, observe that we can use the conic to identify the plane with its dual plane. In this interpretation a triple of points becomes a triple of lines, the polar lines of the points with respect to the conic. Intersecting each line with the conic, we obtain three ordered pairs of points on a conic.

Note that a set of six distinct points on a nonsingular conic can be viewed as the set of Weierstrass points of a hyperelliptic curve C of genus 2. An order on this set defines a symplectic basis of the $\mathbb{F}_2$-symplectic space $\mathrm{Jac}(C)[2]$ of 2-torsion points of its Jacobian variety $\mathrm{Jac}(C)$. The GIT-quotient of the subvariety of $(\mathbb{P}^2)^6$ of ordered points on a conic by the group SL_3 is isomorphic to the Igusa quartic in $\mathbb{P}^4$ (see [3, Chapter 1, Example 3]). A partition of the set of Weierstrass points in three pairs defines a maximal isotropic subspace in $\mathrm{Jac}(C)[2]$. An order of the three pairs chooses a basis in this space. The moduli space of principally polarized abelian surfaces A equipped with a symplectic basis in $A[2]$ is isomorphic to the quotient of the Siegel space $\mathcal{Z}_2 = \{X \in \mathrm{Sym}_4 : \mathrm{Im}(X) > 0\}$ by the group $\Gamma(2) = \{M \in \mathrm{Sp}(4,\mathbb{Z}) : A \equiv I_4 \bmod 2\}$. The moduli space of principally polarized abelian surfaces, together with a choice of a basis in a maximal isotropic subspace of 2-torsion points, is isomorphic to the quotient of $\mathcal{Z}_2$ by the group $\Gamma_1(2) = \{M = \begin{pmatrix} A & B \\ C & D \end{pmatrix} : A - I_2 \equiv C \equiv 0 \bmod 2\}$. Thus, we obtain that our variety $\mathbf{O}_2^3$ is naturally birationally isomorphic to the quotient $\mathcal{Z}_2/\Gamma_1(2)$ and this variety is isomorphic to the quotient of $\mathcal{Z}_2/\Gamma(2)$ by the group $G = \mathfrak{S}_2 \times \mathfrak{S}_2 \times \mathfrak{S}_2$. The Satake compactification of $\mathcal{Z}_2/\Gamma(2)$ is isomorphic to the Igusa quartic. In [12], Mukai shows that the Satake compactification of $\mathcal{Z}_2/\Gamma_1(2)$ is isomorphic to the double cover of $\mathbb{P}^3$ branched along the union of 4-coordinate hyperplanes. It is easy to see that it is birationally isomorphic to the cubic hypersurface defining $\mathbf{O}_2^3$. A remarkable result of Mukai is that the Satake compactifications of $\mathcal{Z}_2/\Gamma(2)$ and $\mathcal{Z}_2/\Gamma_1(2)$ are isomorphic.

Remark 4.3 Assume $m = n + 1$. Fix a volume form on V and use it to identify the linear spaces V^* and $\bigwedge^n V$. This identification is equivariant with respect to the action of $\mathrm{O}(V)$ on V and $\mathrm{O}(V^*)$, where the orthogonal group of V^* is with respect to the dual quadratic form on V^*. Passing to the configuration spaces, we obtain a natural birational involution $F : \mathbf{O}_n^{n+1} \dashrightarrow \mathbf{O}_n^{n+1}$. If G is the Gram matrix of vectors $v_1, \ldots, v_{n+1}$, then the Gram matrix G^* of the vectors $w_i = v_1 \wedge \cdots \wedge v_{i-1} \wedge v_{i+1} \wedge \cdots \wedge v_{n+1} \in V^*$ is equal to the adjugate matrix of G (see [1, Lemma 10.3.2]). In the case $n = 2$, the birational involution corresponds to the involution defined by conjugate triangles (see [1, 2.1.4]). Using the modular interpretation of $\mathbf{O}_2^3$ from the previous example, the involution F corresponds to the Fricke (or Richelot) involution of $\mathcal{Z}_2/\Gamma_1(2)$ (see [12, Theorem 2]).

Example 4.4 Now let us consider the variety O_1^4 of 4-points in $\mathbb{P}^1$ modulo $\mathrm{PO}(2) \cong \mathbb{C}^* \times \mathbb{Z}/2\mathbb{Z}$. It is another threefold. First we get the 5-dimensional toric variety of symmetric matrices of size 4. The coordinate ring is generated by 17 (3+4+6+3+1) determinantal terms:

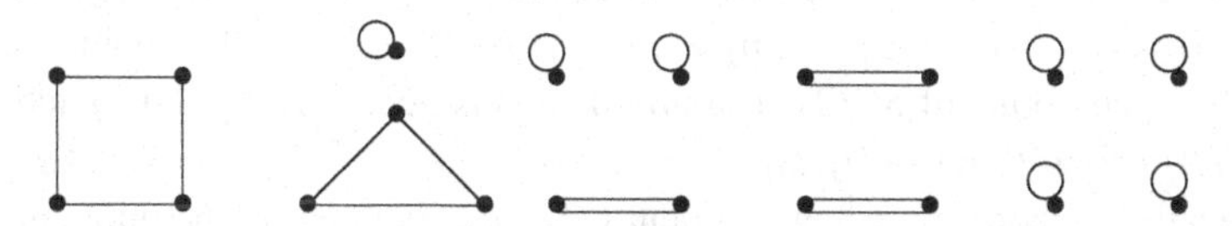

Let $x_1, x_2, x_3, y_1, y_2, y_3, y_4, z_1, \ldots, z_6, u_1, u_2, u_3, v$ be the variables. We have additional equations expressing the condition that the rank of matrices is less than or equal to 2. We can show that the equations are all linear:

$$a_{ij} \det A_{ij} = 0.$$

Their number is equal to 10 but there are three linear dependencies found by expanding the determinant expression along columns.

We may also consider the spaces $\mathbf{SO}_n^m = \mathbb{P}(V)^m // \mathrm{O}^+(V)$, where $\mathrm{O}^+(V) = \mathrm{O}(V) \cap \mathrm{SL}(V)$ is the special orthogonal group. Note that $\mathrm{PO}(V) \cong \mathrm{PO}^+(V)$ if $\dim V$ is odd. Thus we will be interested only in the case when $\dim V$ is even. In this case $\mathrm{PO}^+(V)$ is a subgroup of index 2 in $\mathrm{PO}(V)$, so the variety $\mathbf{SO}_n^m$ is a double cover of $\mathbf{O}_n^m$. We have

$$\mathbf{SO}_n^m = \operatorname{Proj} R^+(n; m),$$

where $R^+(n;m) = \oplus_{d=0}^{\infty}(S^d(V)^{*\otimes m})^{O^+(V)}$. There are more invariants now. The additional invariants in $\mathbb{C}[V^m]^{O^+(V)}$ are the Plücker brackets

$$p_{i_1,\ldots,i_{n+1}} : (v_1,\ldots,v_m) \mapsto v_{i_1} \wedge \ldots \wedge v_{i_{n+1}},$$

where we have fixed a volume form on V. There are additional basic relations (see [19, Chapter 2, Section 17])

$$p_{i_1,\ldots,i_{n+1}} p_{j_1,\ldots,j_{n+1}} - \det([i_\alpha, i_\beta])_{1\le\alpha,\beta\le n+1} = 0, \tag{4.3}$$

$$\sum_{j=1}^{n+1}(-1)^j \sum p_{i_1,\ldots,\hat{i_j},\ldots,i_{n+1}}[i_j, i_{n+2}] = 0. \tag{4.4}$$

The graded part $R^+(n;m)_d$ is spanned by the monomials $p_{I_1}\cdots p_{I_k}[i_1 j_1]\cdots[i_s, j_s]$, where each index $j \in [1,m]$ appears exactly d times. Using the first relation in (4.3), we may assume that at most one Plücker coordinate p_I appears. Also, we see that the product of any two elements in $R^+(n;m)_d$ belongs to $R(N;m)_{2d}$.

5 Points in $\mathbb{P}^1$ and generalized Cayley cubics

The group SO(2) is isomorphic to the 1-dimensional complex torus $\mathbb{C}^*$. We choose projective coordinates in $\mathbb{P}^1$ to identify a quadric in $\mathbb{P}^1$ with the set $Q = \{0,\infty\}$ so that SO(2) acts by $\lambda : [t_0,t_1] \mapsto [\lambda t_0, \lambda^{-1}t_1]$. The points on Q are the fixed points of SO(2). The group O(2) is generated by SO(2) and the transformation $[t_0,t_1] \mapsto [t_1,t_0]$.

Recall that there is a Chow quotient $(\mathbb{P}^1)^m//\mathbb{C}^*$ defined by the quotient fan of the toric variety $(\mathbb{P}^1)^m$ (see [6]).

Lemma 5.1 *Consider $(\mathbb{P}^1)^m$ as a toric variety, the Cartesian product of the toric varieties $\mathbb{P}^1$. Then the Chow quotient $(\mathbb{P}^1)^m//\mathrm{SO}(2)$ is isomorphic to the toric variety $X(A_{m-1})$ associated with the root system of type A_{m-1} defined by the fan in the dual lattice of the root lattice of type A_{m-1} formed by the Weyl chambers.*

Proof The toric variety $(\mathbb{P}^1)^m$ is defined by the complete fan Σ in the lattice $\mathbb{Z}^m$ with 1-skeleton formed by the rays $\mathbb{R}_{\ge0}e_i$ and $\mathbb{R}_{\le0}e_i$, $i = 1,\ldots,m$. The action of SO(2) on the torus $(\mathbb{C}^*)^m \subset (\mathbb{P}^1)^m$ is defined by the surjection of the lattices $\mathbb{Z}^m \to \mathbb{Z}$ given by the map $e_i \to e_1, i = 1,\ldots,m$. Thus the lattice M of characters of the torus $(\mathbb{C}^*)^{m-1}$ acting on $(\mathbb{C}^*)^m/\mathbb{C}^*$ can be identified with the sublattice of $\mathbb{Z}^m$ spanned by the vectors $\alpha_1 = e_1 - e_2,\ldots,\alpha_{m-1} = e_{m-1} - e_m$. This is the root lattice of type A_{m-1}. The dual lattice N is the lattice $\mathbb{Z}^m/\mathbb{Z}e$, where $e = e_1 + \ldots + e_m$. The quotient fan is defined as follows. For any coset $y \in N_{\mathbb{R}} = \mathbb{R}^m/\mathbb{R}e$, one considers the set

$$\mathcal{N}(y) = \{\sigma \in \Sigma : y + \mathbb{R}e \cap \sigma \neq \emptyset\}.$$

A coset $y + \mathbb{R}e \in N_{\mathbb{R}}$ is called admissible if $\mathcal{N}(\phi) \neq \emptyset$. Two admissible cosets $y + \mathbb{R}e$ and $y' + \mathbb{R}e$ are called equivalent if $\mathcal{N}(y) = \mathcal{N}(y')$. The closure of each equivalence class of admissible cosets is a rational polyhedral convex cone in $N_{\mathbb{R}}$ and the set of such cones defines a fan Σ' in $N_{\mathbb{R}}$ which is the quotient fan.

In our case, Σ consists of open faces σ_I of the 2^m m-dimensional cones

$$\sigma_{I,J} = \{(x_1, \dots, x_m) \in \mathbb{R}^m : (-1)^{\delta_I} x_i \geq 0, (-1)^{\delta_J} x_i \leq 0\},$$

where I, J are subsets of $[1, m]$ such that $I \cup J = [1, m]$ and δ_K is a delta-function of a subset K of $[1, m]$. The cones of maximal dimension correspond to pairs of complementary subsets I, J. The k-dimensional cones correspond to the pairs I, J with $\#I \cap J = m - k$.

Let $y = (y_1, \dots, y_m) \in \mathbb{R}^m$ with $y_i \neq y_j, i \neq j$ and let $s \in \mathfrak{S}_m$ be a unique permutation such that $y_{s(1)} > y_{s(2)} > \dots > y_{\sigma(m)}$. Then $y + \mathbb{R}e$ intersects $\sigma_{I,J}$ if and only if $s(I) = \{1, \dots, k\}$ for some $k \leq m$ or $\emptyset$ and $J = [1, m] \setminus I$. Since $\mathfrak{S}_m$ has only one orbit on the set of pairs of complementary subsets of $[1, m]$, we see that the interiors of maximal cones in the quotient fan are obtained from the image of the subset

$$\{y \in \mathbb{R}^m : y \cdot (e_i - e_{i+1}) \geq 0, i = 1, \dots, m - 1\}$$

in $N_{\mathbb{R}}$. This is exactly one of the Weyl chambers in $N_{\mathbb{R}}$. All other cones in the quotient fan are translates of the faces of the closure of this chamber. This proves the assertion. □

It is known that the toric variety $X(A_{m-1})$ is isomorphic to the blow-up of $\mathbb{P}^{m-1}$ of the faces of the coordinate simplex (see, e.g., [2, Lemma 5.1]). Let

$$\tau_{m-1} : \mathbb{P}^{m-1} \dashrightarrow \mathbb{P}^{m-1}, \ [t_0, \dots, t_{m-1}] \mapsto [1/t_0, \dots, 1/t_{m-1}]$$

be the standard Cremona transformation of $\mathbb{P}^{m-1}$. The variety $X(A_{m-1})$ is isomorphic to a minimal resolution of indeterminacy points of the standard involution (see [1, Example 7.2.5]). Equivalently, $X(A_{m-1})$ is isomorphic to the closure of the graph of τ_{m-1} in $\mathbb{P}^{m-1} \times \mathbb{P}^{m-1}$. It is given by the 2×2-minors of the matrix

$$\begin{pmatrix} F_0(x) & F_1(x) & \dots & F_{m-1}(x) \\ y_0 & y_1 & \dots & y_{m-1} \end{pmatrix},$$

where $F_i(x) = (x_0 \cdots x_{m-1})/x_i$. It follows from this formula that the standard involution τ_{m-1} of $X(A_{m-1})$ is induced by the switching involution ι of the factors of $\mathbb{P}^{m-1} \times \mathbb{P}^{m-1}$. The image of composition of the embedding $X(A_{m-1})$ in $\mathbb{P}^{m-1} \times \mathbb{P}^{m-1}$ and the Segre embedding $\mathbb{P}^{m-1} \times \mathbb{P}^{m-1} \hookrightarrow \mathbb{P}^{m^2-1}$ is equal to the intersection of the Segre variety with the linear subspace of dimension $m-1$ defined by

$$t_{00} = t_{11} = \ldots = t_{m-1m-1}, \tag{5.1}$$

where we use the coordinates $t_{ij} = x_i y_j$ in $\mathbb{P}^{m^2-1}$. So $X(A_{m-1})$ is isomorphic to a closed smooth subvariety of $\mathbb{P}^{m(m-1)}$ of degree $\binom{2(m-1)}{m-1}$.

Consider the embedding of $(\mathbb{P}^{m-1} \times \mathbb{P}^{m-1})/\langle\iota\rangle$ in $\mathbb{P}^{\frac{1}{2}(m+2)(m-1)}$ given by the linear system of symmetric divisors of type $(1,1)$. Its image is equal to the secant variety of the Veronese variety $v_2(\mathbb{P}^{m-1})$ isomorphic to the symmetric square $\mathrm{Sym}^2\,\mathbb{P}^{m-1}$ of $\mathbb{P}^{m-1}$. The image of $X(A_{m-1})/\langle\tau_{m-1}\rangle$ in $\mathbb{P}^{\frac{1}{2}(m+2)(m-1)}$ is equal to the intersection of the secant variety with a linear subspace L of codimension $m-1$ given by (5.1). It is known that the singular locus of the secant variety is equal to the Veronese variety. The singular locus of the embedded $X(A_{m-1})/\langle\tau_{m-1}\rangle$ is equal to the intersection of L with the Veronese subvariety $v_2(\mathbb{P}^{m-1})$ and consists of 2^{m-1} points. We have $\dim L = \frac{1}{2}m(m-1)$ and $\deg \mathrm{Sym}^2\,\mathbb{P}^{m-1}) = \frac{1}{2}\binom{2(m-1)}{m-1}$. So $X(A_{m-1})/\langle\tau_{m-1}\rangle$ embeds into $\mathbb{P}^{\frac{1}{2}m(m-1)}$ as a subvariety of degree $\frac{1}{2}\binom{2(m-1)}{m-1}$ with 2^{m-1} singular points locally isomorphic to the singular point of the cone over the Veronese variety $v_2(\mathbb{P}^{m-2})$. We call it the *generalized Cayley cubic* and denote it by Cay_{m-1}.

It follows from above that Cay_{m-1} is isomorphic to the subvariety of the projective space of symmetric $m \times m$ matrices with the conditions that the rank is equal to 2 and the diagonal elements are equal.

In the case when $m = 3$, the variety $X(A_2)$ is a del Pezzo surface of degree 6, the blow-up of three non-collinear points in $\mathbb{P}^2$, and Cay_2 is isomorphic to the Cayley 4-nodal cubic surface in $\mathbb{P}^3$. The variety Cay_3 is a 3-dimensional subvariety of $\mathbb{P}^6$ of degree 10 with eight singular points locally isomorphic to the cone over the Veronese surface.

It is known that the Chow quotient birationally dominates all the GIT-quotients [7, Theorem (0.4.3)]. So we have a $\mathfrak{S}_m$-equivariant birational morphism

$$\Phi_m : X(A_{m-1}) \to \mathbf{SO}_1^m$$

which, after dividing by the involution τ_{m-1}, defines a $\mathfrak{S}_m$-equivariant birational morphism

$$\Phi_m^c : \mathrm{Cay}_{m-1} \to \mathbf{O}_1^m.$$

For example, take $m = 3$. The variety Cay_2 is the Cayley 4-nodal cubic, the morphism Φ_m is an isomorphism. Take $m = 4$. We know from Proposition

2.4 that the variety $\mathbf{SO}_1^4$ has six singular points corresponding to strictly semi-stable points defined by vanishing of two complementary principal matrices of the Gram matrix. They are represented by the point sets of the form (a, a, b, b), where $a, b \in \{0, \infty\}$. The morphism Φ_4 resolves these points with the exceptional divisors equal to the exceptional divisors of $X(A_3) \to \mathbb{P}^3$ over the edges of the coordinate tetrahedron. The morphism Φ_4^c resolves three singular points of $\mathbf{O}_4^1$ and leaves unresolved the eight singular points coming from the fixed points of τ_3. Altogether, the variety $\mathbf{O}_4^1$ has 11 singular points: eight points locally isomorphic to the cones of the Veronese surface and three conical double points. The latter three singular points correspond to strictly semi-stable orbits.

It is known that $X(A_{m-1})$ is isomorphic to the closure of a general maximal torus orbit in PGL_m /B, where B is a Borel subgroup [9, Theorem 1]. Let P be a parabolic subgroup containing B defined by a subset S of the set of simple roots, and W_S be the subgroup of the Weil group $\mathfrak{S}_m$ generated by simple roots in S. Let $\phi_S : \mathrm{PGL}_m /B] \to \mathrm{PGL}_m /P$ be the natural projection. The image of $X(A_{m-1})$ in PGL_m /P is a toric variety $X(A_{m-1})_S$ defined by the fan whose maximal cones are $\mathfrak{S}_m$-translates of the cone $W_S\sigma$, where σ is a fundamental chamber ([4, Theorem 1]). The morphism $\phi_S : X(A_{m-1}) \to X(A_{m-1})_S$ is a birational morphism which is easy to describe.

We believe, but could not find a proof, that for odd m, the morphism Φ_m and Φ_m^c are isomorphisms. If m is even, then the morphism Φ_m is equal to the morphism ϕ_S, where S is the complement of the central vertex of the Dynkin diagram of type A_{m-1}.

6 Rational functions

First we shall prove the rationality of our moduli spaces.

Theorem 6.1 *The varieties $\mathbf{O}_n^m$ are rational varieties.*

Proof The assertion is trivial when $m = n + 1$ because in this case the variety is isomorphic to the toric variety $\mathbb{S}_m$. If $m < n + 1$, a general point set spans $\mathbb{P}(W)$, where W is a subspace of $\mathbb{P}(V)$ of dimension m. Since $\mathrm{O}(V)$ acts transitively on a dense orbit of the Grassmannian $G(m, V)$ (the subspaces containing an orthogonal basis), we may transform a general set to a subset of a fixed $\mathbb{P}(W)$. This shows that the varieties $\mathbf{O}_n^m$ and $\mathbf{O}_{m-1}^m$ are birationally isomorphic. If $m > n+1$, we use the projection map $\mathbf{O}_n^m \dashrightarrow \mathbf{O}_n^{m-1}$ onto the first $m-1$ factors. It is a rational map with general fiber isomorphic to $\mathbb{P}^n$. Its geometric generic fiber is isomorphic to the projective space over the algebraic closure of the field

K of rational functions of $\mathbf{O}_n^{m-1}$. In other words, the generic fiber is a Severi–Brauer variety over K (see [17, Chapter X, Section 6]). The rational map has a rational section $(x_1, \ldots, x_{m-1}) \mapsto (x_1, \ldots, x_{m-1}, x_{m-1})$. Thus the generic fiber is a Severi–Brauer variety with a rational point, hence isomorphic to the projective space over K (see [17, Exercise 1]). Thus the field of rational functions on $\mathbf{O}_n^m$ is a purely transcendental extension of K, and by induction on m, we obtain that $\mathbf{O}_n^m$ is rational. □

We know that the ring $R(n;m)^{(2)}$ is generated by determinantal terms d_σ of the Gram matrix of m points. If we take σ to be a transposition (ab), then the ratio $d_\sigma / d_{(12\ldots m)}$ is equal to

$$R_{ab} = [ab]^2/[aa][bb]. \tag{6.1}$$

More generally, for any cyclic permutation $\sigma = (a_1 \ldots a_k)$ we can do the same to obtain the rational invariant function

$$R_{a_1\ldots a_k} = \frac{[a_1a_2]\cdots[a_{k-1}a_k][a_{k-1}a_k][a_ka_1]}{[a_1a_1]\cdots[a_ka_k]}. \tag{6.2}$$

Writing any permutation as a product of cycles, we see that the field of rational functions on $\mathbf{O}_n^m$ is generated by functions $R_{a_1\ldots a_k}$. Note that

$$R^2_{a_1\ldots a_k} = R_{a_1a_2}\cdots R_{a_ka_1}.$$

We do not know whether a transcendental basis of the field can be chosen among the functions $R_{a_1\ldots a_k}$ or their ratios.

7 Complex spheres

An $(n-1)$-dimensional sphere is given by an equation in $\mathbb{R}^n$ of the form

$$\sum_{i=1}^n (x_i - a_i)^2 = R^2.$$

After homogenizing, we get the equation in $\mathbb{P}^n(\mathbb{R})$:

$$\mathsf{Q} : \sum_{i=1}^n (x_i - a_ix_0)^2 - R^2x_0^2 = 0. \tag{7.1}$$

The hyperplane section $x_0 = 0$ is a sphere in $\mathbb{P}^{n-1}(\mathbb{R})$ with equation

$$\mathsf{Q}_0 : \sum_{i=1}^n x_i^2 = 0. \tag{7.2}$$

The quadric has no real points, and for this reason it is called the imaginary sphere. Now we abandon the real space and replace $\mathbb{R}$ with $\mathbb{C}$. Equation (7.1) defines a *complex sphere*. A coordinate-free definition of a complex sphere is a nonsingular quadric hypersurface Q in $\mathbb{P}^n$ intersecting a fixed hyperplane H_0 along a fixed nonsingular quadric $\mathcal{Q}_0$ in H_0. In the real case, we additionally assume that $\mathrm{Q}_0(\mathbb{R}) = \emptyset$. If we choose coordinates such that $\mathcal{Q}_0$ is given by equation (7.2), then a quadric in $\mathbb{P}^{n+1}(\mathbb{C})$ containing the imaginary sphere has an equation

$$b\left(\sum_{i=1}^{n} x_i^2\right) - 2x_0\left(\sum_{i=0}^{n} a_i x_i\right) = 0.$$

If $b \neq 0$, we may assume that $b = 1$ and rewrite the equation in the form

$$\sum_{i=1}^{n}(x_i - a_i x_0)^2 - (2a_0 + \sum_{i=1}^{n} a_i^2)x_0^2 = 0,$$

so it is a complex sphere. Consider the rational map given by the linear system of quadrics in $\mathbb{P}^n$ containing the fixed quadric Q_0 with equation (7.2). We can choose a basis formed by the quadric Q_0 and the quadrics $V(x_0x_i), i = 0, \ldots, n$. This defines a rational map $\mathbb{P}^n \dashrightarrow \mathbb{P}^{n+1}$ given by the formulas

$$[x_0, \ldots, x_n] \mapsto [t_0, \ldots, t_{n+1}] = \left[\sum_{i=1}^{n} x_i^2, x_0x_1, \ldots, x_0x_n, x_0^2\right].$$

The image of this map is a nonsingular quadric in $\mathbb{P}^{n+1}$ given by the equation $\mathcal{Q} = V(q)$, where

$$q = -t_0t_{n+1} + \sum_{i=1}^{n+1} t_i^2 = 0. \tag{7.3}$$

We call $\mathcal{Q}$ the *fundamental quadric*. The quadratic form q defines a symmetric bilinear form on V whose value on vectors $v, w \in V$ is denoted by $\langle v, w\rangle$. The pre-image of a hyperplane section $\sum A_i t_i = 0$ is a complex sphere, or its degeneration. For example, the sphere corresponding to a hyperplane which is tangent to the quadric has zero radius, and hence it is defined by a singular quadric.

The idea of replacing a quadratic equation of a sphere by a linear equation goes back to Moebius and Chasles in 1850, but was developed by Klein and Lie 20 years later. The spherical geometry, as it is understood in Klein's Erlangen program, becomes isomorphic to the orthogonal geometry. More precisely, the *inversive group* of birational transformations of $\mathbb{P}^n$ sending spheres to spheres or their degenerations is isomorphic to the projective orthogonal group $\mathrm{PO}(n + 2)$.

Let us use the quadric $\mathfrak{Q}$ to define a *polarity duality* between points and hyperplanes in $\mathbb{P}^{n+1}$. If we use the equation of $\mathfrak{Q}$ to define a symmetric bilinear form in $\mathbb{C}^{n+2}$, the polarity is just the orthogonality of lines and hyperplanes with respect to this form. Under the polarity, hyperplanes become points, and hence spheres in $\mathbb{P}^n$ can be identified with points in $\mathbb{P}^{n+1}$.

Explicitly, a point $\alpha = [\alpha_0, \dots, \alpha_{n+1}] \in \mathbb{P}^{n+1}$ defines the sphere

$$S(\alpha) : \alpha_0(\sum_{i=1}^{n} x_i^2) - 2\sum_{i=1}^{n} \alpha_i x_0 x_i + \alpha_{n+1}x_0^2 = 0. \tag{7.4}$$

By definition, its *center* is the point $c = [\alpha_0, \alpha_1, \dots, \alpha_n]$, its *radius square* R^2 is defined by the formula

$$\alpha_0^2 R^2 = \sum_{i=1}^{n} \alpha_i^2 - \alpha_0\alpha_{n+1} = q_0(\alpha_0, \dots, \alpha_{n+1}). \tag{7.5}$$

Computing the discriminant D of the quadratic form in (7.4), we find

$$D = \alpha_0^{n-1}(\alpha_0\alpha_{n+1} - \sum_{i=1}^{n} \alpha_i^2) = \alpha_0^{n-1} q_0. \tag{7.6}$$

This proves the following:

Proposition 7.1 *A complex sphere $S(\alpha)$ is singular if and only if its radius-square is equal to zero, or, equivalently, the point $\alpha \in \mathbb{P}^{n+1}$ lies on the fundamental quadric $\mathfrak{Q}$. The center of a singular complex sphere is its unique singular point.*

Remark 7.2 Spheres of radius zero are points on the fundamental quadric. Thus the spaces $\mathbf{O}_n^m$ contain as its closed subsets the moduli space of m points on the fundamental quadric modulo the automorphism group of the quadric. For example, when $n = 2$, this is the moduli space $P_1^m = (\mathbb{P}^1)^m // \mathrm{SL}(2)$ studied intensively in many papers (see, e.g., [3, 5]).

Many geometrically mutual properties of complex spheres are expressed by vanishing of some orthogonal invariant of point sets in $\mathbb{P}^{n+1}$. We give here only some simple examples.

We define two complex spheres in $\mathbb{P}^n$ to be *orthogonal* to each other if the corresponding points in $\mathbb{P}^{n+1}$ are *conjugate* in the sense that one point lies on the polar hyperplane to another point.

Proposition 7.3 *Two real spheres in $\mathbb{R}^n$ are orthogonal to each other (i.e., the radius-vectors at their intersection points are orthogonal) if and only if the corresponding complex spheres are orthogonal in the sense of the previous definition.*

Proof Let

$$\sum_{i=1}^{n}(x_i - a_i)^2 = r^2, \quad \sum_{i=1}^{n}(x_i - b_i)^2 = r'^2$$

be two orthogonal spheres. Let $(c_1, \ldots, c_n)$ be their intersection point. Then we have

$$0 = \sum_{i=1}^{n}(c_i - a_i)(c_i - b_i) = \sum_{i=1}^{n} c_i^2 - \sum_{i=1}^{n}(a_i + b_i)c_i + \sum_{i=1}^{n} a_i b_i,$$

$$\sum_{i=1}^{n}(c_i - a_i)^2 = r^2, \quad \sum_{i=1}^{n}(c_i - b_i)^2 = r'^2.$$

This gives the equality

$$2\sum_{i=1}^{n} a_i b_i - \sum_{i=1}^{n} a_i^2 - \sum_{i=1}^{n} b_i^2 + r^2 + r'^2 = 0.$$

It gives a necessary and sufficient condition that two spheres intersect orthogonally. It is clear that the condition does not depend on the choice of intersection point. The corresponding complex spheres correspond to points $[1, a_1, \ldots, a_n, \frac{1}{2}(r^2 - \sum a_i^2)]$ and $[1, b_1, \ldots, b_n, \frac{1}{2}(r^2 - \sum b_i^2)]$. The condition that two points $[\alpha_0, \ldots, \alpha_{n+1}]$ and $[\beta_0, \ldots, \beta_{n+1}]$ are conjugate is

$$\alpha_0\beta_{n+1} + \alpha_{n+1}\beta_0 + \sum_{i=1}^{n} \alpha_i\beta_i = 0.$$

So we see that the two conditions agree. □

For convenience of notation, we denote $x = \mathbb{C}v \in \mathbb{P}^n$ by $[v]$. We use the symmetric form $\langle v, w\rangle$ in V defined by the fundamental quadric.

We have learnt the statements of the following two propositions from [10].

Proposition 7.4 *Two complex spheres $S([v])$ and $S([w])$ are tangent at some point if and only if*

$$\det\begin{pmatrix}\langle v, v\rangle & \langle v, w\rangle \\ \langle w, v\rangle & \langle w, w\rangle\end{pmatrix} = 0.$$

Proof Let $\lambda\phi_1 + \mu\phi_2$ be a 1-dimensional space of quadratic forms in V and $V(\lambda\phi_1 + \mu\phi_2)$ be the corresponding pencil of quadrics in $\mathbb{P}^n$. We assume that it contains a nonsingular quadric. Then the equation $\mathrm{discr}(\lambda\phi_1 + \mu\phi_2) = 0$ is a homogeneous form of degree $n + 1$ whose zeros define singular quadrics in the pencil. The quadrics $V(\phi_1)$ and $V(\phi_2)$ are tangent at some point p if and only if p is a singular point of some member of the pencil. It is well known that the corresponding root $[\lambda, \mu]$ of the discriminant equation is of higher multiplicity. If $V(\phi_1) = S([v])$ and $V(\phi_2) = S([w])$ are nonsingular complex spheres, then the pencil $V(\lambda q_1 + \mu q_2)$ corresponds to the line $\overline{x, y}$ in $\mathbb{P}^{n+1}$ spanned by the points x and y. A point $[\lambda v + \mu w]$ on the line defines a singular quadric if and only if

$$D(\lambda v + \mu w) = (\lambda \alpha_{n+1} + \mu \beta_{n+1})^{n-1} q_0(\lambda v + \mu w) = 0.$$

Our condition for quadrics $V(\phi_1)$ and $V(\phi_2)$ to be tangent to each other is that the equation $q_0(\lambda v + \mu w) = 0$ has a double root. We have

$$q_0(\lambda v + \mu w) = \lambda^2 \langle v, v\rangle + 2\lambda\mu\langle v, w\rangle + \mu^2\langle w, w\rangle.$$

Thus the condition becomes

$$\det\begin{pmatrix} \langle v, v\rangle & \langle v, w\rangle \\ \langle w, v\rangle & \langle w, w\rangle \end{pmatrix} = 0.$$

□

Proposition 7.5 *$n + 1$ complex spheres $S([v_i])$ in $\mathbb{P}^n$ have a common point if and only if*

$$\det\begin{pmatrix} \langle w_1, w_1\rangle & \dots & \langle w_1, w_{n+1}\rangle \\ \langle w_2, w_1\rangle & \dots & \langle w_2, w_{n+1} \\ \vdots & \vdots & \vdots \\ \langle w_{n+1}, w_1\rangle & \dots & \langle w_{n+1}, w_{n+1}\rangle \end{pmatrix} = 0,$$

where w_i are the vectors of coordinates of the polar hyperplane of $[v_i]$.

Proof We use the following known identity in the theory of determinants (see, e.g., [1, Lemma 10.3.2]). Let $A = (a_{ij}), B = (b_{ij})$ be two matrices of sizes $k \times m$ and $m \times k$ with $k \leq m$. Let $|A_I|, |B_I|, I = (i_1, \dots, i_k), 1 \leq i_1 < \dots < i_k \leq m$, be maximal minors of A and B. Then

$$|A \cdot B| = \sum_I |A_I||B_I|. \tag{7.7}$$

Let $H_i := \sum_{j=0}^{n+1} a_j^{(i)} t_j = 0$ be the polar hyperplanes of the complex spheres. We may assume that they are linearly independent, i.e., the vectors w_i are linearly independent in V. Otherwise the determinant is obviously equal to zero. Thus the hyperplanes intersect at one point. The spheres have a common point if and only if the intersection point of the hyperplanes H_i lies on the fundamental quadric. Let X be the matrix with rows equal to vectors $w_i = (a_0^{(i)}, \dots, a_{n+1}^{(i)})$. The intersection point has projective coordinates $[C_1, -C_2, \dots, (-1)^{n+1} C_{n+2}]$, where C_j is the maximal minor obtained from X by deleting the jth column. Let G be the symmetric matrix defining the fundamental quadric. We take in the above formula $A = X \cdot G, B = {}^tX$. Then the product $A \cdot B$ is equal to the LHS of the formula in the assertion of the proposition. The RHS is equal to $\pm(C_1C_{n+2} - \sum_{i=2}^{n+1} C_i^2)$. It is equal to zero if and only if the intersection point lies on the fundamental quadric. □

We refer to [18] for many other mutual geometrical properties of circles expressed in terms of invariants of the orthogonal group O(4).

8 Real points

We choose V to be a real vector space equipped with a positive definite inner product $\langle -, - \rangle$. A real point in $\mathbb{P}(V)$ is represented by a nonzero vector $v \in V$. Since $\langle v, v \rangle > 0$, we obtain from Propositions 2.3 and 2.4 that all real point sets $(x_1, \ldots, x_m)$ are stable points. Another nice feature of real point sets is the criterion for vanishing of the Gram functions: $\det G(v_1, \ldots, v_k) = 0$ if and only if $v_1, \ldots, v_k$ are linear dependent vectors in V.

It follows from the FFT and SFT that the varieties $\mathbf{O}_n^m$ are defined over $\mathbb{Q}$. In particular, we may speak about the set $\mathbf{O}_N^m(\mathbb{R})$ of their real points.

Theorem 8.1 *Let V be a real inner-product space. Consider the open subset U of linear independent point sets $(x_1, \ldots, x_m)$ in $\mathbb{P}(V)(\mathbb{R})$. Then the map*

$$U \to \mathbf{O}_n^m(\mathbb{R})$$

is injective.

Proof To show the injectivity of the map, it suffices to show that

$$(g(x_1), \ldots, g(x_m)) = (y_1, \ldots, y_m) \tag{8.1}$$

for $g \in \mathrm{PO}(V_\mathbb{C})$ implies that $(g(x_1), \ldots, g(x_m) = (y_1, \ldots, y_m)$ for some $g' \in \mathrm{PO}(V)$. Choose an orthonormal basis in V to identify V with the Euclidean real space $\mathbb{R}^{n+1}$. The transformation g is represented by a complex orthogonal matrix. If (8.1) holds, we can find some representatives v_i and w_i of points x_i, y_i, respectively, and a matrix $A \in \mathrm{O}(V_\mathbb{C})$ such that $A \cdot v_i = w_i, i = 1, \ldots, m$. This is an inhomogeneous system of linear equations in the entries of A. Since the rank of the matrix $[v_1, \ldots, v_n]$ with columns v_i is maximal, there is a unique solution for A and it is real. Thus g is represented by a transformation from $\mathrm{O}(n + 1, \mathbb{R})$. □

Let us look at the rational invariants $R_{a_1,\ldots,a_k}$. Let $\phi_{ij}, \pi - \phi_{ij}$, denote the angles between basis vectors of the lines $x_i = \mathbb{R}v_i$. Obviously,

$$R_{ij} = \cos^2 \phi_{ij}$$

is well defined and does not depend on the choice of the bases. Also,

$$R_{ij...k} = \cos\phi_{ij} \cdots \cos\phi_{ki}$$

are well defined too. Applying the previous theorem, we see that the cyclic products of the cosines determine uniquely the orbit of a linearly independent point set.

Finally, let us discuss configuration spaces of real spheres. For this we have to choose $V = \mathbb{R}^{n+2}$ to be a real space with quadratic form q_0 of signature $(1, n)$ defined in (7.3). A real sphere with nonzero radius is defined by formula (7.4), where the coefficients $(\alpha_0, \ldots, \alpha_{n+1})$ belong to the set $q_0^{-1}(\mathbb{R}_{>0})$. It consists of two connected components corresponding to the choice of the sign of α_0. Choose the component V^+ where $\alpha_0 > 0$. The image $V^+/\mathbb{R}_{>0}$ of V^+ in the projective space $\mathbb{P}(V^+)$ is, by definition, the *hyperbolic space* $\mathbb{H}^{n+1}$. Each point in $\mathbb{H}^{n+1}$ can be uniquely represented by a unique vector $v = (\alpha_0, \ldots, \alpha_{n+1})$ with

$$\langle v, v\rangle = \sum_{i=1}^{n+1} \alpha_i^2 - \alpha_0\alpha_{n+1} = 1, \quad \alpha_0 > 0.$$

Each $v \in \mathbb{H}^{n+1}$ defines the orthogonal hyperplane

$$H_v = \{x \in \mathbb{H}^{n+1} : \langle x, v\rangle = 0, q(x) = 1\}.$$

The cosine of the angle between the hyperplanes H_v and H_w is defined by

$$\cos\phi = -\langle v, w\rangle.$$

If $|\langle v, w\rangle| > 1$, the hyperplanes are divergent, i.e., they do not intersect in the hyperbolic space. In this case $\cosh(|\langle v, w\rangle|)$ is equal to the distance between the two divergent hyperplanes. If $\langle v, w\rangle = 1$, the hyperplanes are parallel. By Proposition 2.1, the corresponding real spheres $S(v)$ and $S(w)$ are tangent to each other.

References

[1] Dolgachev, I. *Classical Algebraic Geometry. A modern view*. Cambridge: Cambridge University Press, 2012.

[2] Dolgachev, I. and Lunts, V. A character formula for the representation of a Weyl group in the cohomology of the associated toric variety. *J. Algebra* **168** (1994) 741–772.

[3] Dolgachev, I. Ortland, D. Point sets in projective spaces and theta functions. *Astérisque* **165** (1988), 210 pp.

[4] Flaschka, H. Haine, L. Torus orbits in G/P. *Pacific J. Math.* **149** (1991) 251–292.

[5] Howard, B. Millson, J. Snowden, A. and Vakil, R. The equations for the moduli space of n points on the line. *Duke Math. J.* **146** (2009) 175–226.
[6] Kapranov, M. Sturmfels, B. and Zelevinsky, A. Quotients of toric varieties. *Math. Ann.* **290** (1991) 643–655.
[7] Kapranov, M. Chow quotients of Grassmannians. I. I. M. Gelfand Seminar. *Adv. Soviet Math.* **16**(2) (1993) 29–110.
[8] Klein, F. *Vorlesungen über höhere Geometrie*. Dritte Auflage. Bearbeitet und herausgegeben von W. Blaschke. Die Grundlehren der mathematischen Wissenschaften, Band 22. Berlin: Springer-Verlag, 1968.
[9] Klyachko, A. A. Orbits of a maximal torus on a flag space. *Funktsional. Anal. i Prilozhen.* **19** (1985) 77–78 [English transl.: *Functional Analysis and its Applications*, **19** (1985) 65–66].
[10] Kranser, E. The invariant theory of the inversion group geometry upon a quadric surface. *Trans. Amer. Math. Soc.* **1** (1900) 430–498.
[11] Mukai, S. Umemura, H. Minimal rational threefolds. *Algebraic Geometry (Tokyo/Kyoto, 1982)*, pp. 490–518. Lecture Notes in Math., Vol. 1016. Berlin: Springer-Verlag, 1983.
[12] Mukai, S. Igusa quartic and Steiner surfaces. In *Compact Moduli Spaces and Vector Bundles*, pp. 205–210. Contemporary Mathematics, Vol. 564. Providence, RI: American Mathematical Society, 2012.
[13] Petersen, J. Die Theorie der regulären graphs. *Acta Math.* **15** (1891) 193–220.
[14] Procesi, C. *Lie Groups: An approach through invariants and representations*. Berlin: Springer-Verlag, 2007.
[15] Procesi, C. *The Toric Variety Associated to Weyl Chambers*, pp. 153–161. Paris: Hermès, 1990.
[16] Salmon, G. *Lessons Introductory to the Modern Higher Algebra. London*: Hodges and Smith, 1859 (reprinted by Chelsea Publishing Co., 1964).
[17] Serre, J.-P. *Local Fields*. Berlin: Springer-Verlag, 1979.
[18] Study, E. Das Apollonische Problem, *Math. Ann.* **49** (1897) 497–542.
[19] Weyl, H. *The Classical Groups, their Invariants and Representations*. Princeton, NJ: Princeton University Press, 1939.

10

Twenty points in $\mathbb{P}^3$

D. Eisenbud
University of California, Berkeley

R. Hartshorne
University of California, Berkeley

F.-O. Schreyer
Universität des Saarlandes

Abstract

Using the possibility of computationally determining points on a finite cover of a unirational variety over a finite field, we determine all possibilities for direct Gorenstein linkages between general sets of points in $\mathbb{P}^3$ over an algebraically closed field of characteristic 0. As a consequence, we show that a general set of d points is glicci (that is, in the Gorenstein linkage class of a complete intersection) if $d \leq 33$ or $d = 37, 38$. Computer algebra plays an essential role in the proof. The case of 20 points had been an outstanding problem in the area for a dozen years [8].

For Rob Lazarsfeld on the occasion of his 60th birthday

1 Introduction

The theory of liaison (linkage) is a powerful tool in the theory of curves in $\mathbb{P}^3$ with applications, for example, to the question of the unirationality of the moduli spaces of curves (e.g., [3, 26, 29]). One says that two curves $C, D \subset \mathbb{P}^3$ (say, reduced and without common components) are directly linked if their union is a complete intersection, and *evenly linked* if there is a chain of curves $C = C_0, C_1, \ldots, C_{2m} = D$ such that C_i is directly linked to C_{i+1} for all i. The first step in the theory is the result of Gaeta that any two arithmetically

From *Recent Advances in Algebraic Geometry*, edited by Christopher Hacon, Mircea Mustaţă and Mihnea Popa

Cohen–Macaulay curves are evenly linked, and in particular are *in the linkage class of a complete intersection*, usually written *licci*. Much later Rao [23] showed that even linkage classes are in bijection with graded modules of finite length up to shift, leading to an avalanche of results (reported, e.g., in [19, 20]). However, in codimension > 2 linkage yields an equivalence relation that seems to be very fine, and thus not so useful; for example, the scheme consisting of the four coordinate points in $\mathbb{P}^3$ is not licci.

A fundamental paper of Peskine and Szpiro [22] laid the modern foundation for the theory of linkage. They observed that some of the duality used in liaison held more generally in a Gorenstein context, and Schenzel [25] introduced a full theory of Gorenstein liaison. We say that two schemes $X, Y \subset \mathbb{P}^n$ that are reduced and without common components are *directly Gorenstein linked* if their union is arithmetically Gorenstein (for general subschemes the right definition is that $\mathcal{I}_G : \mathcal{I}_X = \mathcal{I}_Y$ and $\mathcal{I}_G : \mathcal{I}_Y = \mathcal{I}_X$). We define *Gorenstein linkage* to be the equivalence relation generated by this notion. This does not change the codimension-2 theory, since, by a result of Serre [27], every Gorenstein scheme of codimension 2 is a complete intersection.

The first serious study of Gorenstein linkage is in the paper [17], where the authors observed that Gorenstein linkage in higher codimensions is analogous to the usual linkage in codimesion 2, and raised the question whether every arithmetically Cohen–Macaulay subscheme of projective space is Gorenstein linked to a complete intersection (*glicci* for short). They verified this in a special case by generalizing Gaeta's theorem from codimension 2 to show that every standard determinantal scheme in any codimension is glicci.

Since any 0 dimensional scheme is arithmetically Cohen–Macaulay, one hopes that every finite set of points in $\mathbb{P}^n$ is glicci. This was verified by the second author in 2001 (see [8]) for general sets of d points in $\mathbb{P}^3$ with $d < 20$, and he proposed the case of 20 general points in $\mathbb{P}^3$ as a "first candidate counterexample." The question whether 20 general points in $\mathbb{P}^3$ is glicci has remained open since then.

Theorem 1.1 *Over an algebraically closed field of characteristic* 0, *a scheme consisting of d general points in $\mathbb{P}^3$ is glicci when $d \leq 33$ and also when $d = 37$ or $d = 38$.*

Further, we determine all pairs of numbers d, e such that there exist "bi-dominant" direct Gorenstein linkage correspondences between the smoothing components (that is, the components containing reduced sets of points) of the Hilbert schemes of degree d subschemes and degree e finite subschemes of $\mathbb{P}^3$ – see Section 2 for a precise statement. All such bi-dominant correspondences are indicated by the edges in the graph shown in Figure 1.

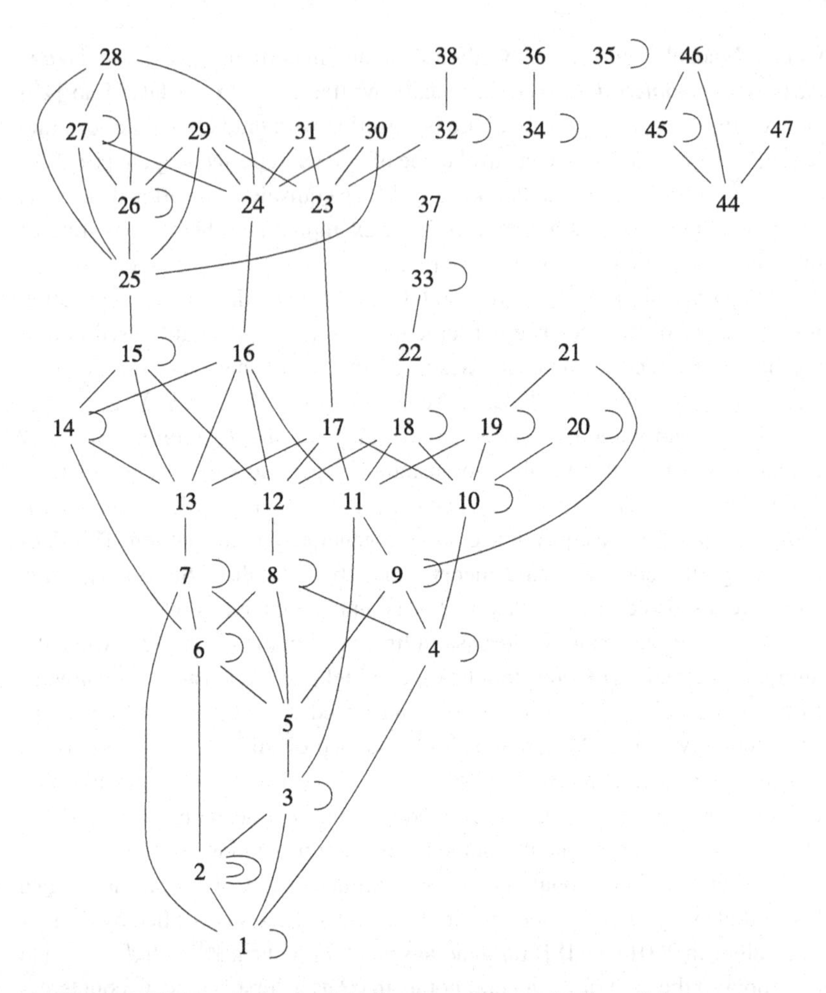

Figure 1 The graph shows all bi-dominant Gorenstein direct linkage correspondences for collections of points in $\mathbb{P}^3$ over an algebraically closed field of characteristic 0. A vertex d represents a set of d points, and an edge $d - e$ represents linkage by an arithmetically Gorenstein set of $d + e$ points with a particular h-vector. Thus two edges are shown where two different h-vectors are possible.

Our approach makes essential use of computation, done in *Macaulay2* [6] by the package GlicciPointsInP3 [5]. It passes by way of characteristic $p > 0$, and we get the same results in all the characteristics we have tested.

Paradoxically, though our method proves that almost all sets of points of the given degrees are glicci, and uses computation, it is not constructive: we prove that a general set of 20 points in $\mathbb{P}^3$ is linked to a set of 10 points by a

Gorenstein scheme of length 30. But if you give us a set of 20 points, we have no way of producing a Gorenstein scheme of length 30 containing it.

This paper is one of the few studies to make explicit use of linkages by Gorenstein schemes that are not divisors of the form $mH-K$ on some arithmetically Cohen–Macaulay scheme of one larger dimension (see Section 6).

2 Basic definitions and outline of the argument

For simplicity, we work throughout this paper over a perfect (but not necessarily algebraically closed) field k. Given a closed subscheme $X \subset \mathbb{P}^n$, we write S_X for the homogeneous coordinate ring of X and $\mathcal{O}_X$ for the structure sheaf of X. Similarly, we write I_X and $\mathcal{I}_X$ for the homogeneous ideal and the ideal sheaf of X.

We write $\mathcal{H}_X$ for the union of the components of the Hilbert scheme of subschemes of $\mathbb{P}^n$ that contain X; and we write $\mathcal{HC}_X$ for the union of the components of the Hilbert scheme of cones in $\mathbb{A}^{n+1}$ that contain the cone over X (see [7]). For example, if X is a reduced set of points then $\mathcal{H}_X$ is smooth at X of dimension nd, but $\mathcal{HC}_X$ may be more complicated.

The reason for considering these cones is the following. While, for a set of general points $X \subset \mathbb{P}^n$, the deformation theory and Hilbert schemes $\mathcal{H}_X$ and $\mathcal{HC}_X$ are naturally isomorphic, this is not so for other subschemes. Also, if G is an arithmetically Gorenstein set of points in $\mathbb{P}^n$, then deformations of G will not in general be arithmetically Gorenstein. On the other hand, the deformations of the cone over G in $\mathcal{HC}_G$ are again cones over arithmetically Gorenstein schemes, and this is the way one normally defines the scheme structure (called PGor(H) in [16]) on the subset of the Hilbert scheme of points in $\mathbb{P}^n$ consisting of arithmetically Gorenstein subschemes with Hilbert function H. In particular, the tangent space to $\mathcal{HC}_G$ at the point corresponding to G is $\mathrm{Hom}_S(I_G, S/I_G)_0$.

We say that a finite scheme of degree d in $\mathbb{P}^n$ has *generic Hilbert function* if the maps $H^0(\mathcal{O}_{\mathbb{P}^n}(\ell)) \to H^0(\mathcal{O}_X(\ell))$ are either injective or surjective for each d. This is the case, for example, when X consists of d general reduced points. In any case, such a generic Hilbert function is determined by d alone. Moreover, any nearby set of points will also have a generic Hilbert function, so the Hilbert scheme $\mathcal{H}_X$ is naturally isomorphic, in a neighborhood of X, to the Hilbert scheme $\mathcal{HC}_X$, and in particular the latter is also smooth and of dimension nd at X.

By [1], any arithmetically Gorenstein scheme G of codimension 3 in $\mathbb{P}^n$ is defined by the Pfaffians of a skew-symmetric matrix of homogeneous forms. From this description and Macaulay's growth conditions, Stanley [28] derived a characterization of the possible Hilbert functions of such schemes, and [4, 16] showed that the family $\mathcal{HC}_G$ is smooth and irreducible. To simplify our discussion, we use the term *Gorenstein Hilbert function* in this paper to refer to a Hilbert function of some arithmetically Gorenstein scheme of codimension 3 in $\mathbb{P}^3$.

Given two positive integers d, e, we ask whether there exists a 0-dimensional arithmetically Gorenstein scheme G of degree $d+e$ in $\mathbb{P}^3$ that "could" provide a direct Gorenstein linkage between a set X of d points with generic Hilbert function and a set Y of e points with generic Hilbert function. It turns out that there is only a finite number of possibilities for the Hilbert function of such a scheme, and we can list them.

If G is a finite reduced Gorenstein scheme in $\mathbb{P}^3$ containing a subscheme X with complement Y, we let

$$\mathcal{HC}_{X\cup Y=G} = \{(X', Y', G') \in \mathcal{HC}_X \times \mathcal{HC}_Y \times \mathcal{HC}_G \mid \\ X' \cup Y' = G' \text{ and } X' \cap Y' = \emptyset\}$$

be the incidence correspondence, and we ask when $\mathcal{HC}_{X\cup Y=G}$ projects dominantly onto $\mathcal{H}_X$ and $\mathcal{H}_Y$ via $\mathcal{HC}_X$ and $\mathcal{HC}_Y$; in this case we say that the correspondence is bi-dominant. Of course for this to happen, the family $\mathcal{HC}_G$ must have dimension at least $\max(3d, 3e)$. We will show that no bi-dominant correspondence is possible unless the numbers d, e are both ≤ 47.

Given d, e and an appropriate Gorenstein Hilbert function, we search for an example of a reduced Gorenstein scheme $G \subset \mathbb{P}^3$ with the given Hilbert function such that G contains a degree d subscheme X with generic Hilbert function whose complement Y also has generic Hilbert function. We do not know how to find such examples directly in characteristic 0. It is perhaps the most surprising part of this work that one can find examples of the type above by computer over a finite field. We can then lift the examples of pairs $X \subset G$ to characteristic 0. The success of the method in finite characteristic is based on the observation that, given integers d, e, a random polynomial of degree $d+e$ over a large finite field "often" has a factor of degree e. This phenomenon is explored in Section 3.

Set

$$\mathcal{HC}_{X\subset G} = \{(X', G') \in \mathcal{HC}_X \times \mathcal{HC}_G \mid X' \subset G'\}.$$

Near a triple (X, Y, G) as above, the natural projection $\mathcal{HC}_{X\cup Y=G} \to \mathcal{HC}_{X\subset G}$ is an isomorphism, with inverse defined from the family, over $\mathcal{HC}_{X\subset G}$, of residual subschemes to X in G, and the universal property of $\mathcal{HC}_{X\cup Y=G}$.

We use a deformation-theoretic argument given in Section 4, together with machine computation, to test whether $\mathcal{HC}_{X\subset G}$ is smooth at the pair (X, G) and projects dominantly onto $\mathcal{H}_X$, and similarly for Y. If our example passes both these tests, it follows that the incidence correspondence $\mathcal{HC}_{X\cup Y=G}$ is bi-dominant.

If $\mathcal{HC}_{X\cup Y=G}$ is bi-dominant then for any dense open set $V \subset \mathcal{HC}_Y$ there exists a dense open set $U \subset \mathcal{HC}_X$ such that each point $X' \in U$ can be directly Gorenstein linked to a point $Y' \in V$. In particular, if a general point $Y' \in \mathcal{HC}_Y$ is glicci, then a general point $X' \in \mathcal{HC}_X$ is glicci as well.

Figure 1 presents a graph (produced by the program Graphviz) of all bi-dominant correspondences. A node numbered d is connected to a node numbered e if, for a general set of d points X, there exists a Gorenstein scheme of degree $d + e$ containing X with complement Y such that the scheme $\mathcal{HC}_{X\cup Y=G}$ dominates both $\mathcal{HC}_X$ and $\mathcal{HC}_Y$. By the remark above, a general set of d points is glicci if d lies in the connected component of 1 in this graph.

Larger numbers of points? The degree of the smallest collection of general points in $\mathbb{P}^3$ that is still not known to be glicci is 34. Here is a possible attack on this case that might extend to larger cases as well.

A general set of 34 points can be linked (using a five-dimensional family of Gorenstein schemes whose existence follows as in the computations below) to a 5-dimensional family $\mathcal{F}$ of sets of 34 points. On the other hand, the schemes in $\mathcal{H}_{34}$ that can be directly Gorenstein linked to 21 points form a subfamily of codimension only 3. Hence it is plausible that the family $\mathcal{F}$ meets this stratum. If this does indeed happen then, since we know that a set of 21 general points is glicci, it would follow that a set of 34 general points is glicci.

This argument makes it very plausible that general sets of 34 points are glicci. A proof along these lines seems to require a good compactification of $\mathcal{HC}_G$ on which one could do intersection theory.

3 Split polynomials over finite fields

In this section we describe the philosophy leading us to believe that the computations underlying this paper could be successful. In a nutshell: $\mathbb{Q}$-rational points on varieties over $\mathbb{Q}$ are very hard to find, but $\mathbb{F}_p$-rational points on varieties over a finite field $\mathbb{F}_p$ are much more accessible. No new result is proven in this section, and none of the results mentioned will be used in the rest of the paper.

By a result of Buchsbaum and Eisenbud [1], the homogeneous ideal of an arithmetically Gorenstein subscheme G of codimension 3 in projective space has a minimal presentation matrix that is homogeneous and skew-symmetric. The degrees of the elements in this matrix (the *degree matrix*) determine the Hilbert function of G. For the arithmetically Gorenstein schemes in an open dense subset of the Hilbert scheme, the degree matrix is also determined (up to permutation of the rows and columns) by the Hilbert function.

Our method thus requires that we find a reduced arithmetically Gorenstein scheme with given degree matrix that contains a union of components of given total degree that is sufficiently general so that its Hilbert function is "generic." We do this by choosing a *random* skew-symmetric matrix with appropriate degrees over a moderately large finite field. We then *hope* that the associated Gorenstein scheme contains a reduced subscheme of the right degree, and we check whether it does by projecting (in a random direction) to a line and factoring the polynomial in one variable that corresponds to the subscheme of the line. Once we have such an example, we can proceed with the computations of tangent spaces described in the next sections.

For instance, to show that a set of 20 points in $\mathbb{P}^3$ is directly Gorenstein linked to a set of 10 Gorenstein points, we will choose a random arithmetically Gorenstein, 0-dimensional scheme $G \subset \mathbb{P}^3$ such that the presentation matrix of I_G is a 9×9 skew-symmetric matrix of linear forms (this choice of a 9×9 matrix of linear forms is determined by considerations to be described later). If M is a sufficiently general matrix of this kind over a polynomial ring in four variables then the cokernel of M will be the homogeneous ideal of a reduced 0-dimensional scheme of degree 30 in $\mathbb{P}^3$. We then decompose G into irreducible components, and search for a combination of components that has total degree 20. We check to see whether the union of these components has generic Hilbert function.

If we carried out this procedure over the rational numbers, we would expect G to be irreducible. However suppose that the ground field k is a moderately large finite field, and we choose such a matrix randomly (say by choosing each coefficient of each linear form in the upper half of the matrix uniformly at random from k). What will be the chance that it contains a subscheme of length 20 defined over k? That is, how many random examples should one expect to investigate before finding a good one?

The answer was surprising to us: Taking k to be a field with 10,007 elements, and making 10,000 random trials, we found that the desired subscheme occurs in 3868 examples – about 38% of the time. For the worst case needed for this paper, where the Gorenstein scheme has degree 90 and the desired subscheme has degree 45, the proportion is about 17% in our experiments.

The proportion of Gorenstein schemes of degree 30 that are reduced and have a subscheme of degree 20 defined over k turns out to be quite close to the proportion of polynomials of degree 30 that have a factor of degree 20 over k. That proportion can be computed explicitly as a rational function in the size of the ground field; for $|k| = 10,007$ it is approximately 0.385426.

Thus, to consider applications of this random search technique it is worthwhile to know something about the proportion of polynomials of degree n in one variable over a finite field $\mathbb{F}_q$ that are square-free and have a factor of given degree k.

This proportion can be computed explicitly (for small n and k). Gauss showed that the number of irreducible monic polynomials in $\mathbb{F}_q[x]$ of degree ℓ is

$$N(\ell, q) = \frac{1}{\ell} \sum_{d|\ell} \mu(\ell/d) q^d,$$

where μ denotes the Möbius function. Thus the number of square-free polynomials of degree n is

$$\sum_{\lambda \vdash n} \prod_{i=1}^{r} \binom{N(\lambda_i, q)}{t_i},$$

where t_i denotes the frequency of λ_i in the partition $\lambda = (\lambda_1^{t_1}, \lambda_2^{t_2}, \ldots, \lambda_r^{t_r})$. (This number, rather amazingly, can also be written as $q^n - q^{n-1}$; for a simple proof see [24].)

The number of square-free polynomials of degree n with a factor of degree k is

$$A(n, k, q) = \sum_{\substack{\lambda \vdash n \\ \text{with subpartition of size } k}} \prod_{i=1}^{r} \binom{N(\lambda_i, q)}{t_i}.$$

For small n and k the polynomial in q can be evaluated explicitly. For example

$$A(6, 3, q) = \frac{29}{80} q^6 - \frac{11}{16} q^5 + \frac{5}{16} q^4 - \frac{5}{16} q^3 + \frac{13}{40} q^2$$

and

$$A(30, 20, q)/q^{30} \doteq 0.385481 - 0.550631 q^{-1} + O(q^{-2}).$$

Since

$$\lim_{q \to \infty} N(\ell, q)/q^\ell = \frac{1}{\ell},$$

the relative size of the contribution of a partition λ converges to

$$\lim_{q\to\infty}\prod_{i=1}^{r}\binom{N(\lambda_i,q)}{t_i}/q^n = \frac{1}{\prod_{i=1}^{r} t_i!\lambda_i^{t_i}} = |C_\lambda|/n!,$$

which is also the relative size of the conjugacy class C_λ in the symmetric group S_n. Thus the sum

$$p(n,k) = \sum_{\substack{\lambda\vdash n \\ \text{with subpartition of size } k}} |C_\lambda|/n!$$

can serve as an approximation for $A(n,k,q)/q^n$ for large q.

For fixed k, Cameron (unpublished) proved that the limit $\lim_{n\to\infty} p(n,k)$ exists and is positive. For example,

$$\lim_{n\to\infty} p(n,1) = 1 - exp(-1)$$

was established by Montmort around 1708 [21].

Indeed, over a finite field $\mathbb{F}$ with q elements the fraction of polynomials with a root in $\mathbb{F}$ is about 63% nearly independently of q and n. Experimentally we find that

$$A(n,1,q)/q^n \doteq 0.632121 - 0.81606q^{-1} + O(q^{-2}).$$

4 Deformation theory

We are interested in pairs of schemes $X \subset G$ such that the projection

$$\mathcal{HC}_{X\subset G} \to \mathcal{HC}_X$$

is dominant, meaning geometrically that each small deformation of X is still contained in a small deformation of G that is still arithmetically Gorenstein. We will check this condition by showing that the map of tangent spaces is surjective at some smooth point of $\mathcal{HC}_{X\subset G}$.

We thus begin by recalling the construction of the tangent space to $\mathcal{HC}_{X\subset G}$. Though the case of interest to us has to do with finite schemes in $\mathbb{P}^n$, the (well-known) result is quite general:

Lemma 4.1 *Suppose that $X \subset G$ are closed subschemes of a scheme Z, and let*

$$T_{X/Z} = H^0\mathcal{H}om_Z(\mathcal{I}_X,\mathcal{O}_X),\quad T_{G/Z} = H^0\mathcal{H}om_Z(\mathcal{I}_G,\mathcal{O}_G)$$

be the tangent spaces to the functors of embedded flat deformations of X *and* G *in* Z. *The functor of pairs of embedded flat deformations of* X *and* G *in* Z *that preserve the inclusion relation* $X \subset G$ *has Zariski tangent space* $T_{(X\subset G)/Z}$ *at* $X \subset G$ *equal to* $H^0\mathcal{T}_{(X\subset G)/Z}$, *where* $\mathcal{T} := \mathcal{T}_{(X\subset G)/Z}$ *is defined by the fibered product diagram*

$$\begin{array}{ccc} \mathcal{H}om_Z(\mathcal{I}_G, \mathcal{O}_G) & \longrightarrow & \mathcal{H}om_Z(\mathcal{I}_G, \mathcal{O}_X) \\ \uparrow & & \uparrow \\ \mathcal{T} & \longrightarrow & \mathcal{H}om_Z(\mathcal{I}_X, \mathcal{O}_X) \end{array}$$

In particular, if the restriction map $H^0\mathcal{H}om(\mathcal{I}_X, \mathcal{O}_X) \to H^0\mathcal{H}om(\mathcal{I}_G, \mathcal{O}_X)$ *is an isomorphism then* $T_{(X\subset G)/Z} \cong T_{G/Z}$.

Proof (See also [11, Ex. 6.8].) It suffices to prove the lemma in the affine case, so we suppose that $Z = \operatorname{Spec} R$ and that $X \subset G$ are defined by ideals $I \supset J$ in R. The first-order deformation of I corresponding to a homomorphism $\phi : I \to R/I$ is the ideal

$$I_\phi := \{i + \epsilon\phi(i) \mid i \in I\} + \epsilon I \subset R[\epsilon]/(\epsilon^2),$$

and similarly for J_ψ, so we have

$$\mathcal{T} = \{(\psi : J \to R/J, \phi : I \to R/I) \mid \psi(j) \equiv \phi(j) (\mathrm{mod}\ I) \text{ for all } j \in J\}.$$

If $\psi(j) \equiv \phi(j) (\mathrm{mod}\ I)$ for all $j \in J$ then every element $j + \epsilon\psi(j)$ is obviously in I_ϕ. Conversely, if $j + \epsilon\psi(j) = i + \epsilon\phi(i) + \epsilon i'$ with $i' \in I$, then $i = j$, so $\psi(j) = \phi(j) + i'$. This proves that $\mathcal{T}$ is the fibered product. The last statement of the lemma is an immediate consequence. □

Recall from Section 2 that if X is reduced and has generic Hilbert function, then $\mathcal{HC}_X$ and the Hilbert scheme $\mathcal{H}_X$ coincide in a neighborhood of X. Moreover, $\mathcal{HC}_X$ and $\mathcal{H}_X$ are irreducible.

Theorem 4.2 *Let* $G \subset \mathbb{P}^n$ *be a finite scheme such that the cone over* G *is a smooth point on* $\mathcal{HC}_G$. *Suppose that* $X \subset G$ *is a union of some of the components of* G *that are reduced, and that* X *has generic Hilbert function. Let* $d = \deg X$. *If*

$$\dim_k \mathrm{Hom}_S(I_G, S_G)_0 - \dim_k \mathrm{Hom}_S(I_G, I_X/I_G)_0 = nd,$$

then the projection map $\mathcal{HC}_{X\subset G} \to \mathcal{H}_X$ *is dominant.*

Proof Consider the diagram with exact row

$$\begin{array}{ccccccc} 0 & \longrightarrow & \mathrm{Hom}_S(I_G, I_X/I_G) & \longrightarrow & \mathrm{Hom}_S(I_G, S_G) & \longrightarrow & \mathrm{Hom}_S(I_G, S_X) \\ & & & & & & \uparrow \phi \\ & & & & & & \mathrm{Hom}_S(I_X, S_X) \end{array}$$

We begin by computing the dimension of $\mathrm{Hom}_S(I_X, S_X)_0$, the degree 0 part of $\mathrm{Hom}_S(I_X, S_X)$. We may interpret this space as the space of first-order infinitesimal deformations of X as a cone – that is, as the tangent space to $\mathcal{HC}_X$ at the point X. The computation of this space commutes with base change, and since we have assumed that the ground field k is perfect, the base change to $\overline{k}$ remains reduced. Thus to compute $\dim \mathrm{Hom}_S(I_X, S_X)_0$ we may assume that X consists of d distinct k-rational points.

The sheaf $\mathcal{H}om_S(\mathcal{I}_X, \mathcal{O}_X)$ is the sheafification of $\mathrm{Hom}_S(I_X, S_X)$, so there is a natural map

$$\alpha : \mathrm{Hom}_S(I_X, S_X)_0 \to H^0(\mathcal{H}om_S(\mathcal{I}_X, \mathcal{O}_X)).$$

The source of α may be identified with the tangent space to a Hilbert scheme of cones near X, while the target may be identified with the Hilbert scheme of collections of points near X, and α is the map induced by forgetting the cone structure. Since, by assumption, X has generic Hilbert function, these Hilbert schemes coincide and α is an isomorphism. Thus $\dim \mathrm{Hom}_S(I_X, S_X)_0 = nd$.

Though the map $\phi : \mathrm{Hom}_S(I_X, S_X) \to \mathrm{Hom}_S(I_G, S_X)$ in the diagram above is generally not an isomorphism, we will next show that it induces an isomorphism between the components of degree 0. Using the fact that S_X is reduced, so that $\mathrm{Hom}_S(I_G, S_X)$ has depth ≥ 1, we see that the natural map

$$\beta : \mathrm{Hom}_S(I_G, S_X)_0 \to H^0\mathcal{H}om(\mathcal{I}_G, \mathcal{O}_X)$$

is an injection. On the other hand, any section of $\mathcal{H}om(\mathcal{I}_G, \mathcal{O}_X)$ is supported on X. Because X is a union of the components of G, this implies that

$$H^0\mathcal{H}om(\mathcal{I}_G, \mathcal{O}_X) = H^0\mathcal{H}om(\mathcal{I}_X, \mathcal{O}_X).$$

Together with the equality $\mathrm{Hom}_S(I_X, S_X)_0 = H^0\mathcal{H}om(\mathcal{I}_X, \mathcal{O}_X)$, this implies that the map

$$\phi_0 : \mathrm{Hom}_S(I_X, S_X)_0 \to \mathrm{Hom}_S(I_G, S_X)_0$$

is an isomorphism, as claimed.

Let $\operatorname{cone} X \subset \operatorname{cone} G \subset \mathbb{A}^{n+1}$ be the cones over X and G respectively. We may apply Lemma 4.1, which tells us that the space of first-order deformations of the pair $\operatorname{cone} X \subset \operatorname{cone} G$ is the fibered product of $\operatorname{Hom}_S(I_X, S_X)$ and $\operatorname{Hom}_S(I_G, S_G)$ over $\operatorname{Hom}_S(I_G, S_X)$. Since we wish to look only at deformations as cones, we take the degree 0 parts of these spaces, and we see that the tangent space to $\mathcal{HC}_{X\subset G}$ is the fibered product of $\operatorname{Hom}_S(I_X, S_X)_0$ and $\operatorname{Hom}_S(I_G, S_G)_0$ over $\operatorname{Hom}_S(I_G, S_X)_0$. Since ϕ_0 is an isomorphism, the tangent space to $\mathcal{HC}_{X\subset G}$ is isomorphic, via the projection, to $\operatorname{Hom}_S(I_G, S_G)_0$, the tangent space to $\mathcal{HC}_G$ at G.

Since $\operatorname{cone} X$ consists of a subset of the irreducible components of $\operatorname{cone} G$, and X has generic Hilbert function, it follows that the map $\mathcal{HC}_{X\subset G} \to \mathcal{HC}_G$ is surjective. Since $\mathcal{HC}_G$ is smooth at G, and the map of tangent spaces is an isomorphism, it follows that $\mathcal{HC}_{X\subset G}$ is smooth at the pair $(X \subset G)$.

To prove that the other projection map $\mathcal{HC}_{X\subset G} \to \mathcal{HC}_X$ is dominant, it now suffices to show that the map on tangent spaces

$$T_{X\subset G}\mathcal{HC}_{X\subset G} = T_G\mathcal{HC}_G \to T_X\mathcal{HC}_X = \operatorname{Hom}_S(I_G, S_X)_0$$

is onto or, equivalently, that the right-hand map in the sequence

$$0 \to \operatorname{Hom}_S(I_G, I_X/I_G)_0 \to \operatorname{Hom}_S(I_G, S_G)_0 \to \operatorname{Hom}_S(I_G, S_X)_0$$

is surjective. Since the right-hand vector space has dimension nd, this follows from our hypothesis on dimensions. □

Corollary 4.3 *Let $G \subset \mathbb{P}^n$ be a finite scheme. Suppose that $X \subset G$ is a union of some of the components of G that are reduced, and that X has generic Hilbert function. Let $d = \deg X$. If*

$$\dim_k \operatorname{Hom}_S(I_G, I_X/I_G)_0 = \dim_G \mathcal{HC}_G - nd,$$

then $\mathcal{HC}_G$ and $\mathcal{HC}_{X\subset G}$ are smooth in G and $X \subset G$ respectively and the projection map $\mathcal{HC}_{X\subset G} \to \mathcal{HC}_X$ is dominant.

Proof Since

$$\begin{aligned}\dim_G \mathcal{HC}_G &\le \dim T_G\mathcal{HC}_G = \dim \operatorname{Hom}_S(I_G, S_G)_0\\ &\le \dim \operatorname{Hom}_S(I_G, I_X/I_G)_0 + \dim \operatorname{Hom}_S(I_G, S_X)_0\\ &= \dim \operatorname{Hom}_S(I_G, I_X/I_G)_0 + nd,\end{aligned}$$

equality holds by our assumption and $\mathcal{HC}_G$ is smooth at G. Now the theorem applies. □

5 Computational approach

To classify the possible bi-dominant direct Gorenstein linkage correspondences we make use of h-vectors [28], which are defined as follows. See [14] for further details.

Let $R = S/I$ be a homogeneous Cohen–Macaulay factor ring of a polynomial ring S with $\dim R = k$. The *h-vector* of R (or of X in case $I = I_X$) is defined to be the kth difference of the Hilbert function of R. If $\ell_1, \ldots, \ell_k$ is an R-sequence of linear forms, then $R/(\ell_1, \ldots, \ell_k)$ is artinian, and its Hilbert function is equal to the h-vector of R. It follows in particular that the h-vector consists of a finite sequence of positive integers followed by zeros. Over a small field such an R-sequence may not exist, but the h-vector does not change under extension of scalars, so the conclusion remains true. We often specify an h-vector by giving just the list of nonzero values.We can make a similar construction for any Cohen–Macaulay module. If X is a finite scheme then the sum of the terms in the h-vector is the degree of X. The h-vector of a Gorenstein ideal is symmetric.

From the definition it follows at once that the h-vector of a set of points X with general Hilbert function is equal to the Hilbert function of the polynomial ring in three variables except (possibly) for the last nonzero term, and thus has the form

$$1, 3, 6, \ldots, \binom{s+1}{2}, a, 0, \ldots \text{ with } 0 \leq a < \tbinom{s+2}{2}$$

where s is the least degree of a surface containing X. For example, the h-vector of a general collection of 21 points in $\mathbb{P}^3$ is $\{1, 3, 6, 10, 1\}$.

We will make use of the well-known observation that if I_X and I_Y in S are directly Gorenstein linked via I_G then the h-vector of X, plus some shift of the reverse of the h-vector of Y, is equal to the h-vector of G (see [14, 2.14]). One way to see this is to reduce all three of I_X, I_Y, I_G modulo a general linear form ℓ. The relation of linkage is preserved, and $\omega_X/ \cong I_Y/I_G$ (up to a shift). Since the Hilbert function of ω_X mod ℓ is the reverse of the Hilbert function of S/I_X mod ℓ, this gives the desired relation.

For example, our computations show that a general collection of 21 points in $\mathbb{P}^3$, with h-vector $\{1, 3, 6, 10, 1\}$ as above, is directly Gorenstein linked to a collection of nine points with general Hilbert function, and thus h-vector $\{1, 3, 5\}$. The Gorenstein ideal that links them will have h-vector $\{1, 3, 6, 10, 6, 3, 1\}$, and we have

$$\begin{array}{rl} & 1\;3\;6\;10\;1 \\ + & \qquad\;\;\; 5\;3\;1 \\ = & 1\;3\;6\;10\;6\;3\;1 \end{array}$$

The additivity of h-vectors in 0-dimensional Gorenstein liaison can be traced back to Macaulay (see [18, p. 112]) as Tony Iarrobino pointed out to us.

We return to the problem of linking general sets of points. Consider the irreducible component $\mathcal{H}_d \subset Hilb_d(\mathbb{P}^3)$ whose general point corresponds to a collection of d distinct points. Consider an (irreducible) direct Gorenstein linkage correspondence

$$\mathcal{HC}_{X\cup Y=G} \to \mathcal{H}_d \times \mathcal{H}_e.$$

Recall that $\mathcal{HC}_{X\cup Y=G}$ is said to be bi-dominant if it dominates both $\mathcal{H}_d$ and $\mathcal{H}_e$. The h-vector of a general point $G = X \cup Y$ of a bi-dominant correspondence is rather special. Most of the following proposition can be found in [14, 7.2]; we repeat it for the reader's convenience.

Proposition 5.1 *Let $G = X \cup Y$ be a general point of a bi-dominant correspondence. Then the h-vector of G is one of the following:*

(I) $\{1, 3, 6, \ldots, \binom{s+1}{2}, \binom{s+1}{2} + c, \binom{s+1}{2}, \ldots, 3, 1\}$ *with* $0 \le c \le s+1$, *or*
(II) $\{1, 3, 6, \ldots, \binom{s+1}{2}, \binom{s+1}{2} + c, \binom{s+1}{2} + c, \binom{s+1}{2}, \ldots, 3, 1\}$ *with* $0 \le c \le s+1$.

Proof We may assume that $d \ge e$. A collection of d general points has generic Hilbert function. Thus X has h-vector of shape

$$h_X = \{1, 3, \ldots, \binom{t+1}{2}, a\},$$

where $d = \binom{t+2}{3} + a$ is the unique expression with $0 \le a < \binom{t+2}{2}$. Similarly, we have $h_Y = \{1, 3, \ldots, \binom{t'+1}{2}, a'\}$, where $e = \binom{t'+1}{3} + a'$. On the other hand, since G is arithmetically Gorenstein, the h-vector of G is symmetric. As explained above, the difference $h_G - h_X$ coincides, after a suitable shift, with the h-vector of Y read backwards. Since the h-vector of G coincides with the h-vector of X up to position t, we have only the possibilities

1. $t = t'$ and $a = a' = 0$ with $h_G = \{1, 3, \ldots, \binom{t+1}{2}, \binom{t+1}{2}, \ldots, 3, 1\}$ of type *(II)* with $s = t - 1$ and $c = s + 1$.
2. $t = t' + 1$ and $a = a' = 0$ with $h_G = \{1, 3, \ldots, \binom{s+2}{2}, \binom{s+1}{2}, \ldots, 3, 1\}$ of type *(I)* with $s = t - 1 = t'$ and $c = s + 1$.
3. $t = t' + 1$ and $a + a' = \binom{s+2}{2}$ with $h_G = \{1, 3, \ldots, \binom{s+2}{2}, \binom{s+2}{2}, \ldots, 3, 1\}$ of type *(II)* with $s = t - 1 = t'$ and $c = s + 1$.
4. $t = t' + 2$ and $a + a' = \binom{s+1}{2}$ with $h_G = \{1, 3, \ldots, \binom{s+2}{2}, \binom{s+1}{2}, \ldots, 3, 1\}$ of type *(I)* with $s = t - 1 = t' + 1$ and $c = s + 1$.
5. $s = t = t'$ and $h_G = \{1, 3, \ldots, \binom{s+1}{2}, a + a', \binom{s+1}{2}, \ldots, 3, 1\}$ of type *(I)*.
6. $s = t = t'$, $a = a'$ with $h_G = \{1, 3, \ldots, \binom{s+1}{2}, a, a, \binom{s+1}{2}, \ldots, 3, 1\}$ of type *(II)*.

By Stanley's theorem [28, 4.2] the difference function of the first half of h_G is non-negative, so we have $a + a' \geq \binom{s+1}{2}$ and $a = a' \geq \binom{s+1}{2}$ respectively in the last two cases. □

Proposition 5.2 *Let G be an arithmetically Gorenstein set of points in $\mathbb{P}^3$ with h-vector h as in Proposition 5.1, and let $g(h)$ be the dimension of $\mathcal{HC}_G$.*

- *In case I, $g(h) = 4s(s+1) + 4c - 1$.*
- *In case II, $g(h) = \frac{9}{2}s(s+1) + \frac{1}{2}c(c+13) - cs - 1$.*

Proof Case I is type 3 of [14, 7.2]. Case II is proved analogously, by induction on s, using [14, 5.3] starting with the cases of h-vectors $\{1, 1, 1, 1\}$ and $\{1, 2, 2, 1\}$. Note that when $c \geq 2$, case II will at some point reduce to type 2 of [14, 7.2]. □

Corollary 5.3 *There are only finitely many bi-dominant correspondences.*

Proof If $\mathcal{HC}_h \to \mathcal{HC}_d$ is dominant then we must have $g(h) \geq \dim \mathcal{H}_d = 3d$. But examining the Hilbert functions in the different cases we find that $d \geq \binom{s+2}{3}$, which is cubic in s, while the functions $g(h)$ are quadratic in s, so the inequality cannot hold for large s. Calculation shows that $s \leq 5$, and that $d = 47$ is the maximal degree possible [5]. (See also [14, 7.2, 7.3].) □

Theorem 5.4 *The bi-dominant correspondences are precisely those indictated in Figure 1.*

Proof To prove existence of a bi-dominant correspondence it suffices to find a smooth point $(X, Y) \in \mathcal{HC}_{X \cup Y = G}$ and to verify that both maps on tangent spaces

$$T_{X \subset G}\mathcal{HC}_{X \subset G} \to T_X\mathcal{H}_d \text{ and } T_{Y \subset G}\mathcal{HC}_{Y \subset G} \to T_Y\mathcal{H}_e$$

are surjective. We test [5] each of the finitely many triples consisting of an h-vector $h = \{h_0, \ldots, h_n\}$ and integers (d, e) satisfying $g(h) \geq \max(3d, 3e)$ and $\sum h_i = d + e$ and subject to the condition that h can be expressed as the sum of the h-vector of a general set of d points and the reverse of the h-vector of a general set of e points, as follows:

1. Using the probabilistic method of Section 2, find a pair $X \subset G$ over a finite field $\mathbb{F}_p$.
2. Let Y be the scheme defined by $I_Y = I_G : I_X$. Test whether G is reduced, and whether X and Y have generic Hilbert functions.
3. Test whether

$$\dim \operatorname{Hom}(I_G, I_X/I_G)_0 = \dim_G \mathcal{HC}_G - 3d$$

and

$$\dim\ \mathrm{Hom}(I_G, I_Y/I_G)_0 = \dim_G \mathcal{HC}_G - 3e.$$

If the example $X \subset G$ and Y passes the tests in steps 2 and 3 then, by Corollary 4.3, $(X, Y) \in \mathcal{HC}_{X\cup Y=G}$ is a point on a bi-dominant correspondence over $\mathbb{F}_p$. Since we may regard our example as the reduction mod p of an example defined over some number field, this shows the existence of a bi-dominant correspondence in characteristic 0.

There are nine pairs (d, h), involving six different h-vectors, where the procedure above did not, in our experiments, lead to a proof of bi-dominance. They are given in the following table:

Degrees	h-Vector	Buchsbaum–Eisenbud matrix
7	$\{1,3,3,3,1\}$	$S^2(-3) \oplus S^3(-6) \to S^2(-5) \oplus S^3(-2)$
7	$\{1,3,3,3,3,1\}$	$S^2(-3) \oplus S^3(-5) \to S^2(-4) \oplus S^3(-2)$
13, 14, 15	$\{1,3,6,6,6,3,1\}$	$S^3(-4) \oplus S^4(-6) \to S^3(-5) \oplus S^4(-3)$
16	$\{1,3,6,6,6,6,3,1\}$	$S^3(-4) \oplus S^4(-7) \to S^3(-6) \oplus S^4(-3)$
17	$\{1,3,6,7,7,6,3,1\}$	$S(-4) \oplus S(-5) \oplus S^3(-7) \to$ $S(-6) \oplus S(-5) \oplus S^3(-3)$
25, 26	$\{1,3,6,10,10,10,6,3,1\}$	$S^4(-5) \oplus S^5(-7) \to S^4(-6) \oplus S^5(-4)$

It remains to show that, in these numerical cases, there really is *no* bi-dominant family. In each of these cases the Buchsbaum–Eisenbud matrix (the skew-symmetric presentation matrix of the I_G) has a relatively large block of zeros, since the maps between the first summands of the free modules shown in the table are zero for degree reasons. (In case $d = 17$ the map from the first two summands in the source to the first two summands in the target is zero, as the matrix is skew symmetric.) Thus, among the Pfaffians of this matrix are the minors of an $n \times (n + 1)$ matrix, for a certain value of n. These minors generate the ideal of an arithmetically Cohen–Macaulay (ACM) curve. In the given cases, the general such curve will be smooth. Thus, in these cases, the Gorenstein points lie on smooth ACM curves of degree c, the maximal integer in the h-vector of G (so $c \in \{3, 6, 7, 10\}$). For example if $c = 3$ there are seven points, but a twisted cubic curve can contain at most six general points. More generally, for a curve C moving in its Hilbert scheme $\mathcal{H}_C$ to contain d general points we must have $2d \le \dim \mathcal{H}_C$. In all cases listed above, $\dim \mathcal{H}_C = 4c$ and $2d \le 4c$ is not satisfied. □

The method discussed above can be used to show more generally that having certain h-vectors forces a 0-dimensional Gorenstein scheme to be a divisor on an ACM curve. Here is a special case:

Proposition 5.5 *If Z is a 0-dimensional AG scheme with h-vector h of type I with $c = 0$ or type II with $c = 0, 1$ in Proposition 5.1, then Z is a divisor in a class of the form $mH - K$ on some ACM curve whose h-vector is the first half of h.*

We sketch an alternative proof.

Proof For the cases with $c = 0$ this is a consequence of [14, 3.4] (see also [15, Theorem 5.77a]). For type II with $c = 1$ we use induction on s. Using results 5.3 and 5.5 of [14] we compare h to the h-vector h' defined there, which is the same thing with s replaced by $s - 1$, and we compute that the dimension of $\mathcal{HC}_h$ is equal to the dimension of the family of those $Z \sim mH - K$ on a C, computed as $\dim \mathcal{H}_{h_C} + \dim_C |mH - K|$. The induction starts with $h = \{1, 3, 4, 4, 3, 1\}$, where the corresponding scheme Z is a complete intersection of type $(2, 2, 3)$ and the result is obvious. □

Remark 5.6 In some cases the projection $\mathcal{HC}_{X \subset G} \to \mathcal{H}_X$ is finite. This happens for the following degrees d and h-vectors of G:

Degree	h-Vector
7	$\{1, 3, 3, 1\}$
17	$\{1, 3, 6, 7, 6, 3, 1\}$
21	$\{1, 3, 6, 10, 6, 3, 1\}$
25	$\{1, 3, 6, 10, 10, 6, 3, 1\}$
29	$\{1, 3, 6, 10, 12, 10, 6, 3, 1\}$
32	$\{1, 3, 6, 10, 12, 12, 10, 6, 3, 1\}$
33	$\{1, 3, 6, 10, 15, 10, 6, 3, 1\}$
38	$\{1, 3, 6, 10, 15, 15, 10, 6, 3, 1\}$
45	$\{1, 3, 6, 10, 15, 19, 15, 10, 6, 3, 1\}$

It would be interesting to compute the degree of the projection in these cases. When G is a complete intersection the projection is one-to-one.

Corollary 5.7 *A general collection of d points in $\mathbb{P}^3$ over an algebraically closed field of characteristic 0 is glicci if $1 \leq d \leq 33$ or $d = 37$ or 38.*

Proof Since the correspondences are bi-dominant, a general collection of d points will be Gorenstein linked to a general collection of degree e. Thus we may repeat, and the result follows, because these degrees form a connected component of the graph in Figure 1. □

6 Strict Gorenstein linkage

One way to obtain an arithmetically Gorenstein (AG) subscheme of any projective space $\mathbb{P}^n$ is to take an ACM subscheme S satisfying the condition, called G_1, of being Gorenstein in codimension 1, and a divisor X on it that is linearly equivalent to $mH-K$, where H is the hyperplane class and K is the canonical divisor of S (see [17, 5.4]). A slight variation of this construction allows one to reduce the condition G_1 to G_0 (Gorenstein in codimension 0; see [10, 3.3]). A direct linkage using one of these AG schemes is called a *strict* direct Gorenstein link, and the equivalence relation generated by these is called *strict Gorenstein linkage* [10].

Nearly all of the proofs in the literature that certain classes of schemes are glicci use this more restrictive notion of Gorenstein linkage. (See [20] for a survey, and [8–10, 12–14, 17] for some of the results). In fact, the one paper we are aware of that actually makes use of the general notion for this purpose is [2, 7.1], which uses general Gorenstein linkages to show that any AG subscheme of $\mathbb{P}^n$ is glicci.

By contrast, some of the direct linkages established in this paper cannot be strict direct Gorenstein links. We do not know whether such links can be achieved by a sequence of strict Gorenstein links, but one can show that if this is possible then some of the links must be to larger sets of points, and some of the intermediate sets of points must fail to be general.

Proposition 6.1 *A general arithmetically Gorenstein scheme of 30 points in $\mathbb{P}^3$ cannot be written as a divisor of the form $mH-K$ on any ACM curve $C \subset \mathbb{P}^3$, where H is the hyperplane class and K the canonical class of C. The linkages 20–10 and 21–9 in Figure 1 are not strict direct Gorenstein links.*

Proof The h-vector of a Gorenstein scheme Z of 30 points, of which 20 or 21 are general, is necessarily $h = \{1, 3, 6, 10, 6, 3, 1\}$. If Z lay on an ACM curve in the class $mH - K$, then the h-vector of the curve would be $\{1, 2, 3, 4\}$ ([14, 3.1]). This is a curve of degree 10 and genus 11. The Hilbert scheme of such curves has dimension 40, so such a curve can contain at most 20 general points. On the contrary, our Theorem 1.1 shows that there are Gorenstein schemes with h-vector h containing 21 general points. In particular, the linkage 21–9 is not strict.

If the link 20–10 were a strict Gorenstein link, then a set of 20 general points X would lie in an AG scheme of 30 points in the class $5H - K$ on a curve C as above. Since the Hilbert scheme of X and the incidence correspondence $\mathcal{H}_{X \subset C}$ both have dimension 60, a general X would be contained in a general, and thus smooth and integral, curve C. But the family of pairs $Z \subset C$ of this type has dimension only 59 [14, 6.8] and thus there is no such Z containing 20 general points. □

Some of the direct Gorenstein links in Figure 1 can be obtained by direct strict Gorenstein links (e.g., the cases $d \leq 19$ are treated in [9]). However, for $d = 20,\ 24 \leq d \leq 33$, and $d = 37, 38$ this is not the case.

Acknowledgments We are grateful to the Mathematical Sciences Research Institute in Berkeley, CA, for providing such an exciting environment during the Commutative Algebra Program, 2012–13, where a chance meeting led to the progress described here. The work was performed while the first and third authors were in residence at MSRI, which was supported by the National Science Foundation under grant no. 0932078 000. The first author was partially supported by National Science Foundation grant no. 1001867. The third author was supported at MSRI by a Simons Visiting Professorship.

We are grateful to Peter Cameron, Persi Diaconis, Ira Gessel, Robert Guralnick, Brendan Hassett, and Yuri Tschinkel for helping us to understand various aspects of this work.

References

[1] Buchsbaum, D. A., Eisenbud, D. Algebra structures for finite free resolutions, and some structure theorems for ideals of codimension 3. *Amer. J. Math.* **99**(3) (1977) 447–485.

[2] Casanellas, M., Drozd, E., Hartshorne, R. Gorenstein liaison and ACM sheaves. *J. Reine Angew. Math.* **584** (2005) 149–171.

[3] Chang, M. C., Ran, Z. Unirationality of the moduli spaces of curves of genus 11, 13 (and 12). *Invent. Math.* **76** (1984) 41–54.

[4] Diesel, S.J. Irreducibility and dimension theorems for families of height 3 Gorenstein algebras. *Pacific J. Math.* **172**(2) (1996) 365–397.

[5] Eisenbud, D., Schreyer, F.-O. GlicciPointsInP3.m2 – Experiments establishing direct Gorenstein linkages for finite subsets of $\mathbb{P}^3$. A Macaulay2 [6] package, available at http://www.math.uni-sb.de/ag/schreyer/home/computeralgebra.htm.

[6] Grayson, D. R., Stillman, M. E. Macaulay2, a software system for research in algebraic geometry. Available at http://www.math.uiuc.edu/Macaulay2/.

[7] Haiman, M., Sturmfels, B. Multigraded Hilbert schemes. *J. Alg. Geom.* **13**(4) (2004) 72–769.

[8] Hartshorne, R. Experiments with Gorenstein liaison. Dedicated to Silvio Greco on the occasion of his 60th birthday *(Catania, 2001). Le Matematiche (Catania)* **55**(2) (2002) 305–318.

[9] Hartshorne, R. Some examples of Gorenstein liaison in codimension three. *Collect. Math.* **53** (2002) 21–48.

[10] Hartshorne, R. Generalized divisors and biliaison. *Illinois J. Math.* **51** (2007) 83–98.
[11] Hartshorne, R. *Deformation Theory*. Graduate Texts in Mathematics, No. 257. New York: Springer-Verlag, 2010.
[12] Hartshorne, R., Martin-Deschamps, M., Perrin, D. Un théorème de Rao pour les familles de courbes gauches. *J. Pure Appl. Algebra* **155** (2001) 53–76.
[13] Hartshorne, R., Migliore, J., Nagel, U. Liaison addition and the structure of a Gorenstein liaison class. *J. Algebra* **319**(8) (2008) 3324–3342.
[14] Hartshorne, R., Sabadini, I., Schlesinger, E. Codimension 3 arithmetically Gorenstein subschemes of projective N-space. *Ann. Inst. Fourier (Grenoble)* **58**(6) (2008) 2037–2073.
[15] Iarrobino, A., Kanev, V. *Power Sums, Gorenstein Algebras, and Determinantal Loci.* Lecture Notes in Mathematics, No. 1721. Berlin: Springer-Verlag, 1999.
[16] Kleppe, J. O. The smoothness and the dimension of PGor(H) and of other strata of the punctual Hilbert scheme. *J. Algebra* **200**(2) (1998) 606–628.
[17] Kleppe, J. O., Migliore, J. C., Miró-Roig, R., Nagel, U., Peterson, C. Gorenstein liaison, complete intersection liaison invariants and unobstructedness. *Mem. Amer. Math. Soc.* **154** (2001).
[18] Macaulay, F. S. On the resolution of a given modular equation into primary systems. *Math. Ann.* **74** (1913) 66–121.
[19] Martin-Deschamps, M., Perrin, D. Sur la classification des courbes gauches. *Astérisque* No. 184–185 (1990), 208 pp.
[20] Migliore, J. C. *Introduction to Liaison Theory and Deficiency Modules*. Progress in Mathematics, No. 165. Boston, MA: Birkhäuser, 1998.
[21] de Montmort, P. R. *Essay d'analyse sur les jeux de hazard.* Paris: Jacque Quillau. Seconde Edition, Revue et augmentée de plusieurs Lettres, 1713.
[22] Peskine, C., Szpiro, L. Liaison des variétés algébriques. I. *Invent. Math.* **26** (1974) 271–302.
[23] Rao, A. P. Liaison among curves in $\mathbb{P}^3$. *Invent. Math.* **50**(3) (1978/79) 205–217.
[24] Reifegerste, A. Enumeration of special sets of polynomials over finite fields. *Finite Fields Appl.* **5**(2) (1999) 112–156.
[25] Schenzel, P. Notes on liaison and duality. *J. Math. Kyoto Univ.* **22**(3) (1982/83) 485–498.
[26] Schreyer, F.-O. Computer aided unirationality proofs of moduli spaces. In: Farkas, G., Morrison, I. (eds), *Handbook of Moduli*, Vol. 3, International Press, Somerville, Massachusetts, USA (2013), 257–280.
[27] Serre, J.-P. Modules projectifs et espaces fibrés à fibre vectorielle. 1958 Séminaire P. Dubreil, M.-L. Dubreil-Jacotin et C. Pisot, 1957/58, Fasc. 2, Exposé 23, 18 pp. Paris: Secrétariat mathématique.
[28] Stanley, R. P. Hilbert functions of graded algebras. *Adv. Math.* **28** (1978) 57–83.
[29] Verra, A. The unirationality of the moduli spaces of curves of genus 14 or lower. *Compos. Math.* **141**(6) (2005) 1425–1444.

11

The Betti table of a high-degree curve is asymptotically pure

D. Erman[a]
University of Wisconsin

Dedicated to Rob Lazarsfeld on the occasion of his 60th birthday

1 Introduction

Syzygies can encode subtle geometric information about an algebraic variety, with the most famous examples coming from the study of smooth algebraic curves. Though little is known about the syzygies of higher-dimensional varieties, Ein and Lazarsfeld have shown that at least the asymptotic behavior is uniform [1]. More precisely, given a projective variety $X \subseteq \mathbb{P}^n$ embedded by the very ample bundle A, Ein and Lazarsfeld ask: which graded Betti numbers are nonzero for X re-embedded by dA? They prove that, asymptotically in d, the answer (or at least the main term of the answer) only depends on the dimension of X.

Boij–Söderberg theory [4] provides refined invariants of a graded Betti table, and it is natural to ask about the asymptotic behavior of these Boij–Söderberg decompositions. In fact, this problem is explicitly posed by Ein and Lazarsfeld [1, Problem 7.4], and we answer their question for smooth curves in Theorem 3.

Fix a smooth curve C and a sequence $\{A_d\}$ of increasingly positive divisors on C. We show that, as $d \to \infty$, the Boij–Söderberg decomposition of the Betti table of C embedded by $|A_d|$ is increasingly dominated by a single pure diagram that depends only on the genus of the curve. The proof combines an explicit computation about the numerics of pure diagrams with known facts about when an embedded curve satisfies Mark Green's N_p-condition.

[a] Research of the author partially supported by the Simons Foundation.

From *Recent Advances in Algebraic Geometry*, edited by Christopher Hacon, Mircea Mustaţă and Mihnea Popa

2 Setup

We work over an arbitrary field $\mathbf{k}$. Throughout, we will fix a smooth curve C of genus g and a sequence $\{A_d\}$ of line bundles of increasing degree. Since we are interested in asymptotics, we assume that for all d, $\deg A_d \geq 2g+1$. Let $r_d := \dim H^0(C, A_d) - 1 = \deg A_d - g$ so that the complete linear series $|A_d|$ embeds $C \subseteq \mathbb{P}^{r_d}$. For each d, we consider the homogeneous coordinate ring $R(C, A_d) := \oplus_{e\geq 0} H^0(C, eA_d)$ of this embedding. We may then consider $R(C, A_d)$ as a graded module over the polynomial ring $\operatorname{Sym}(H^0(C, A_d))$.

If $\mathbf{F} = [\mathbf{F}_0 \leftarrow \mathbf{F}_1 \leftarrow \cdots \leftarrow \mathbf{F}_n \leftarrow 0]$ is a minimal graded free resolution of $R(C, A_d)$, then we will use $\beta_{i,j}(\mathcal{O}_C, A_d)$ to denote the number of minimal generators of $\mathbf{F}_i$ of degree j. Equivalently, we have

$$\beta_{i,j}(\mathcal{O}_C, A_d) = \dim_{\mathbf{k}} \operatorname{Tor}_i^{\operatorname{Sym}(H^0(C,A_d))}(R(C, A_d), \mathbf{k})_j.$$

We define the **graded Betti table** $\beta(\mathcal{O}_C, A_d)$ as the vector with coordinates $\beta_{i,j}(\mathcal{O}_C, A_d)$ in the vector space $\mathbf{V} = \bigoplus_{i=0}^{n} \bigoplus_{j\in\mathbb{Z}} \mathbb{Q}$.

We use the standard Macaulay2 notation for displaying Betti tables, where

$$\beta = \begin{pmatrix} \beta_{0,0} & \beta_{1,1} & \beta_{2,2} & \cdots \\ \beta_{0,1} & \beta_{1,2} & \beta_{2,3} & \cdots \\ \beta_{0,2} & \beta_{1,3} & \beta_{2,4} & \cdots \\ \vdots & \vdots & \vdots & \ddots \end{pmatrix}.$$

Boij–Söderberg theory focuses on the rational cone spanned by all graded Betti tables in $\mathbf{V}$. The extremal rays of this cone correspond to certain **pure diagrams**, and hence every graded Betti table can be written as a positive rational sum of pure diagrams; this decomposition is known as a **Boij–Söderberg decomposition**. For a good introduction to the theory, see either [5] or [6]. We introduce only the notation and results that we need.

For a given d and some $i \in [0, g]$, we define the (degree) sequence $\mathbf{e} = \mathbf{e}(i, d) := (0, 2, 3, 4, \ldots, r_d - i, r_d - i + 2, r_d - i + 3, \ldots, r_d + 1) \in \mathbb{Z}^{r_d - 1}$, and we define the pure diagram $\pi_{i,d} \in \mathbf{V}$ by the formula

$$\beta_{p,q}(\pi_{i,d}) = \begin{cases} (r_d - 1)! \cdot \left(\prod_{\ell \neq p} \frac{1}{|\mathbf{e}_\ell - \mathbf{e}_p|} \right) & \text{if } p \in [0, r_d - 1] \text{ and } q = \mathbf{e}_p \\ 0 & \text{else.} \end{cases} \tag{1}$$

Note that the shape of $\pi_{i,d}$ is the following, where $*$ indicates a nonzero entry:

$$\pi_{i,d} = \begin{array}{c} \\ \\ \\ \end{array} \begin{array}{cccccccc} 0 & 1 & \ldots & r_d - i - 1 & r_d - i & \ldots & r_d - 1 \\ \end{array}$$

$$\pi_{i,d} = \begin{pmatrix} * & 0 & \ldots & 0 & 0 & \ldots & 0 \\ 0 & * & \ldots & * & 0 & \ldots & 0 \\ 0 & 0 & \ldots & 0 & * & \ldots & * \end{pmatrix}.$$

It turns out that these are the only pure diagrams that appear in the Boij–Södergberg decomposition of the Betti tables $\beta(C, A_d)$ (see Lemma 2 below).

We next recall the notion of a (reduced) Hilbert numerator, which will be central to our proof. If $S = \mathbf{k}[x_0, \ldots, x_n]$ is a polynomial ring, and M is a graded S-module, then the **Hilbert series** of a finitely generated, graded module M is the power series $\mathrm{HS}_M(t) := \sum_{i \in \mathbb{Z}} \dim_{\mathbf{k}} M_i \cdot t^i \in \mathbb{Q}[[t]]$. The Hilbert series can be written uniquely as a rational function of the form

$$\mathrm{HS}_M(t) = \frac{\mathrm{HN}_M(t)}{(1-t)^{\dim M}}$$

and we define the **Hilbert numerator** of M as the polynomial $\mathrm{HN}_M(t)$. The **multiplicity** of M is $\mathrm{HN}_M(1)$.

As is standard in Boij–Söderberg theory, we allow formal rescaling of Betti tables by rational numbers. Note the Hilbert numerator is invariant under modding out by a regular linear form or adjoining an extra variable; also, the Hilbert numerator is computable entirely in terms of the graded Betti table (see [2, Section 1]). Similar statements hold for the codimension of a module. Thus we may and do formally extend the notions of Hilbert numerator, codimension, and multiplicity to all elements of the vector space **V**.

Lemma 1 *For any i, d, the diagram $\pi_{i,d}$ has multiplicity* 1.

Proof By (1) we have $\beta_{0,0}(\pi_{i,d}) = \frac{(r_d-1)!}{2 \cdot 3 \cdots (r_d-i)\cdot(r_d-i+2)\cdots(r_d+1)}$. Up to a positive scalar multiple, the diagram $\pi_{i,d}$ equals the graded Betti table of a Cohen–Macaulay module by [4, Theorem 0.1]. Then by Huneke and Miller's multiplicity computation for Cohen–Macaulay modules with a pure resolution[1] [9, Proof of Theorem 1.2], it follows that the multiplicity of $\pi_{i,d}$ equals

$$\begin{aligned}\beta_{0,0}(\pi_{i,d}) \cdot \frac{2 \cdot 3 \cdots (r_d - i) \cdot (r_d - i + 2) \cdots (r_d + 1)}{(r_d - 1)!} &= \beta_{0,0}(\pi_{i,d}) \cdot (\beta_{0,0}(\pi_{i,d}))^{-1} \\ &= 1.\end{aligned}$$

□

3 Main result and proof

To make sensible comparisons between the graded Betti tables $\beta(\mathcal{O}_C, A_d)$ for different values of d, we will rescale by the degree of the curve so that we are always considering Betti tables of (formal) multiplicity equal to 1. Namely, we define

[1] Strictly speaking, Huneke and Miller's computation is for graded algebras. But by including a $\beta_{0,0}$ factor, the argument goes through unchanged for a graded Cohen–Macaulay module generated in degree 0 and with a pure resolution.

$$\overline{\beta}(\mathcal{O}_C, A_d) := \frac{1}{\deg A_d} \cdot \beta(\mathcal{O}_C, A_d).$$

The Boij–Söderberg decomposition of $\overline{\beta}(\mathcal{O}_C, A_d)$ has a relatively simple form.

Lemma 2 *For any d, the Boij–Söderberg decomposition of $\overline{\beta}(\mathcal{O}_C, A_d)$ has the form*

$$\overline{\beta}(\mathcal{O}_C, A_d) = \sum_{i=0}^{g} c_{i,d} \cdot \pi_{i,d}, \tag{2}$$

where $c_{i,d} \in \mathbb{Q}_{\geq 0}$ and $\sum_i c_{i,d} = 1$.

The above lemma shows that the number of potential pure diagrams appearing in the decomposition of $\overline{\beta}(C, A_d)$ is at most $g + 1$. Note that the precise number of summands with a nonzero coefficient is closely related to Green and Lazarsfeld's gonality conjecture [8, Conjecture 3.7], and hence will vary even among curves of the same genus. However our main result, which we now state, shows that this variance plays a minor role in the asymptotics.

Theorem 3 *The Betti table $\overline{\beta}(\mathcal{O}_C, A_d)$ converges to the pure diagram $\pi_{g,d}$ in the sense that*

$$c_{i,d} \to \begin{cases} 0 & i \neq g \\ 1 & i = g \end{cases} \quad \text{as} \quad d \to \infty.$$

In particular, the limiting pure diagram only depends on the genus of the curve. A nearly equivalent statement of the theorem is: asymptotically in d, the main term of the Boij–Söderberg decomposition of the (unscaled) Betti table $\beta(C, A_d)$ is the $\pi_{g,d}$ summand.

Proof of Lemma 2 Since the homogeneous coordinate ring of $C \subseteq \mathbb{P}^{r_d}$ is Cohen–Macaulay (see [2, Section 8A] for a proof and the history of this fact), it follows from [4, Theorem 0.2] that $\overline{\beta}(\mathcal{O}_C, A_d)$ can be written as a positive rational sum of pure diagrams of codimension $r_d - 1$. Since $C \subseteq \mathbb{P}^{r_d}$ satisfies the N_p condition for $p = r_d - g - 1$ by [7, Theorem 4.a.1], it follows that the shape of $\beta(\mathcal{O}_C, A_d)$ is

$$\begin{array}{cccccccc} 0 & 1 & 2 & \dots & r_d - g - 1 & r_d - g & \dots & r_d - 1 \end{array}$$
$$\begin{pmatrix} * & - & \dots & - & - & - & \dots & - \\ - & * & * & \dots & * & * & \dots & * \\ - & - & - & \dots & - & * & \dots & * \end{pmatrix}.$$

Thus the pure diagrams $\pi_{i,d}$ for $i = 0, 1, \ldots, g$ are the only diagrams that can appear in the Boij–Söderberg decomposition of $\overline{\beta}(C, A_d)$, and so we may write

$$\overline{\beta}(O_C, A_d) = \sum_{i=0}^{g} c_{i,d} \cdot \pi_{i,d}$$

with $c_{i,d} \in \mathbb{Q}_{\geq 0}$. The (formal) multiplicity of $\overline{\beta}(C, A_d)$ is 1 by construction, and the same holds for the $\pi_{i,d}$ by Lemma 1, so it follows that $\sum c_{i,d} = 1$. □

Lemma 4 *The Hilbert numerator of the pure diagram* $\pi_{i,d}$ *is*

$$\mathrm{HN}_{\pi_{i,d}}(t) = \left(\frac{r_d - i + 1}{r_d(r_d + 1)}\right)t^0 + \left(\frac{(r_d - 1)(r_d - i + 1)}{r_d(r_d + 1)}\right)t^1 + \left(\frac{i}{r_d + 1}\right)t^2.$$

The Hilbert numerator of the rescaled Betti table $\overline{\beta}(O_C, A_d)$ *is*

$$\left(\frac{1}{r_d + g}\right)t^0 + \left(\frac{r_d - 1}{r_d + g}\right)t^1 + \left(\frac{g}{r_d + g}\right)t^2.$$

Proof We prove the first statement by direct computation. Since $\pi_{i,d}$ represents, up to scalar multiple, the Betti table of a Cohen–Macaulay module M, we may assume by Artinian reduction that the module M has finite length. For a finite-length module, the Hilbert numerator equals the Hilbert series. Since the Betti table $\pi_{i,d}$ has two rows, it follows that the Castelnuovo–Mumford regularity of M equals 2 (except for $\pi_{0,d}$, which has regularity 1). The coefficient of t^0 is thus the value of the Hilbert function in degree 0, which is the 0th Betti number of the pure diagram $\pi_{i,d}$. By (1), this equals

$$\beta_{0,0}(\pi_{i,d}) = \frac{(r_d - 1)!}{2 \cdot 3 \cdots (r_d - i) \cdot (r_d - i + 2) \cdots (r_d + 1)} = \frac{r_d + 1 - i}{r_d(r_d + 1)}.$$

Similarly, the coefficient of t^2 is given by the bottom-right Betti number of $\pi_{i,d}$, which is

$$\beta_{r_d - 1, r_d + 1}(\pi_{i,d}) = \frac{i}{r_d + 1}.$$

Finally, since $\pi_{i,d}$ has multiplicity 1 by Lemma 1, it follows that $\mathrm{HN}_{\pi_{i,d}}(1) = 1$ and hence the coefficient of t^1 equals 1 minus the coefficients of t^0 and t^2:

$$1 - \left(\frac{r_d + 1 - i}{r_d(r_d + 1)}\right) - \left(\frac{i}{r_d + 1}\right) = \frac{r_d(r_d + 1) - (r_d - i + 1) - i \cdot r_d}{r_d(r_d + 1)}$$
$$= \frac{(r_d - 1)(r_d - i + 1)}{r_d(r_d + 1)}.$$

For the Hilbert numerator of $\overline{\beta}(O_C, A_d)$ statement, we note that $\deg A_d = r_d + g$, yielding

$$\overline{\beta}(O_C, A_d) = \tfrac{1}{r_d + g} \cdot \beta(O_C, A_d).$$

As above, we can compute the t^0 and t^2 coefficients via the first and last entries in the Betti table, and these are thus $\frac{1}{r_d+g}$ and $\frac{g}{r_d+g}$ respectively (see [2, Section 8A], for instance). Since $\overline{\beta}(O_C, A_d)$ has multiplicity 1, the t^1 coefficient is again 1 minus the t^0 and t^2 coefficients, and the statement follows immediately. □

Proof of Theorem 3 Note that $r_d \to \infty$ as $d \to \infty$. We rewrite the Hilbert numerator of $\pi_{i,d}$ as

$$\mathrm{HN}_{\pi_{i,d}}(t) = \left(\frac{1}{r_d} + \epsilon_{0,i,d}\right)t^0 + \left(1 + \frac{1}{r_d} + \epsilon_{1,i,d}\right)t^1 + \left(\frac{i}{r_d} + \epsilon_{2,i,d}\right)t^2,$$

where $r_d\epsilon_{j,i,d} \to 0$ as $d \to \infty$ for all $j = 0, 1, 2$ and $i = 0, \ldots, g$. For instance

$$\epsilon_{0,i,d} = \frac{r_d - i + 1}{r_d(r_d + 1)} - \frac{1}{r_d} = \frac{-i}{r_d(r_d + 1)}.$$

We may similarly rewrite the Hilbert numerator of $\overline{\beta}(O_C, A_d)$ as

$$\left(\tfrac{1}{r_d} + \delta_{0,d}\right)t^0 + \left(1 - \tfrac{g+1}{r_d} + \delta_{1,d}\right)t^1 + \left(\tfrac{g}{r_d} + \delta_{2,d}\right)t^2$$

where for $j = 0, 1, 2$ we have $r_d\delta_{j,d} \to 0$ as $d \to \infty$.

Since the Hilbert numerator is additive with respect to the Betti table decomposition of (2), combining the above computations with our Boij–Söderberg decomposition from (2), we see that the t^2 coefficient of the Hilbert numerator of $\overline{\beta}(O_C, A_d)$ may be written as

$$\tfrac{g}{r_d} + \delta_{2,d} = \sum_{i=0}^{g} c_{i,d} \cdot \left(\tfrac{i}{r_d} + \epsilon_{2,i,d}\right).$$

We multiply through by r_d and take the limit as $d \to \infty$. Since $r_d\delta_{j,d}$ and $r_d\epsilon_{j,i,d}$ both go to 0 as $d \to \infty$, this yields

$$g = \lim_{d\to\infty} \sum_{i=0}^{g} c_{i,d} \cdot i.$$

But $c_{i,d} \geq 0$ and $\sum_i c_{i,d} = 1$. Hence, as $d \to \infty$, we obtain $c_{i,d} \to 0$ for all $i \neq g$ and $c_{g,d} \to 1$. □

Remark If X is a variety with $\dim X > 1$, then our argument fails in several important ways. To begin with, Ein and Lazarsfeld's nonvanishing syzygy results from [1] show that the number of potential pure diagrams for the Boij–Söderberg decomposition of $\beta(X, A_d)$ is unbounded.

Moreover, in the case of curves, the Hilbert numerator of the embedded curves converged to the Hilbert numerator of one of the potential pure diagrams; the N_p condition then implied that this pure diagram had the largest degree sequence of any potential pure diagram. Our result then followed by the semicontinuous behavior of the Hilbert numerators of pure diagrams (for a related semicontinuity phenomenon, see [3, monotonicity principle, p. 758]). Ein and Lazarsfeld's asymptotic nonvanishing results imply that, even for $\mathbb{P}^2$, the limit of the Hilbert numerator will fail to correspond to an extremal potential pure diagram, and so the semicontinuity does not obviously help.

Acknowledgments The questions considered in this paper arose in conversations with Rob Lazarsfeld, and in addition I learned a tremendous amount about these topics from him. It is a great pleasure to thank Rob Lazarsfeld for his influence and his superb mentoring. I also thank Lawrence Ein and David Eisenbud for helpful insights and conversations related to this paper. I thank Christine Berkesch, Frank-Olaf Schreyer, and the referee for comments that improved the chapter.

References

[1] Ein, L. and Lazarsfeld, R. Asymptotic syzygies of algebraic varieties, 2011. arXiv:1103.0483.

[2] Eisenbud, D. The Geometry of Syzygies, Graduate Texts in Mathematics, Vol. 229. New York: Springer-Verlag, 2005.

[3] Eisenbud, D., Erman, D. and Schreyer, F. O. Filtering free resolutions. *Compos. Math.* **149** (2013), 754–772.

[4] Eisenbud, D. and Schreyer, F. O. Betti numbers of graded modules and cohomology of vector bundles. *J. Amer. Math. Soc.* **22**(3) (2009) 859–888.

[5] Eisenbud, D. and Schreyer, F. O. Betti numbers of syzygies and cohomology of coherent sheaves. In *Proceedings of the International Congress of Mathematicians*, Hyderabad, India, 2010.

[6] Fløystad, G. Boij-Söderberg theory: Introduction and survey. Progress in Commutative Algebra 1, No. 154. Berlin: de Gruyter, 2012.

[7] Green, M. L. Koszul cohomology and the geometry of projective varieties. *J. Diff. Geom.* **19** (1984) 125–171.

[8] Green, M. and Lazarsfeld, R. On the projective normality of complete linear series on an algebraic curve. *Invent. Math.* **83** (1985) 73–90.

[9] Huneke, C. and Miller, M. A note on the multiplicity of Cohen–Macaulay algebras with pure resolutions. *Canad. J. Math.* **37**(6) (1985) 1149–1162.

12

Partial positivity: Geometry and cohomology of q-ample line bundles

D. Greb
Ruhr-Universität Bochum

A. Küronya
Budapest University of Technology and Economics

Abstract

We give an overview of partial positivity conditions for line bundles, mostly from a cohomological point of view. Although the current work is to a large extent expository in nature, we present some minor improvements on the existing literature and a new result: a Kodaira-type vanishing theorem for effective q-ample Du Bois divisors and log canonical pairs.

To Rob Lazarsfeld on the occasion of his 60th birthday

1 Introduction

Ampleness is one of the central notions of algebraic geometry, possessing the extremely useful feature that it has geometric, numerical, and cohomological characterizations. Here we will concentrate on its cohomological side. The fundamental result in this direction is the theorem of Cartan–Serre–Grothendieck (see [Laz04, Theorem 1.2.6]): for a complete projective scheme X, and a line bundle $\mathcal{L}$ on X, the following are equivalent to $\mathcal{L}$ being ample:

1. There exists a positive integer $m_0 = m_0(X, \mathcal{L})$ such that $\mathcal{L}^{\otimes m}$ is very ample for all $m \geq m_0$.
2. For every coherent sheaf $\mathcal{F}$ on X, there exists a positive integer $m_1 = m_1(X, \mathcal{F}, \mathcal{L})$ for which $\mathcal{F} \otimes \mathcal{L}^{\otimes m}$ is globally generated for all $m \geq m_1$.
3. For every coherent sheaf $\mathcal{F}$ on X, there exists a positive integer $m_2 = m_2(X, \mathcal{F}, \mathcal{L})$ such that

From *Recent Advances in Algebraic Geometry*, edited by Christopher Hacon, Mircea Mustaţă and Mihnea Popa

$$H^i\left(X, \mathcal{F} \otimes \mathcal{L}^{\otimes m}\right) = \{0\}$$

for all $i \geq 1$ and all $m \geq m_2$.

We will focus on the direction pointed out by Serre's vanishing theorem, part 3 above, and concentrate on line bundles with vanishing cohomology above a certain degree.

Historically, the first result in this direction is due to Andreotti and Grauert [AG62]. They prove that given a compact complex manifold X of dimension n, and a holomorphic line bundle $\mathcal{L}$ on X equipped with a Hermitian metric whose curvature is a $(1, 1)$-form with at least $n - q$ positive eigenvalues at every point of X, then for every coherent sheaf $\mathcal{F}$ on X, there exists a natural number $m_0(\mathcal{L}, \mathcal{F})$ such that

$$H^i\left(X, \mathcal{F} \otimes \mathcal{L}^{\otimes m}\right) = \{0\} \qquad \text{for all } m \geq m_0(\mathcal{L}, \mathcal{F}) \text{ and for all } i > q. \quad (1)$$

In [DPS96], Demailly, Peternell, and Schneider posed the question of under what circumstances the converse does hold. That is, they asked: assume that for every coherent sheaf $\mathcal{F}$ there exists $m = m_0(\mathcal{L}, \mathcal{F})$ such that the vanishing (1) holds. Does $\mathcal{L}$ admit a Hermitian metric with the expected number of positive eigenvalues? In dimension two, Demailly [Dem11] proved an asymptotic version of this converse to the Andreotti–Grauert theorem using tools related to asymptotic cohomology; subsequently, Matsumura [Mat13] gave a positive answer to the question for surfaces. However, there exist higher-dimensional counterexamples to the converse Andreotti–Grauert problem in the range $\frac{\dim X}{2} - 1 < q < \dim X - 2$, constructed by Ottem [Ott12].

We study line bundles with the property of the conclusion of the Andreotti–Grauert theorem; let X be a complete scheme of dimension n, $0 \leq q \leq n$ an integer. A line bundle $\mathcal{L}$ is called *naively q-ample*, or simply *q-ample* if for every coherent sheaf $\mathcal{F}$ on X there exists an integer $m_0 = m_0(\mathcal{L}, \mathcal{F})$ for which

$$H^i\left(X, \mathcal{F} \otimes \mathcal{L}^{\otimes m}\right) = \{0\}$$

for $m \geq m_0$ and for all $i > q$.

It is a consequence of the Cartan–Grothendieck–Serre result discussed above that 0-ampleness reduces to the usual notion of ampleness. In [DPS96], the authors studied naive q-ampleness along with various other notions of partial cohomological positivity. Part of their approach is to look at positivity of restrictions to elements of a complete flag, and they use it to give a partial vanishing theorem similar to that of Andreotti–Grauert.

In a beautiful paper [Tot13], Totaro proves that the competing partial positivity concepts are in fact all equivalent in characteristic 0, thus laying down the foundations for a very satisfactory theory. His result goes as follows: let X

be a projective scheme of dimension n, $\mathcal{A}$ a sufficiently ample line bundle on X, $0 \leq q \leq n$ a natural number. Then for all line bundles $\mathcal{L}$ on X, the following are equivalent:

1. $\mathcal{L}$ is naively q-ample.
2. $\mathcal{L}$ is uniformly q-ample, that is, there exists a constant $\lambda > 0$ such that
$$H^i\left(X, \mathcal{L}^{\otimes m} \otimes \mathcal{A}^{\otimes -j}\right) = \{0\}$$
for all $i > q$, $j > 0$, and $m \geq \lambda j$.
3. There exists a positive integer $m_1 = m_1(\mathcal{L}, \mathcal{A})$ such that
$$H^{q+i}\left(X, \mathcal{L}^{\otimes m_1} \otimes \mathcal{A}^{\otimes -(n+i)}\right) = \{0\}$$
for all $1 \leq i \leq n - q$.

As an outcome, Totaro can prove that on the one hand, for a given q, the set of q-ample line bundles forms an open cone in the Néron–Severi space, and on the other hand, q-ampleness is an open property in families as well.

In his recent article [Ott12], Ottem works out other basic properties of q-ample divisors and employs them to study subvarieties of higher codimension. We will give an overview of his results in Section 2.4.

Interestingly enough, prior to [DPS96], Sommese [Som78] defined a geometric version of partial ampleness by studying the dimensions of the fibers of the morphism associated with a given line bundle. In the case of semi-ample line bundles, where Sommese's notion is defined, he proves his condition to be equivalent to naive q-ampleness. Although more limited in scope, Sommese's geometric notion extends naturally to vector bundles as well.

One of the major technical vanishing theorems for ample divisors that does not follow directly from the definition is Kodaira's vanishing theorem if X is a smooth projective variety, defined over an algebraically closed field of characteristic 0, and $\mathcal{L}$ is an ample line bundle on X, then
$$H^j\left(X, \omega_X \otimes \mathcal{L}\right) = \{0\} \quad \text{for all } j > 0.$$

We refer the reader to [Kod53] for the original analytic proof, to [DI87] for a subsequent algebraic proof, to [Ray78] for a counterexample in characteristic p, and to [Kol95, Chapter 9] as well as to [EV92] for a general discussion of vanishing theorems. It is a natural question to ask whether an analogous vanishing holds for q-ample divisors. While one of the ingredients of classical proofs of Kodaira vanishing, namely Lefschetz' hyperplane section theorem, has been generalized to the q-ample setup by Ottem [Ott12, Corollary 5.2], at the same time he gives a counterexample to the Kodaira vanishing theorem for q-ample divisors, which we recall in Section 2.4. In Ottem's example, the chosen q-ample divisor is not pseudo-effective. In Section 3 we show that there is a good reason for this: we prove that q-Kodaira vanishing holds for reduced effective divisors which are not too singular; more precisely, we prove two versions of q-Kodaira vanishing which are related to log canonicity and the Du Bois condition.

Theorem (= Theorem 3.4) *Let X be a normal proper variety, $D \subset X$ a reduced effective (Cartier) divisor such that*

1. *$\mathcal{O}_X(D)$ is q-ample;*
2. *$X \setminus D$ is smooth, and D (with its reduced subscheme structure) is Du Bois.*

Then, we have

$$H^j(X, \omega_X \otimes \mathcal{O}_X(D)) = \{0\} \quad \textit{for all } j > q$$

as well as

$$H^j(X, \mathcal{O}_X(-D)) = \{0\} \quad \textit{for all } j < \dim X - q.$$

Theorem (= Theorem 3.8) *Let X be a proper Cohen–Macaulay variety, $\mathcal{L}$ a q-ample line bundle on X, and D_i different irreducible Weil divisors on X. Assume that $\mathcal{L}^m \cong \mathcal{O}_X(\sum d_j D_j)$ for some integers $1 \leq d_j < m$. Set $m_j := m/\gcd(m, d_j)$. Assume furthermore that the pair $(X, \sum(1 - \frac{1}{m_j})D_j)$ is log canonical. Then, we have*

$$H^j(X, \omega_X \otimes \mathcal{L}) = \{0\} \quad \textit{for all } j > q$$

as well as

$$H^j\left(X, \mathcal{L}^{-1}\right) = \{0\} \quad \textit{for all } j < \dim X - q.$$

Global conventions If not mentioned otherwise, we work over the complex numbers, and all divisors are assumed to be Cartier.

2 Overview of the theory of q-ample line bundles

2.1 Vanishing of cohomology groups and partial ampleness

Starting with the pioneering work [DPS96] of Demailly, Peternell, and Schneider related to the Andreotti–Grauert problem, there has been a certain interest in studying line bundles with vanishing cohomology above a given degree. Just as big line bundles are a generalization of ample ones along its geometric side, these so-called q-ample bundles focus on a weakening of the cohomological characterization of ampleness. In general, there exist competing definitions of various flavors, which were shown to be equivalent in characteristic 0 in [Tot13].

Definition 2.1 (Definitions of partial ampleness) Let X be a complete scheme of dimension n over an algebraically closed field of arbitrary characteristic, $\mathcal{L}$ an invertible sheaf on X, q a natural number.

1. The invertible sheaf $\mathcal{L}$ is called *naively q-ample* if for every coherent sheaf $\mathcal{F}$ on X there exists a natural number $m_0 = m_0(\mathcal{L}, \mathcal{F})$ having the property that
$$H^i\left(X, \mathcal{F} \otimes \mathcal{L}^{\otimes m}\right) = \{0\} \quad \text{for all } i > q \text{ and } m \geq m_0.$$
2. Fix a very ample invertible sheaf $\mathcal{A}$ on X. We call $\mathcal{L}$ *uniformly q-ample* if there exists a constant $\lambda = \lambda(\mathcal{A}, \mathcal{L})$ such that
$$H^i\left(X, \mathcal{L}^{\otimes m} \otimes \mathcal{A}^{\otimes -j}\right) = \{0\} \quad \text{for all } i > q,\ j > 0, \text{ and } m \geq \lambda \cdot j.$$
3. Fix a Koszul-ample invertible sheaf $\mathcal{A}$ on X. We say that $\mathcal{L}$ is *q-T-ample* if there exists a positive integer $m_1 = m_1(\mathcal{A}, \mathcal{L})$ satisfying
$$H^{q+1}\left(X, \mathcal{L}^{\otimes m_1} \otimes \mathcal{A}^{\otimes -(n+1)}\right) = H^{q+2}\left(X, \mathcal{L}^{\otimes m_1} \otimes \mathcal{A}^{\otimes -(n+2)}\right)$$
$$\ldots = H^n\left(X, \mathcal{L}^{\otimes m_1} \otimes \mathcal{A}^{\otimes -2n+q}\right) = \{0\}.$$

An integral Cartier divisor is called *naively q-ample/uniformly q-ample/q-T-ample* if the invertible sheaf $\mathcal{O}_X(D)$ has the appropriate property.

Remark 2.2 (Koszul-ampleness) We recall that a connected locally finite[1] graded ring $R_\bullet = \oplus_{i=0}^{\infty} R_i$ is called N-Koszul for a positive integer N if the field $k = R_0$ has a resolution
$$\ldots \longrightarrow M_1 \longrightarrow M_0 \longrightarrow k \longrightarrow 0$$
as a graded $R_\bullet$-module, where for all $i \geq N$ the module M_i is free and generated in degree i.

In turn, a very ample line bundle $\mathcal{A}$ on a projective scheme X (taken to be connected and reduced to arrange that the ring of its regular functions $k \stackrel{\text{def}}{=} \mathcal{O}_X(X)$ is a field) is called N-Koszul if the section ring $R(X, \mathcal{A})$ is N-Koszul. The line bundle $\mathcal{A}$ is said to be *Koszul-ample* if it is N-Koszul with $N = 2 \dim X$. It is important to point out that Castelnuovo–Mumford regularity with respect to a Koszul-ample line bundle has favorable properties.

[1] In this context locally finite means that $\dim R_i < \infty$ for all graded pieces.

If $\mathcal{A}$ is an arbitrary ample line bundle on X, then Backelin [Bac86] showed that there exists a positive integer $k_0 \in \mathbb{N}$ having the property that $\mathcal{A}^{\otimes k}$ is Koszul-ample for all $k \geq k_0$.

Remark 2.3 The idea of naive q-ampleness is the immediate extension of the Grothendieck–Cartan–Serre vanishing criterion for ampleness. Uniform q-ampleness first appeared in [DPS96]; the term q-T-ampleness was coined by Totaro in [Tot13, Section 7], extending the idea of Castelnuovo–Mumford regularity.

A line bundle is ample if and only if it is naively 0-ample, while all line bundles are $n = \dim X$-ample.

Example 2.4 One source of examples of q-ample divisors comes from ample vector bundles: according to [Ott12, Proposition 4.5], if $\mathcal{E}$ is an ample vector bundle of rank $r \leq \dim X$ on a scheme X, $s \in H^0(X, \mathcal{E})$, then $Y \stackrel{\text{def}}{=} Z(s) \subseteq X$ is an ample subvariety. By definition, this means that $\mathcal{O}_{X'}(E)$ is $(r-1)$-ample, where $\pi\colon X' \to X$ is the blow-up of X along Y with exceptional divisor E, cf. Definition 2.48.

Remark 2.5 A straightforward sufficient condition for (naive) q-ampleness can be obtained by studying restrictions of $\mathcal{L}$ to general complete intersection subvarieties. More specifically, the following claim is shown in [Kür10, Theorem A]: let X be a projective variety over the complex numbers, L a Cartier divisor, $A_1, \ldots, A_q$ very ample Cartier divisors on X such that $L|_{E_1 \cap \cdots \cap E_q}$ is ample for general $E_j \in |A_j|$. Then, for any coherent sheaf $\mathcal{F}$ on X there exists an integer $m(L, A_1, \ldots, A_q, \mathcal{F})$ such that

$$H^i(X, \mathcal{F} \otimes \mathcal{O}_X(mL + N + \sum_{j=1}^{q} k_j A_j)) = \{0\}$$

for all $i > q$, $m \geq m(L, A_1, \ldots, A_q, \mathcal{F})$, $k_j \geq 0$, and all nef divisors N. In particular, the conditions of the above claim ensure that L is q-ample.

Remark 2.6 There are various implications among the three definitions over an arbitrary field. As was verified in [DPS96, Proposition 1.2] via an argument resolving coherent sheaves by direct sums of ample line bundles,[2] a uniformly q-ample line bundle is necessarily naively q-ample. On the other hand, naive q-ampleness implies q-T-ampleness by definition.

[2] Note that the resulting resolution is not guaranteed to be finite; see [Laz04, Example 1.2.21] for a discussion of this possibility.

Remark 2.7 The idea behind the definition of q-T-ampleness is to reduce the question of q-ampleness to the vanishing of finitely many cohomology groups. It is not known whether the three definitions coincide in positive characteristic.

The following result of Totaro compares the three different approaches to q-ampleness in characteristic 0.

Theorem 2.8 (Totaro; [Tot13, Theorem 8.1]) *Over a field of characteristic* 0, *the three definitions of partial ampleness are equivalent.*

The proof uses methods from positive characteristics to generalize a vanishing result of Arapura [Ara06, Theorem 5.4], at the same time it relies on earlier work of Orlov and Kawamata on resolutions of the diagonal via Koszul-ample line bundles.

Remark 2.9 (Resolution of the diagonal) One of the main building blocks of [Tot13] is an explicit resolution of the diagonal as a sheaf on $X \times X$ depending on an ample line bundle $\mathcal{A}$ on X. This is used as a tool to prove an important result on the regularity of tensor products of sheaves (see Theorem 2.10 below, itself an improvement over a statement of Arapura [Ara06, Corollary 1.12]), which in turn is instrumental in showing that q-T-ampleness implies uniform q-ampleness, the nontrivial part of Totaro's theorem on the equivalence of the various definitions of partial vanishing.

The resolution in question – which exists over an arbitrary field – had first been constructed by Orlov [Orl97, Proposition A.1] under the assumption that $\mathcal{A}$ is sufficiently ample, and subsequently improved by Kawamata [Kaw04] by making the more precise assumption that the coordinate ring $R(X, \mathcal{A})$ is a Koszul algebra.

Totaro reproves Kawamata's result under the weaker hypothesis that for $\mathcal{A}$, the section ring $R(X, \mathcal{A})$ is N-Koszul for some positive integer N (for the most of [Tot13] one will set $N = 2 \dim X$).

To construct the Kawamata–Orlov–Totaro resolution of the diagonal, we will proceed as follows. First, we define a sequence of k-vector spaces B_i by setting

$$B_i \stackrel{\text{def}}{=} \begin{cases} k & \text{if } i = 0, \\ H^0(X, \mathcal{A}) & \text{if } i = 1, \\ \ker\left(B_{i-1} \otimes H^0(X, \mathcal{A}) \longrightarrow B_{i-2} \otimes H^0\left(X, \mathcal{A}^2\right)\right) & \text{if } i \geq 2. \end{cases}$$

Note that $\mathcal{A}$ is N-Koszul precisely if the following sequence of graded $R(X, \mathcal{A})$-modules cooked up from the B_is is exact:

$$B_N \otimes R(X, \mathcal{A})(-N) \longrightarrow \ldots \longrightarrow B_1 \otimes R(X, \mathcal{A})(-1) \longrightarrow R(X, \mathcal{A}) \longrightarrow k \longrightarrow 0 .$$

Next, set

$$\mathcal{R}_i \stackrel{\text{def}}{=} \begin{cases} \mathcal{O}_X & \text{if } i = 0\,, \\ \ker\left(B_i \otimes \mathcal{O}_X \longrightarrow B_{i-1} \otimes \mathcal{A}\right) & \text{if } i > 0\,. \end{cases}$$

Totaro's claim [Tot13, Theorem 2.1] goes as follows: if $\mathcal{A}$ is an N-Koszul line bundle on X, then there exists an exact sequence

$$\mathcal{R}_{N-1} \boxtimes \mathcal{A}^{-N+1} \longrightarrow \cdots \longrightarrow \mathcal{R}_1 \boxtimes \mathcal{A}^{-1} \longrightarrow \mathcal{R}_0 \boxtimes \mathcal{O}_X \longrightarrow \mathcal{O}_\Delta \longrightarrow 0\,,$$

where $\boxtimes$ denotes the external tensor product on $X \times X$, and $\Delta \subset X \times X$ stands for the diagonal.

The above construction leads to a statement of independent interest about the regularity of tensor products of sheaves, which had already appeared in some form in [Ara06].

Theorem 2.10 (Totaro; [Tot13, Theorem 3.4]) *Let X be a connected and reduced projective scheme of dimension n, $\mathcal{A}$ a $2n$-Koszul line bundle, $\mathcal{E}$ a vector bundle, $\mathcal{F}$ a coherent sheaf on X. Then,*

$$\operatorname{reg}(\mathcal{E} \otimes \mathcal{F}) \le \operatorname{reg}(\mathcal{E}) + \operatorname{reg}(\mathcal{F})\,.$$

Remark 2.11 In the case of $X = \mathbb{P}^n_{\mathbb{C}}$ the above theorem is a simple application of Koszul complexes (see [Laz04, Proposition 1.8.9], for instance).

Remark 2.12 (Positive characteristic methods) Another crucial point in the proof of the equivalence of the various definitions of q-ampleness is a vanishing result in positive characteristic originating in the work of Arapura in the smooth case [Ara06, Theorem 5.3] that was extended to possibly singular schemes by Totaro [Tot13, Theorem 5.1] exploiting a flatness property of the Frobenius over arbitrary schemes over fields of prime cardinality.

The statement is essentially as follows: let X be a connected and reduced projective scheme of dimension n over a field of positive characteristic p, $\mathcal{A}$ a Koszul-ample line bundle on X, q a natural number. Let $\mathcal{L}$ be a line bundle on X satisfying

$$H^{q+1}\left(X, \mathcal{L} \otimes \mathcal{A}^{\otimes(-n-1)}\right) = H^{q+2}\left(X, \mathcal{L} \otimes \mathcal{A}^{\otimes(-n-2)}\right) = \ldots = \{0\}\,.$$

Then, for any coherent sheaf $\mathcal{F}$ on X one has

$$H^i\left(X, \mathcal{L}^{\otimes p^m} \otimes \mathcal{F}\right) = \{0\} \text{ for all } i > q \text{ and } p^m \ge \operatorname{reg}_{\mathcal{A}}(\mathcal{F}).$$

2.2 Basic properties of q-ampleness

From now on we return to our blanket assumption and work over the complex number field; we will immediately see that the equivalence of the various definitions brings all sorts of perks.

First, we point out that q-ampleness enjoys many formal properties analogous to ampleness. The following statements have been part of the folklore; for precise proofs we refer the reader to [Ott12, Proposition 2.3] and [DPS96, Lemma 1.5].

Lemma 2.13 *Let X be a projective scheme, $\mathcal{L}$ a line bundle on X, q a natural number. Then, the following hold:*

1. *$\mathcal{L}$ is q-ample if and only if $\mathcal{L}|_{X_{\text{red}}}$ is q-ample on X_{red}.*
2. *$\mathcal{L}$ is q-ample precisely if $\mathcal{L}|_{X_i}$ is q-ample on X_i for every irreducible component X_i of X.*
3. *For a finite morphism $f\colon Y \to X$, if $\mathcal{L}$ on X is q-ample then so is $f^*\mathcal{L}$. Conversely, if f is surjective as well, then the q-ampleness of $f^*\mathcal{L}$ implies the q-ampleness of $\mathcal{L}$.*

The respective proofs of the ample case go through with minimal modifications. Another feature surviving in an unchanged form is the fact that to check (naive) q-ampleness we can restrict our attention to line bundles.

Lemma 2.14 *Let X be a projective scheme, $\mathcal{L}$ a line bundle, $\mathcal{A}$ an arbitrary ample line bundle on X. Then, $\mathcal{L}$ is q-ample precisely if there exists a natural number $m_0 = m_0(\mathcal{A}, \mathcal{L})$ having the property that*

$$H^i\left(X, \mathcal{L}^{\otimes m} \otimes \mathcal{A}^{\otimes -k}\right) = \{0\}$$

for all $i > q$, $k \geq 0$, and $m \geq m_0 k$.

Proof Follows immediately by decreasing induction on q from the fact that every coherent sheaf $\mathcal{F}$ on X has a possibly infinite resolution by finite direct sums of non-positive powers of the ample line bundle $\mathcal{A}$ ([Laz04, Example 1.2.21]). □

Ample line bundles are good to work with for many reasons, but the fact that they are open both in families and in the Néron–Severi space contributes considerably. As it turns out, the same properties are valid for q-ample line bundles as well.

Theorem 2.15 (Totaro; [Tot13, Theorem 9.1]) *Let $\pi\colon X \to B$ be a flat projective morphism of schemes (over $\mathbb{Z}$) with connected fibers, $\mathcal{L}$ a line bundle* *on X, q a natural number. Then, the subset of points b of B having the property that $\mathcal{L}|_{X_b}$ is q-ample is Zariski open.*

Sketch of proof This is one point where q-T-ampleness plays a role, since in that formulation one only needs to check vanishing for a finite number of cohomology groups.

Assume that $\mathcal{L}|_{X_b}$ is q-T-ample for a given point $b \in B$; let U be an affine open neighborhood on $b \in B$, and $\mathcal{A}$ a line bundle on $\pi^{-1}(U) \subseteq X$, whose restriction to X_b is Koszul-ample. Since Koszul-ampleness is a Zariski-open property, $\mathcal{A}|_{X_{b'}}$ is again Koszul-ample for an open subset of points $b' \in U$; without loss of generality we can assume that this holds on the whole of U.

We will use the line bundle $\mathcal{A}|_{X_{b'}}$ to check q-T-ampleness of $\mathcal{L}_{X_{b'}}$ in an open neighborhood on $b \in U$. By the q-T-ampleness of $\mathcal{L}|_{X_b}$ there exists a positive integer m_0 satisfying

$$H^{q+1}\left(X_b, \mathcal{L}^{\otimes m_0} \otimes \mathcal{A}^{\otimes -n-1}\right) = \ldots = H^n\left(X_b, \mathcal{L}^{\otimes m_0} \otimes \mathcal{A}^{\otimes -2n+q}\right) = \{0\} .$$

It follows from the semicontinuity theorem that the same vanishing holds true for points in an open neighborhood of b. □

In a different direction, Demailly, Peternell, and Schneider proved that uniform q-ampleness is open in the Néron–Severi space. To make this precise we need the fact that uniform q-ampleness is a numerical property; once this is behind us, we can define q-ampleness for numerical equivalence classes of $\mathbb{R}$-divisors.

Remark 2.16 Note that a line bundle $\mathcal{L}$ is q-ample if and only if $\mathcal{L}^{\otimes m}$ is q-ample for some positive integer m. Therefore it makes sense to talk about q-ampleness of $\mathbb{Q}$-Cartier divisors; a $\mathbb{Q}$-divisor D is said to be q-ample if it has a multiple mD that is integral and $\mathcal{O}_X(mD)$ is q-ample.

Theorem 2.17 (q-ampleness is a numerical property) *Let D and D' be numerically equivalent integral Cartier divisors on an irreducible complex projective variety X, q a natural number. Then,*

$$D \text{ is } q\text{-ample} \quad \Leftrightarrow \quad D' \text{ is } q\text{-ample}.$$

Demailly, Peternell, and Schneider [DPS96, Proposition 1.4] only prove this claim for smooth projective varieties. The proof in [DPS96] cites the completeness of $\mathrm{Pic}^0(X)$, hence it is far from obvious how to extend it. Here we give a proof that is valid under the more general given hypothesis. Instead of dealing with uniform q-ampleness, we use the naive formulation.

Proof Let $\mathcal{N}'$ be a numerically trivial line bundle, $\mathcal{L}$ a q-ample line bundle on X. This means that for a given coherent sheaf $\mathcal{F}$, we have

$$H^i\left(X, \mathcal{L}^{\otimes m} \otimes \mathcal{F}\right) = \{0\} \quad \text{for } i > q \text{ and } m_0 = m_0(\mathcal{L}, \mathcal{F}).$$

We need to prove that

$$H^i\left(X, (\mathcal{L} \otimes \mathcal{N}')^{\otimes m} \otimes \mathcal{F}\right) = \{0\}$$

holds for all $i > q$, and for suitable $m \geq m_1(\mathcal{L}, \mathcal{N}', \mathcal{F})$. To this end, we will study the function

$$f_i^{(m)} : \mathcal{N} \mapsto h^i\left(X, \mathcal{L}^{\otimes m} \otimes \mathcal{N} \otimes \mathcal{F}\right)$$

as a function on the closed points of the subscheme $\mathcal{X}$ of the Picard scheme that parameterizes numerically trivial line bundles on X, which is a scheme of finite type by the boundedness of numerically trivial line bundles [Laz04, Theorem 1.4.37]. We know that

$$f_i^{(m)}(\mathcal{O}_X) = h^i\left(X, \mathcal{L}^{\otimes m} \otimes \mathcal{F}\right) = \{0\}$$

for $i > q$, and $m \geq m_0$. By the semicontinuity theorem, $f_i^{(m)}$ attains the same value on a dense open subset of $\mathcal{X}$.

By applying Noetherian induction and semicontinuity on the irreducible components of the complement (on each of which we apply the q-ampleness of $\mathcal{L}$ for coherent sheaves of the shape $\mathcal{F} \otimes \mathcal{N}$, $\mathcal{N}$ numerically trivial) we will eventually find a value $m_0' = m_0'(\mathcal{L}, \mathcal{F})$ such that

$$f_i^{(m)} \equiv 0$$

for all $i > q$ and $m \geq m_0'$.

But this implies that

$$H^i\left(X, (\mathcal{L} \otimes \mathcal{N}')^{\otimes m} \otimes \mathcal{F}\right) = H^i\left(X, \mathcal{L}^{\otimes m} \otimes ((\mathcal{N}')^{\otimes m}) \otimes \mathcal{F}\right) = \{0\}$$

for $m \geq m_0'$, since the required vanishing holds for an arbitrary numerically trivial divisor in place of $(\mathcal{N}')^{\otimes m}$. □

Remark 2.18 As a result, we are in a position to extend the definition of q-ampleness elements of $N^1(X)_{\mathbb{Q}}$: if α is a numerical equivalence class of $\mathbb{Q}$-divisors, then we will call it q-ample if one (equivalently all) of its representatives are q-ample.

Remark 2.19 Since Fujita's vanishing theorem also holds over algebraically closed fields of positive characteristic (see [Fuj83] or [Laz04, Remark 1.4.36]), the boundedness of numerically trivial line bundles holds again in this case by

[Laz04, Proposition 1.4.37] and we can conclude that naive q-ampleness is invariant with respect to numerical equivalence in that situation as well.

Definition 2.20 (q-ampleness for $\mathbb{R}$-divisors) An $\mathbb{R}$-divisor D on a complex projective variety is *q-ample* if

$$D = D' + A ,$$

where D' is a q-ample $\mathbb{Q}$-divisor, A an ample $\mathbb{R}$-divisor.

The result that q-ample $\mathbb{R}$-divisors form an open cone in $N^1(X)_{\mathbb{R}}$ was proved in [DPS96]. Here we face the same issue as with Theorem 2.17: in the article [DPS96] only the smooth case is considered, and the proof given there does not seem to generalize to general varieties. Here we present a proof of the general case.

Definition 2.21 Given $\alpha \in N^1(X)_{\mathbb{R}}$, we set

$$q(\alpha) \stackrel{\text{def}}{=} \min\{q \in \mathbb{N} \mid \alpha \text{ is } q\text{-ample}\} .$$

Theorem 2.22 (Upper-semicontinuity of q-ampleness) *Let X be an irreducible projective variety over the complex numbers. Then, the function*

$$q : N^1(X)_{\mathbb{R}} \longrightarrow \mathbb{N}$$

is upper-semicontinuous. In particular, for a given $q \in \mathbb{N}$, the set of q-ample classes forms an open cone.

In order to be able to prove this result, we need some auxiliary statements. To this end, Demailly, Peternell, and Schneider introduce the concept of height of coherent sheaves with respect to a given ample divisor. Roughly speaking, the height of a coherent sheaf tells us, what multiples of the given ample divisor we need to obtain a linear resolution.

Definition 2.23 (Height) Let X be an irreducible projective variety, $\mathcal{F}$ a coherent sheaf, $\mathcal{A}$ an ample line bundle. Consider the set $\mathcal{R}$ of all resolutions

$$\ldots \to \bigoplus_{1 \le l \le m_k} \mathcal{A}^{\otimes - d_{k,l}} \to \ldots \to \bigoplus_{1 \le l \le m_0} \mathcal{A}^{\otimes - d_{0,l}} \to \mathcal{F} \to 0$$

of $\mathcal{F}$ by non-positive powers of $\mathcal{A}$ (that is, $d_{k,l} \geq 0$). Then,

$$\mathrm{ht}_{\mathcal{A}}(\mathcal{F}) \stackrel{\text{def}}{=} \min_{\mathcal{R}} \max_{0 \le k \le \dim X, 1 \le l \le m_k} d_{k,l} .$$

Remark 2.24 One could define the height by looking at resolutions without truncating, that is, by

$$\widetilde{\mathrm{ht}}_{\mathcal{A}}(\mathcal{F}) \stackrel{\text{def}}{=} \min_{\mathcal{R}} \max_{0 \le k, 1 \le l \le m_k} d_{k,l} .$$

On a general projective variety there might be sheaves that do not possess finite locally free resolutions at all, and it can happen that the height of a sheaf is infinite if we do not truncate resolutions.

A result of Arapura [Ara04, Corollary 3.2] gives effective estimates on the height of a coherent sheaf in terms of its Castelnuovo–Mumford regularity.

Lemma 2.25 *Let X be an irreducible projective variety, $\mathcal{A}$ an ample and globally generated line bundle, $\mathcal{F}$ a coherent sheaf on X. Given a natural number k, there exist vector spaces V_i for $1 \le i \le k$ and a resolution*

$$V_k \otimes \mathcal{A}^{\otimes -r_{\mathcal{F}} - kr_X} \to \ldots \to V_1 \otimes \mathcal{A}^{\otimes -r_{\mathcal{F}} - r_X} \to V_0 \otimes \mathcal{A}^{\otimes -r_{\mathcal{F}}} \to \mathcal{F} \to 0 ,$$

where

$$r_{\mathcal{F}} \stackrel{def}{=} reg_{\mathcal{A}}(\mathcal{F}) \quad and \quad r_X \stackrel{def}{=} \max\{1, reg_{\mathcal{A}}(\mathcal{O}_X)\} .$$

Proof Without loss of generality we can assume that $\mathcal{F}$ is 0-regular by replacing $\mathcal{F}$ by $\mathcal{F} \otimes \mathcal{A}^{r_{\mathcal{F}}}$, thus we can assume $r_{\mathcal{F}} = 0$. Consequently, $\mathcal{F}$ is globally generated by Mumford's theorem; set $V_0 \stackrel{\text{def}}{=} H^0(X, \mathcal{F})$, and

$$\mathcal{K}_0 \stackrel{\text{def}}{=} \mathcal{F} \otimes \mathcal{A}^{\otimes -r_X} , \; \mathcal{K}_1 \stackrel{\text{def}}{=} \ker\left(V_0 \otimes \mathcal{O}_X \twoheadrightarrow \mathcal{K}_0 \otimes \mathcal{A}^{\otimes r_X}\right) .$$

A quick cohomology computation [Ara06, Lemma 3.1] shows that $\mathcal{K}_1$ is r_X-regular, hence we can repeat the above process for $\mathcal{K}_1$ in place of $\mathcal{F}$. This leads to a sequence of vector spaces V_i, and sheaves $\mathcal{K}_i$ which fit into the exact sequences

$$0 \longrightarrow \mathcal{K}_{i+1} \longrightarrow V_i \otimes \mathcal{O}_X \longrightarrow \mathcal{K}_i \otimes \mathcal{A}^{\otimes r_X} \longrightarrow 0$$

or, equivalently,

$$0 \longrightarrow \mathcal{K}_{i+1} \otimes \mathcal{A}^{\otimes -ir_X} \longrightarrow V_i \otimes \mathcal{A}^{\otimes -ir_X} \longrightarrow \mathcal{K}_i \otimes \mathcal{A}^{\otimes (1-i)r_X} \longrightarrow 0 .$$

We obtain the statement of the lemma by combining these sequences into the required resolution. □

Corollary 2.26 *With notation as above, the height of an $r_{\mathcal{F}}$-regular coherent sheaf $\mathcal{F}$ is*

$$\mathrm{ht}_{\mathcal{A}}(\mathcal{F}) \le r_{\mathcal{F}} + r_X \cdot \dim X .$$

Proposition 2.27 (Properties of height) *Let X be an irreducible projective variety of dimension n, $\mathcal{A}$ an ample line bundle. Then, the following hold:*

1. *For coherent sheaves $\mathcal{F}_1$ and $\mathcal{F}_2$ we have*

$$\mathrm{ht}_{\mathcal{A}}(\mathcal{F}_1 \otimes \mathcal{F}_2) \le \mathrm{ht}_{\mathcal{A}}(\mathcal{F}_1) + \mathrm{ht}_{\mathcal{A}}(\mathcal{F}_2) .$$

2. *There exists a positive constant $M = M(X, \mathcal{A})$ having the property that*

$$\mathrm{ht}_{\mathcal{A}}(\mathcal{N}) \leq M$$

for all numerically trivial line bundles $\mathcal{N}$ on X.

Proof The first statement is an immediate consequence of the fact that the tensor product of appropriate resolutions of $\mathcal{F}_1$ and $\mathcal{F}_2$ is a resolution of $\mathcal{F}_1 \otimes \mathcal{F}_2$ of the required type.

The second claim is a consequence of the fact that numerically trivial divisors on a projective variety are parameterized by a quasi-projective variety. Indeed, it follows by Lemma 2.28 and the Noetherian property of the Zariski topology that there exists a constant M' satisfying

$$\mathrm{reg}_{\mathcal{A}}(\mathcal{N}) \leq M'$$

for all numerically trivial line bundles $\mathcal{N}$. By the Corollary of Lemma 2.25,

$$\mathrm{ht}_{\mathcal{A}}(\mathcal{N}) \leq M \stackrel{\mathrm{def}}{=} M' + r_X \cdot \dim X ,$$

as required. □

Lemma 2.28 (Upper-semicontinuity of Castelnuovo–Mumford regularity) *Let X be an irreducible projective variety, $\mathcal{A}$ an ample and globally generated line bundle on X. Given a flat family of line bundles $\mathcal{L}$ on X parameterized by a quasi-projective variety T, the function*

$$T \ni t \mapsto \mathrm{reg}_{\mathcal{A}}(\mathcal{L}_t)$$

is upper-semicontinuous.

Proof Since Castelnuovo–Mumford regularity is checked by the vanishing of finitely many cohomology groups of line bundles, the statement follows from the semicontinuity theorem for cohomology. □

Example 2.29 (Height and regularity on projective spaces) Here we discuss the relationship between height and regularity with respect to $\mathcal{O}(1)$ on an n-dimensional projective space $\mathbb{P}$. We claim that

$$\mathrm{ht}(\mathcal{F}) = \mathrm{reg}(\mathcal{F}) + n$$

holds for an arbitrary coherent sheaf $\mathcal{F}$ on $\mathbb{P}$.

For the inequality

$$\mathrm{ht}(\mathcal{F}) \leq \mathrm{reg}(\mathcal{F}) + n$$

observe that by [Laz04, Proposition 1.8.8] there must exist a long exact sequence

$$\dots \longrightarrow \bigoplus \mathcal{O}_{\mathbb{P}}(-\mathrm{reg}(\mathcal{F}) - 1) \longrightarrow \bigoplus \mathcal{O}_{\mathbb{P}}(-\mathrm{reg}(\mathcal{F})) \longrightarrow \mathcal{F} \longrightarrow 0\,,$$

hence we are done by the definition of height.

To see the reverse inequality, let

$$\dots \longrightarrow \mathcal{F}_2 \longrightarrow \mathcal{F}_1 \longrightarrow \mathcal{F}_0 \longrightarrow \mathcal{F} \longrightarrow 0$$

be a resolution of $\mathcal{F}$ with

$$\mathcal{F}_k = \bigoplus_{i=1}^{k} \mathcal{O}_{\mathbb{P}}(-d_{k,i})\,.$$

Since $d_{k,i} \leq \mathrm{ht}(\mathcal{F})$ by definition, we obtain that $\mathcal{F}_k$ is $\mathrm{ht}(\mathcal{F}) - k$ regular for all $n \geq k \geq 0$, therefore

$$\mathrm{reg}(\mathcal{F}) \leq \mathrm{ht}(\mathcal{F}) - n$$

according to [Laz04, Example 1.8.7].

Remark 2.30 (Height and Serre vanishing with estimates) The introduction of the height of a coherent sheaf leads to an effective version of Serre's vanishing theorem. With the notation of the Introduction, we have

$$m_0(\mathcal{A}, \mathcal{F}) \leq m_0(\mathcal{A}, \mathcal{A}) + \mathrm{ht}_{\mathcal{A}}(\mathcal{F})\,.$$

Lemma 2.31 (Demailly–Peternell–Schneider; [DPS96, Proposition 1.2]) *Let $\mathcal{L}$ be a uniformly q-ample line bundle on X with respect to an ample line bundle $\mathcal{A}$ for a given constant $\lambda = \lambda(\mathcal{A}, \mathcal{L})$. Given a coherent sheaf $\mathcal{F}$ on X,*

$$H^i\left(X, \mathcal{L}^{\otimes m} \otimes \mathcal{F}\right) = \{0\}$$

for all $i > q$ and $m \geq \lambda \cdot (\mathrm{ht}_{\mathcal{A}}(\mathcal{F}) + 1)$.

Proof Let

$$\dots \to \bigoplus_{1 \leq l \leq m_k} \mathcal{A}^{\otimes - d_{k,l}} \to \dots \to \bigoplus_{1 \leq l \leq m_0} \mathcal{A}^{\otimes - d_{0,l}} \to \mathcal{F} \to 0$$

be a resolution where the value of $\mathrm{ht}_{\mathcal{A}}(\mathcal{F})$ is attained, and write $\mathcal{F}_k$ for the image sheaf of the kth differential in the above sequence (note that $\mathcal{F}_0 = \mathcal{F}$). Chopping up the resolution of $\mathcal{F}$ into short exact sequences yields

$$0 \longrightarrow \mathcal{F}_{k+1} \longrightarrow \bigoplus_{1\le l\le m_k} \mathcal{A}^{\otimes -d_{k,l}} \longrightarrow \mathcal{F}_k \longrightarrow 0$$

for all $0 \le k \le \dim X$. By the uniform q-ampleness assumption on $\mathcal{L}$ we obtain that

$$H^i\left(X, \mathcal{L}^{\otimes m} \otimes \mathcal{A}^{\otimes -d_{k,l}}\right) = \{0\} \quad \text{for all } i > q \text{ and } m \ge \lambda(d_{k,l}+1).$$

By induction on k we arrive at

$$H^i\left(X, \mathcal{L}^{\otimes m} \otimes \mathcal{F}\right) \simeq \ldots \simeq H^{i+k}\left(X, \mathcal{L}^{\otimes m} \otimes \mathcal{F}_k\right) \simeq H^{i+k+1}\left(X, \mathcal{L}^{\otimes m} \otimes \mathcal{F}_{k+1}\right)$$

for all $i > q$ and $m \ge \lambda(\mathrm{ht}_{\mathcal{A}}(\mathcal{F}) + 1)$. The statement of the lemma follows by taking $k = \dim X$. □

The idea for the following modification of the proof of [DPS96, Proposition 1.4] was suggested to us by Burt Totaro.

Proof of Theorem 2.22 Fix an integral ample divisor A, as well as integral Cartier divisors $B_1, \ldots, B_\rho$ whose numerical equivalence classes form a basis of the rational Néron–Severi space. Let D be an integral uniformly q-ample divisor (for a constant $\lambda = \lambda(D, A)$), D' a $\mathbb{Q}$-Cartier divisor, and write

$$D' \equiv D + \sum_{i=1}^{\rho} \lambda_i B_i$$

for rational numbers λ_i. Let k be a positive integer clearing all denominators, then

$$kD' = kD + \sum_{i=1}^{\rho} k\lambda_i B_i + N$$

for a numerically trivial (integral) divisor N. We want to show that

$$H^i(X, mkD' - pA) = \{0\}$$

whenever $m \ge \lambda(D', A) \cdot p$ for a suitable positive constant λ. By Lemma 2.31 applied with

$$\mathcal{L} = \mathcal{O}_X(D), \ \mathcal{A} = \mathcal{O}_X(A), \text{ and } \mathcal{F} = \mathcal{O}_X\Big(\sum_{i=1}^{\rho} mk\lambda_i B_i + mN - pA\Big),$$

this will happen whenever

$$m \ge \lambda(D, A) \cdot \mathrm{ht}_A\Big(\sum_{i=1}^{\rho} mk\lambda_i B_i + mN - pA\Big).$$

Observe that

$$\mathrm{ht}_A\Big(\sum_{i=1}^{\rho} mk\lambda_i B_i + mN - pA\Big) = \sum_{i=1}^{\rho} \mathrm{ht}_A(mk\lambda_i B_i) + \mathrm{ht}_A(mN) + \mathrm{ht}_A(-pA)$$
$$\leq \sum_{i=1}^{\rho} mk|\lambda_i| \cdot \max\{\mathrm{ht}_A(B_i), \mathrm{ht}_A(-B_i)\} + M + p\,,$$

where M is the constant from Proposition 2.27; note that $\mathrm{ht}_A(-pA) = p$ for $p \geq 0$. Therefore, if the λ_is are close enough to zero so that

$$\lambda \cdot \sum_{i=1}^{\rho} k|\lambda_i| \cdot \max\{\mathrm{ht}_A(B_i), \mathrm{ht}_A(-B_i)\} < \frac{1}{2}\,,$$

then it suffices to require

$$m \geq 2\lambda(M + p)\,,$$

and D' will be q-ample. This shows the upper-semicontinuity of uniform q-ampleness. □

Remark 2.32 If D_1 is a q-ample and D' is an r-ample divisor, then their sum $D + D'$ can only be guaranteed to be $q + r$-ample; this bound is sharp, as shown in [Tot13, Section 8]. As a consequence, the cone of q-ample $\mathbb{R}$-divisor classes is not necessarily convex. We denote this cone by $\mathrm{Amp}^q(X)$.

Interestingly enough, if we restrict our attention to semi-ample divisors, then Sommese proves in [Som78] (see also Corollary 2.45 below) that the sum of q-ample divisors retains this property.

It is an interesting question how to characterize the cone of q-ample divisors for a given integer q. If $q = 0$, then the Cartan–Serre–Grothendieck theorem implies that $\mathrm{Amp}^0(X)$ equals the ample cone.

Totaro describes the $(n-1)$-ample cone with the help of duality theory.

Theorem 2.33 (Totaro; [Tot13, Theorem 10.1]) *For an irreducible projective variety X we have*

$$\mathrm{Amp}^{n-1}(X) = N^1(X)_{\mathbb{R}} \setminus (-\overline{\mathrm{Eff}(X)})\,.$$

Corollary 2.34 (1-ampleness on surfaces) *If X is a surface, then a divisor D on X is* 1*-ample if and only if $(D \cdot A) > 0$ for some ample divisor A on X.*

Remark 2.35 The cone of q-ample divisors on a $\mathbb{Q}$-factorial projective toric variety has been shown to be polyhedral (more precisely, to be the interior of the union of finitely many rational polyhedral cones) by Broomhead and

Prendergast-Smith [BPS12, Theorem 3.3]. Nevertheless, an explicit combinatorial description in terms of the fan of the underlying toric variety along the lines of [HKP06] is not yet known.

Totaro links partial positivity to the vanishing of higher asymptotic cohomology. Generalizing the main result of [dFKL07] (see also [Kür06] for terminology), he asks the following question:

Question 2.36 (Totaro) Let D be an $\mathbb{R}$-divisor class on a complex projective variety, $0 \leq q \leq n$ an integer. Assume that $\widehat{h}^i(X, D') = 0$ for all $i > q$ and all $D' \in N^1(X)_{\mathbb{R}}$ in a neighborhood of D. Is is true that D is q-ample?

Remark 2.37 Broomhead and Prendergast-Smith [BPS12, Theorem 5.1] answered Totaro's question positively for toric varieties.

It is expected the q-ampleness should have more significance in the case of big line bundles. A first move in this direction comes from the following Fujita-type vanishing statement (see [Fuj83] or [Laz04, Theorem 1.4.35] for Fujita's original statement).

Theorem 2.38 ([Kür10], Theorem C) *Let X be a complex projective scheme, L a big Cartier divisor, $\mathcal{F}$ a coherent sheaf on X. Then there exists a positive integer $m_0(L, \mathcal{F})$ such that*

$$H^i(X, \mathcal{F} \otimes \mathcal{O}_X(mL + D)) = \{0\}$$

for all $i > \dim \mathbf{B}_+(L)$, $m \geq m_0(L, \mathcal{F})$, and all nef divisors D on X.

Remark 2.39 (Augmented base loci on schemes) The augmented base locus of a $\mathbb{Q}$-Cartier divisor L is defined in [ELM⁺06] via

$$\mathbf{B}_+(L) \stackrel{\text{def}}{=} \bigcap_A \mathbf{B}(L - A),$$

where A runs through all ample $\mathbb{Q}$-Cartier divisors. As opposed to the stable base locus of a divisor, the augmented base locus is invariant with respect to numerical equivalence of divisors. The augmented base locus of a $\mathbb{Q}$-divisor L is empty precisely if L is ample.

Although it is customary to define the stable base locus and the augmented base locus of a divisor in the setting of projective varieties, as pointed out in [Kür10, Section 3], these notions make perfect sense on more general schemes.

For an invertible sheaf $\mathcal{L}$ on an arbitrary scheme X, let us denote by $\mathcal{F}_{\mathcal{L}}$ the quasi-coherent subsheaf of $\mathcal{L}$ generated by $H^0(X, \mathcal{L})$. Then we can set

$$\mathfrak{b}(\mathcal{L}) \stackrel{\text{def}}{=} \operatorname{ann}_{\mathcal{O}_X}(\mathcal{L}/\mathcal{F}_{\mathcal{L}}),$$

and define $\mathrm{Bs}(\mathcal{L})$ to be the closed subscheme corresponding to $\mathfrak{b}(\mathcal{L})$. Furthermore, we define

$$\mathbf{B}(\mathcal{L}) \stackrel{\text{def}}{=} \bigcap_{m=1}^{\infty} \mathrm{Bs}(\mathcal{L}^{\otimes m})_{\mathrm{red}} \subseteq X$$

as a closed subset of the topological space associated with X. All basic properties of the stable base locus are retained (see again [Kür10, Section 3]), in particular, if X is complete and algebraic over $\mathbb{C}$ (by which we mean separated, and of finite type over $\mathbb{C}$), then we recover the original definition of stable base loci.

Assuming X to be projective and algebraic over $\mathbb{C}$, we define the augmented base locus of a $\mathbb{Q}$-Cartier divisor L via

$$\mathbf{B}_{+}(L) \stackrel{\text{def}}{=} \bigcap_{A} \mathbf{B}(L - A),$$

where A runs through all ample $\mathbb{Q}$-Cartier divisors. Again, basic properties are preserved, and in the case of projective varieties we recover the original definition.

An interesting further step in this direction is provided by Brown's work, where he connects q-ampleness of a big line bundle to its behavior when restricted to its augmented base locus.

Theorem 2.40 (Brown; [Bro12, Theorem 1.1]) *Let $\mathcal{L}$ be a big line bundle on a complex projective scheme X, denote by $\mathbf{B}_{+}(\mathcal{L})$ the augmented base locus of $\mathcal{L}$. For a given integer $0 \le q \le n$, $\mathcal{L}$ is q-ample if and only if $\mathcal{L}|_{\mathbf{B}_{+}(\mathcal{L})}$ is q-ample.*

We give a very rough outline of the proof of Brown's result. First, if $\mathcal{L}$ is q-ample on X and $Y \subseteq X$ denotes $\mathbf{B}_{+}(\mathcal{L})$ with the reduced induced scheme structure, then the projection formula and the preservation of cohomology groups under push-forward by closed immersions imply that $\mathcal{L}|_{\mathbf{B}_{+}(\mathcal{L})}$ is q-ample as well.

The other implication comes from the following useful observation from [Bro12], a restriction theorem for line bundles that are not q-ample [Bro12, Theorem 2.1]: let $\mathcal{L}$ be a line bundle on a reduced projective scheme X, which is not q-ample, and let $\mathcal{L}'$ be a line bundle on X with a nonzero section s having the property that $\mathcal{L}^{\otimes a} \otimes \mathcal{L}'^{\otimes -b}$ is ample for some positive integers a, b. Then, $\mathcal{L}|_{Z(s)}$ is not q-ample.

2.3 Sommese's geometric q-ampleness

In this section, we will discuss Sommese's geometric notion of q-ampleness, and relate it to the more cohomologically oriented discussion in the previous sections.

Definition 2.41 (Sommese; [Som78, Definition 1.3]) Let X be a projective variety, $\mathcal{L}$ a line bundle on X. We say that $\mathcal{L}$ is *geometrically q-ample* for a natural number q if

1. $\mathcal{L}$ is semi-ample, i.e., $\mathcal{L}^{\otimes m}$ is globally generated for some natural number $m \geq 1$;
2. the maximal fiber dimension of $\phi_{|\mathcal{L}^{\otimes m}|}$ is at most q.

More generally, Sommese defines a vector bundle $\mathcal{E}$ over X to be geometrically q-ample if $\mathcal{O}_{\mathbb{P}(\mathcal{E})}(1)$ is geometrically q-ample, and goes on to prove many interesting results for vector bundles (see [Som78, Proposition 1.7 or 1.12], for instance). In this paper we will only treat the line bundle case.

Remark 2.42 (Iitaka fibration) We briefly recall the semi-ample or Iitaka fibration associated with a semi-ample line bundle $\mathcal{L}$ on a normal projective variety X [Laz04, Theorem 2.1.27]: there exists an algebraic fiber space (a surjective projective morphism with connected fibers) $\phi\colon X \to Y$ with the property that for any sufficiently large and divisible $m \in \mathbb{N}$ one has

$$\phi_{|\mathcal{L}^{\otimes m}|} = \phi \text{ and } Y_m = Y\,,$$

where Y_m denotes the image of X under $\phi_{|\mathcal{L}^{\otimes m}|}$.

In addition there exists an ample line bundle $\mathcal{A}$ on X such that

$$\phi^*\mathcal{A} = \mathcal{L}^{\otimes k}$$

for a suitable positive integer k.

Remark 2.43 If X is a normal variety then Sommese's conditions are equivalent to requiring that $\mathcal{L}$ is semi-ample and its semi-ample fibration has fiber dimension at most q. In the case when X is not normal, it is a priori not clear if the set of integers q for which $\mathcal{L}$ is q-ample depends on the choice of m; a posteriori this follows from Sommese's theorem. Nevertheless, for this reason the definition via the Iitaka fibration is cleaner in the case of normal varieties.

We summarize Sommese's results in this direction.

Theorem 2.44 (Sommese; [Som78, Proposition 1.7]) *Let X be a projective variety, $\mathcal{L}$ a semi-ample line bundle over X with Iitaka fibration $\phi_{\mathcal{L}}$. For a natural number q, the following are equivalent:*

(i) The line bundle $\mathcal{L}$ is geometrically q-ample.

(ii) The maximal dimension of an irreducible subvariety $Z \subseteq X$ with the property that $\mathcal{L}|_Z$ is trivial is at most q.

(iii) If $\psi\colon Z \to X$ is a morphism from a projective variety Z such that $\phi^*\mathcal{L}$ is trivial, then $\dim Z \leq q$.

(iv) The line bundle $\mathcal{L}$ is naively q-ample, that is, for every coherent sheaf $\mathcal{F}$ on X there exists a natural number $m_0 = m_0(\mathcal{L}, \mathcal{F})$ with the property that

$$H^i\left(X, \mathcal{F} \otimes \mathcal{L}^{\otimes m}\right) = 0 \quad \textit{for all } i > q \textit{ and } m \geq m_0.$$

Proof All equivalences are treated in [Som78]. Here we describe the equivalence between (i) and (iv), that is, we prove that for semi-ample line bundles geometric and cohomological q-ampleness agree.

Let $m_0 \geq 1$ be an integer for which $\mathcal{L}^{\otimes m_0}$ is globally generated, and let $\phi\colon X \to Y \subseteq \mathbb{P}$ denote the associated morphism. Then there exists an ample line bundle $\mathcal{A}$ on Y and a positive integer k such that $\mathcal{L}^{\otimes k} = \phi^*\mathcal{A}$. Fix a coherent sheaf $\mathcal{F}$ on X, and consider the Leray spectral sequence

$$H^p\left(Y, R^r\phi_*(\mathcal{F} \otimes \mathcal{L}^{\otimes m})\right) \Longrightarrow H^{p+r}\left(X, \mathcal{F} \otimes \mathcal{L}^{\otimes m}\right).$$

Let us write $m = sk + t$ with $0 \leq t < k$ and $s \geq 0$ integers. For the cohomology groups on the LHS

$$\begin{aligned} H^p\left(Y, R^r\phi_*(\mathcal{F} \otimes \mathcal{L}^{\otimes m})\right) &= H^p\left(Y, R^r\phi_*(\mathcal{F} \otimes (\mathcal{L}^{\otimes k})^{\otimes s} \otimes \mathcal{L}^{\otimes t})\right) \\ &= H^p\left(Y, R^r\phi_*(\mathcal{F} \otimes \mathcal{L}^{\otimes t}) \otimes \mathcal{A}^{\otimes s}\right) \end{aligned}$$

by the projection formula. Serre's vanishing theorem yields

$$H^p\left(Y, R^r\phi_*(\mathcal{F} \otimes \mathcal{L}^{\otimes t}) \otimes \mathcal{A}^{\otimes s}\right) = 0 \ \text{ for all } p \geq 1,\ s \gg 0 \text{ and all } 0 \leq t < k,$$

hence

$$H^0\left(Y, R^r\phi_*(\mathcal{F} \otimes \mathcal{L}^{\otimes m})\right) \simeq H^r\left(X, \mathcal{F} \otimes \mathcal{L}^{\otimes m}\right)$$

for $m \gg 0$. On the other hand, if the maximal fiber dimension of ϕ is q, then $R^r\phi_*(\mathcal{F} \otimes \mathcal{L}^{\otimes m}) = 0$ for all $r > q$ and therefore

$$H^r\left(X, \mathcal{F} \otimes \mathcal{L}^{\otimes m}\right) = 0 \ \text{ for all } m \gg 0 \text{ and } r > q.$$

Consequently, $\mathcal{L}$ is naively q-ample as claimed.

For the other implication assume that $\mathcal{L}$ is not geometrically q-ample, hence has a fiber $F \subseteq X$ of dimension $f > q$. Starting from here one constructs a coherent sheaf $\mathcal{F}$ on F having the property that $R^f\phi_*\mathcal{F}$ is a skyscraper sheaf, which, by the Leray spectral sequence above, would imply that

$$H^f\left(X, \mathcal{F} \otimes \mathcal{L}^{\otimes ks}\right) \neq 0 \ \text{ for all } s \geq 1.$$

Without loss of generality we can assume that F is irreducible, otherwise we replace it by one of its top-dimensional irreducible components. Let $\pi\colon \widetilde{F} \to F$ denote a resolution of singularities of F, and consider the Grauert–Riemen–Schneider canonical sheaf

$$\mathcal{F} \stackrel{\text{def}}{=} \mathcal{K}_{\widetilde{F}/F} \stackrel{\text{def}}{=} \pi_*\omega_{\widetilde{F}} .$$

By Grauert–Riemenschneider [GR70]

$$R^f(\phi|_F)_*\mathcal{F} \simeq H^f(F, \mathcal{F}) \neq 0$$

as $\phi|_F$ maps F to a point. □

Corollary 2.45 *Let D_1 and D_2 be geometrically q-ample divisors on a smooth variety. Then so is $D_1 + D_2$.*

Proof This is [Som78, Corollary 1.10.2]. □

Additionally, Kodaira–Akizuki–Nakano vanishing continues to hold for the expected range of cohomology groups and degrees of differential forms.

Theorem 2.46 (Kodaira–Akizuki–Nakano for geometrically q-ample bundles) *Let $\mathcal{L}$ be a geometrically q-ample line bundle on a smooth projective variety X. Then,*

$$H^i\left(X, \wedge^j\Omega_X \otimes \mathcal{L}\right) = \{0\} \quad \textit{for all } i + j > \dim X + q.$$

*In particular, the following q-*Kodaira vanishing *holds:*

$$H^i(X, \omega_X \otimes \mathcal{L}) = \{0\} \quad \textit{for all } i > q.$$

Proof This is proven in [Som78, Proposition 1.12]. □

Remark 2.47 Sommese's version of the Kodaira–Akizuki–Nakano vanishing was later shown by Esnault and Viehweg to hold for an even larger range of values for i and j, see [EV89] and [EV92, Corollary 6.6].

In Example 2.54 and Section 3 below we discuss the question of whether Kodaira vanishing still continues to hold when one drops the semi-ampleness condition, i.e., for general q-ample line bundles.

2.4 Ample subschemes and a Lefschetz hyperplane theorem for q-ample divisors

Building upon the theory of q-ample line bundles and Hartshorne's classical work [Har70], Ottem [Ott12] defines the notion of an ample subvariety (or subscheme) and goes on to verify that ample subvarieties share many of the significant algebro-geometric and topological properties of their codimension-1 counterparts. One of the highlights of his work is a Lefschetz-type hyperplane theorem, which we will use for a proof of Kodaira's vanishing theorem for effective q-ample Du Bois divisors in Section 3 below. Here we briefly recall the theory obtained in [Ott12].

Definition 2.48 (Ample subschemes) Let X be a projective scheme, $Y \subseteq X$ a closed subscheme of codimension r, $\pi\colon \widetilde{X} \to X$ the blow-up of Y with exceptional divisor E. Then, Y is called *ample* if E is $(r-1)$-ample on $\widetilde{X}$.

The idea behind this notion is classical: it has been known for a long time that positivity properties of Y can often be read off from the geometry of the complement $X - Y \simeq \widetilde{X} - E$. In spite of this, the concept has not been defined until recently.

Example 2.49 As can be expected, linear subspaces of projective spaces are ample.

Remark 2.50 (Cohomological dimension of the complement of an ample subscheme) An important geometric feature of ample divisors is that their complement is affine. In terms of cohomology, this is equivalent to requiring

$$H^i(X, \mathcal{F}) = \{0\} \quad \text{for all } i > 0,\ \mathcal{F} \text{ coherent sheaf on } X.$$

If we denote as customary the cohomological dimension of a subset $Y \subseteq X$ by $\operatorname{cd}(Y)$, then we can phrase Ottem's generalization [Ott12, Proposition 5.1] to the q-ample case as follows: if $U \subseteq X$ is an open subset of a projective scheme X having the property that $X \setminus U$ is the support of a q-ample divisor, then

$$\operatorname{cd}(U) \leq q\,.$$

The observation on cohomological dimensions of complements leads to the following Lefschetz-type statement:

Theorem 2.51 (Generalized Lefschetz hyperplane theorem, Ottem; [Ott12, Corollary 5.2]) *Let D be an effective q-ample divisor on a projective variety X with smooth complement. Then, the restriction morphism*

$$H^i(X, \mathbb{Q}) \longrightarrow H^i(D, \mathbb{Q}) \ \textit{is} \ \begin{cases} \textit{an isomorphism for } 0 \leq i \leq n-q-1, \\ \textit{injective for } i = n-q-1. \end{cases}$$

We give Ottem's proof to show the principles at work.

Proof Via the long exact sequence for relative cohomology and Lefschetz duality, the statement reduces to the claim that

$$H^i(X \setminus D, \mathbb{C}) = \{0\} \quad \text{for } i > n + q.$$

This latter follows from the Frölicher spectral sequence

$$E_1^{st} = H^s\left(X \setminus D, \Omega^t_{X\setminus D}\right) \Longrightarrow H^{s+t}(X \setminus D, \mathbb{C}),$$

as

$$H^s\left(X \setminus D, \Omega^t_{X\setminus D}\right) = \{0\} \quad \text{for all } s + t > n + q$$

by Remark 2.50. □

Corollary 2.52 (Lefschetz hyperplane theorem for ample subvarieties; [Ott12, Corollary 5.3]) *Let Y be an ample local complete intersection subscheme in a smooth complex projective variety X. Then, the restriction morphism*

$$H^i(X, \mathbb{Q}) \longrightarrow H^i(Y, \mathbb{Q}) \ \text{ is } \begin{cases} \text{an isomorphism} & \text{if } i < \dim Y, \\ \text{injective} & \text{if } i = \dim Y. \end{cases}$$

The following is a summary of properties of smooth ample subschemes:

Theorem 2.53 (Properties of smooth ample subschemes; [Ott12, Corollary 5.6 and Theorem 6.6]) *Let X be a smooth projective variety, $Y \subseteq X$ a non-singular ample subscheme of dimension $d \geq 1$. Then, the following hold:*

1. *The normal bundle $N_{Y/X}$ of Y is an ample vector bundle.*
2. *For every irreducible* ($\dim X - d$)*-dimensional subvariety $Z \subseteq X$ one has* $(Y \cdot Z) > 0$. *In particular, Y meets every divisor.*
3. *The Lefschetz hyperplane theorem holds for rational cohomology on Y:*

$$H^i(X, \mathbb{Q}) \longrightarrow H^i(Y, \mathbb{Q}) \ \text{ is } \begin{cases} \text{an isomorphism} & \text{if } i < \dim Y, \\ \text{injective} & \text{if } i = \dim Y. \end{cases}$$

4. *Let $\widehat{X}$ denote the completion of X with respect to Y. For any coherent sheaf $\mathcal{F}$ on X one has*

$$H^i(X, \mathcal{F}) \longrightarrow H^i\left(\widehat{X}, \mathcal{F}\right) \ \text{ is } \begin{cases} \text{an isomorphism} & \text{if } i < \dim Y, \\ \text{injective} & \text{if } i = \dim Y. \end{cases}$$

5. *The inclusion $Y \hookrightarrow X$ induces a surjection*

$$\pi_1(Y) \twoheadrightarrow \pi_1(X)$$

on the level of fundamental groups.

Proof We will only discuss point 5, since it is the only statement that is slightly different from its original source. Because Y is smooth over a reduced base, it is automatically reduced and by smoothness again, it is irreducible exactly when it is connected. But point 3 in the case of $i = 0$ implies that Y is connected. The rest follows from Ottem's proof. □

Based on the fact that a Lefschetz-type theorem holds for q-ample divisors, as seen in Theorem 2.51, and that this forms one of the ingredients of the proof of the Kodaira vanishing theorem in the classical setup, cf. [Laz04, Section 4.2], as well as on the fact that q-Kodaira vanishing continues to hold for *geometrically* q-ample divisors, Theorem 2.46, one might hope that there is a q-Kodaira vanishing theorem for q-ample divisors. The following example shows that this is not true in general.

Example 2.54 (Counterexample to Kodaira vanishing for non-pseudo-effective q-ample divisors, Ottem; [Ott12, Section 9]) Let $G = SL_3(\mathbb{C})$, $B \leqslant G$ be the Borel subgroup consisting of upper triangular matrices, and consider the homogeneous space G/B. By the Bott–Borel–Weil theorem and a brief computation, Ottem shows the existence of a non-pseudo-effective line bundle $\mathcal{L}$ on G/B, which is 1-ample, but for which the cohomology group $H^2(G/B, \omega_{G/B} \otimes \mathcal{L})$ does not vanish.

It turns out, however, that by putting geometric restrictions on the q-ample divisor in question, one can in fact prove Kodaira-style vanishing theorems. We will do this in the next section.

3 q-Kodaira vanishing for Du Bois divisors and log canonical pairs

This section contains the proofs of various versions of Kodaira's vanishing theorem for q-ample divisors. First we present the argument in the smooth case, where the reasoning is particularly transparent and simple.

Theorem 3.1 *Let X be a smooth projective variety, D a smooth reduced effective q-ample divisor on X. Then,*

$$H^i(X, \mathcal{O}_X(K_X + D)) = \{0\} \quad \textit{for all } i > q.$$

Proof By Serre duality, it suffices to show

$$H^i(X, \mathcal{O}_X(-D)) = \{0\} \quad \text{for all } i \leq n - q - 1\,.$$

To this end, follow the proof of Kodaira vanishing in [Laz04, Section 4.2] with minor modifications. Ottem's generalized Lefschetz hyperplane theorem for effective q-ample divisors, Theorem 2.51, asserts that

$$H^i(X, \mathbb{C}) \longrightarrow H^i(D, \mathbb{C})$$

is an isomorphism for $0 \leq i < n - q - 1$ and an injection for $i = n - q - 1$.

The Hodge decomposition then gives rise to homomorphisms

$$r_{k,l}\colon H^l\left(X, \Omega_X^k\right) \longrightarrow H^l\left(D, \Omega_D^k\right) \tag{1}$$

for which $r_{k,l}$ is an isomorphism for $0 \leq k + l < n - q - 1$, and an injection for $k + l = n - q - 1$.

Consequently, one has that $r_{o,l}\colon H^l(X, \mathcal{O}_X) \to H^l(D, \mathcal{O}_D)$ is an isomorphism for $l < n - q - 1$, and is injective for $l = n - q - 1$.

Finally, consider the exact sequence

$$0 \to \mathcal{O}_X(-D) \to \mathcal{O}_X \to \mathcal{O}_D \to 0\,.$$

The properties of $r_{0,l}$ applied to the associated long exact sequence imply that

$$H^i(X, \mathcal{O}_X(-D)) = \{0\} \quad \text{for all } i \leq n - q - 1$$

as we wished. □

Remark 3.2 Our proof here should also be compared with the discussion in [EV92, Section 4], where bounds on the cohomological dimension of the complement of a smooth divisor are also used to derive vanishing theorems of the type considered here.

Remark 3.3 Note that we have used only part of the information provided by the homomorphisms (1). The remaining instances lead to an Akizuki–Nakano-type vanishing result for effective reduced smooth q-ample divisors; cf. the discussion in [Laz04, Section 4.2]. Note however that a generalization to the singular setup cannot be expected, as already for ample divisors on Kawamata log terminal varieties the natural generalization of Kodaira–Akizuki–Nakano vanishing does not hold in general; we refer the reader to [GKP12, Section 4] for a discussion and for explicit counterexamples.

One of the original contributions of our work is the observation that the argument in the proof of Theorem 3.1 can be modified to go through in the Du Bois case, in particular, for log canonical pairs on smooth projective varieties. Our discussion here is very much influenced by "Kollár's principle," cf. [Rei97, Section 3.12], that vanishing occurs when a coherent cohomology group has a topological interpretation, see [Kol86, Section 5]. For this, we depend heavily on the discussion of vanishing theorems in [Kol95].

Theorem 3.4 (q-Kodaira vanishing for reduced effective Du Bois divisors) *Let X be a normal proper variety, $D \subset X$ a reduced effective (Cartier) divisor such that*

1. *$\mathcal{O}_X(D)$ is q-ample,*
2. *$X \setminus D$ is smooth,*
3. *D (with its reduced subscheme structure) is Du Bois.*

Then, we have

$$H^j(X, \omega_X \otimes \mathcal{O}_X(D)) = \{0\} \quad \textit{for all } j > q \tag{2}$$

as well as

$$H^j(X, \mathcal{O}_X(-D)) = \{0\} \quad \textit{for all } j < \dim X - q. \tag{3}$$

We refer the reader to [KS11] and [Kol95, Chapter 12] for introductions to the theory of Du Bois singularities, as well as to [Kov12] for a simple characterization of the Du Bois property for projective varieties, related to the properties of Du Bois singularities used here.

Proof Since D is effective and q-ample, and $X \setminus D$ is smooth, Theorem 2.51 states that the restriction morphism

$$H^j(X, \mathbb{C}) \xrightarrow{\Phi_j} H^j(D, \mathbb{C})$$

is an isomorphism for $0 \leq j < n - q - 1$, and injective for $j = n - q - 1$. If $F^\bullet$ denotes Deligne's Hodge filtration on $H^j(D, \mathbb{C})$ with associated graded pieces $Gr^\bullet_F$, then without any assumption on D, the natural map $\alpha_j \colon H^j(D, \mathbb{C}) \to Gr^0_F(H^j(D, \mathbb{C}))$ factors as

$$H^j(D, \mathbb{C}) \xrightarrow{\beta_j} H^j(D, \mathcal{O}_D) \xrightarrow{\gamma_j} Gr^0_F(H^j(D, \mathbb{C})), \qquad \alpha_j = \gamma_j \circ \beta_j \tag{4}$$

for each j.

Moreover, since D is assumed to be Du Bois, the map γ_j is an isomorphism, see [Kov12, Section 1]. By abuse of notation, the Hodge filtration on $H^j(X, \mathbb{C})$ will likewise be denoted by $F^\bullet$. By standard results of Hodge theory (e.g., see [PS08, Theorem 5.33.iii)], the map Φ_j is a morphism of Hodge structures; in particular it is compatible with the filtrations $F^\bullet$ on $H^j(X,\mathbb{C})$ and $H^j(D,\mathbb{C})$. Hence, for each j, we obtain a natural commutative diagram

$$\begin{array}{ccc} H^j(X,\mathbb{C}) & \xrightarrow{\Phi_j} & H^j(D,\mathbb{C}) \\ \downarrow & & \downarrow \beta_j \\ H^j(X,\mathcal{O}_X) & \xrightarrow{\phi_j} & H^j(D,\mathcal{O}_D) \\ \downarrow \cong & & \cong \downarrow \gamma_j \\ Gr^0_F(H^j(X,\mathbb{C})) & \longrightarrow & Gr^0_F(H^j(D,\mathbb{C})) \end{array}$$

Since Φ_j is an isomorphism for $0 \le j < n-q-1$, and injective for $j = n-q-1$, Hodge theory [PS08, Corollaries 3.6 and 3.7] implies that the same is true for ϕ_j; i.e., ϕ_j is an isomorphism for $0 \le j < n-q-1$, and injective for $j = n-q-1$. Looking at the long exact cohomology sequence associated with

$$0 \to \mathcal{O}_X(-D) \to \mathcal{O}_X \to \mathcal{O}_D \to 0$$

we conclude that

$$H^j(X,\mathcal{O}_X(-D)) = \{0\} \quad \text{for } 0 \le j \le n-q-1, \tag{5}$$

as claimed in equation (3).

Now, as $X \setminus D$ is smooth and D is Du Bois, X itself has rational singularities by a result of Schwede [Sch07, Theorem 5.1], see also [KS11, Section 12]. In particular, X is Cohen–Macaulay, [KM98, Theorem 5.10], and hence we may apply Serre duality [KM98, Theorem 5.71] to equation (5) to obtain

$$H^j(X,\omega_X \otimes \mathcal{O}_X(D)) = \{0\} \text{ for all } j > q,$$

as claimed in equation (2). □

Corollary 3.5 (*q*-Kodaira vanishing for reduced log canonical pairs) *Let X be a smooth projective variety, $D \subset X$ a reduced effective (Cartier) divisor such that*

1. *$\mathcal{O}_X(D)$ is q-ample,*
2. *the pair (X,D) is log canonical.*

Then, we have

$$H^j(X,\omega_X \otimes \mathcal{O}_X(D)) = \{0\} \quad \text{for all } j > q \tag{6}$$

as well as

$$H^j(X,\mathcal{O}_X(-D)) = \{0\} \quad \text{for all } j < \dim X - q. \tag{7}$$

Proof Since D is a union of log canonical centers of the pair (X, D), it is Du Bois by [KK10, Theorem 1.4]. Hence, the claim follows immediately from Theorem 3.4. □

Example 3.6 To give an example of a q-ample divisor that satisfies the assumptions of Theorems 3.4 and 3.1 above, and for which the desired vanishing does not follow directly from Sommese's results, let $Z \subset X$ be a smooth ample subscheme of pure codimension r in a projective manifold X, and $\mathcal{O}_{\hat{X}}(E)$ the corresponding $(r-1)$-ample line bundle on the blow-up $\hat{X}$ of X along Z, cf. Example 2.4. Then, $\mathcal{O}_{\hat{X}}(E)$ is clearly not semi-ample, and hence not geometrically q-ample in the sense of Definition 2.41. However, E is effective, reduced, and smooth, and hence fulfills the assumptions of Theorems 3.4 and 3.1. In this special case, the desired vanishing can also be derived from the results presented in [EV92, Section 4].

Finally, we will prove a version of the above for line bundles that are only $\mathbb{Q}$-effective. We start with the following slight generalization of [Kol95, Theorem 12.10]:

Proposition 3.7 *Let X be a normal and proper variety, $\mathcal{L}$ a rank 1 reflexive sheaf on X, and D_i different irreducible Weil divisors on X. Assume that $\mathcal{L}^{[m]} := (\mathcal{L}^{\otimes m})^{**} \cong \mathcal{O}_X(\sum d_j D_j)$ for some integers $1 \le d_j < m$. Set $m_j := m/\gcd(m, d_j)$. Assume furthermore that the pair $\left(X, \sum(1 - \frac{1}{m_j})D_j\right)$ is log canonical. Then, for every $i \ge 0$ and $n_j \ge 0$, the natural map*

$$H^i\left(X, \mathcal{L}^{[-1]}(-\textstyle\sum n_j D_j)\right) \to H^i\left(X, \mathcal{L}^{[-1]}\right)$$

is surjective.

Proof Let $p\colon Y \to X$ be the normalization of the cyclic cover corresponding to the isomorphism $\mathcal{L}^{[m]} \cong \mathcal{O}_X(\sum d_j D_j)$. By [Kol92, Proposition 20.2], we have

$$K_Y = p^*(K_X + \textstyle\sum(1 - \frac{1}{m_j})D_j),$$

and therefore Y is log canonical by [Kol92, Proposition 20.3]. Hence, Y is Du Bois by [KK10, Theorem 1.4], and the natural map $H^j(Y, \mathbb{C}) \to H^j(Y, \mathcal{O}_Y)$ is surjective, cf. the discussion in the proof of Theorem 3.4. Consequently, the assumptions of [Kol95, Theorem 9.12] are fulfilled. This implies the claim. □

We are now in a position to prove a version of q-Kodaira vanishing that works for $\mathbb{Q}$-effective line bundles.

Theorem 3.8 (q-Kodaira vanishing for effective log canonical $\mathbb{Q}$-divisors) *Let X be a proper, normal, Cohen–Macaulay variety, $\mathcal{L}$ a q-ample line bundle on X, and D_i different irreducible Weil divisors on X. Assume that $\mathcal{L}^m \cong \mathcal{O}_X(\sum d_j D_j)$ for some integers $1 \le d_j < m$. Set $m_j := m/\gcd(m, d_j)$. Assume furthermore that the pair $(X, \sum(1 - \frac{1}{m_j})D_j)$ is log canonical. Then, we have*

$$H^j(X, \omega_X \otimes \mathcal{O}_X(D)) = \{0\} \quad \textit{for all } j > q \tag{8}$$

as well as

$$H^j(X, \mathcal{O}_X(-D)) = \{0\} \quad \textit{for all } j < \dim X - q. \tag{9}$$

Proof With Proposition 3.7 at hand, the proof is the same as in the klt case, cf. [Kol95, Chapter 10]. By Proposition 3.7, for all $i \ge 0$ and for all $k \in \mathbb{N}^{>0}$ such that $k - 1$ is divisible by m, we obtain a surjection

$$H^i\left(X, \mathcal{L}^{-k}\right) \twoheadrightarrow H^i\left(X, \mathcal{L}^{-1}\right).$$

Since X is Cohen–Macaulay, by Serre duality [KM98, Theorem 5.71], this surjection is dual to an injection

$$H^{n-i}(X, \omega_X \otimes \mathcal{L}) \hookrightarrow H^{n-i}\left(X, \omega_X \otimes \mathcal{L}^k\right) \text{ for all } k \text{ as above.} \tag{10}$$

As $\mathcal{L}$ is q-ample, there exists a $k \gg 0$ such that

$$H^{n-i}\left(X, \omega_X \otimes \mathcal{L}^k\right) = \{0\} \text{ for all } n - i > q.$$

Hence, owing to the injection (10), we obtain

$$H^j(X, \omega_X \otimes \mathcal{L}) = \{0\} \text{ for all } j > q,$$

as claimed in equation (8). The dual vanishing (9) then follows from a further application of Serre duality. □

Remark 3.9 For related work discussing ample divisors on (semi)-log canonical varieties, the reader is referred to [KSS10].

Remark 3.10 (Necessity of assumptions on the singularities) To see that some assumption on the singularities of the pair (X, D) is necessary in Theorems 3.4 and 3.8, we note that Kodaira vanishing may fail already for ample line bundles on Gorenstein varieties with worse than log canonical singularities, see [BS95, Example 2.2.10]. Moreover, we note that for the dual form (9) of Kodaira vanishing the Cohen–Macaulay condition is strictly necessary. If X is a projective variety with ample (Cartier) divisor D for which (9) holds (with $q = 0$), then X is Cohen–Macaulay by [KM98, Corollary 5.72].

Acknowledgments Both authors received partial support from the DFG-Forschergruppe 790 "Classification of Algebraic Surfaces and Compact Complex Manifolds," as well as by the DFG-Graduiertenkolleg 1821 "Cohomological Methods in Geometry." The first author gratefully acknowledges additional support by the Baden-Württemberg-Stiftung through the "Eliteprogramm für Postdoktorandinnen und Postdoktoranden." The second author was in addition partially supported by the OTKA grants 77476 and 81203 of the Hungarian Academy of Sciences.

References

[AG62] Andreotti, A. and Grauert, H. Théorème de finitude pour la cohomologie des espaces complexes. *Bull. Soc. Math. France* **90** (1962) 193–259.

[Ara04] Arapura, D. Frobenius amplitude and strong vanishing theorems for vector bundles. *Duke Math. J.* **121**(2) (2004) 231–267. With an appendix by D. S. Keeler.

[Ara06] Arapura, D. Partial regularity and amplitude. *Amer. J. Math.* **128**(4) (2006) 1025–1056.

[Bac86] Backelin, J. On the rates of growth of the homologies of Veronese subrings. In *Algebra, Algebraic Topology and their Interactions (Stockholm, 1983)*. Lecture Notes in Mathematics, Vol. 1183. Berlin: Springer-Verlag, 1986, pp. 79–100.

[BPS12] Broomhead, N. and Prendergast-Smith, A. Partially ample line bundles on toric varieties, 2012 arXiv:1202.3065v1.

[Bro12] Brown, M. V. Big q-ample line bundles. *Compos. Math.* **148**(3) (2012) 790–798.

[BS95] Beltrametti, M. C. and Sommese, A. J. *The Adjunction Theory of Complex Projective Varieties*. de Gruyter Expositions in Mathematics, Vol. 16, Berlin: Walter de Gruyter & Co., 1995.

[Dem11] Demailly, J.-P. A converse to the Andreotti–Grauert theorem, *Ann. Fac. Sci. Toulouse Math. (6)* **20**(S2) (2011) 123–135.

[dFKL07] de Fernex, T., Küronya, A., and Lazarsfeld, R. Higher cohomology of divisors on a projective variety. *Math. Ann.* **337**(2) (2007) 443–455.

[DI87] Deligne, P. and Illusie, L. Relèvements modulo p^2 et décomposition du complexe de de Rham. *Invent. Math.* **89**(2) (1987) 247–270.

[DPS96] Demailly, J.-P., Peternell, T., and Schneider, M. Holomorphic line bundles with partially vanishing cohomology. Proceedings of the Hirzebruch 65 Conference on Algebraic Geometry *(Ramat Gan, 1993). Israel Mathematics Conference Proceedings*, Vol. 9, Bar-Ilan University, Ramat Gan, 1996, pp. 165–198.

[ELM+06] Ein, L., Lazarsfeld, R., Mustaţă, M., Nakamaye, M., and Popa, M. Asymptotic invariants of base loci. *Ann. Inst. Fourier (Grenoble)* **56**(6) (2006) 1701–1734.

[EV89] Esnault, H. and Viehweg, E. Vanishing and nonvanishing theorems. Actes du Colloque de Théorie de Hodge *(Luminy, 1987)*. *Astérisque*, Vol. 179–180. Paris: Société Mathématique de France, 1989, pp. 97–112.

[EV92] Esnault, E. and Viehweg, E. Lectures on Vanishing Theorems. DMV Seminar, Vol. 20. Basel: Birkhäuser Verlag, 1992.

[Fuj83] Fujita, T. Vanishing theorems for semipositive line bundles. *Algebraic Geometry (Tokyo/Kyoto, 1982)*. Lecture Notes in Mathematics, Vol. 1016. Berlin: Springer-Verlag, 1983, pp. 519–528.

[GKP12] Greb, D., Kebekus, S., and Peternell, T. Reflexive differential forms on singular spaces. Geometry and cohomology. *J. Reine Angew. Math.* (2012). DOI: 10.1515/crelle-2012-0097.

[GR70] Grauert, H. and Riemenschneider, O. Verschwindungssätze für analytische Kohomologiegruppen auf komplexen Räumen, *Invent. Math.* **11** (1970) 263–292.

[Har70] Hartshorne, R. *Ample Subvarieties of Algebraic Varieties*. Notes written in collaboration with C. Musili. Lecture Notes in Mathematics, Vol. 156. Berlin: Springer-Verlag, 1970.

[HKP06] Hering, M., Küronya, A., and Payne, S. Asymptotic cohomological functions of toric divisors. *Adv. Math.* **207**(2) (2006) 634–645.

[Kaw04] Kawamata, Y. Equivalences of derived categories of sheaves on smooth stacks. *Amer. J. Math.* **126**(5) (2004) 1057–1083.

[KK10] Kollár, J. and Kovács, S. J. Log canonical singularities are Du Bois. *J. Amer. Math. Soc.* **23**(3) (2010) 791–813.

[KM98] Kollár, J. and Mori, S. *Birational Geometry of Algebraic Varieties*. Cambridge Tracts in Mathematics, Vol. 134. Cambridge: Cambridge University Press, 1998.

[Kod53] Kodaira, K. On a differential-geometric method in the theory of analytic stacks. *Proc. Nat. Acad. Sci. USA* **39** (1953) 1268–1273.

[Kol86] Kollár, J. Higher direct images of dualizing sheaves. II. *Ann. Math. (2)* **124** (1986) 171–202.

[Kol92] Kollár, J. Flips and abundance for algebraic threefolds. *Astérisque*, Vol. 211. Paris: Société Mathématique de France, 1992.

[Kol95] Kollár, J. *Shafarevich Maps and Automorphic Forms*. M. B. Porter Lectures. Princeton, NJ: Princeton University Press, 1995.

[Kov12] Kovács, S. J. The intuitive definition of Du Bois singularities. Geometry and Arithmetic, EMS Series Congress Report. European Mathematical Society, Zürich, 2012, pp. 257–266.

[KS11] Kovács, S. J. and Schwede, K. Hodge theory meets the minimal model program: A survey of log canonical and Du Bois singularities. In *Topology of Stratified Spaces*, Mathematical Sciences Research Institute Publication, Vol. 58. Cambridge: Cambridge University Press, 2011, pp. 51–94.

[KSS10] Kovács, S. J., Schwede, K., and Smith, K. E. The canonical sheaf of Du Bois singularities. *Adv. Math.* **224**(4) (2010) 1618–1640.

[Kür06] Küronya, A. Asymptotic cohomological functions on projective varieties. *Amer. J. Math.* **128**(6) (2006) 1475–1519.

[Kür10] Küronya, A. Positivity on subvarieties and vanishing theorems for higher cohomology. *Ann. Inst. Fourier (Grenoble)* **63**(5) (2013) 1717–1737.

[Laz04] Lazarsfeld, R. *Positivity in Algebraic Geometry. I.* Ergebnisse der Mathematik und ihrer Grenzgebiete. 3. Folge. A Series of Modern Surveys in Mathematics, Vol. 48. Berlin: Springer-Verlag, 2004.

[Mat13] Matsumura, S.-I. Asymptotic cohomology vanishing and a converse to the Andreotti–Grauert theorem on a surface. *Ann. Inst. Fourier (Grenoble)* **63**(6) (2013) 2199–2221.

[Orl97] Orlov, D. O. Equivalences of derived categories and *K*3 surfaces. *J. Math. Sci. (New York)* **84**(5) (1997) 1361–1381.

[Ott12] Ottem, J. C. Ample subvarieties and *q*-ample divisors. *Adv. Math.* **229**(5) (2012) 2868–2887.

[PS08] Peters, C. A. M. and Steenbrink, J. H. M. *Mixed Hodge Structures*. Ergebnisse der Mathematik und ihrer Grenzgebiete. 3. Folge. A Series of Modern Surveys in Mathematics, Vol. 52. Berlin: Springer-Verlag, 2008.

[Ray78] Raynaud, M. Contre-exemple au "vanishing theorem" en caractÃ©ristique $p > 0$. C. P. Ramanujama tribute, Tata Institute of Fundamental Research Studies in Mathematics, Vol. 8. Berlin: Springer-Verlag, 1978, pp. 273–278.

[Rei97] Reid, M. Chapters on algebraic surfaces. *Complex Algebraic Geometry (Park City, UT, 1993)*. IAS/Park City Mathematical Series, Vol. 3. Providence, RI: American Mathematical Society, 1997, pp. 3–159.

[Sch07] Schwede, K. A simple characterization of Du Bois singularities. *Compos. Math.* **143**(4) (2007), 813–828.

[Som78] Sommese, A. J. Submanifolds of abelian varieties. *Math. Ann.* **233**(3) (1978) 229–256.

[Tot13] Totaro, B. Line bundles with partially vanishing cohomology, *J. Eur. Math. Soc.* **15**(3) (2013) 731–754.

13

Generic vanishing fails for singular varieties and in characteristic $p > 0$

C. D. Hacon[a]
University of Utah

S. J. Kovács[b]
University of Washington

To Rob Lazarsfeld on the occasion of his 60th birthday

1 Introduction

In recent years there has been considerable interest in understanding the geometry of irregular varieties, i.e., varieties admitting a nontrivial morphism to an abelian variety. One of the central results in the area is the following, conjectured by M. Green and R. Lazarsfeld (cf. [GL91, 6.2]) and proven in [Hac04] and [PP09].

Theorem 1.1 *Let $\lambda : X \to A$ be a generically finite (onto its image) morphism from a compact Kähler manifold to a complex torus. If $\mathscr{L} \to X \times \mathrm{Pic}^0(A)$ is the universal family of topologically trivial line bundles, then*

$$R^i \pi_{\mathrm{Pic}^0(A)*} \mathscr{L} = 0 \qquad \text{for } i < n.$$

At first sight the above result appears to be quite technical, however it has many concrete applications (see, e.g., [CH11], [JLT11], and [PP09]). In this paper we will show that Theorem 1.1 does not generalize to characteristic $p > 0$ or to singular varieties in characteristic 0.

[a] Partially supported by NSF research grants DMS-0757897 and DMS-1300750 and a grant from the Simons Foundation.
[b] Partially supported by NSF grants DMS-0856185, DMS-1301888, and the Craig McKibben and Sarah Merner Endowed Professorship in Mathematics at the University of Washington.

Notation 1.2 Let A be an abelian variety over an algebraically closed field k, $\widehat{A}$ its dual abelian variety, $\mathscr{P}$ the normalized Poincaré bundle on $A \times \widehat{A}$, and $p_{\widehat{A}}: A \times \widehat{A} \to \widehat{A}$ the projection. Let $\lambda: X \to A$ be a projective morphism, $\pi_{\widehat{A}}: X \times \widehat{A} \to \widehat{A}$ the projection, and $\mathscr{L} := (\lambda \times \mathrm{id}_{\widehat{A}})^* \mathscr{P}$ where $(\lambda \times \mathrm{id}_{\widehat{A}}): X \times \widehat{A} \to A \times \widehat{A}$ is the product morphism.

Theorem 1.3 *Let k be an algebraically closed field. Then, using Notation 1.2, there exists a projective variety X over k such that*

- *if* $\operatorname{char} k = p > 0$, *then X is smooth and*
- *if* $\operatorname{char} k = 0$, *then X has isolated Gorenstein log canonical singularities*

and a separated projective morphism to an abelian variety $\lambda: X \to A$ which is generically finite onto its image such that

$$R^i \pi_{\widehat{A}*} \mathscr{L} \neq 0 \qquad \text{for some } 0 \leq i < n.$$

Remark 1.4 Owing to the birational nature of the statement, Theorem 1.1 generalizes trivially to the case of X having only rational singularities. Arguably, Gorenstein log canonical singularities are the simplest examples of singularities that are not rational. Therefore, the characteristic 0 part of Theorem 1.3 may be interpreted as saying that generic vanishing does not extend to singular varieties in a nontrivial way.

Remark 1.5 Note that Theorem 1.3 seems to contradict the main result of [Par03].

2 Preliminaries

Let A be a g-dimensional abelian variety over an algebraically closed field k, $\widehat{A}$ its dual abelian variety, p_A and $p_{\widehat{A}}$ the projections of $A \times \widehat{A}$ onto A and $\widehat{A}$, and $\mathscr{P}$ the normalized Poincaré bundle on $A \times \widehat{A}$. We denote by $\mathbf{R}\widehat{S}: \mathbf{D}(A) \to \mathbf{D}(\widehat{A})$ the usual Fourier–Mukai functor given by $\mathbf{R}\widehat{S}(\mathscr{F}) = \mathbf{R}p_{\widehat{A}*}(p_A^* \mathscr{F} \otimes \mathscr{P})$ (cf. [Muk81]). There is a corresponding functor $\mathbf{R}S: \mathbf{D}(\widehat{A}) \to \mathbf{D}(A)$ such that

$$\mathbf{R}S \circ \mathbf{R}\widehat{S} = (-1_A)^*[-g] \qquad \text{and} \qquad \mathbf{R}\widehat{S} \circ \mathbf{R}S = (-1_{\widehat{A}})^*[-g].$$

Definition 2.1 An object $F \in \mathbf{D}(A)$ is called *WIT-i* if $R^j\widehat{S}(F) = 0$ for all $j \neq i$. In this case we use the notation $\widehat{F} = R^i\widehat{S}(F)$.

Notice that if F is a WIT-i coherent sheaf (in degree 0), then $\widehat{F}$ is a WIT-$(g-i)$ coherent sheaf (in degree i) and $F \simeq (-1_A)^* R^{g-i} S(\widehat{F})$.

One easily sees that if F and G are arbitrary objects, then

$$\mathrm{Hom}_{\mathbf{D}(A)}(F,G) = \mathrm{Hom}_{\mathbf{D}(\widehat{A})}(\mathbf{R}\widehat{S}F, \mathbf{R}\widehat{S}G).$$

An easy consequence (cf. [Muk81, 2.5]) is that if F is a WIT-i sheaf and G is a WIT-j sheaf (or if F is a WIT-i locally free sheaf and G is a WIT-j object – not necessarily a sheaf), then

$$\begin{aligned}\mathrm{Ext}^k_{\mathscr{O}_A}(F,G) &\simeq \mathrm{Hom}_{\mathbf{D}(A)}(F,G[k]) \\ &\simeq \mathrm{Hom}_{\mathbf{D}(\widehat{A})}(\mathbf{R}\widehat{S}F, \mathbf{R}\widehat{S}G[k]) \\ &= \mathrm{Hom}_{\mathbf{D}(\widehat{A})}(\widehat{F}[-i], \widehat{G}[k-j]) \simeq \mathrm{Ext}^{k+i-j}_{\mathscr{O}_{\widehat{A}}}(\widehat{F},\widehat{G}). \quad (1)\end{aligned}$$

Let L be any ample line bundle on $\widehat{A}$, then $\mathbf{R}S(L) = R^0S(L) = \widehat{L}$ is a vector bundle on A of rank $h^0(L)$. For any $x \in A$, let $t_x\colon A \to A$ be the translation by x and let $\phi_L\colon \widehat{A} \to A$ be the isogeny determined by $\phi_L(\widehat{x}) = t^*_{\widehat{x}}L \otimes L^\vee$, then $\phi^*_L(\widehat{L}) = \bigoplus_{h^0(L)} L^\vee$.

Let $\lambda\colon X \to A$ be a projective morphism of normal varieties, and $\mathscr{L} = (\lambda \times \mathrm{id}_{\widehat{A}})^*\mathscr{P}$. We let $\mathbf{R}\Phi\colon \mathbf{D}(X) \to \mathbf{D}(\widehat{A})$ be the functor defined by $\mathbf{R}\Phi(F) = \mathbf{R}\pi_{\widehat{A}*}(\pi^*_X F \otimes \mathscr{L})$, where π_X and $\pi_{\widehat{A}}$ denote the projections of $X \times \widehat{A}$ onto the first and second factor. Note that

$$\begin{aligned}\mathbf{R}\Phi(F) &= \mathbf{R}\pi_{\widehat{A}*}(\pi^*_X F \otimes \mathscr{L}) \\ &\simeq^1 \mathbf{R}p_{\widehat{A}*}\mathbf{R}(\lambda \times \mathrm{id}_{\widehat{A}})_*(\pi^*_X F \otimes (\lambda \times \mathrm{id}_{\widehat{A}})^*\mathscr{P}) \\ &\simeq^2 \mathbf{R}p_{\widehat{A}*}\left(\mathbf{R}(\lambda \times \mathrm{id}_{\widehat{A}})_*(\pi^*_X F) \otimes \mathscr{P}\right) \\ &\simeq^3 \mathbf{R}p_{\widehat{A}*}(p^*_A \mathbf{R}\lambda_* F \otimes \mathscr{P}) \simeq \mathbf{R}\widehat{S}(\mathbf{R}\lambda_* F), \quad (2)\end{aligned}$$

where $\simeq^1$ follows by composition of derived functors [Har66, II.5.1], $\simeq^2$ follows by the projection formula [Har66, II.5.6], and $\simeq^3$ follows by flat base change [Har66, II.5.12].

We also define $\mathbf{R}\Psi\colon \mathbf{D}(\widehat{A}) \to \mathbf{D}(X)$ by $\mathbf{R}\Psi(F) = \mathbf{R}\pi_{X*}(\pi^*_{\widehat{A}}F \otimes \mathscr{L})$. Notice that if F is a locally free sheaf, then $\pi^*_{\widehat{A}}F \otimes \mathscr{L}$ is also a locally free sheaf. In particular, for any $i \in \mathbb{Z}$, we have that

$$R^i\Psi(F) \simeq R^i\pi_{X*}(\pi^*_{\widehat{A}}F \otimes \mathscr{L}). \quad (3)$$

We will need the following fact (which is also proven during the proof of Theorem B of [PP11]):

Lemma 2.2 *Let L be an ample line bundle on $\widehat{A}$, then*

$$\mathbf{R}\Psi(L^\vee) = R^g\Psi(L^\vee) = \lambda^*\widehat{L^\vee}.$$

Proof Since L is ample, $H^i(\widehat{A}, L^\vee \otimes \mathscr{L}_x) = H^i(\widehat{A}, L^\vee \otimes \mathscr{P}_{\lambda(x)}) = 0$ for $i \neq g$, where $\mathscr{P}_{\lambda(x)} = \mathscr{P}|_{\lambda(x)\times\widehat{A}}$ and $\mathscr{L}_x = \mathscr{L}|_{x\times\widehat{A}}$ are isomorphic. By cohomology and base change, $\mathbf{R}\Psi(L^\vee) = R^g\Psi(L^\vee)$ (resp. $\widehat{L^\vee}$) is a vector bundle of rank $h^g(\widehat{A}, L^\vee)$ on X (resp. on A).

The natural transformation $\mathrm{id}_{A\times\widehat{A}} \to (\lambda \times \mathrm{id}_{\widehat{A}})_*(\lambda \times \mathrm{id}_{\widehat{A}})^*$ induces a natural morphism

$$\widehat{L^\vee} = R^g p_{A*}(p_{\widehat{A}}^* L^\vee \otimes \mathscr{P}) \to R^g p_{A*}(\lambda \times \mathrm{id}_{\widehat{A}})_*(\pi_{\widehat{A}}^* L^\vee \otimes \mathscr{L}).$$

Let $\sigma = p_A \circ (\lambda \times \mathrm{id}_{\widehat{A}}) = \lambda \circ \pi_X$. By the Grothendieck spectral sequence associated with $p_{A*} \circ (\lambda \times \mathrm{id}_{\widehat{A}})_*$ there exists a natural morphism

$$R^g p_{A*}(\lambda \times \mathrm{id}_{\widehat{A}})_*(\pi_{\widehat{A}}^* L^\vee \otimes \mathscr{L}) \to R^g \sigma_*(\pi_{\widehat{A}}^* L^\vee \otimes \mathscr{L}),$$

and similarly by the Grothendieck spectral sequence associated with $\lambda_* \circ \pi_{X*}$ there exists a natural morphism

$$R^g \sigma_*(\pi_{\widehat{A}}^* L^\vee \otimes \mathscr{L}) \to \lambda_* R^g \pi_{X*}(\pi_{\widehat{A}}^* L^\vee \otimes \mathscr{L}).$$

Combining the above three morphisms gives a natural morphism

$$\widehat{L^\vee} \to \lambda_* R^g \pi_{X*}(\pi_{\widehat{A}}^* L^\vee \otimes \mathscr{L}) = \lambda_* R^g \Psi(L^\vee),$$

and hence by adjointness a natural morphism

$$\eta \colon \lambda^* \widehat{L^\vee} \to R^g\Psi(L^\vee).$$

For any point $x \in X$, by cohomology and base change, the induced morphism on the fiber over x is an isomorphism:

$$\begin{aligned}\eta_x \colon \lambda^* \widehat{L^\vee} \otimes \kappa(x) \simeq H^g(\lambda(x) \times \widehat{A}, L^\vee \otimes \mathscr{P}_{\lambda(x)}) \\ \xrightarrow{\simeq} H^g(x \times \widehat{A}, L^\vee \otimes \mathscr{L}_x) \simeq R^g\Psi(L^\vee) \otimes \kappa(x).\end{aligned}$$

Therefore η_x is an isomorphism for all $x \in X$ and hence η is an isomorphism. □

3 Examples

Notation 3.1 Let $T \subseteq \mathbb{P}^n$ be a projective variety. The cone over T in $\mathbb{A}^{n+1}$ will be denoted by $C(T)$. In other words, if $T \simeq \operatorname{Proj} S$, then $C(T) \simeq \operatorname{Spec} S$.

Linear equivalence between (Weil) divisors is denoted by $\sim$ and strict transform of a subvariety T by the inverse of a birational morphism σ is denoted by $\sigma_*^{-1} T$.

Example 3.2 Let k be an algebraically closed field, $V \subseteq \mathbb{P}^n$ and $W \subseteq \mathbb{P}^m$ two smooth projective varieties over k, and $p \in V$ a closed point. Let $x_0, \dots, x_n$ and $y_0, \dots, y_m$ be homogeneous coordinates on $\mathbb{P}^n$ and $\mathbb{P}^m$ respectively.

Consider the embedding $V \times W \subset \mathbb{P}^N$ induced by the Segre embedding of $\mathbb{P}^n \times \mathbb{P}^m$. We may choose homogeneous coordinates z_{ij} for $i = 0, \dots, n$ and $j = 0, \dots, m$ on $\mathbb{P}^N$, and in these coordinates $\mathbb{P}^n \times \mathbb{P}^m$ is defined by the equations $z_{\alpha\gamma}z_{\beta\delta} - z_{\alpha\delta}z_{\beta\gamma}$ for all $0 \le \alpha, \beta \le n$ and $0 \le \gamma, \delta \le m$.

Next let $H \subset W$ such that $\{p\} \times H \subset \{p\} \times W$ is a hyperplane section of $\{p\} \times W$ in $\mathbb{P}^N$. Let $Y = C(V \times W) \subset \mathbb{A}^{N+1}$ and $Z = C(V \times H) \subset Y$, and let $v \in Z \subset Y$ denote the common vertex of Y and Z. If $\dim W = 0$, then $H = \emptyset$. In this case let $Z = \{v\}$, the vertex of Y. Finally, let $\mathfrak{m}_v$ denote the ideal of v in the affine coordinate ring of Y. It is generated by all the variables z_{ij}.

Proposition 3.3 *Let $f\colon X \to Y$ be the blowing up of Y along Z. Then f is an isomorphism over $Y \setminus \{v\}$ and the scheme-theoretic pre-image of v (whose support is the exceptional locus) is isomorphic to V:*

$$f^{-1}(v) \simeq V.$$

Proof As Z is of codimension 1 in Y and $Y \setminus \{v\}$ is smooth, it follows that $Z \setminus \{v\}$ is a Cartier divisor in $Y \setminus \{v\}$ and hence f is indeed an isomorphism over $Y \setminus \{v\}$.

To prove the statement about the exceptional locus of f, first assume that $V = \mathbb{P}^n$, $W = \mathbb{P}^m$, $p = [1 : 0 : \cdots : 0]$, and $\{p\} \times H = (z_{0m} = 0) \cap (\{p\} \times W)$. Then $H = (y_m = 0) \subseteq W$ and hence $I = I(Z)$, the ideal of Z in the affine coordinate ring of Y, is generated by $\{z_{im} | i = 0, \dots, n\}$. Then by the definition of blowing up, $X = \operatorname{Proj} \oplus_{d \ge 0} I^d$ and $f^{-1}v \simeq \operatorname{Proj} \oplus_{d \ge 0} I^d / I^d \mathfrak{m}_v$.

Notice that $I^d / I^d \mathfrak{m}_v$ is a k-vector space generated by the degree-d monomials in the variables $\{z_{im} | i = 0, \dots, n\}$. It follows that the graded ring $\oplus_{d \ge 0} I^d / I^d \mathfrak{m}_v$ is nothing else but $k[z_{im} | i = 0, \dots, n]$ and hence $f^{-1}v \simeq \mathbb{P}^n = V$, so the claim is proved in this case.

Next consider the case when $V \subseteq \mathbb{P}^n$ is arbitrary, but $W = \mathbb{P}^m$. In this case the calculation is similar, except that we have to account for the defining equations of V. They show up in the definition of the coordinate ring of Y in the following way. If a homogeneous polynomial $g \in k[x_0, \dots, x_n]$ vanishes on V (i.e., $g \in I(V)_h$), then define $g_\gamma \in k[z_{ij}]$ for any $0 \le \gamma \le m$ by replacing x_α with $z_{\alpha\gamma}$ for each $0 \le \alpha \le n$. Then $\{g_\gamma | 0 \le \gamma \le m, g \in I(V)_h\}$ generates the ideal of Y in the affine coordinate ring of $C(\mathbb{P}^n \times \mathbb{P}^m)$. It follows that the above computation goes through in the same way, except that the variables $\{z_{im} | i = 0, \dots, n\}$ on the exceptional $\mathbb{P}^n$ are subject to the equations $\{g_m | g \in I(V)_h\}$. However, this simply means that the exceptional locus of f, i.e., $f^{-1}v$, is cut out from $\mathbb{P}^n$ by these equations and hence it is isomorphic to V.

Finally, consider the general case. The way W changes the setup is the same as what we described for V. If a homogeneous polynomial $h \in k[y_0, \dots, y_m]$ vanishes on W (i.e., $h \in I(W)_h$), then define $h_\alpha \in k[z_{ij}]$ for any $0 \le \alpha \le n$ by replacing y_γ with $z_{\alpha\gamma}$ for each $0 \le \gamma \le m$. Then $\{h_\alpha | 0 \le \alpha \le n, h \in I(W)_h\}$ generates the ideal of Y in the affine coordinate ring of $C(V \times \mathbb{P}^m)$.

However, in this case, differently from the case of V, we do not get any additional equations. Indeed, we chose the coordinates so that $H = (y_m = 0)$ and hence $y_m \notin I(W)$, which means that we may choose the rest of the coordinates such that $[0 : \cdots : 0 : 1] \in W$. This implies that no polynomial in the ideal of W may have a monomial term that is a constant multiple of a power of y_m. It follows that, since $I = I(Z)$ is generated by the elements $\{z_{im} | i = 0, \dots, n\}$, any monomial term of any polynomial in the ideal of Y in the affine coordinate ring of $C(V \times \mathbb{P}^m)$ that lies in I^d for some $d > 0$ also lies in $I^d \mathfrak{m}_v$. Therefore, these new equations do not change the ring $\oplus I^d / I^d \mathfrak{m}_v$ and so $f^{-1}v$ is still isomorphic to V. □

Notation 3.4 We will use the notation introduced in Proposition 3.3 for X, Y, Z, and f. We will also use $X_\mathbb{P}$, $Y_\mathbb{P}$, $Z_\mathbb{P}$, and $f_\mathbb{P} \colon X_\mathbb{P} \to Y_\mathbb{P}$ to denote the same objects in the case $W = \mathbb{P}^m$, i.e., $Y_\mathbb{P} = C(V \times \mathbb{P}^m)$, $Z_\mathbb{P} = C(V \times H)$, where $H \subset \mathbb{P}^m$ is such that $\{p\} \times H \subset \{p\} \times \mathbb{P}^m$ is a hyperplane section of $\{p\} \times \mathbb{P}^m$ in $\mathbb{P}^N$.

Corollary 3.5 *$f_\mathbb{P}$ is an isomorphism over $Y_\mathbb{P} \setminus \{v\}$ and the scheme-theoretic pre-image of v (whose support is the exceptional locus) via $f_\mathbb{P}$ is isomorphic to V:*

$$f_\mathbb{P}^{-1} v \simeq V.$$

Proof This was proven as an intermediate step in Proposition 3.3, and is also straightforward by taking $W = \mathbb{P}^m$. □

Proposition 3.6 *Assume that V and W are both positive dimensional, $W \subseteq \mathbb{P}^m$ is a complete intersection, and the embedding $V \times \mathbb{P}^r \subset \mathbb{P}^N$ for any linear subvariety $\mathbb{P}^r \subseteq \mathbb{P}^m$ induced by the Segre embedding of $\mathbb{P}^n \times \mathbb{P}^m$ is projectively normal. Then X is Gorenstein.*

Proof First note that the projective normality assumption implies that $Y_\mathbb{P} = C(V \times \mathbb{P}^m)$ is normal and hence we may consider divisors and their linear equivalences.

Let $H' \subset \mathbb{P}^m$ be an arbitrary hypersurface (different from H and not necessarily linear). Observe that $H' \sim d \cdot H$ with $d = \deg H'$, so $V \times H' \sim d \cdot (V \times H)$, and hence $C(V \times H') \sim d \cdot C(V \times H)$ as divisors on $Y_\mathbb{P}$.

Since $f_{\mathbb{P}}$ is a small morphism it follows that the strict transforms of these divisors on $X_{\mathbb{P}}$ are also linearly equivalent: $f_*^{-1}C(V \times H') \sim d \cdot f_*^{-1}C(V \times H)$ (where by abuse of notation we let $f = f_{\mathbb{P}}$). By the basic properties of blowing up, the (scheme-theoretic) pre-image of $C(V \times H)$ is a Cartier divisor on X which coincides with $f_*^{-1}C(V\times H)$ (as f is small). However, then $f_*^{-1}C(V\times H')$ is also a Cartier divisor and hence it is Gorenstein if and only if $X_{\mathbb{P}}$ is. Note that $f_*^{-1}C(V\times H')$ is nothing else but the blow-up of $C(V\times H')$ along $C(V\times(H'\cap H))$.

By assumption W is a complete intersection, so applying the above argument for the intersection of the hypersurfaces cutting out W shows that X is Gorenstein if and only if $X_{\mathbb{P}}$ is Gorenstein. In other words, it is enough to prove the statement with the additional assumption that $W = \mathbb{P}^m$. In particular, we have $X = X_{\mathbb{P}}$, etc.

In this case the same argument as above shows that the statement holds for m if and only if it holds for $m - 1$, so we only need to prove it for $m = 1$. In that case $H \in \mathbb{P}^1$ is a single point. Choose another point $H' \in \mathbb{P}^1$. As above, $f_*^{-1}C(V \times H')$ is a Cartier divisor in X and it is the blow-up of $C(V \times H')$ along the intersection $C(V \times H') \cap C(V \times H)$.

We claim that this intersection is just the vertex of $C(V)$. To see this, view $Y = Y_{\mathbb{P}} = C(V \times \mathbb{P}^1)$ as a subscheme of $C(\mathbb{P}^n \times \mathbb{P}^1)$. Inside $C(\mathbb{P}^n \times \mathbb{P}^1)$ the cones $C(\mathbb{P}^n \times H)$ and $C(\mathbb{P}^n \times H')$ are just linear subspaces of dimension $n + 1$ whose scheme-theoretic intersection is the single reduced point v. Therefore we have that

$$C(V \times H') \cap C(V \times H) \subseteq C(\mathbb{P}^m \times H') \cap C(\mathbb{P}^m \times H) = \{v\},$$

proving the same for this intersection.

Finally then $f_*^{-1}C(V \times H')$, the blow-up of $C(V \times H')$ along the intersection $C(V \times H') \cap C(V \times H)$, is just the blow-up of $C(V)$ at its vertex and hence it is smooth and in particular Gorenstein. This completes the proof. □

Lemma 3.7 *Let $V \subseteq \mathbb{P}^n$ and $W \subseteq \mathbb{P}^m$ be two normal complete intersection varieties of positive dimension. Assume that either* $\dim V + \dim W > 2$ *or if* $\dim V = \dim W = 1$, *then $n = m = 2$. The embedding $V \times W \subset \mathbb{P}^N$ induced by the Segre embedding of $\mathbb{P}^n \times \mathbb{P}^m$ is then projectively normal.*

Proof It follows easily from the definition of the Segre embedding that it is itself projectively normal and hence it is enough to prove that

$$H^0(\mathbb{P}^n \times \mathbb{P}^m, \mathscr{O}_{\mathbb{P}^N}(d)|_{\mathbb{P}^n\times\mathbb{P}^m}) \to H^0(V \times W, \mathscr{O}_{\mathbb{P}^N}(d)|_{V\times W}) \tag{1}$$

is surjective for all $d \in \mathbb{N}$.

We prove this by induction on the combined number of hypersurfaces cutting out V and W. When this number is 0, then $V = \mathbb{P}^n$ and $W = \mathbb{P}^m$ so we are done.

Otherwise, assume that $\dim V \leq \dim W$ and if $\dim V = \dim W = 1$ then $\deg V = e \geq \deg W$. Let $V' \subseteq \mathbb{P}^n$ be a complete intersection variety of dimension $\dim V+1$ such that $V = V' \cap H'$, where $H' \subset \mathbb{P}^n$ is a hypersurface of degree e. Then $V \times W \subset V' \times W$ is a Cartier divisor with ideal sheaf $\mathscr{I} \simeq \pi_1^* \mathscr{O}_{V'}(-e)$, where $\pi_1 : V' \times W \to V'$ is the projection to the first factor. It follows that for every $d \in \mathbb{N}$ there exists a short exact sequence

$$0 \to \mathscr{O}_{\mathbb{P}^N}(d)|_{V' \times W} \otimes \pi_1^* \mathscr{O}_{V'}(-e) \to \mathscr{O}_{\mathbb{P}^N}(d)|_{V' \times W} \to \mathscr{O}_{\mathbb{P}^N}(d)|_{V \times W} \to 0,$$

and hence an induced exact sequence of cohomology

$$H^0(V' \times W, \mathscr{O}_{\mathbb{P}^N}(d)|_{V' \times W}) \to H^0(V \times W, \mathscr{O}_{\mathbb{P}^N}(d)|_{V \times W}) \\ \to H^1(V' \times W, \pi_1^* \mathscr{O}_{V'}(d-e) \otimes \pi_2^* \mathscr{O}_W(d)),$$

where $\pi_2 : V' \times W \to W$ is the projection to the second factor.

Since by assumption V' is a complete intersection variety of dimension at least 2, it follows that $H^1(V', \mathscr{O}_{V'}(d-e)) = 0$.

If $\dim W > 1$, then it follows similarly that $H^1(W, \mathscr{O}_W(d)) = 0$.

If $\dim W = 1$, then since $0 < \dim V \leq \dim W$ we also have $\dim V = 1$. By assumption V and W are normal and hence regular, and in this case we assumed earlier that $\deg V = e \geq \deg W$. It follows that as long as $e > d$, then $H^0(V', \mathscr{O}_{V'}(d-e)) = 0$ and if $e \leq d$, then $d \geq \deg W$ and hence $H^1(W, \mathscr{O}_W(d)) = 0$.

In both cases we obtain that by the Künneth formula (cf. [EGAIII$_2$, (6.7.8)], [Kem93, 9.2.4]),

$$H^1(V' \times W, \pi_1^* \mathscr{O}_{V'}(d-e) \otimes \pi_2^* \mathscr{O}_W(d)) = 0$$

and hence

$$H^0(V' \times W, \mathscr{O}_{\mathbb{P}^N}(d)|_{V' \times W}) \to H^0(V \times W, \mathscr{O}_{\mathbb{P}^N}(d)|_{V \times W})$$

is surjective. By induction we may assume that

$$H^0(\mathbb{P}^n \times \mathbb{P}^m, \mathscr{O}_{\mathbb{P}^N}(d)|_{\mathbb{P}^n \times \mathbb{P}^m}) \to H^0(V' \times W, \mathscr{O}_{\mathbb{P}^N}(d)|_{V' \times W})$$

is surjective, so it follows that the desired map in (1) is surjective as well and the statement is proven. □

Corollary 3.8 *Let $V \subseteq \mathbb{P}^n$ and $W \subseteq \mathbb{P}^m$ be two positive-dimensional normal complete intersection varieties and assume that if* $\dim V = 1$, *then* $n = 2$. *X is then Gorenstein.*

Proof Follows by combining Proposition 3.6 and Lemma 3.7. Note that in Proposition 3.6 the embedding $V \times W \hookrightarrow \mathbb{P}^N$ does not need to be projectively normal, only $V \times \mathbb{P}^r \hookrightarrow \mathbb{P}^N$ does, which indeed follows from Lemma 3.7. □

Example 3.9 Let k be an algebraically closed field. We will construct a birational projective morphism $f\colon X \to Y$ such that X is Gorenstein (and log canonical) and $R^1 f_*\omega_X \neq 0$.

Let $E_1, E_2 \subseteq \mathbb{P}^2$ be two smooth projective cubic curves. Consider the construction in Example 3.2 with $V = E_1$, $W = E_2$. As in that construction let $f\colon X \to Y$ be the blow-up of $Y = C(E_1 \times E_2)$ along $Z = C(E_1 \times H)$, where $H \subseteq E_2$ is a hyperplane section. The common vertex of Y and Z will still be denoted by $v \in Z \subset Y$. The map f is an isomorphism over $Y \setminus \{v\}$ and $f^{-1}v \simeq E_1$ by Proposition 3.3.

Proposition 3.10 *Both X and Y are smooth in codimension* 1 *with trivial canonical divisor and X is Gorenstein and hence Cohen–Macaulay.*

Proof By construction $Y \setminus \{v\} \simeq X \setminus f^{-1}v$ is smooth, so the first statement follows. Furthermore, $Y \setminus \{v\} \simeq X \setminus f^{-1}v$ is an affine bundle over $E_1 \times E_2$, so by the choice of E_1 and E_2, the canonical divisor of $Y \setminus \{v\} \simeq X \setminus f^{-1}v$ is trivial. However, the complement of this set has codimension at least 2 in both X and Y and hence their canonical divisors are trivial as well. Since $E_1, E_2 \subset \mathbb{P}^2$ are hypersurfaces, X is Gorenstein by Corollary 3.8. □

Let E denote $f^{-1}v$, so we have that $E \simeq E_1$ and there is a short exact sequence

$$0 \to \mathscr{I}_E \to \mathscr{O}_X \to \mathscr{O}_E \to 0.$$

Pushing this forward via f we obtain a homomorphism $\phi\colon R^1 f_*\mathscr{O}_X \to R^1 f_*\mathscr{O}_E$. Since the maximum dimension of any fiber of f is 1, we have $R^2 f_*\mathscr{I}_E = 0$. It follows that $R^1 f_*\omega_X = R^1 f_*\mathscr{O}_X \neq 0$, because $R^1 f_*\mathscr{O}_E \neq 0$ (it is a sheaf supported on v of length $h^1(\mathscr{O}_E) = 1$).

Example 3.11 Let k be an algebraically closed field of characteristic $p \neq 0$. Then there exists a birational morphism $f\colon X \to Y$ of varieties (defined over k) such that X is smooth of dimension 7 and $R^i f_*\omega_X \neq 0$, for some $i \in \{1, 2, 3, 4, 5\}$.

Let Z be a smooth 6-dimensional variety and L a very ample line bundle such that $H^1(Z, \omega_Z \otimes L) \neq 0$ (such varieties exist by [LR97]). By Serre vanishing, $H^i(Z, \omega_Z \otimes L^j) = 0$ for all $i > 0$ and $j \gg 0$. Let m be the largest positive integer such that $H^i(Z, \omega_Z \otimes L^m) \neq 0$ for some $i > 0$.

After replacing L by L^m we may assume that there exists a $q > 0$ such that $H^q(Z, \omega_Z \otimes L) \neq 0$, but $H^i(Z, \omega_Z \otimes L^j) = 0$ for all $i > 0$ and $j \geq 2$. Note that $q < 6$, because $H^6(Z, \omega_Z \otimes L)$ is dual to $H^0(Z, L^{-1}) = 0$.

Let Y be the cone over the embedding of Z given by L, $f: X \to Y$ the blow-up of the vertex $v \in Y$, and $E = f^{-1}v$ the exceptional divisor of f. Note that $E \simeq Z$ and $\omega_E(-jE) \simeq \omega_Z \otimes L^j$ for any j.

For $j \geq 1$ consider the short exact sequence

$$0 \to \omega_X(-jE) \to \omega_X(-(j-1)E) \to \omega_E(-jE) \to 0.$$

Claim $R^i f_* \omega_X(-E) = 0$ for all $i > 0$ and $R^i f_* \omega_X = 0$ for all $i > 0$, such that $H^i(Z, \omega_Z \otimes L) = 0$.

Proof of claim As $-E$ is f-ample we have, by Serre vanishing again, that $R^i f_* \omega_X(-jE) = 0$ for all $i > 0$ and some $j > 0$. If either $j > 1$ or $j = 1$ and $H^i(Z, \omega_Z \otimes L) = 0$, then $R^i f_* \omega_E(-jE) = H^i(Z, \omega_Z \otimes L^j) = 0$ by the choice of L. Therefore, the exact sequence

$$0 = R^i f_* \omega_X(-jE) \to R^i f_* \omega_X(-(j-1)E) \to R^i f_* \omega_E(-jE) = 0$$

gives that $R^i f_* \omega_X(-(j-1)E) = 0$. The claim follows by induction. □

From the above claim it follows that

$$0 = R^q f_* \omega_X(-E) \to R^q f_* \omega_X \to R^q f_* \omega_E(-E) \to R^{q+1} f_* \omega_X(-E) = 0.$$

Since $R^q f_* \omega_E(-E) = H^q(Z, \omega_Z \otimes L) \neq 0$, we obtain that $R^q f_* \omega_X \neq 0$ as claimed. □

Remark 3.12 The above example is certainly well known (see, e.g., [CR11b, 4.7.2]) and one can easily construct examples in dimension ≥ 3 (using, e.g., the results of [Ray78] and [Muk79]). We have chosen to include the above example because of its elementary nature.

Proposition 3.13 *There exists a variety T and a generically finite projective separable morphism to an abelian variety $\lambda: T \to A$ defined over an algebraically closed field k such that:*

- *if* char $k = 0$, *then T is Gorenstein (and hence Cohen–Macaulay) with a single isolated log canonical singularity and $R^1 \lambda_* \omega_T \neq 0$;*
- *if* char $k = p > 0$, *then T is smooth and $R^i \lambda_* \omega_T \neq 0$ for some $i > 0$.*

Proof First assume that char $k = 0$ and let $f: X \to Y$ be as in Example 3.9. We may assume that X and Y are projective. Let $X' \to X$ and $Y' \to Y$ be birational morphisms that are isomorphisms near $f^{-1}(v)$ and v respectively such that there is a birational morphism $f': X' \to Y'$ and a generically finite morphism $g: Y' \to \mathbb{P}^n$. Let $v' \in Y'$ be the inverse image of $v \in Y$ and $p \in \mathbb{P}^n$

its image. We may assume that there is an open subset $\mathbb{P}^n_0 \subset \mathbb{P}^n$ such that $g|_{Y'_0}$ is finite, where $Y'_0 = g^{-1}(\mathbb{P}^n_0)$. Note that if we let X'_0 be the inverse image of Y'_0 and $g' = g \circ f'$, then we have $R^i g'_* \omega_{X'_0} = g_* R^i f'_* \omega_{X'_0}$.

Let A be an n-dimensional abelian variety, $A' \to A$ a birational morphism of smooth varieties, and $A' \to \mathbb{P}^n$ a generically finite morphism. We may assume that there are points $a' \in A'$ and $a \in A$ such that $(A', a') \to (A, a)$ is locally an isomorphism and $(A', a') \to (\mathbb{P}^n, p)$ is locally étale.

Let U be the normalization of the main component of $X' \times_{\mathbb{P}^n} A'$ and $h: U \to X'$ the corresponding morphism. We let $E \subset (f' \circ h)^{-1}(v') \subset U$ be the component corresponding to $(v', a') \in Y' \times_{\mathbb{P}^n} A'$. Then, the morphism $(U, E) \to (Y' \times_{\mathbb{P}^n} A', (v', a')) \to (A, a)$ is étale locally (on the base) isomorphic to $(X, f^{-1}(v)) \to (Y, v) \to (\mathbb{P}^n, p)$.

Let $\nu: T \to U$ be a birational morphism such that ν is an isomorphism over a neighborhood of $E \subset U$ and $T \setminus \nu^{-1}(E)$ is smooth. Let $\lambda: T \to A$ be the induced morphism. It is clear from what we have observed above that $\lambda(E)$ is one of the components of the support of $R^1 \lambda_* \omega_T \neq 0$ and T has the required singularities.

Assume now that char $k = p > 0$ and let $f: X \to Y$ be a birational morphism of varieties such that X is smooth and $R^i f_* \omega_X \neq 0$ for some $i > 0$. This i will be fixed for the rest of the proof. The existence of such morphisms is well known (see Remark 3.12), and Example 3.11 is explicit in dimension 7. Further, let A be an abelian variety of the same dimension as X and Y and set $n = \dim A = \dim X = \dim Y$. There are embeddings $Y \subset \mathbb{P}^{m_1}$, $A \subset \mathbb{P}^{m_2}$, and $\mathbb{P}^{m_1} \times \mathbb{P}^{m_2} \subset \mathbb{P}^M$. Let H be a very ample divisor on $\mathbb{P}^M$ and $U \subset Y \times A$ the intersection of n general members $H_1, \dots, H_n \in |H|$ with $Y \times A$. By choice, the induced maps $h: U \to Y$ and $a: U \to A$ are generically finite, U intersects $v \times A$ transversely so that $V = U \cap (v \times A)$ is a finite set of reduced points, and $U \setminus V$ is smooth by Bertini's theorem (cf. [Har77, II.8.18] and its proof). It follows that any singular point $u \in U$ is a point in V and (U, u) is locally isomorphic to (Y, v). We claim that a is finite in a neighborhood of $u \in U$. Consider any contracted curve, i.e., any curve $C \subset U \cap (Y \times a(u))$. We must show that $u \notin C$. Let $\nu: T \to U$ be the blow-up of U along V and $\tilde{C}$ the strict transform of C on T. We let $\mu: \mathrm{Bl}_V \mathbb{P}^M \to \mathbb{P}^M$, $E = \mu^{-1}(u) \cong \mathbb{P}^{M-1}$ and we denote by $h_i = \mu_*^{-1} H_i|_E$ the corresponding hyperplanes. To verify the claim it suffices to check that $\nu^{-1}(u) \cap \tilde{C} = \emptyset$. But this is now clear, as $\nu^{-1}(u) \cong Z \subset \mathbb{P}^{M-1}$ and the h_i are general hyperplanes so that $Z \cap h_1 \cap \ldots \cap h_n = \emptyset$ as Z is $(n-1)$-dimensional.

Let $\lambda = a \circ \nu: T \to A$ be the induced morphism. By construction, the support of the sheaf $R^i \nu_* \omega_T$ is V. Since a is finite on a neighborhood of $u \in U$, it follows that $0 \neq a_* R^i \nu_* \omega_T \subset R^i \lambda_* \omega_T$ and hence $R^i \lambda_* \omega_T \neq 0$ for the same $i > 0$. □

4 Main result

Proposition 4.1 *Assume that $\lambda\colon X \to A$ is generically finite onto its image, where X is a projective Cohen–Macaulay variety and A is an abelian variety. If* $\mathrm{char}(k) = p > 0$*, then we assume that there is an ample line bundle L on A whose degree is not divisible by p. If $R^i\pi_{\widehat{A}*}\mathscr{L} = 0$ for all $i < n$, then $R^i\lambda_*\omega_X = 0$ for all $i > 0$.*

Proof By Theorem A of [PP11], $R^i\Phi(\mathscr{O}_X) = R^i\pi_{\widehat{A}*}\mathscr{L} = 0$ for all $i < n$ is equivalent to

$$H^i(X, \omega_X \otimes R^g\Psi(L^\vee)) = 0 \qquad \forall\, i > 0,$$

where L is sufficiently ample on $\widehat{A}$ and $R^g\Psi(L^\vee) = \lambda^*\widehat{L^\vee}$ (cf. Lemma 2.2). It is easy to see that this in turn is equivalent to

$$H^i(X, \omega_X \otimes \lambda^*(\widehat{t^*_{\widehat{a}}L^\vee})) = 0 \qquad \forall\, i > 0,\ \forall\, \widehat{a} \in \widehat{A},$$

where L is sufficiently ample on $\widehat{A}$. By [Muk81, 3.1], we have $\widehat{t^*_{\widehat{a}}L^\vee} = \widehat{L^\vee} \otimes P_{-\widehat{a}}$ and hence $H^i(X, \omega_X \otimes \lambda^*(\widehat{L^\vee} \otimes P_{-\widehat{a}})) = 0$. Thus, by cohomology and base change, we have that

$$\mathbf{R}\widehat{S}(\mathbf{R}\lambda_*\omega_X \otimes \widehat{L^\vee}) =^{(2)} \mathbf{R}\Phi(\omega_X \otimes \lambda^*\widehat{L^\vee}) = R^0\Phi(\omega_X \otimes \lambda^*\widehat{L^\vee}).$$

In particular, $\mathbf{R}\lambda_*\omega_X \otimes \widehat{L^\vee}$ is WIT-0. □

Claim 4.2 For any ample line bundle M on A, we have that

$$H^i(X, \omega_X \otimes \lambda^*(\widehat{L^\vee} \otimes M \otimes P_{-\widehat{a}})) = 0 \qquad \forall\, i > 0,\ \forall\, \widehat{a} \in \widehat{A}.$$

Proof We follow the argument in [PP03, 2.9]. For any $P = P_{-\widehat{a}}$,

$$\begin{aligned}
H^i(X, \omega_X \otimes \lambda^*(\widehat{L^\vee} \otimes M \otimes P)) &= R^i\Gamma(X, \omega_X \otimes \lambda^*(\widehat{L^\vee} \otimes M \otimes P)) \\
&=^{\mathrm{P.F.}} R^i\Gamma(A, \mathbf{R}\lambda_*\omega_X \otimes \widehat{L^\vee} \otimes M \otimes P) = \mathrm{Ext}^i_{D(A)}((M \otimes P)^\vee, \mathbf{R}\lambda_*\omega_X \otimes \widehat{L^\vee}) \\
&=^{(1)} \mathrm{Ext}^{i+g}_{D(\widehat{A})}(R^g\widehat{S}((M \otimes P)^\vee), R^0\Phi(\omega_X \otimes \lambda^*\widehat{L^\vee})) \\
&= H^{i+g}(\widehat{A}, R^0\Phi(\omega_X \otimes \lambda^*\widehat{L^\vee}) \otimes R^g\widehat{S}((M \otimes P)^\vee)^\vee) = 0 \qquad i > 0.
\end{aligned}$$

(The third equality follows as $M \otimes P$ is free, the fifth follows since $R^g\widehat{S}(M \otimes P)^\vee$ is free, and the last one since $i + g > g = \dim \widehat{A}$.) □

Let $\phi_L\colon \widehat{A} \to A$ be the isogeny induced by $\phi_L(\widehat{x}) = t^*_{\widehat{x}}L \otimes L^\vee$, then $\phi^*_L\widehat{L^\vee} = L^{\oplus h^0(L)}$. We may assume that the characteristic does not divide the degree of L, so that ϕ_L is separable. Let $X' = X \times_A \widehat{A}$, $\phi\colon X' \to X$, and $\lambda'\colon X' \to \widehat{A}$ be the induced morphisms. Note that $\phi_*\mathscr{O}_{X'} = \lambda^*(\phi_{L*}\mathscr{O}_{\widehat{A}}) = \lambda^*(\oplus P_{\alpha_i})$, where the α_i are the elements in $K \subset \widehat{A}$, the kernel of the induced homomorphism $\phi_L : \widehat{A} \to A$. By the above equation and a flat base change,

$$H^i(X', \omega_{X'} \otimes \lambda'^* \phi_L^*(\widehat{L^\vee} \otimes M)) = \bigoplus_{\alpha \in K} H^i(X, \omega_X \otimes \lambda^*(\widehat{L^\vee} \otimes M \otimes P_\alpha)) = 0$$

for all $i > 0$. But then $H^i(X', \omega_{X'} \otimes \lambda'^*(L \otimes \phi_L^* M)) = 0$ for all $i > 0$. Note that if M is sufficiently ample on A then so is $L \otimes \phi_L^* M$ on $\widehat{A}$. It follows by an easy (and standard) spectral sequence argument that $R^i \lambda'_* \omega_{X'} = 0$ for $i > 0$. Since ω_X is a summand of $\phi_* \omega_{X'} = \mathbf{R}\phi_* \omega_{X'}$, and $\mathbf{R}\lambda_* \mathbf{R}\phi_* \omega_{X'} = \mathbf{R}\phi_{L*} \mathbf{R}\lambda'_* \omega_{X'}$, it follows that $R^i \lambda_* \omega_X$ is a summand of $R^i \lambda_* \phi_* \omega_{X'} = \phi_{L*} R^i \lambda'_* \omega_{X'}$ and hence $R^i \lambda_* \omega_X = 0$ for all $i > 0$. □

Proof of Theorem 1.3 Immediate from Propositions 3.13 and 4.1. □

Acknowledgments The authors would like to thank G. Pareschi, A. Langer, and M. Popa for useful discussions and comments.

References

[CH11] Chen, J. A. and Hacon, C. D. Kodaira dimension of irregular varieties. *Invent. Math.* **186**(3) (2011) 481–500.

[CR11b] Chatzistamatiou, A. and Rülling, K. Hodge Witt cohomology and Witt-rational singularities. arXiv:1104.2145v1.

[GL91] Green, M. and Lazarsfeld, R. Higher obstructions to deforming cohomology groups of line bundles. *J. Amer. Math. Soc.* **4** (1991) 87–103.

[EGAIII$_2$] Grothendieck, A. Éléments de géométrie algébrique, III (seconde partie). *Inst. Hautes Études Sci. Publ. Math.* **17** (1963) 91.

[Hac04] Hacon, C. A derived category approach to generic vanishing. *J. Reine Angew. Math.* **575** (2004) 173–187.

[Har66] Hartshorne, R. Residues and Duality. Lecture notes of a seminar on the work of A. Grothendieck given at Harvard University 1963/64. With an appendix by P. Deligne. Lecture Notes in Mathematics, No. 20. Berlin: Springer-Verlag, 1966.

[Har77] Hartshorne, R. Algebraic Geometry. Graduate Texts in Mathematics, No. 52. New York: Springer-Verlag, 1977.

[JLT11] Jiang, Z., Lahoz, M., and Tirabassi, S. On the Iitaka fibration of varieties of maximal Albanese dimension. arXiv:1111.6279.

[Kem93] Kempf, G. R. Algebraic Varieties. London Mathematical Society Lecture Note Series, Vol. 172. Cambridge: Cambridge University Press, 1993.

[LR97] Lauritzen, N. and Rao, A. P. Elementary counterexamples to Kodaira vanishing in prime characteristic. *Proc. Indian Acad. Sci.* **107** (1997) 21–25.

[Muk79] Mukai, S. On counterexamples for the Kodaira vanishing theorem and the Yau inequality in positive characteristics. In Symposium on Algebraic Geometry, Kinosaki, 1979, p. 923 (in Japanese).

[Muk81] Mukai, S. Duality between $D(X)$ and $D(\widehat{X})$ with its application to Picard sheaves. *Nagoya Math. J.* **81** (1981) 153–175.

[Par03] Pareschi, G.: Generic vanishing, Gaussian maps, and Fourier–Mukai transform. math.arXiv:0310026v1.

[PP03] Pareschi, G. and Popa, M.: Regularity on abelian varieties I. *J. Amer. Math. Soc.* **16**(2) (2003) 285–302.

[PP09] Pareschi, G. and Popa, M. Strong generic vanishing and a higher dimensional Castelnuovo–de Franchis inequality. *Duke Math. J.* **150**(2) (2009) 269–285.

[PP11] Pareschi, G. and Popa, M.: GV-sheaves, Fourier–Mukai transform, and generic vanishing. *Amer. J. Math.* **133** (2011) 235–271.

[Ray78] Raynaud, M. Contre-exemple au "vanishing theorem" en caractéristique $p > 0$. C. P. Ramanujam—A Tribute. Tata Institute of Fundamental Research Studies in Mathematics, Vol. 8. Berlin: Springer-Verlag, 1978, pp. 273–278.

14

Deformations of elliptic Calabi–Yau manifolds

J. Kollár
Princeton University

Abstract

We investigate deformations and characterizations of elliptic Calabi–Yau varieties, building on earlier works of Wilson and Oguiso. We show that if the second cohomology of the structure sheaf vanishes, then every deformation is again elliptic. More generally, all non-elliptic deformations derive from abelian varieties or K3 surfaces. We also give a numerical characterization of elliptic Calabi–Yau varieties under some positivity assumptions on the second Todd class. These results lead to a series of conjectures on fibered Calabi–Yau varieties.

To Robert Lazarsfeld on the occasion of his sixtieth birthday

The aim of this paper is to answer some questions about Calabi–Yau manifolds that were raised during the workshop *String Theory for Mathematicians,* which was held at the Simons Center for Geometry and Physics.

F-theory posits that the "hidden dimensions" constitute a Calabi–Yau 4-fold X that has an elliptic structure with a section. That is, there are morphisms $g: X \to B$ whose general fibers are elliptic curves and $\sigma: B \to X$ such that $g \circ \sigma = 1_B$ (see [Vaf96, Don98]). In his lecture, Donagi asked the following:

Question 1 Is every small deformation of an elliptic Calabi–Yau manifold also an elliptic Calabi–Yau manifold?

Question 2 Is there a good numerical characterization of elliptic Calabi–Yau manifolds?

From *Recent Advances in Algebraic Geometry*, edited by Christopher Hacon, Mircea Mustaţă and Mihnea Popa

Clearly, an answer to Question 2 should give a solution of Question 1. The answers to these problems are quite sensitive to which variant of the definition of Calabi–Yau manifolds one uses. For instance, a general deformation of the product of an Abelian variety and of an elliptic curve has no elliptic fiber space structure and every elliptic K3 surface has non-elliptic deformations. We prove in Section 5 that these are essentially the only such examples, even for singular Calabi–Yau varieties (Theorem 31). In the smooth case, the answer is especially simple.

Theorem 3 *Let X be an elliptic Calabi–Yau manifold such that $H^2(X, \mathcal{O}_X) = 0$. Then every small deformation of X is also an elliptic Calabi–Yau manifold.*

In dimension 3 this was proved in [Wil94, Wil98].

Our results on Question 2 are less complete. Let $L_B \in H^2(B, \mathbb{Q})$ be an ample cohomology class and set $L := g^* L_B$. We interpret Question 2 to mean: *Characterize pairs (X, L) that are elliptic fiber spaces.* Following [Wil89, Ogu93], one is led to the following:

Conjecture 4 *A Calabi–Yau manifold X is elliptic iff there is a $(1,1)$-class $L \in H^2(X, \mathbb{Q})$ such that $(L \cdot C) \geq 0$ for every algebraic curve $C \subset X$, $(L^{\dim X}) = 0$ and $(L^{\dim X - 1}) \neq 0$.*

For threefolds, the more general results of [Ogu93, Wil94] imply Conjecture 4 if L is effective or $(L \cdot c_2(X)) \neq 0$. As in the earlier works, in higher dimensions we study the interrelation of L and of the second Chern class $c_2(X)$. By a result of [Miy88], $(L^{n-2} \cdot c_2(X)) \geq 0$ and we distinguish two cases.

- (Main case) If $(L^{n-2} \cdot c_2(X)) > 0$ then Conjecture 4 is solved in Corollary 11. We also check that all elliptic Calabi–Yau manifolds with a section belong to this class (46).
- (Isotrivial case) If $(L^{n-2} \cdot c_2(X)) = 0$ then there is an elliptic curve E, a finite subgroup $G \subset \mathrm{Aut}(E)$, and a Calabi–Yau manifold Y with a G-action such that $X \cong (E \times Y)/G$ (see Theorem 43). If, in addition, $H^2(X, \mathcal{O}_X) = 0$ then by Theorem 39 every deformation of X is obtained by deforming E and Y.

  However, I have not been able to prove that the numerical conditions of Conjecture 4 guarantee the existence of an elliptic structure.

Following [Ogu93] and [MP97, Lecture 10], the plan is to put both questions in the more general framework of the abundance conjecture [Rei83, 4.6]; see Conjectures 50 and 51 for the precise formulation.

This approach suggests that the key is to understand the rate of growth of $h^0(X, L^m)$. If (X, L) is elliptic, then $h^0(X, L^m)$ grows like $m^{\dim X - 1}$. Given a pair

(X, L), the most important deformation-invariant quantity is the holomorphic Euler characteristic

$$\chi(X, L^m) = h^0(X, L^m) - h^1(X, L^m) + h^2(X, L^m) \cdots$$

The difficulty is that in our case $h^0(X, L^m)$ and $h^1(X, L^m)$ both grow like $m^{\dim X - 1}$ and they cancel each other out. That is,

$$\chi(X, L^m) = O(m^{\dim X - 2}).$$

For the main series $\chi(X, L^m)$ does grow like $m^{\dim X - 2}$, which implies that $h^0(X, L^m)$ grows at least like $m^{\dim X - 2}$.

For the isotrivial series the order of growth of $\chi(X, L^m)$ is even smaller; in fact, $\chi(X, L^m)$ can be identically zero.

Several of the ideas of this paper can be traced back to other sources. Sections 2–4 owe a lot to [Kaw85a, Ogu93, Wil94, Fuj11]; Sections 5 and 6 to [Hor76, KL09]; Sections 7 and 8 to [Kol93, Nak99] and to some old results of Matsusaka. Ultimately the origin of many of these methods is the work of Kodaira on elliptic surfaces [Kod63, Section 12]. (See [BPV84, Sections V.7–13] for a more modern treatment.)

1 Calabi–Yau fiber spaces

For many reasons it is of interest to study proper morphisms with connected fibers $g' \colon X' \to B$ whose general fibers are birational to Calabi–Yau varieties. A special case of the minimal model conjecture, proved by [Lai11, HX13], implies that every such fiber space is birational to a projective morphism with connected fibers $g \colon X \to B$ where X has terminal singularities and its canonical class K_X is relatively trivial, at least rationally. That is, there is a Cartier divisor F on B such that $mK_X \sim g^*F$ for some $m > 0$.

We will work with varieties with log terminal singularities, or later even with klt pairs (X, Δ), but I will state the main results for smooth varieties as well. See [KM98, Section 2.3] for the definitions and basic properties of the singularities we use. Note also that, even if one is primarily interested in smooth Calabi–Yau varieties X, the natural setting is to allow at least canonical singularities on X and at least log terminal singularities on the base B of the elliptic fibration.

Definition 5 In this paper a *Calabi–Yau variety* is a projective variety X with log terminal singularities such that $K_X \sim_{\mathbb{Q}} 0$, that is, mK_X is linearly equivalent to 0 for some $m > 0$. By [Kaw85b] this is equivalent to assuming that $(K_X \cdot C) = 0$ for every curve $C \subset X$.

Note that we allow a rather broad definition of Calabi–Yau varieties. This is very natural for algebraic geometry but less so for physical considerations.

A *Calabi–Yau fiber space* is a proper morphism with connected fibers $g: X \to B$ onto a normal variety where X has log terminal singularities and $K_{X_g} \sim_{\mathbb{Q}} 0$ where $X_g \subset X$ is a general fiber.

We say that $g: X \to B$ is an *elliptic* (or *Abelian*, etc.) fiber space if in addition general fibers are elliptic curves (or Abelian varieties, etc.). Our main interest is in the elliptic case, but in Sections 7 and 8 we also study the general setting.

Let X be a projective, log terminal variety and L a $\mathbb{Q}$-Cartier $\mathbb{Q}$-divisor (or divisor class) on X. We say that (X, L) is a Calabi–Yau fiber space if there is a Calabi–Yau fiber space $g: X \to B$ and an ample $\mathbb{Q}$-Cartier $\mathbb{Q}$-divisor L_B on B such that $L \sim_{\mathbb{Q}} g^* L_B$.

In general, a divisor L is called *semi-ample* if it is the pull-back of an ample divisor by a morphism and *nef* if $(L \cdot C) \geq 0$ for every irreducible curve $C \subset X$. Every semi-ample divisor is nef, but the converse usually fails. However, the hope is that for Calabi–Yau varieties nef and semi-ample are essentially equivalent; see Conjectures 50 and 51.

We say that a Calabi–Yau fiber space $g: X \to B$ is *relatively minimal* if $K_X \sim_{\mathbb{Q}} g^* F$ for some $\mathbb{Q}$-Cartier $\mathbb{Q}$-divisor F on B. This condition is automatic if X itself is Calabi–Yau. (These are called crepant log structures in [Kol13b].)

If $g: X \to B$ is a relatively minimal Calabi–Yau fiber space and X has canonical (resp. log terminal) singularities, then every other relatively minimal Calabi–Yau fiber space $g': X' \to B$ that is birational to $g: X \to B$ also has canonical (resp. log terminal) singularities.

By [Nak88], if X has log terminal singularities then B has rational singularities; more precisely, there is an effective divisor D_B such that (B, D_B) is klt.

6 (Elliptic threefolds) Elliptic threefolds have been studied in detail. The papers [Wil89, Gra91, Nak91, Gra93, DG94, Gra94, Gro94, Wil94, Gro97, Nak02a, Nak02b, CL10, HK11, Klo13] give rather complete descriptions of their local and global structure. However, neither Question 1 nor 2 was fully answered for threefolds.

By contrast, not even the local structure of elliptic fourfolds is understood. Double covers of the $\mathbb{P}^1$-contractions described in [AW98] give some rather surprising examples; there are probably much more complicated ones as well.

Definition 7 Let $g: X \to B$ be a morphism between normal varieties. A divisor $D \subset X$ is called *horizontal* if $g(D) = B$, *vertical* if $g(D) \subset B$ has codimension ≥ 1, and *exceptional* if $g(D)$ has codimension ≥ 2 in B.

If g is birational the latter coincides with the usual notion of exceptional divisors, but the above version makes sense even if $\dim X > \dim B$. (If g is birational then there are no horizontal divisors.)

8 (Birational models of Calabi–Yau fiber spaces) We see in Lemma 18 that if X is smooth (or $\mathbb{Q}$-factorial), g is a Calabi–Yau fiber space, and $D \subset X$ is exceptional then D is not g-nef. Thus, by [Lai11, HX13] the $(X, \epsilon D)$ minimal model program over B (cf. [KM98, Section 3.7]) contracts D. Thus every Calabi–Yau fiber space $g_2 : X_2 \to B_2$ is birational to a relatively minimal Calabi–Yau fiber space $g_1 : X_1 \to B_1 = B_2$ that has no exceptional divisors. Furthermore, let $B \to B_1$ be a small $\mathbb{Q}$-factorialization. Let L_B be an effective, ample divisor and L_1 its pull-back to X_1 by the rational map $X_1 \dashrightarrow B$. Applying [Lai11, HX13] and Theorem 14 to $(X_1, \epsilon L_1)$, we get a birational model $g : X \to B$ where B is also $\mathbb{Q}$-factorial. (In general, such a model is not unique.) Thus, in birational geometry, it is reasonable to focus on the study of relatively minimal Calabi–Yau fiber spaces $g : X \to B$ without exceptional divisors, where X and B are $\mathbb{Q}$-factorial and X is log terminal.

Let $\phi : X_1 \dashrightarrow X_2$ be a birational equivalence of two relatively minimal Calabi–Yau fiber spaces $g_i : X_i \to B$. Thus ϕ is an isomorphism between dense open sets $\phi : X_1^0 \cong X_2^0$. If the X_i are smooth (or they have terminal singularities) then we can choose these sets such that their complements $X_i \setminus X_i^0$ have codimension ≥ 2 (cf. [KM98, 3.52.2]). More generally, this holds if there are no exceptional divisors E_i of discrepancy 0 over X_i such that the center of E_i on X_i is disjoint from X_i^0.

Even in the smooth case, ϕ can be a rather complicated composite of flops.

From the point of view of F-theory it is especially interesting to study the examples $g' : X' \to B$ with a section $\sigma' : B \to X'$, where X' itself is Calabi–Yau. In this case the so-called *Weierstrass model* is a relatively minimal model without exceptional divisors that can be constructed explicitly as follows.

Let L_B be an ample divisor on B. Then $\sigma'(B)+mg'^*L_B$ is nef and big on X' for $m \gg 1$, hence a large multiple of it is base point free (cf. [KM98, Section 3.2]). This gives a morphism $h : X' \to X$, where X is still Calabi–Yau (usually with canonical singularities) and $g : X \to B$ has a section $\sigma : B \to X$ whose image is g-ample. Thus every fiber of g has dimension 1 and so $g : X \to B$ has no exceptional divisors.

Furthermore, $R^1 h_* \mathcal{O}_{X'} = 0$ which implies that every deformation of X' comes from a deformation of X (see 53).

The next result says that once $g : X \to B$ looks like a relatively minimal Calabi–Yau fiber space outside a subset of codimension ≥ 2, then it is a relatively minimal Calabi–Yau fiber space.

Proposition 9 *Let $g\colon X \to B$ be a projective fiber space with X log terminal. Assume the following:*

(1) There are no g-exceptional divisors (Definition 7).
(2) There is a closed subset $Z \subset B$ of codimension ≥ 2 such that K_X is numerically trivial on the fibers over $B \setminus Z$.
(3) B is $\mathbb{Q}$-factorial.

Then $g\colon X \to B$ is a relatively minimal Calabi–Yau fiber space.

Proof First note that, as a very special case of Theorem 14, there is a $\mathbb{Q}$-Cartier $\mathbb{Q}$-divisor F_1 on $B \setminus Z$ such that

$$K_X|_{X \setminus g^{-1}(Z)} \sim_{\mathbb{Q}} g^* F_1.$$

Since B is $\mathbb{Q}$-factorial, F_1 extends to a $\mathbb{Q}$-Cartier $\mathbb{Q}$-divisor F on B.

Thus every point $b \in B$ has an open neighborhood $b \in U_b \subset B$ and an integer $m_b > 0$ such that

$$\mathcal{O}_X(m_b K_X)|_{g^{-1}(U_b \setminus Z)} \cong g^* \mathcal{O}_{U_b}(m_b F|_{U_b}) \cong g^* \mathcal{O}_{U_b} \cong \mathcal{O}_{g^{-1}(U_b)}.$$

By (1), $g^{-1}(Z)$ has codimension ≥ 2 in $g^{-1}(U_b)$ and hence the constant 1 section of $\mathcal{O}_{g^{-1}(U_b \setminus Z)}$ extends to a global section of $\mathcal{O}_X(m_b K_X)|_{g^{-1}(U_b)}$ that has neither poles nor zeros. Thus

$$\mathcal{O}_X(m_b K_X)|_{g^{-1}(U_b)} \cong \mathcal{O}_{g^{-1}(U_b)}.$$

Since this holds for every $b \in B$, we conclude that $K_X \sim_{\mathbb{Q}} g^* F$. □

2 The main case

The next theorem gives a characterization of the main series of elliptic Calabi–Yau fiber spaces. (For the log version, see 54.) The proof is quite short but it relies on auxiliary results that are proved in the next two sections.

Theorem 10 *Let X be a projective variety of dimension n with log terminal singularities and L a Cartier divisor on X. Assume that K_X is nef and $(L^{n-2} \cdot \mathrm{td}_2(X)) > 0$, where $\mathrm{td}_2(X)$ is the second Todd class of X (24). Then (X, L) is a relatively minimal, elliptic fiber space iff*

(1) L is nef,
(2) $L - \epsilon K_X$ is nef for $0 \leq \epsilon \ll 1$,
(3) $(L^n) = 0$, and
(4) (L^{n-1}) is nonzero in $H^{2n-2}(X, \mathbb{Q})$.

Note that if (X, L) is a relatively minimal elliptic fiber space then L is semi-ample (Definition 5) and, as we see in 13 below, the only hard part of Theorem 10 is to show that conditions (1)–(4) imply L is semi-ample. In particular, Theorem 10 also holds over fields that are not algebraicaly closed.

This immediately yields the following partial answer to Question 1:

Corollary 11 *Let X be a smooth, projective variety of dimension n and L a Cartier divisor on X. Assume that $K_X \sim_{\mathbb{Q}} 0$ and $(L^{n-2} \cdot c_2(X)) > 0$. Then (X, L) is an elliptic fiber space iff*

(1) L is nef,
(2) $(L^n) = 0$, and
(3) (L^{n-1}) is nonzero in $H^{2n-2}(X, \mathbb{Q})$. □

Definition 12 Let Y be a projective variety and D a Cartier divisor on X. If $m > 0$ is sufficiently divisible, then, up to birational equivalence, the map given by global sections of $\mathcal{O}_Y(mD)$

$$Y \dashrightarrow I(Y, D) \overset{bir}{\sim} I_m(Y, D) \hookrightarrow \mathbb{P}(H^0(Y, \mathcal{O}_Y(mD))) \quad \text{is independent of } m.$$

It is called the *Iitaka fibration* of (Y, D). The *Kodaira dimension* of D (or of (Y, D)) is $\kappa(D) = \kappa(Y, D) := \dim I(Y, D)$.

If D is nef, the *numerical dimension* of D (or of (Y, D)), denoted by $\nu(D)$ or $\nu(Y, D)$, is the largest natural number r such that the self-intersection $(D^r) \in H^{2r}(Y, \mathbb{Q})$ is nonzero. Equivalently, $(D^r \cdot H^{n-r}) > 0$ for some (or every) ample divisor H.

It is easy to see that $\kappa(D) \leq \nu(D)$. This was probably first observed by Matsusaka as a corollary of his theory of variable intersection cycles (see [Mat72] or [LM75, p. 515]).

13 (Proof of Theorem 10) First note that $\kappa(L) \geq n - 2$ by Lemma 25. We will also need this for some perturbations of L.

By Theorem 10, (2) and (3) we have $0 = (L^n) \geq \epsilon(L^{n-1} \cdot K_X) \geq 0$, thus $(L^{n-1} \cdot K_X) = 0$.

Set $L_m := L - \frac{1}{m} K_X$. For $m \gg 1$ we see that L_m is nef, $(L_m^{n-2} \cdot \mathrm{td}_2(X)) > 0$, and (L_m^{n-1}) is nonzero in $H^{2n-2}(X, \mathbb{Q})$. Note that $mL = K_X + mL_m$, hence

$$m^n(L^n) = \textstyle\sum_{i=0}^{n} m^{n-i}(K_X^i \cdot L_m^{n-i}).$$

Since K_X and L_m are both nef, all the terms on the RHS are ≥ 0. Their sum is zero by assumption, hence $(K_X^i \cdot L_m^{n-i}) = 0$ for every i. Thus Lemma 25 also applies to L_m and we get that $\kappa(L_m) \geq n - 2$.

We can now apply Proposition 15 with $\Delta = 0$, $D := 2mL_m$, and $K_X + 2mL_m = 2mL_{2m}$ to conclude that $\nu(L_m) \le \kappa(L_{2m})$. Since we know that $\nu(L_m) = \dim X - 1$, we conclude that $\kappa(L_{2m}) = \dim X - 1$.

Finally, use Theorem 14 with $S =$ (point), $2mL$ instead of L and $a = 1$ to obtain that some multiple of L is semi-ample. That is, there is a morphism with connected fibers $g\colon X \to B$ and an ample $\mathbb{Q}$-divisor L_B such that $L \sim_{\mathbb{Q}} g^* L_B$. Note that $(L^{\dim B}) \neq 0$ but $(L^{\dim B+1}) = 0$ so, comparing with Theorem 10, (3) and (4), we see that $\dim B = \dim X - 1$. By the adjunction formula, the canonical class of the general fiber is proportional to $(L^{n-1} \cdot K_X) = 0$, thus $g\colon X \to B$ is an elliptic fiber space. □

We have used the following theorem due to [Kaw85a] and [Fuj11]:

Theorem 14 *Let (X, Δ) be an irreducible, projective, klt pair and $g\colon X \to S$ a morphism with generic fiber X_g. Let L be a $\mathbb{Q}$-Cartier $\mathbb{Q}$-divisor on X. Assume that*

(1) L and $L - K_X - \Delta$ are g-nef and
(2) $\nu((L - K_X - \Delta)|_{X_g}) = \kappa((L - K_X - \Delta)|_{X_g}) = \nu(((1 + a)L - K_X - \Delta)|_{X_g}) = \kappa(((1 + a)L - K_X - \Delta)|_{X_g})$ for some $a > 0$.

Then there is a factorization $g\colon X \xrightarrow{h} B \xrightarrow{\pi} S$ and a π-ample $\mathbb{Q}$-Cartier $\mathbb{Q}$-divisor L_B on B such that $L \sim_{\mathbb{Q}} h^* L_B$. □

3 Adjoint systems of large Kodaira dimension

The following is modeled on [Ogu93, 2.4]:

Proposition 15 *Let (X, Δ) be a projective, klt pair such that $K_X + \Delta$ is pseudo-effective, that is, its cohomology class is a limit of effective classes. Let D be an effective, nef, $\mathbb{Q}$-Cartier $\mathbb{Q}$-divisor on X such that $\kappa(K_X + \Delta + D) \ge \dim X - 2$.*
Then $\nu(D) \le \kappa(K_X + \Delta + D)$.

Proof There is nothing to prove if $\kappa(K_X + \Delta + D) = \dim X$. Thus assume that $\kappa(K_X + \Delta + D) \le \dim X - 1$ and let $g\colon X \dashrightarrow B$ be the Iitaka fibration (cf. [Laz04, 2.1.33]). After some blow-ups we may assume in addition that g is a morphism and X, B are smooth.

The generic fiber of g is a smooth curve or surface (S, Δ_S) such that $K_S + \Delta_S$ is pseudo-effective. Since abundance holds for curves and surfaces [Kol92, Section 11], this implies that $\kappa(K_S + \Delta_S) \ge 0$. Furthermore, by Iitaka's theorem (cf. [Laz04, 2.1.33]), $\kappa(K_S + \Delta_S + D|_S) = 0$.

If D is disjoint from S then, by Lemma 17, (2), $\nu(D) \leq \dim B = \kappa(K_X+\Delta+D)$ and we are done. Otherwise $D|_S$ is an effective, nonzero, nef divisor on S. We obtain a contradiction by proving that $\kappa(K_S + \Delta_S + D|_S) \geq 1$.

If S is a curve, then $\deg D|_S > 0$ and hence $\kappa(K_S + \Delta_S + D|_S) \geq \kappa(D|_S) = 1$. If S is a surface, then $\kappa(K_S + \Delta_S + D|_S) \geq 1$ is proved in Lemma 16. □

Lemma 16 *Let (S, Δ_S) be a projective, klt surface such that $\kappa(K_S + \Delta_S) \geq 0$. Let D be a nonzero, effective, nef $\mathbb{Q}$-divisor. Then $\kappa(K_S + \Delta_S + D) \geq 1$.*

Proof Since $\kappa(K_S + \Delta_S + D) \geq \kappa(K_S + \Delta_S)$ we only need to consider the case when $\kappa(K_S + \Delta_S) = 0$. Let $\pi\colon (S, \Delta_S) \to (S^m, \Delta_S^m)$ be the minimal model. It is obtained by repeatedly contracting curves that have negative intersection number with $K_S + \Delta_S$. These curves also have negative intersection number with $K_S + \Delta_S + \epsilon D$ for $0 < \epsilon \ll 1$. Thus

$$\pi\colon (S, \Delta_S + \epsilon D) \to (S^m, \Delta_S^m + \epsilon D^m)$$

is also the minimal model and $(S^m, \Delta^m + \epsilon D^m)$ is klt for $0 < \epsilon \ll 1$. By the Hodge index theorem, every effective divisor contracted by π has negative self-intersection, thus D cannot be π-exceptional. So D^m is again a nonzero, effective, nef $\mathbb{Q}$-divisor.

Since abundance holds for klt surface pairs (cf. [Kol92, Section 11]), we see that $K_{S^m} + \Delta^m \sim_{\mathbb{Q}} 0$ and $\kappa(K_{S^m} + \Delta^m + \epsilon D^m) \geq 1$. Since D is effective, we obtain that $\kappa(K_S + \Delta_S + D) \geq \kappa(K_S + \Delta_S + \epsilon D) = \kappa(K_{S^m} + \Delta^m + \epsilon D^m) \geq 1$. □

Lemma 17 *Let $g\colon X \to B$ be a proper morphism with connected general fiber X_g. Let D be an effective, nef, $\mathbb{Q}$-Cartier $\mathbb{Q}$-divisor on X. Then*

(1) either $D|_{X_g}$ is a nonzero nef divisor,
(2) or D is disjoint from X_g and $(D^{\dim B+1}) = 0$. Thus $\nu(D) \leq \dim B$.

Proof We are done if $D|_{X_g}$ is nonzero. If it is zero then D is vertical, hence there is an ample divisor L_B such that $g^*L_B \sim D + E$ where E is effective. Then

$$(g^*L_B^r) - (D^r) = \textstyle\sum_{i=0}^{r-1}(E \cdot g^*L_B^i \cdot D^{r-1-i})$$

shows that $(D^r) \leq (g^*L_B^r)$. Since $((g^*L_B)^{\dim B+1}) = g^*(L_B^{\dim B+1}) = 0$, we conclude that $(D^{\dim B+1}) = 0$. □

Lemma 18 *Let $g\colon X \to B$ be a proper morphism with connected fibers and D an effective, exceptional, $\mathbb{Q}$-Cartier $\mathbb{Q}$-divisor on X. Then D is not g-nef.*

Proof Let $|H|$ be a very ample linear system on X and $S \subset X$ the intersection of $\dim X - 2$ general members of $|H|$. Then $g|_S : S \to B$ is generically finite over its image and $D \cap S$ is $g|_S$-exceptional. By the Hodge index theorem we conclude that $(D^2 \cdot H^{\dim X-2}) < 0$, a contradiction. □

4 Asymptotic estimates for cohomology groups

19 Let X be a smooth variety and $g: X \to B$ a Calabi–Yau fiber space of relative dimension m over a smooth curve B. Assume that $K_{X_g} \sim 0$, where X_g denotes a general fiber. It is easy to see that the sheaves $R^m g_* \mathcal{O}_X$ and $g_* \omega_{X/B}$ are line bundles and dual to each other. For elliptic surfaces these sheaves were computed by Kodaira. His results were clarified and extended to higher dimensions by [Fuj78]. We will need the following consequences of their results.

The degree of $g_* \omega_{X/B}$ is ≥ 0 and can be written as a sum of two terms. One is a global term (determined by the j-invariant of the fibers in the elliptic case) which is zero iff $g: X \to B$ is *generically isotrivial*, that is, g is an analytically locally trivial fiber bundle over a dense open set $B^0 \subset B$. The other is a local term, supported at the points where the local monodromy of the local system $R^m g_* \mathbb{Q}_{X^0}$ is nontrivial. There is a precise formula for the local term, but we only need to understand what happens with generically isotrivial families. For these the local term is positive iff the local monodromy has eigenvalue $\neq 1$ on $g_* \omega_{X^0/B^0} \subset \mathcal{O}_{B^0} \otimes_{\mathbb{Q}} R^m g_* \mathbb{Q}_{X^0}$.

Over higher-dimensional bases, $R^m g_* \mathcal{O}_X$ and $g_* \omega_{X/B}$ are rank 1 sheaves, and the above considerations describe their codimension 1 behavior. In particular, we see the following:

(1) $c_1(g_* \omega_{X/B})$ is linearly equivalent to a sum of effective $\mathbb{Q}$-divisors. It is zero only if $g: X \to B$ is isotrivial over a dense open set B^0 and the local monodromy around each irreducible component of $B \setminus B^0$ has eigenvalue $= 1$ on $g_* \omega_{X^0/B^0} \subset \mathcal{O}_{B^0} \otimes_{\mathbb{Q}} R^m g_* \mathbb{Q}_{X^0}$.

(2) $c_1(R^m g_* \mathcal{O}_X) = -c_1(g_* \omega_{X/B})$.

Frequently $c_1(g_* \omega_{X/B})$ is denoted by $\Delta_{X/B}$.

Corollary 20 *Let $g: X \to B$ be an elliptic fiber space of dimension n and L a line bundle on B. Then*

$$\begin{aligned} \chi(X, g^* L^m) &= \tfrac{(L^{n-2} \cdot \Delta_{X/B})}{(n-2)!} m^{n-2} + O(m^{n-3}) \quad \textit{and} \\ h^i(X, g^* L^m) &= O(m^{n-3}) \quad \textit{for } i \geq 2. \end{aligned}$$

Proof By the Leray spectral sequence,

$$\chi(X, g^* L^m) = \textstyle\sum_i (-1)^i \chi(B, L^m \otimes R^i g_* \mathcal{O}_X).$$

For $i \geq 2$ the support of $R^i g_* \mathcal{O}_X$ has codimension ≥ 2 in B, hence its cohomologies contribute only to the $O(m^{n-3})$ term.

Since g has connected fibers, $g_* \mathcal{O}_X \cong \mathcal{O}_B$ and $c_1(R^1 g_* \mathcal{O}_X) \sim_{\mathbb{Q}} -\Delta_{X/B}$ by 19, item (2). We conclude by applying Lemma 23 to both terms. □

21 Similar formulas apply to arbitrary Calabi–Yau fiber spaces $g\colon X \to B$ with general fiber F. If L is ample on B then, for $m \gg 1$, we have

$$H^i(X, g^*L^m) = H^0(B, L^m \otimes R^i g_* \mathcal{O}_X) = \chi(B, L^m \otimes R^i g_* \mathcal{O}_X). \tag{21.1}$$

Setting $k = \dim B$, Lemma 23 computes $H^i(X, g^*L^m)$ as

$$\frac{m^k}{k!} h^i(F, \mathcal{O}_F)(L^k) + \frac{m^{k-1}}{(k-1)!}\Big(L^{k-1} \cdot (c_1(R^i g_* \mathcal{O}_X) - \tfrac{h^i(F,\mathcal{O}_F)}{2} K_B)\Big) + O(m^{k-2}).$$

These imply that

$$\chi(X, g^*L^m) = \chi(F, \mathcal{O}_F) \cdot \frac{m^k}{k!}(L^k) + O(m^{k-1}). \tag{21.2}$$

If $\chi(F, \mathcal{O}_F) \neq 0$ then this describes the asymptotic behavior of $\chi(X, g^*L^m)$. However, if $\chi(F, \mathcal{O}_F) = 0$, which happens for Abelian fibers, then we have to look at the next term, which gives that

$$\chi(X, g^*L^m) = \frac{m^{k-1}}{(k-1)!}\Big(L^{k-1} \cdot \textstyle\sum_{i=1}^{\dim F}(-1)^i c_1(R^i g_* \mathcal{O}_X)\Big) + O(m^{k-2}). \tag{21.3}$$

If F is an elliptic curve then the sum on the RHS has only one nonzero term. For higher-dimensional Abelian fibers there are usually several nonzero terms and sometimes they cancel each other.

This is one reason why elliptic fibers are easier to study than higher-dimensional Abelian fibers. The other difficulty with higher-dimensional fibers is that the Euler characteristic only tells us that $h^0 + h^2 + h^4 + \cdots$ grows as expected. Proving that $h^0 \neq 0$ would need additional arguments.

The next result, while stated in all dimensions, is truly equivalent to Kodaira's formula [BPV84, V.12.2].

Corollary 22 *Let $g\colon X \to B$ be a relatively minimal elliptic fiber space of dimension n and L a line bundle on B. Then $(L^{n-2} \cdot \Delta_{X/B}) = (g^*L^{n-2} \cdot \mathrm{td}_2(X))$.*

Proof Expanding the Riemann–Roch formula $\chi(X, g^*L) = \int_X \mathrm{ch}(g^*L) \cdot \mathrm{td}(X)$ and taking into account that $(g^*L^n) = (g^*L^{n-1} \cdot K_X) = 0$ gives

$$\chi(X, g^*L^m) = \frac{(g^*L^{n-2} \cdot \mathrm{td}_2(X))}{(n-2)!} \cdot m^{n-2} + O(m^{n-3}).$$

Comparing this with Corollary 20 yields the claim. □

We used several versions of the asymptotic Riemann–Roch formula.

Lemma 23 *Let Y be a normal, projective variety of dimension n, L a line bundle on X, and F a coherent sheaf of rank r that is locally free in codimension 1. Then*

$$\chi(Y, L^m \otimes F) = \frac{(L^n)\cdot r}{n!} m^n + \frac{(L^{n-1}\cdot (c_1(F) - \frac{r}{2}K_Y))}{(n-1)!} m^{n-1} + O(m^{n-2}). \quad \square$$

24 (Riemann–Roch with rational singularities) The Todd classes of a singular variety X are not always easy to compute, but if X has rational singularities then there is a straightforward formula in terms of the Chern classes of any resolution $h\colon X' \to X$.

By definition, rational singularity means that $R^i h_* \mathcal{O}_{X'} = 0$ for $i > 0$. Thus $\chi(X, L) = \chi(X', h^*L)$ for any line bundle L on X. By the projection formula this implies that $\chi(X, L) = \int_X \operatorname{ch}(L)\cdot h_* \operatorname{td}(X')$ and in fact $\operatorname{td}(X) = h_* \operatorname{td}(X')$ (cf. [Ful98, Theorem 18.2].) In particular, we see that the second Todd class of X is

$$\operatorname{td}_2(X) = h_*\Big(\frac{c_1(X')^2 + c_2(X')}{12}\Big).$$

The following numerical version of Corollary 20 was used in the proof of Theorem 10.

Lemma 25 *Let X be a normal, projective variety of dimension n. Let L be a nef line bundle on X such that* $(L^n) = (L^{n-1}\cdot K_X) = 0$ *but* $(L^{n-1}) \neq 0$. *Then*

$$h^0(X, L^m) - h^1(X, L^m) = \frac{(L^{n-2}\cdot \operatorname{td}_2(X))}{(n-2)!}\cdot m^{n-2} + O(m^{n-3}).$$

Proof The assumptions $(L^n) = (L^{n-1}\cdot K_X) = 0$ imply that the RHS equals $\chi(X, L^m)$. Thus the equality follows if $h^i(X, L^m) = O(m^{n-3})$ for $i \geq 2$. The latter is a special case of Lemma 26. $\square$

Lemma 26 *Let X be a projective variety of dimension n and F a torsion-free coherent sheaf on X. Let L be a nef line bundle on X and set* $d = \nu(X, L)$. *Then*

$$\begin{aligned} h^i(X, F\otimes L^m) &= O(m^d) \quad \text{for } i = 0, \dots, n-d \text{ and}\\ h^{n-j}(X, F\otimes L^m) &= O(m^{j-1}) \quad \text{for } j = 0, \dots, d-1. \end{aligned}$$

Note the key feature of the estimate: the order of growth of H^i is m^d for $i \leq n-d$, then for $i = n-d+1$ it drops by 2 to m^{d-2}, and then it drops by 1 for each increase of i. This strengthens [Laz04, 1.4.40] but the proof is essentially the same.

Proof We use induction on $\dim X$. By Fujita's theorem (cf. [Laz04, 1.4.35]) we can choose a general very ample divisor A on X such that

$$h^i(X, F\otimes \mathcal{O}_X(A)\otimes L^m) = 0 \quad \text{for all } i \geq 1 \text{ and } m \geq 1.$$

We get an exact sequence

$$0 \to F \otimes L^m \to F \otimes O_X(A) \otimes L^m \to G \otimes L^m \to 0,$$

where G is a torsion-free coherent sheaf on A. For $i \geq 1$ its long cohomology sequence gives surjections (even isomorphisms for $i \geq 2$)

$$H^{i-1}(A, G \otimes L^m) \twoheadrightarrow H^i(X, F \otimes L^m).$$

By induction this shows the claim except for $i = 0$.

One can realize F as a subsheaf of a sum of line bundles, thus it remains to prove that $H^0(X, F \otimes L^m) = O(m^d)$ when $F \cong O_X(H)$ is a very ample line bundle. The exact sequence

$$0 \to L^m \to O_X(H) \otimes L^m \to O_H(H|_H) \otimes L^m \to 0$$

finally reduces the problem to $\kappa(L) \leq \nu(L)$, which was discussed in Definition 12. □

5 Deforming morphisms

Here we answer Question 1, but first two technical issues need to be discussed: the distinction between étale and quasi-étale covers and the existence of non-Calabi–Yau deformations. Both appear only for singular Calabi–Yau varieties.

Definition 27 Following [Cat07], a finite morphism $\pi\colon U \to V$ is called *quasi-étale* if there is a closed subvariety $Z \subset V$ of codimension ≥ 2 such that π is étale over $V \setminus Z$.

If V is a normal variety, then there is a one-to-one correspondence between quasi-étale covers of V and finite, étale covers of $V \setminus \operatorname{sing} V$.

In particular, if X is a Calabi–Yau variety then there is a quasi-étale morphism $X_1 \to X$ such that $K_{X_1} \sim 0$.

Among all such covers $X_1 \to X$ there is a unique smallest one, called the *index 1 cover* of X, which is Galois with cyclic Galois group. We denote it by $X^{\mathrm{ind}} \to X$.

28 (Deformation theory) For a general introduction, see [Har10]. By a deformation of a proper scheme (or analytic space) X we mean a flat, proper morphism $g\colon \mathbf{X} \to (0 \in S)$ to a pointed scheme (or analytic space) together with a fixed isomorphism $X_0 \cong X$.

By a deformation of a morphism of proper schemes (or analytic spaces) $f\colon X \to Y$ we mean a morphism $\mathbf{f}\colon \mathbf{X} \to \mathbf{Y}$ where $\mathbf{X}$ is a deformation of X, $\mathbf{Y}$ is a deformation of Y, and $\mathbf{f}|_{X_0} = f$.

When we say that an assertion holds for all *small deformations* of X, this means that for every deformation $g: \mathbf{X} \to (0 \in S)$ there is an étale (or analytic) neighborhood $(0 \in S') \to (0 \in S)$ such that the assertion holds for $g': \mathbf{X} \times_S S' \to (0 \in S')$.

29 (Deformations of Calabi–Yau varieties) Let X be a Calabi–Yau variety. If X is smooth (or has canonical singularities) then every small deformation of X is again a Calabi–Yau variety. This, however, fails in general; see Example 47, where X is a surface with quotient singularities.

Dealing with such unexpected deformations is a basic problem in the moduli theory of higher-dimensional varieties; see [Kol13a, Section 4], [HK10, Section 14B], or [AH11] for a discussion and solutions. For Calabi–Yau varieties one can use a global trivialization of the canonical bundle to get a much simpler answer.

We say that a deformation $g: \mathbf{X} \to (0 \in S)$ of X over a reduced, local space S is a *Calabi–Yau deformation* if the following equivalent conditions hold:

(1) Every fiber of g is a Calabi–Yau variety.
(2) The deformation can be lifted to a deformation $g^{\mathrm{ind}}: \mathbf{X}^{\mathrm{ind}} \to (0 \in S)$ of X^{ind}, the index 1 cover of X.

Thus, studying Calabi–Yau deformations of Calabi–Yau varieties is equivalent to studying deformations of Calabi–Yau varieties whose canonical class is Cartier. As we noted, for the latter every deformation is automatically a Calabi–Yau deformation. Thus we do not have to deal with this issue at all.

Theorem 30 *Let X be a Calabi–Yau variety and $g: X \to B$ an elliptic fiber space. Then at least one of the following holds:*

(1) The morphism g extends to every small Calabi–Yau deformation of X.
(2) There is a quasi-étale cover $\tilde{X} \to X$ such that the Stein factorization $\tilde{g}: \tilde{X} \to \tilde{B}$ of $\tilde{X} \to B$ is one of the following:
   (a) $(\tilde{g}: \tilde{X} \to \tilde{B}) \cong (p_1: \tilde{B} \times (\text{elliptic curve}) \to \tilde{B})$ where p_1 is the first projection or
   (b) $(\tilde{g}: \tilde{X} \to \tilde{B}) \cong (p_1: \tilde{Z} \times (\text{elliptic K3}) \to \tilde{Z} \times \mathbb{P}^1)$ where $\tilde{Z}$ is a Calabi–Yau variety of dimension $\dim X - 2$ and p_1 is the product of the first projection with the elliptic pencil map of the K3 surface.

Proof As noted in 29, we may assume that $K_X \sim 0$.

By [KMM92] there is a unique map (up to birational equivalence) $h: B \dashrightarrow Z$ whose general fiber F is rationally connected and whose target Z is not uniruled by [GHS03]. (See [Kol96, Chapter IV] for a detailed treatment or [AK03] for an introduction.) Next apply [KL09, Theorem 14] to $X \dashrightarrow Z$ to conclude that there is a finite, quasi-étale cover $\tilde{X} \to X$, a product decomposition $\tilde{X} \cong Y \times \tilde{Z}$, and a generically finite map $\tilde{Z} \dashrightarrow Z$ that factors $\tilde{X} \dashrightarrow Z$.

If $\dim Z = \dim B$ then we are in case (2a). If $\dim Z = \dim B - 1$ then the generic fiber of $\tilde{B} \to \tilde{Z}$ is $\mathbb{P}^1$. Furthermore, $\dim Y = 2$, hence either Y is an elliptic K3 surface and we are in case (2b) or Y is an Abelian surface that has an elliptic pencil and after a further cover we are again in case (2a).

It remains to prove that if $\dim F \geq 2$ then the assertion of (1) holds. By Theorem 35 it is sufficient to check that

$$\operatorname{Hom}_B(\Omega_B, R^1 g_* O_X) = 0.$$

Note that 19 and $\omega_X \sim O_X$ imply that $R^1 g_* O_X \cong (g_* \omega_{X/B})^{-1} \cong \omega_B$, at least over the smooth locus of B. Since $R^1 g_* O_X$ is reflexive by [Kol86a, 7.8], the isomorphism holds everywhere. Thus

$$\mathcal{H}om_B(\Omega_B, R^1 g_* O_X) \cong \mathcal{H}om_B(\Omega_B, \omega_B) \cong (\Omega_B^{\dim B-1})^{**},$$

where $(\ )^{**}$ denotes the double dual or reflexive hull. By taking global sections we get that

$$\operatorname{Hom}_B(\Omega_B, R^1 g_* O_X) = H^0(B, (\Omega_B^{\dim B-1})^{**}).$$

Let $B' \to B$ be a resolution of singularities such that $B' \to Z$ is a morphism and $F' \subset B'$ a general fiber. Since F' is rationally connected, it is covered by rational curves $C \subset F'$ such that

$$T_{F'}|_C \cong \sum O_C(a_i) \quad \text{where } a_i > 0\ \forall i;$$

see [Kol96, IV.3.9]. Thus $T_{B'}|_C$ is a sum of line bundles $O_C(a_i)$ where $a_i > 0$ for $\dim F$ summands and $a_i = 0$ for the rest. Since $\dim F \geq 2$ we conclude that

$$\wedge^{\dim B-1} T_{B'}|_C \cong \sum O_C(b_i) \quad \text{where } b_i > 0 \text{ for every } i.$$

By duality this gives that $H^0(B', \Omega_{B'}^{\dim B-1}) = 0$. Finally we use that B has log terminal singularities by [Nak88] and so [GKKP11] shows that

$$\operatorname{Hom}_B(\Omega_B, R^1 g_* O_X) = H^0(B, (\Omega_B^{\dim B-1})^{**}) = H^0(B', \Omega_{B'}^{\dim B-1}) = 0.$$

□

We are now ready to answer Question 1.

Theorem 31 *Let X be an elliptic Calabi–Yau variety such that $H^2(X, O_X) = 0$. Then every small Calabi–Yau deformation of X is also an elliptic Calabi–Yau variety.*

Proof Let $g: X \to B$ be an elliptic Calabi–Yau variety. By Theorem 30 every small Calabi–Yau deformation of X is also an elliptic Calabi–Yau variety except possibly when there is a quasi-étale cover $\tilde{X} \to X$ such that

(1) either $\tilde{X} \cong \tilde{Z} \times$ (elliptic curve)
(2) or $\tilde{X} \cong \tilde{Z} \times$ (elliptic K3).

In both cases, $\tilde{X}$ can have non-elliptic deformations but we show that these do not correspond to a deformation of X. Here we use that $H^2(X, O_X) = 0$.

Let $\pi: \mathbf{X} \to (0 \in S)$ be a flat deformation of X over a local scheme S. Let L be the pull-back of an ample line bundle from B to X. Since $H^2(X, O_X) = 0$, L lifts to a line bundle $\mathbf{L}$ on $\mathbf{X}$ (cf. [Gro62, p. 236–16]) thus we get a line bundle $\tilde{\mathbf{L}}$ on $\tilde{\mathbf{X}}$. We need to show that a large multiple of $\mathbf{L}$ is base-point-free over S; then it gives the required morphism $\mathbf{g}: \mathbf{X} \to \mathbf{B}$. One can check base-point-freeness of some multiple after a finite surjection, thus it is enough to show that some multiple of $\tilde{\mathbf{L}}$ is base-point-free over S.

The first case (more generally, deformations of products with Abelian varieties) is treated in Lemma 40.

In the K3 case note first that every small deformation of $\tilde{X}$ is of the form $\tilde{\mathbf{Z}} \times_S \tilde{\mathbf{F}}$, where $\tilde{\mathbf{F}} \to S$ is a flat family of K3 surfaces. This is a trivial case of Theorem 35; see 53 for an elementary argument. Hence we only need to show that the restriction of $\tilde{\mathbf{L}}$ to $\tilde{\mathbf{F}}$ is base-point-free over S. Equivalently, that the elliptic structure of the central K3 surface $\tilde{F}$ is preserved by our deformation. The restriction of $\tilde{\mathbf{L}}$ to every fiber of $\tilde{\mathbf{F}} \to S$ gives a nonzero, nef line bundle with self-intersection 0, hence an elliptic pencil. □

32 (Deformation of sections) Let $g: X \to B$ be an elliptic Calabi–Yau fiber space with a section $S \subset X$. Let us assume first that S is a Cartier divisor in X. (This is automatic if X is smooth.) Then S is g-nef, g-big, and $S \sim_{\mathbb{Q},g} K_X + S$ hence $R^i g_* O_X(S) = 0$ for $i > 0$ (cf. [KM98, Section 2.5]). Thus $H^i(X, O_X(S)) = H^i(B, g_* O_X(S))$ for every i. In order to compute $g_* O_X(S)$ we use the exact sequence

$$0 \to O_B = g_* O_X \xrightarrow{\alpha} g_* O_X(S) \to g_* O_S(S|_S).$$

A degree-1 line bundle over an elliptic curve has only one section, thus α is an isomorphism over an open set where the fiber is a smooth elliptic curve. Since $g_* O_S(S|_S) \cong O_S(S|_S)$ is torsion-free we conclude that $g_* O_X(S) \cong O_B$. Thus

$$H^1(X, O_X(S)) = H^1(B, O_B) \subset H^1(X, O_X).$$

If $H^2(X, O_X) = 0$ then the line bundle $O_X(S)$ lifts to every small deformation of X and if $H^1(X, O_X) = 0$ then the unique section of $O_X(S)$ also lifts.

The situation is quite different if the section is not assumed Cartier. For instance, let $X_0 \subset \mathbb{P}^2 \times \mathbb{P}^2$ be a general hypersurface of multidegree $(3,3)$ containing $S := \mathbb{P}^2 \times \{p\}$ for some point p. Then X_0 is a Calabi–Yau variety and the first projection shows that it is elliptic with a section. Note that X_0 is singular, it has nine ordinary nodes along S.

By contrast, if $X_t \subset \mathbb{P}^2 \times \mathbb{P}^2$ is a smooth hypersurface of multidegree $(3,3)$ then the restriction map $\operatorname{Pic}(\mathbb{P}^2 \times \mathbb{P}^2) \to \operatorname{Pic}(X_t)$ is an isomorphism by the Lefschetz hyperplane theorem. Thus the degree of every divisor $D \subset X_t$ on the general fiber of the first projection $X_t \to \mathbb{P}^2$ is a multiple of 3. Therefore $X_t \to \mathbb{P}^2$ does not even have a rational section.

As an aside, we consider the general question of deforming morphisms $g\colon X \to Y$ whose target is not uniruled.

There are some obvious examples when not every deformation of X gives a deformation of $g\colon X \to Y$. For example, let A_1, A_2 be positive-dimensional Abelian varieties and $g\colon A_1 \times A_2 \to A_2$ the second projection. A general deformation of $A_1 \times A_2$ is a simple Abelian variety which has no maps to lower-dimensional Abelian varieties. One can now get more complicated examples by replacing $A_1 \times A_2$ by say a complete intersection subvariety or by a cyclic cover. The next result says that this essentially gives all examples.

Theorem 33 *Let X be a projective variety with rational singularities, Y a normal variety, and $g\colon X \to Y$ a surjective morphism with connected fibers. Assume that Y is not uniruled. Then at least one of the following holds:*

(1) Every small deformation of X gives a deformation of $(g\colon X \to Y)$.
(2) There is a quasi-étale cover $\tilde{Y} \to Y$, a normal variety Z, and positive-dimensional Abelian varieties A_1, A_2 such that the lifted morphism $\tilde{g}\colon \tilde{X} := X \times_Y \tilde{Y} \to \tilde{Y}$ factors as

$$\begin{array}{ccc} \tilde{X} & \to & Z \times A_2 \times A_1 \\ \tilde{g} \downarrow & & \downarrow \\ \tilde{Y} & \cong & Z \times A_2 \end{array}$$

Proof By Theorem 35 every deformation of X gives a deformation of $g\colon X \to Y$ if

$$\operatorname{Hom}_Y(\Omega_Y, R^1 g_* O_X) = 0. \tag{3}$$

Thus we need to show that if Theorem 33, (3) fails then we get a structural description as in (2).

Assuming that there is a nonzero map $\phi\colon \Omega_Y \to R^1 g_* O_X$, let $E \subset R^1 g_* O_X$ denote its image. Our first aim is to prove that E becomes trivial after a quasi-étale base change.

Let $C \subset Y$ be a high-degree, general, complete intersection curve. First we show that $E|_C$ is stable and has degree 0.

Since Y is not uniruled, $\Omega_Y|_C$ is semi-positive by [Miy88] (see also [Kol92, Section 9]). Thus $E|_C$ is also semi-positive.

Let $Y^0 \subset Y$ be a dense open set and $X^0 := g^{-1}(Y^0)$ such that $g^0 : X^0 \to Y^0$ is smooth. Set $C^0 := Y^0 \cap C$. By [Ste76], $(R^1 g_* O_X)|_C$ is the (lower) canonical extension of the top quotient of the variation of Hodge structures $R^1 g^0_* \mathbb{Q}_{X^0}|_{C^0}$. (Note that [Ste76] works with ω_{X^0/Y^0} but the proof is essentially the same; see [Kol86b, pp. 177–179].) Thus $(R^1 g_* O_X)|_C$ is semi-negative by [Ste76] and so is $E|_C$. Thus $E|_C$ is stable of degree 0, hence it corresponds to a unitary representation ρ of $\pi_1(C)$.

By [Gri70, Section 5], ρ is a subrepresentation of the monodromy representation on $R^1 g^0_* \mathbb{Q}_{X^0}|_{C^0}$ and by [Del71, Theorem 4.2.6], it is even a direct summand $\mathbb{E}$. Since we have a polarized variation of Hodge structures, the monodromy representation on $\mathbb{E}$ has finite image. Thus E becomes trivial after a quasi-étale base change and then it corresponds to a direct factor of the relative Albanese variety of $X_1 := Y_1 \times_Y X$, giving the Abelian variety A_1.

Furthermore, in this case $T_{Y_1}|_C = \mathcal{H}om_{Y_1}(\Omega_{Y_1}, O_{Y_1})|_C$ has a global section. Since T_{Y_1} is reflexive, the Enriques–Severi–Zariski lemma (as proved, though not as claimed in [Har77, III.7.8]) implies that $H^0(Y_1, T_{Y_1}) \neq 0$. Therefore $\dim \mathrm{Aut}(Y_1) > 0$. Since Y_1 is not uniruled, $\mathrm{Aut}^0(Y_1)$ has no linear algebraic subgroups, thus the connected component $\mathrm{Aut}^0(Y_1)$ is an Abelian variety A_2. By Proposition 34, A_2 becomes a direct factor after a suitable étale cover $\tilde{Y} \to Y_1 \to Y$. □

The following result was known in [Ser01, Ses63]; see [Bri10] for the general theory.

Proposition 34 *Let $W \to S$ be a flat, projective morphism with normal fibers over a field of characteristic 0 and $A \to S$ an Abelian scheme acting faithfully on W. Then there is a flat, projective morphism with normal fibers $Z \to S$ and an A-equivariant étale morphism $A \times_S Z \to W$.* □

The following is a combination of [Hor76, Theorem 8.1] and the method of [Hor76, Theorem 8.2] in the smooth case and [BHPS12, Proposition 3.10] in general.

Theorem 35 *Let $f : X \to Y$ be a morphism of proper schemes over a field such that $f_* O_X = O_Y$ and $\mathrm{Hom}_Y(\Omega_Y, R^1 f_* O_X) = 0$.*

Then for every small deformation $\mathbf{X}$ of X there is a small deformation $\mathbf{Y}$ of Y such that f lifts to $\mathbf{f} : \mathbf{X} \to \mathbf{Y}$. □

6 Smoothings of very singular varieties

One can frequently construct smooth varieties by first exhibiting some very singular, even reducible schemes with suitable numerical invariants and then smoothing them. For such Calabi–Yau examples, see [KN94]. Thus it is of interest to know when an elliptic fiber space structure is preserved by a smoothing. In some cases, when Theorem 31 does not apply, the following result, relying on Corollary 11, provides a quite satisfactory answer.

Proposition 36 *Let X be a projective, reduced, Gorenstein scheme of pure dimension n such that ω_X is numerically trivial and $H^2(X, \mathcal{O}_X) = 0$. Let $g: X \to B$ be a morphism whose general fibers (over every irreducible component of B) are curves of arithmetic genus 1. Assume also that every irreducible component of X dominates an irreducible component of B.*

Let L_B be an ample line bundle on B and assume that $\chi(X, g^*L_B^m)$ is a polynomial of degree $\dim X - 2$. *Then every smoothing (and every log terminal deformation) of X is an elliptic fiber space.*

Warning Note that we do not claim that g lifts to every deformation of X. In Example 49 X has smoothings, which are elliptic, and also other singular deformations that are not elliptic.

Proof As before, $H^2(X, \mathcal{O}_X) = 0$ implies that g^*L_B lifts to every small deformation [Gro62, p. 236–16]. Thus we have a deformation $h: (\mathbf{X}, \mathbf{L}) \to (0 \in S)$ of $(X_0, L_0) \cong (X, L = g^*L_B)$.

We claim that $\mathbf{L}$ is h-nef and $K_{\mathbf{X}}$ is trivial on the fibers of h. This is a somewhat delicate point since being nef is not an open condition in general [Les12]. We get around this problem as follows.

Let (X_{gen}, L_{gen}) be a generic fiber. (Note the difference between generic and general.) First we show that L_{gen} is nef and $K_{X_{gen}} \equiv 0$. Indeed, assume that $(L_{gen} \cdot C_{gen}) < 0$ for some curve C_{gen}. Let $C_0 \subset X_0$ be a specialization of C_{gen}. Then $(L_0 \cdot C_0) = (L_{gen} \cdot C_{gen}) < 0$ gives a contradiction. A similar argument shows that $(K_{X_{gen}} \cdot C_{gen}) = 0$ for every curve C_{gen}.

Next, the deformation invariance of $\chi(X, g^*L_B^m)$ and Riemann–Roch (cf. Corollary 22 and 24) show that

$$(L_{gen}^{n-2} \cdot c_2(X_{gen})) = (n-2)! \cdot (\text{coefficient of } m^{n-2} \text{ in } \chi(X, g^*L_B^m)).$$

Therefore $(L_{gen}^{n-2} \cdot c_2(X_{gen})) > 0$ and, as we noted after Theorem 10, this implies that $|mL_{gen}|$ is base-point-free for some $m > 0$.

Thus there is a dense Zariski open subset $S^0 \subset S$ such that $|mL_s|$ is base-point-free for $s \in S^0$, hence (X_s, L_s) is an elliptic fiber space for $s \in S^0$. We repeat the argument for the generic points of $S \setminus S^0$ and conclude by Noetherian induction. □

It may be useful to see how to modify the above proof to work in the analytic case when there are no generic points.

The (Barlet or Douady) space of curves in $h\colon \mathbf{X} \to (0 \in S)$ has only countably many irreducible components, thus there are countably many closed subspaces $S_i \subsetneq S$ such that every curve $C_s \subset X_s$ is deformation equivalent to a curve $C_0 \subset X_0$ for $s \notin \cup S_i$. In particular, L_s is nef and $K_{X_s} \equiv 0$ whenever $s \notin \cup S_i$. Thus (X_s, L_s) is an elliptic fiber space for $s \notin \cup S_i$.

By semicontinuity, there are closed subvarieties $T_m \subsetneq S$ such that

$$h_*O_{\mathbf{X}}(m\mathbf{L}) \otimes \mathbb{C}_s = H^0(X_s, O_{X_s}(mL_s)) \quad \text{for } s \notin T_m.$$

Thus if $s \notin \cup_i S_i \bigcup \cup_m T_m$ and $O_{X_s}(m_0 L_s)$ is generated by global sections then

$$\phi_{m_0}\colon h^*(h_*O_{\mathbf{X}}(m_0\mathbf{L})) \to O_{\mathbf{X}}(m_0\mathbf{L})$$

is surjective along X_s. Thus there is a dense Zariski open subset $S^0 \subset S$ such that ϕ_{m_0} is surjective for all $s \in S^0$. Now we can finish by Noetherian induction as before.

7 Calabi–Yau orbibundles

The techniques of this section are mostly taken from [Kol93, Section 6] and [Nak99].

Definition 37 A Calabi–Yau fiber space $g\colon X \to B$ is called an *orbibundle* if it can be obtained by the following construction.

Let $\tilde{B}$ be a normal variety, F a Calabi–Yau variety, and $\tilde{X} := \tilde{B} \times F$. Let G be a finite group, $\rho_B\colon G \to \mathrm{Aut}(\tilde{B})$ and $\rho_F\colon G \to \mathrm{Aut}(F)$ two faithful representations. Set

$$(g\colon X \to B) := (\tilde{X}/G \to \tilde{B}/G);$$

it is a generically isotrivial Calabi–Yau fiber space with general fiber F.

(It would seem more natural to require the above property only locally on B. We see in Theorem 43 that in the algebraic case the two versions are equivalent. However, for complex manifolds, the local and global versions are different.)

For any normal variety Z with non-negative Kodaira dimension, the connected component $\mathrm{Aut}^0(Z)$ of $\mathrm{Aut}(Z)$ is an Abelian variety; we call its elements translations. The quotient Aut(Z)/ $\mathrm{Aut}^0(Z)$ is the discrete part of the automorphism group.

For G acting on F, let $G_t := \rho_F^{-1}\,\mathrm{Aut}^0(F) \subset G$ be the normal subgroup of translations and set $X^d := \tilde{X}/G_t$. Then $G_d := G/G_t$ acts on X^d and $X = X^d/G_d$. Thus every orbibundle comes with two covers:

$$\begin{array}{ccccc} X & \stackrel{\tau_X}{\longleftarrow} & X^d & \stackrel{\pi_X}{\longleftarrow} & \tilde{X} \\ g\downarrow & & g^d\downarrow & & \tilde{g}\downarrow \\ B & \stackrel{\tau_B}{\longleftarrow} & B^d & \stackrel{\pi_B}{\longleftarrow} & \tilde{B} \end{array} \tag{37.1}$$

We see during the proof of Theorem 43 that the cover $X \leftarrow X^d$ corresponding to the discrete part of the monodromy representation is uniquely determined by $g\colon X \to B$. By contrast, the $X^d \leftarrow \tilde{X}$ part is not unique. Its group of deck transformations is $G_t \subset \mathrm{Aut}^0(F)$, hence Abelian. It is not even clear that there is a natural "smallest" choice of $X^d \leftarrow \tilde{X}$.

If $F = A$ is an Abelian variety, then $g^d\colon X^d \to B^d$ is a Seifert bundle where an orbibundle $g^s\colon X^s \to B^s$ is called a *Seifert bundle* if $F = A$ is an Abelian variety and G acts on A by translation. Note that in this case the A-action on $\tilde{B}\times A$ descends to an A-action on X^s and $B^s = X^s/A$. Thus the reduced structure of every fiber is a smooth Abelian variety isogenous to A.

Lemma 38 *Notation as above. Then*

(1) π_X and τ_X are étale in codimension 1 (that is, quasi-étale),
(2) π_X and τ_X are étale in codimension 2 if one of the following holds:

  (a) G acts freely on F outside a codimension ≥ 2 subset or
  (b) $K_F \sim 0$ and $\Delta_{X/B} = 0$.

Proof The first claim is clear since both ρ_F, ρ_B are faithful.

Since ρ_F, ρ_B are faithful, τ_X fails to be étale in codimension 2 iff some $1 \neq g \in G$ fixes a divisor $\tilde{D}_B \subset \tilde{B}$ and also a divisor $D_F \subset F$. This is excluded by (2a).

Next we check that (2b) implies (2a). At a general point $p \in D_F$ choose local g-equivariant coordinates $x_1, \ldots, x_m$ such that $D_F = (x_1 = 0)$. Thus $\rho_F(g)^*$ acts on x_1 nontrivially but it fixes $x_2, \ldots, x_m$. Let ω_0 be a nonzero section of ω_F. Locally near p we can write

$$\omega_0 = f \cdot dx_1 \wedge \cdots \wedge dx_m,$$

thus $\rho_F(g)^*$ acts on $H^0(F, \omega_F)$ with the same eigenvalue as on x_1.

Thus, by 19, (1), the image of $\tilde{D}_X$ gives a positive contribution to $\Delta_{X/B}$. This contradicts $\Delta_{X/B} = 0$. □

There are some obvious deformations of X obtained by deforming $\tilde{B}$ and F in a family $\{(\tilde{B}_t, F_t)\}$ such that the representations ρ_B, ρ_F lift to $\rho_{B,t}\colon G \to \mathrm{Aut}(\tilde{B}_t)$ and $\rho_{F,t}\colon G \to \mathrm{Aut}(F_t)$.

In general, not every deformation of X arises this way. For instance, let $\tilde{B}$ and $F = A$ be elliptic curves and X the Kummer surface of $\tilde{B} \times A$. The obvious deformations of X form a 2-dimensional family obtained by deforming $\tilde{B}$ and A. Thus a general deformation of X is not obtained this way and it is not even elliptic. Even worse, a general elliptic deformation of X is also not Kummer; thus not every deformation of the morphism $(g\colon X \to B)$ is obtained by the quotient construction.

Theorem 39 *Let $g\colon X \to B$ be a Calabi–Yau orbibundle with general fiber F. Assume that X has log terminal singularities, $H^2(X, \mathcal{O}_X) = 0$, $\kappa(X) \geq 0$, $K_F \sim 0$, and $\Delta_{X/B} = 0$. Then every flat deformation of X arises from a flat deformation of $(\tilde{B}, F, \rho_B, \rho_F)$.*

Proof Let L_B be an ample line bundle on B and set $L := g^* L_B$.

Let $h\colon \mathbf{X} \to (0 \in S)$ be a deformation of $X_0 \cong X$. In the sequel we will repeatedly replace S by a smaller analytic (or étale) neighborhood of 0 if necessary.

Since $H^2(X, \mathcal{O}_X) = 0$, L lifts to a line bundle $\mathbf{L}$ on $\mathbf{X}$ by [Gro62, p. 236–16].

Since $K_F \sim 0$ and $\Delta_{X/B} = 0$, Lemma 38 implies that $\pi\colon \tilde{X} \to X$ is étale in codimension 2. Since X is log terminal, so is $\tilde{X}$, hence it is Cohen–Macaulay (see, e.g., [KM98, 5.10 and 5.22]). Thus, by [Kol95, Corollary 12.7], the cover π lifts to a cover $\Pi\colon \tilde{\mathbf{X}} \to \mathbf{X}$.

Finally we show that the product decomposition $\tilde{X} \cong \tilde{B} \times F$ lifts to a product decomposition

$$\tilde{\mathbf{X}} \cong \tilde{\mathbf{B}} \times_S \mathbf{F}$$

where $\tilde{\mathbf{B}} \to S$ is a flat deformation of $\tilde{B}$ and $\mathbf{F} \to S$ is a family of Calabi–Yau varieties over S. After a further étale cover of $\tilde{F} \to F$ we may assume that $\tilde{F} \cong Z \times A$, where $H^1(Z, \mathcal{O}_Z) = 0$ and A is an Abelian variety. Set $\hat{X} := \tilde{B} \times Z \times A$; then $\hat{X} \to \tilde{X}$ lifts to a deformation $\widehat{\mathbf{X}} \to \tilde{\mathbf{X}} \to S$.

First we use Lemma 40 and Proposition 34 to show that the product decomposition $\hat{X} \cong (\tilde{B} \times Z) \times A$ lifts to a product decomposition

$$\widehat{\mathbf{X}} \cong \widehat{\mathbf{BZ}} \times_S \mathbf{A}$$

where $\widehat{\mathbf{BZ}} \to S$ is a flat deformation of $\tilde{B} \times Z$ and $\mathbf{A} \to S$ is a family of Abelian varieties over S. The deformation of the product $\tilde{B} \times Z$ is much easier to understand; we discuss it in 53. □

Lemma 40 *Let $Y \to S$ be a flat, proper morphism whose fibers are normal and L a line bundle on Y. Let $0 \in S$ be a point such that*

(1) Y_0 is not birationally ruled,
(2) an Abelian variety $A_0 \subset \mathrm{Aut}^0(Y_0)$ acts faithfully on Y_0,
(3) L_0 is nef, L_0 is numerically trivial on the A_0-orbits but not numerically trivial on general A'_0-orbits for any $A_0 \subsetneq A'_0 \subset \mathrm{Aut}^0(Y_0)$.

Then, possibly after shrinking S, there is an Abelian scheme $A \to S$ extending A_0 such that A acts faithfully on Y.

Proof By [Mat68, p. 217] (see also [Kol85, p. 392]), possibly after shrinking S, $g^a\colon \mathrm{Aut}^0(Y/S) \to S$ is a smooth Abelian scheme, where $\mathrm{Aut}^0(Y/S)$ denotes the identity component of the automorphism scheme $\mathrm{Aut}(Y/S)$. The fibers are normal, hence $Y \to S$ is smooth over a dense subset of every fiber. Since a smooth morphism has sections étale locally, we may assume after an étale base change that there is a section $Z \subset Y$. Acting on Z gives a morphism $\rho_Z\colon \mathrm{Aut}^0(Y/S) \to Y$. Then $\rho_Z^* L$ is a nef line bundle on $\mathrm{Aut}^0(Y/S)$. The kernel of the cup-product map

$$c_1(\rho_Z^* L)\colon R^1 g^a_* \mathbb{Q} \to R^3 g^a_* \mathbb{Q}$$

is a variation of sub-Hodge structures, hence it corresponds to a smooth Abelian subfamily $A \subset \mathrm{Aut}^0(Y/S)$. By (3), this is the required extension of A_0.

The quotient then exists by [Ses63]. □

We will also need to understand the class group of an orbibundle.

41 (Divisors on orbibundles) We use the notation of Definitions 37 and 42.

By [BGS11, 5.3] (see also [HK11, CL10] for the elliptic case), the class group of the product $\tilde{B} \times F$ is

$$\mathrm{Cl}(\tilde{B} \times F) = \mathrm{Cl}(\tilde{B}) + \mathrm{Cl}(F) + \mathrm{Hom}(\mathrm{Alb}^{\mathrm{rat}}(\tilde{B}), \mathrm{Pic}^0(F)). \tag{41.1}$$

This comes with a natural G-action and, up to torsion, the class group of the quotient is

$$\mathrm{Cl}(B) + \mathrm{Cl}(F)^G + \mathrm{Hom}(\mathrm{Alb}^{\mathrm{rat}}(\tilde{B}), \mathrm{Pic}^0(F))^G. \tag{41.2}$$

Here $\mathrm{Cl}(F)^G + \mathrm{Hom}(\mathrm{Alb}^{\mathrm{rat}}(\tilde{B}), \mathrm{Pic}^0(F))^G$ can be identified with the class group of the generic fiber of g. If $\tilde{B}$ has rational singularities, then $\mathrm{Alb}^{\mathrm{rat}}(\tilde{B}) = \mathrm{Alb}(\tilde{B})$. Thus the extra component $\mathrm{Hom}(\mathrm{Alb}(\tilde{B}), \mathrm{Pic}^0(F))$ corresponds to divisors that are pulled back from $\mathrm{Alb}(\tilde{B}) \times F$, hence they are Cartier.

We will use the following variant of these observations:

Claim 41.3 Let $g\colon X \to B$ be an orbibundle such that X has log terminal singularities. Then the natural map

$$\mathrm{Cl}(B)/\operatorname{Pic}(B) + (\mathrm{Cl}(F)/\operatorname{Pic}(F))^G \to \mathrm{Cl}(X)/\operatorname{Pic}(X)$$

is an isomorphism modulo torsion. In particular, if B and the generic fiber of g are $\mathbb{Q}$-factorial, then so is X.

Proof By Lemma 38, $\tau_X\colon X^d \to X$ is étale in codimension 1, hence X^d also has log terminal singularities. As noted in Definition 5, this implies that B^d has rational singularities.

Let us now study more carefully the RHS of (41.2). Let $G_t \subset G$ denote the subgroup of translations. Then

$$\operatorname{Hom}(\mathrm{Alb}^{\mathrm{rat}}(\tilde{B}), \operatorname{Pic}^0(F))^G \subset \operatorname{Hom}(\mathrm{Alb}^{\mathrm{rat}}(\tilde{B}), \operatorname{Pic}^0(F))^{G_t}.$$

Since translations act trivially on $\operatorname{Pic}^0(F)$, the latter can be identified (up to torsion) as

$$\begin{aligned}\operatorname{Hom}(\mathrm{Alb}^{\mathrm{rat}}(\tilde{B}), \operatorname{Pic}^0(F))^{G_t} \otimes \mathbb{Q} &\cong \operatorname{Hom}(\mathrm{Alb}^{\mathrm{rat}}(\tilde{B})^{G_t}, \operatorname{Pic}^0(F)) \otimes \mathbb{Q}\\ &\cong \operatorname{Hom}(\mathrm{Alb}^{\mathrm{rat}}(B^d), \operatorname{Pic}^0(F)) \otimes \mathbb{Q}\\ &\cong \operatorname{Hom}(\mathrm{Alb}(B^d), \operatorname{Pic}^0(F)) \otimes \mathbb{Q}.\end{aligned}$$

Thus this extra term gives only $\mathbb{Q}$-Cartier divisors on X^d and hence also on X. □

The following local example shows that it is not enough to assume that B has rational singularities. Set $\tilde{B} = (u^3 + v^3 + w^3 = 0) \subset \mathbb{A}^3$ and $E = (x^3 + y^3 + z^3 = 0) \subset \mathbb{P}^2$. On both factors, $\mathbb{Z}/3$ acts by weights $(0, 0, 1)$. Then $B = \tilde{B}/\frac{1}{3}(0, 0, 1) \cong \mathbb{A}^2$ is even smooth, but

$$X = \tilde{B} \times E/\tfrac{1}{3}(0, 0, 1) \times (0, 0, 1)$$

is not $\mathbb{Q}$-factorial. For instance, the closure of the graph of the natural projection $\tilde{B} \dashrightarrow E$ gives a non-$\mathbb{Q}$-Cartier divisor on X.

Definition 42 (Albanese varieties) For a smooth projective variety V let $\mathrm{Alb}(V)$ denote the Albanese variety, that is, the target of the universal morphism from V to an Abelian variety. (See [BPV84, Section I.13] or [Gro62, p. 236–16] for introductions.)

There are two ways to generalize this concept to normal varieties.

The above definition yields what we again call the *Albanese variety* $\mathrm{Alb}(V)$. Alternatively, the *rational Albanese* variety $\mathrm{Alb}^{\mathrm{rat}}(V)$ is defined as the target of the universal rational map from V to an Abelian variety. One can identify $\mathrm{Alb}^{\mathrm{rat}}(V) = \mathrm{Alb}(V')$, where $V' \to V$ is any resolution of singularities.

It is easy to see that if V has log terminal (more generally rational) singularities, then $\mathrm{Alb}^{\mathrm{rat}}(V) = \mathrm{Alb}(V)$.

8 Generically isotrivial Calabi–Yau fiber spaces

In this section we prove that all generically isotrivial Calabi–Yau fiber spaces are essentially Calabi–Yau orbibundles.

Theorem 43 *Let $g\colon X \to B$ be a projective, generically isotrivial, Calabi–Yau fiber space with general fiber F. Then*

(1) $g\colon X \to B$ is birational to an orbibundle $(g^{\mathrm{orb}}\colon X^{\mathrm{orb}} \to B)$.
(2) $g\colon X \to B$ is isomorphic to $(g^{\mathrm{orb}}\colon X^{\mathrm{orb}} \to B)$ if

- *(a) X is $\mathbb{Q}$-factorial, log terminal,*
- *(b) $g\colon X \to B$ is relatively minimal, without exceptional divisors,*
- *(c) B is $\mathbb{Q}$-factorial, and*
- *(d) one of the following holds:*
  - *(i) $K_F \sim 0$ and $\Delta_{X/B} = 0$, or*
  - *(ii) there is a closed subset $Z_B \subset B$ of codimension ≥ 2 such that $g\colon X \to B$ is locally an orbibundle over $B \setminus Z_B$.*

Proof Let $B^0 \subset B$ be a Zariski open subset over which $X^0 \to B^0$ is isotrivial with general fiber F. This gives a well-defined representation

$$\rho\colon \pi_1(B^0) \to \mathrm{Aut}(F)/\,\mathrm{Aut}^0(F).$$

Let $B^{(d,0)} \to B^0$ be the corresponding étale, Galois cover with group G_d and $B^d \to B$ its extension to a (usually ramified) Galois cover of B with group G_d. This gives the well-defined cover in (37.1).

The trivialization of the translation part is more subtle and it depends on additional choices.

A general $\mathrm{Aut}^0(F)$-orbit $A_F \subset F$ defines an isotrivial Abelian family $X^{(d,0)} \supset A_X^{(d,0)} \to B^{(d,0)}$. By assumption there is a g-ample line bundle L on X. It pulls back to a relatively ample line bundle L_A on $A_X^{(d,0)}$. We may assume that its degree on the general fiber is at least 3. Let $T^{(d,0)} \subset A_X^{(d,0)}$ be the subscheme as in 44. Since L_A is G_d-invariant, $T^{(d,0)}$ is G_d-equivariant hence it defines a monodromy representation of $\pi_1(B^{(d,0)}) \to \mathrm{Aut}^0(F)$. Let $\Gamma \subset \pi_1(B^{(d,0)})$ be a finite-index subgroup that is normal in $\pi_1(B^0)$ and $\tilde{B}^0 \to B^0$ the corresponding étale, Galois cover with group $G = \pi_1(B^0)/\Gamma$. Let $\tilde{B} \to B$ denote its extension to a (usually ramified) Galois cover of B with group G.

By pull-back we obtain an isotrivial, Abelian fiber space $\tilde{A}^0_X \to \tilde{B}^0$ with a trivialization of the m-torsion points. For $m \geq 3$ this implies that $\tilde{A}^0_X \cong \tilde{B}^0 \times A$. (This is quite elementary, cf. [ACG11, p. 513].) Thus the same pull-back also trivializes $X^0 \to B^0$. We can compactify $\tilde{X}^0$ as $\tilde{X} := \tilde{B} \times F$.

The G-action on $\tilde{X}$ can be given as

$$\gamma\colon (\tilde{b}, c) \mapsto (\rho_B(\gamma) \cdot \tilde{b}, \rho_{F,\tilde{b}}(\gamma) \cdot c).$$

Note that $\rho_{F,\tilde{b}}$ preserves the m-torsion points and the automorphisms of an Abelian torsor that preserve any finite non-empty set form a discrete group. Thus in fact $\rho_{F,\tilde{b}}$ is independent of $\tilde{b}$ and hence the G-action on $\tilde{X}$ is given by

$$\gamma\colon (\tilde{b}, c) \mapsto (\rho_B(\gamma) \cdot \tilde{b}, \rho_F(\gamma) \cdot c)$$

for some isomorphism $\rho_B\colon G \cong \mathrm{Gal}(\tilde{B}/B)$ and homomorphism $\rho_F\colon G \to \mathrm{Aut}(F)$. We can replace $\tilde{B}$ by $\tilde{B}/\ker \rho_F$, hence we may assume that ρ_F is faithful.

Thus we have $g^{\mathrm{orb}} : X^{\mathrm{orb}} := \tilde{X}/G \to B$ and a birational map

$$\phi\colon X \dashrightarrow X^{\mathrm{orb}} \quad \text{such that} \quad g = g^{\mathrm{orb}} \circ \phi.$$

Assume next that conditions (2a–d) hold. First we use (2d) to prove that ϕ extends to an isomorphism over codimension 1 points of B. Then we use conditions (2a–c) to show that ϕ is an isomorphism everywhere.

In order to understand the codimension 1 behavior, we can take a transversal curve section (or localize at a codimension 1 point). Thus we may assume that $B = (0 \in D)$ is a unit disc (or the spectrum of a DVR) and $g\colon X \to D$ is isotrivial on $D \setminus \{0\}$. Thus X^{orb} is of the form

$$X^{\mathrm{orb}} \cong (F \times D)/(\rho, e^{2\pi i/m}),$$

where ρ is an automorphism of order m of F.

In case (2d(i)) ρ acts trivially on $H^0(F, \omega_F)$ by 19, (1), thus the canonical class of $F_0 := F/\langle\rho\rangle$ is trivial and so F_0 has canonical singularities. By inversion of adjunction [Kol13b, Theorem 4.9], the pair (X^{orb}, F_0) is also canonical. It is clear that $a(E, X^{\mathrm{orb}}) > a(E, X^{\mathrm{orb}}, F_0) \geq 0$ for every divisor over X^{orb} dominating $0 \in D$; cf. [KM98, 2.27]. Thus, as we noted in 8, ϕ restricts to a birational map of the central fibers $\phi_0\colon X_0 \dashrightarrow F_0$.

Now let H be a relatively ample divisor on $X \to D$. Then $\phi_* H$ is a divisor on X^{orb} that is Cartier and ample outside F_0. Since every curve on F_0 is the specialization of a curve in F, we see that $\phi_* H$ is $\mathbb{Q}$-Cartier and ample everywhere. Thus ϕ is an isomorphism by Lemma 45.

In case (2d(ii)), let $g_i\colon X_i \to D$ be two orbibundles and $\phi\colon X_1 \dashrightarrow X_2$ a birational map that is an isomorphism over $D^0 := D \setminus \{0\}$. Let $\Gamma \subset X_1 \times_D X_2$

denote the closure of the graph of ϕ. We need to prove that the coordinate projections $\Gamma \to X_i$ are finite. It is enough to check this after a finite base change. Thus we may assume that $X_i \cong F \times D$. Then ϕ can be identified with a map $D^0 \to \mathrm{Aut}(F)$ and this extends to $D \to \mathrm{Aut}(F)$ since every connected component of $\mathrm{Aut}(F)$ is proper. Thus ϕ is an isomorphism.

Now we return to the general case. We have shown that there is a subset $Z_B \subset B$ of codimension ≥ 2 such that $\phi \colon X \dashrightarrow X^{\mathrm{orb}}$ is an isomorphism over $B \setminus Z_B$. By assumption (2b), the pre-image $g^{-1}(Z_B)$ has codimension ≥ 2 and the pre-image $(g^{\mathrm{orb}})^{-1}(Z_B)$ has codimension ≥ 2 by construction. Since X is $\mathbb{Q}$-factorial, so is the generic fiber of g, hence X^{orb} is $\mathbb{Q}$-factorial by Claim 41.3. Thus the assumptions of Lemma 45, part (2) are satisfied and hence ϕ is an isomorphism. □

44 (Multisections of Abelian families) Let E be a smooth projective curve of genus 1 and L a line bundle of degree m on E. If $m = 1$ then L has a unique section, thus we can associate a point $p \in E$ to L. If $m \geq 2$, then sections define a linear equivalence class $|L|$ of m points. If we fix a point $0 \in E$ to be the origin, then we can add these m points together and get a well-defined point of E associated with L. This, however, depends on the choice of the origin.

To get something invariant, let us look at the points $p \in E$ such that $m \cdot p \in |L|$. There are m^2 such points, together forming a translate of the subgroup of m-torsion points. This construction also works in families.

Let $g \colon X \to B$ be a smooth, projective morphism whose fibers E_b are curves of genus 1. Let L be a line bundle on X that has degree m on each fiber. Then there is a closed subscheme $T \subset X$ such that $g|_T : T \to B$ is étale of degree m^2 and every fiber $T_b \subset E_b$ is a translate of the subgroup of m-torsion points.

There is a similar construction for higher-dimensional Abelian varieties. For clarity, I say *Abelian torsor* when talking about an Abelian variety without a specified origin.

Thus let A be an Abelian torsor of dimension d and L an ample line bundle on A. It has a first Chern class $\tilde{c}_1(L)$ in the Chow group and we get $\tilde{c}_1(L)^d$ as an element of the Chow group of 0-cycles. (It is important to use the Chow group, the Chern class in cohomology is not sufficient.) Let its degree be m.

Fix a base point $0 \in A$. This defines a map from the Chow group of 0-cycles to $(A, 0)$; let $\alpha(\tilde{c}_1(L)^d)$ denote the image.

Finally let $T \subset A$ be the set of points $t \in A$ such that $m \cdot t = \alpha(\tilde{c}_1(L)^d)$. This T is a translate of the subgroup of m-torsion points. As before, the key point is that T is independent of the choice of the base point $0 \in A$. Indeed, if we change 0 by a translation by $c \in A$ then $\alpha(\tilde{c}_1(L)^d)$ is changed by translation by $m \cdot c$ so T is changed by translation by c.

Furthermore, if (A_b, L_b) is a family of polarized Abelian torsors that varies analytically (or algebraically) with b then $T_b \subset A_b$ is a family of subschemes that also vary analytically (or algebraically) with b. Thus we obtain the following.

Let $g: X \to B$ be a smooth, projective morphism whose fibers are Abelian torsors. Then there is a closed subscheme $T \subset X$ such that $g|_T: T \to B$ is étale and every fiber $T_b \subset A_b$ is a translate of the subgroup of m-torsion points (where $\deg T/B = m^{2d}$).

Lemma 45 *Let $g_i: X_i \to B$ be projective fiber spaces, the X_i normal and $\phi: X_1 \dashrightarrow X_2$ a rational map. Assume that there are closed subsets $Z_i \subset X_i$ such that $\operatorname{codim}_{X_i} Z_i \geq 2$ and ϕ induces an isomorphism $X_1 \setminus Z_1 \cong X_2 \setminus Z_2$. Let H_1 be a g_1-ample divisor on X_1 and set $H_2 := \phi_* H_1$.*

(1) ϕ is an isomorphism iff H_2 is g_2-ample.

(2) If ϕ induces an isomorphism of the generic fibers, X_2 is $\mathbb{Q}$-factorial and every curve $C \subset X_2$ contracted by g_2 is $\mathbb{Q}$-homologous to a curve in a general fiber, then H_2 is g_2-ample.

Proof The first claim is a lemma of Matsusaka and Mumford [MM64]; see [KSC04, 5.6] or [Kol10, Exercise 75] for the variant used here.

It follows from assumption (2) that H_2 is $\mathbb{Q}$-Cartier and strictly positive on the cone of curves, hence it is g_2-ample. □

46 (F-theory examples) Let X be a smooth, projective variety and $g: X \to B$ a relatively minimal elliptic fiber space with a section $\sigma: B \to X$. Since X is smooth, so is B.

Assume that $\Delta_{X/S} = 0$. Then, by Lemma 38, it can have only multiple smooth fibers at codimension-1 points, but then the section shows that there are no multiple fibers. Thus there is an open subset $B^0 \subset B$ such that $\operatorname{codim}_B(B \setminus B^0) \geq 2$ and $X^0 \to X$ is a fiber bundle with fiber a pointed elliptic curve $(E, 0)$. Thus X^0 is given by the data

$$(B^0, E, \rho: \pi_1(B^0) \to \operatorname{Aut}(E, 0)).$$

Note that $\pi_1(B^0) = \pi_1(B)$ since B is smooth and $\operatorname{codim}_B(B \setminus B^0) \geq 2$. Thus X is birational to a fiber bundle $g': X' \to B$ given by the data

$$(B, E, \rho: \pi_1(B) \to \operatorname{Aut}(E, 0)).$$

All the fibers of g' are elliptic curves, but the exceptional locus of a flip or a flop is always covered by rational curves (cf. [Kol96, VI.1.10]). Thus in fact

$X \cong X'$, hence $g\colon X \to B$ is a locally trivial fiber bundle. The image of the monodromy representation $\rho\colon \pi_1(B) \to \operatorname{Aut}(E, 0)$ is usually $\mathbb{Z}/2$, but for elliptic curves with extra automorphisms it can also be $\mathbb{Z}/3, \mathbb{Z}/4$, or $\mathbb{Z}/6$.

It is easy to write down examples where $K_X \sim 0$ and $H^i(X, \mathcal{O}_X) = 0$ for $0 < i < \dim X$. However, $\pi_1(X)$ is always infinite, so such an X cannot be a "true" Calabi–Yau manifold.

By Theorem 39, if $H^2(X, \mathcal{O}_X) = 0$ then every small deformation of X is obtained by deforming B and, if the image of ρ is $\mathbb{Z}/2$, also deforming E.

9 Examples

The first example is an elliptic Calabi–Yau surface with quotient singularities that has a flat smoothing which is neither Calabi–Yau nor elliptic.

Example 47 We start with a surface S_F^* which is the quotient of the square of the Fermat cubic curve by $\mathbb{Z}/3$:

$$S_F^* \cong (u_1^3 = v_1^3 + w_1^3) \times (u_2^3 = v_2^3 + w_2^3)/\tfrac{1}{3}(1,0,0;1,0,0).$$

To describe the deformation, we need a different representation of it.

In $\mathbb{P}^3$ consider two lines $L_1 = (x_0 = x_1 = 0)$ and $L_2 = (x_2 = x_3 = 0)$. The linear system $\left|\mathcal{O}_{\mathbb{P}^3}(2)(-L_1 - L_2)\right|$ is spanned by the four reducible quadrics $x_i x_j$ for $i \in \{0, 1\}$ and $j \in \{2, 3\}$. They satisfy a relation $(x_0x_2)(x_1x_3) = (x_0x_3)(x_1x_2)$. Thus we get a morphism

$$\pi\colon B_{L_1+L_2}\mathbb{P}^3 \to \mathbb{P}^1 \times \mathbb{P}^1,$$

which is a $\mathbb{P}^1$-bundle whose fibers are the birational transforms of lines that intersect both of the L_i.

Let $S \subset \mathbb{P}^3$ be a cubic surface such that $\mathbf{p} := S \cap (L_1 + L_2)$ is six distinct points. Then we get $\pi_S : B_{\mathbf{p}}S \to \mathbb{P}^1 \times \mathbb{P}^1$.

In general, none of the lines connecting two points of $\mathbf{p}$ is contained in S. Thus in this case π_S is a finite triple cover.

Both of the lines L_i determine an elliptic pencil on $B_{\mathbf{p}}S$ but if we move the six points $\mathbf{p}$ into general position, we lose both elliptic pencils.

At the other extreme we have the Fermat-type surface

$$S_F := (x_0^3 + x_1^3 = x_2^3 + x_3^3) \subset \mathbb{P}^3.$$

We can factor both sides and write its equation as $m_1m_2m_3 = n_1n_2n_3$. The nine lines $L_{ij} := (m_i = n_j = 0)$ are all contained in S_F. Let $L'_{ij} \subset B_{\mathbf{p}}S_F$ denote their birational transforms. Then the self-intersections $(L'_{ij} \cdot L'_{ij})$ equal -3 and

π_{S_F} contracts these nine curves L'_{ij}. Thus the Stein factorization of π_{S_F} gives a triple cover $S^*_F \to \mathbb{P}^1 \times \mathbb{P}^1$ and S^*_F has nine singular points of type $\mathbb{A}^2/\frac{1}{3}(1,1)$. We see furthermore that

$$-3K_{S_F} \sim \sum_{ij} L_{ij} \quad \text{and} \quad -3K_{B_P S_F} \sim \sum_{ij} L'_{ij}.$$

Thus $-3K_{S^*_F} \sim 0$.

To see that this is the same S^*_F, note that the morphism of the original S^*_F to $\mathbb{P}^1 \times \mathbb{P}^1$ is given by

$$(u_1{:}v_1{:}w_1) \times (u_2{:}v_2{:}w_2) \mapsto (v_1{:}w_1) \times (v_2{:}w_2)$$

and the rational map to the cubic surface is given by

$$(u_1{:}v_1{:}w_1) \times (u_2{:}v_2{:}w_2) \mapsto (v_2 u_1 u_2^2{:}u_1 u_2^2{:}v_1 u_2^3{:}u_2^3).$$

Varying S gives a flat deformation whose central fiber is S^*_F, a surface with quotient singularities and torsion canonical class and whose general fiber is a cubic surface blown up at six general points, hence rational and without elliptic pencils.

The next example gives local models of generically isotrivial elliptic orbibundles that have a crepant resolution.

Example 48 Let $Z \subset \mathbb{P}^N$ be an anticanonically embedded Fano variety and $X \subset \mathbb{A}^{N+1}_{\mathbf{x}}$ the cone over Z. Let $0 \in E$ be an elliptic curve with a marked point. Consider the elliptic fiber space

$$Y := X \times E/(-1,-1) \to X/(-1).$$

We claim that Y has a crepant resolution.

First we blow up the vertex of X. We get $B_0 X \to X$ with exceptional divisor $F \cong Z$. Note further that $B_0 X \to X$ is crepant. The involution lifts to $B_0 X \times E/(-1,-1)$. The fixed point set of this action is $F \times \{0\}$; a smooth subvariety of codimension 2. Thus $B_0 X \times E/(-1,-1)$ is resolved by blowing up the singular locus.

The next example shows that for surfaces with normal crossing singularities, a deformation may lose the elliptic structure.

Example 49 Let $S \subset \mathbb{P}^1 \times \mathbb{P}^2$ be a smooth surface of bi-degree $(1,3)$. The first projection $\pi\colon S \to \mathbb{P}^1$ is an elliptic fiber space. The other projection $\tau\colon S \to \mathbb{P}^2$ exhibits it as the blow-up of $\mathbb{P}^2$ at nine base points of an elliptic pencil. Let $F_1, \dots, F_9 \subset S$ denote the nine exceptional curves. Thus S is an elliptic dP_9. In particular, specifying $\pi\colon S \to \mathbb{P}^1$ plus a fiber of π is equivalent to a pair

$(E \subset \mathbb{P}^2)$ plus nine points $P_1, \dots, P_9 \in E$ such that $P_1 + \cdots + P_9 \sim O_{\mathbb{P}^2}(3)|_E$. The elliptic pencils are given by $\pi^* O_{\mathbb{P}^1}(1) \cong \tau^* O_{\mathbb{P}^2}(3)(-F_1 - \cdots - F_9)$.

Let us now vary the points on E in a family $P_i(t) : t \in \mathbb{C}$. The line bundle giving the elliptic pencil deforms as $\tau^* O_{\mathbb{P}^2}(3)(-F_1(t) - \cdots - F_9(t))$ but the elliptic pencil deforms only if $P_1(t) + \cdots + P_9(t) \sim O_{\mathbb{P}^2}(3)|_E$ holds for every t.

Let $X \subset \mathbb{P}^2 \times \mathbb{P}^2$ be a smooth threefold of bi-degree $(1, 3)$. The first projection $\pi : X \to \mathbb{P}^2$ is an elliptic fiber space.

If $C \subset \mathbb{P}^2$ is a conic, its pre-image $X_C \to C$ is an elliptic K3 surface. If C is general then X_C is smooth.

If $C = L_1 \cup L_2$ is a pair of general lines then $X_C = S_1 \cup S_2$ is a singular K3 surface which is a union of two smooth dP_9 that intersect along a smooth elliptic curve E.

We can thus think of X_C as obtained from two pairs $(E^i \subset \mathbb{P}^2)$ $(i = 1, 2)$ with an isomorphism $\phi : E^1 \to E^2$ by blowing up nine points $P^i_j \subset E^i$ $(j = 1, \dots, 9)$ and gluing the resulting surfaces along the birational transforms of E^1 and E^2.

Let us now vary the points on both curves $P^1_i(t)$ and $P^2_i(t)$. We get two families $S_1(t), S_2(t)$ and this induces a deformation $X_C(t) = S_1(t) \cup S_2(t)$.

Although the line bundle $\pi^* O_C(1)$ giving the elliptic pencil $X_C \to C$ deforms on both of the $S_i(t)$, in general we do not get a line bundle on $X_C(t)$ unless

$$P^1_1(t) + \cdots + P^1_9(t) \sim \phi^*(P^2_1(t) + \cdots + P^2_9(t))$$

holds for every t. We can thus arrange that $\pi^* O_C(1)$ deforms along $X_C(t)$ but we lose the elliptic pencil.

10 General conjectures

A straightforward generalization of Conjecture 4 is the following (cf. [Ogu93] and [MP97, Lecture 10]):

Conjecture 50 (Strong abundance for Calabi–Yau manifolds) *Let X be a Calabi–Yau manifold and $L \in H^2(X, \mathbb{Q})$ a $(1, 1)$-class such that $(L \cdot C) \geq 0$ for every algebraic curve $C \subset X$. Then there is a unique morphism with connected fibers $g : X \to B$ onto a normal variety B and an ample $L_B \in H^2(B, \mathbb{Q})$ such that $L = g^* L_B$.*

The usual abundance conjecture assumes that L is effective, but this may not be necessary.

One expects Conjecture 50 to get harder as $\dim X - \dim B$ increases. The easiest case, when $\dim X - \dim B = 1$, corresponds to Questions 1 and 2.

From the point of view of higher-dimensional birational geometry, it is natural to consider a more general setting.

A *log Calabi–Yau fiber space* is a proper morphism with connected fibers $g\colon (X,\Delta) \to B$ onto a normal variety where (X,Δ) is klt (or possibly lc) and $(K_X+\Delta)|_{X_g} \sim_{\mathbb{Q}} 0$, where $X_g \subset X$ is a general fiber.

Let (X,Δ) be a proper klt pair such that $K_X+\Delta$ is nef and $g\colon (X,\Delta) \to B$ a relatively minimal Calabi–Yau fiber space. Let L_B be an ample $\mathbb{Q}$-divisor on B and set $L := g^*L_B$. Then $L-\epsilon(K_X+\Delta)$ is nef for $0 \le \epsilon \ll 1$. The converse fails in some rather simple cases, for instance when $X = B \times E$ for an elliptic curve E and we twist L by a degree-0 non-torsion line bundle on E.

It is natural to expect that the above are essentially the only exceptions.

Conjecture 51 *Let (X,Δ) be a proper klt pair such that $K_X+\Delta$ is nef and $H^1(X,\mathcal{O}_X)=0$. Let L be a Cartier divisor on X such that $L-\epsilon(K_X+\Delta)$ is nef for $0 \le \epsilon \ll 1$.*

Then there is a relatively minimal log Calabi–Yau fiber space structure $g\colon (X,\Delta) \to B$ and an ample $\mathbb{Q}$-divisor L_B on B such that $L \sim_{\mathbb{Q}} g^*L_B$.

If $L-\epsilon(K_X+\Delta)$ is effective then Conjecture 51 is implied by the abundance conjecture. Note also that Example 49 shows that Conjecture 51 fails if (X,Δ) is log canonical.

Conjecture 52 *Let $g_0\colon (X_0,\Delta_0) \to B_0$ be a relatively minimal log Calabi–Yau fiber space where (X_0,Δ_0) is a proper klt pair and $H^2(X_0,\mathcal{O}_{X_0})=0$.*

Let (X,Δ) be a klt pair and $h\colon (X,\Delta) \to (0 \in S)$ a flat proper morphism whose central fiber is (X_0,Δ_0).

Then, after passing to an analytic or étale neighborhood of $0 \in S$, there is a proper, flat morphism $B \to (0 \in S)$ whose central fiber is B_0 such that g_0 extends to a log Calabi–Yau fiber space $g\colon (X,\Delta) \to B$.[1]

53 Although Conjecture 52 looks much more general than Theorem 31, it seems that Abelian fibrations comprise the only unknown case.

Indeed, let X_0, B_0 be projective varieties with rational singularities and $g_0\colon X_0 \to B_0$ a morphism with connected general fiber F_0. Assume that $H^1(F_0,\mathcal{O}_{F_0})=0$. Then $R^1(g_0)_*\mathcal{O}_{X_0}$ is a torsion sheaf. On the contrary, it is reflexive by [Kol86a, 7.8]. Thus $R^1(g_0)_*\mathcal{O}_{X_0}=0$.

We could use Theorem 35, but there is an even simpler argument. Let L_{B_0} be a sufficiently ample line bundle on B_0 and set $L_0 := g_0^*L_{B_0}$. Then $H^1(X_0,L_0)=0$ by (21.1). Thus, if $h\colon X \to (0 \in S)$ is a deformation of X_0 such that L_0 lifts to a line bundle L on X then every section of L_0 lifts to a

[1] Recent work of Katzarkov, Kontsevich, and Pantev establishes this in case X_0 is smooth.

section of L (after passing to an analytic or étale neighborhood of $0 \in S$). Thus Conjecture 52 holds in this case.

Furthermore, the method of Theorem 30 suggests that the most difficult case is Abelian pencils over $\mathbb{P}^1$.

Note also that it is easy to write down examples of Abelian Calabi–Yau fiber spaces $f\colon X \to B = \mathbb{P}^1$ such that $\operatorname{Hom}_B(\Omega_B, R^1 f_* O_X) \neq 0$, thus Theorem 35 does not seem to be sufficient to prove Conjecture 52.

54 (Log elliptic fiber spaces) As before, $g\colon (X, \Delta) \to B$ is a log elliptic fiber space iff $(L^{\dim X}) = 0$ but $(L^{\dim X-1}) \neq 0$. There are three cases to consider.

(1) If $(L^{\dim X-1} \cdot \Delta) > 0$ then Riemann–Roch shows that $h^0(X, L^m)$ grows like $m^{\dim X-1}$ and we get Conjecture 51 as in Theorem 10. In this case the general fiber of g is $F \cong \mathbb{P}^1$ and $(F \cdot \Delta) = 2$.
(2) If $(L^{\dim X-1} \cdot \Delta) = 0$ but $(L^{n-2} \cdot \mathrm{td}_2(X)) > 0$ then the proof of Theorem 10 works with minor changes.
(3) The hard and unresolved case is again when $(L^{\dim X-1} \cdot \Delta) = 0$ and $(L^{n-2} \cdot \mathrm{td}_2(X)) = 0$, so $\chi(X, L^m) = O(m^{\dim X-3})$.

Acknowledgments I thank G. Di Cerbo, R. Donagi, O. Fujino, A. Langer, R. Lazarsfeld, K. Oguiso, Y.-C. Tu, and C. Xu for helpful discussions, comments and references. Partial financial support was provided by the NSF under grant number DMS-07-58275.

References

[ACG11] Arbarello, E., Cornalba, M., and Griffiths, P. A. Geometry of Algebraic Curves, Vol. II. Grundlehren der Mathematischen Wissenschaften, Vol. 268. Heidelberg: Springer-Verlag, 2011. With a contribution by J. D. Harris.

[AH11] Abramovich, D. and Hassett, B. Stable varieties with a twist. In *Classification of Algebraic Varieties*. EMS Series of Congress Reports, European Mathematical Society, Zürich, 2011, pp. 1–38.

[AK03] Araujo, C. and Kollár, J. Rational curves on varieties. *Higher Dimensional Varieties and Rational Points (Budapest, 2001)*. Bolyai Society Mathematics Studies, Vol. 12. Berlin: Springer-Verlag, 2003, pp. 13–68.

[AW98] Andreatta, M. and Wiśniewski, J. A. On contractions of smooth varieties. *J. Alg. Geom.* **7**(2) (1998) 253–312.

[BGS11] Boissière, S. Gabber, O. and Serman, O. Sur le produit de variétés localement factorielles ou Q-factorielles. ArXiv:1104.1861.

[BHPS12] Bhatt, B. Ho, W. Patakfalvi, Z. and Schnell, C. Moduli of products of stable varieties. ArXiv:1206.0438.

[BPV84] Barth, W. Peters, C. and Van de Ven, A. *Compact Complex Surfaces*. Ergebnisse der Mathematik und ihrer Grenzgebiete (3), Vol. 4. Berlin: Springer-Verlag, 1984.

[Bri10] Brion, M. *Some Basic Results on Actions of Nonaffine Algebraic Groups. Symmetry and spaces*. Progress in Mathematics, Vol. 278. Boston: Inc. Birkhäuser, 2010, pp. 1–20.

[Cat07] Catanese, F. Q.E.D. for algebraic varieties. *J. Diff. Geom.* **77** (2007) 43–75. With an appendix by S. Rollenske.

[CL10] Cogolludo-Agustin, J. I. and Libgober, A. Mordell-Weil groups of elliptic threefolds and the Alexander module of plane curves. ArXiv:1008.2018.

[Del71] Deligne, P. Théorie de Hodge, II. *Inst. Hautes Études Sci. Publ. Math.* (1971) 5–57.

[DG94] Dolgachev, I. and Gross, M. Elliptic threefolds. I. Ogg-Shafarevich theory. *J. Alg. Geom.* **3** (1994) 39–80.

[Don98] Donagi, R. Y. ICMP lecture on heterotic/F-theory duality. ArXiv:hep-th/9802093.

[Fuj78] Fujita, T. On Kähler fiber spaces over curves. *J. Math. Soc. Japan* **30**(4) (1978) 779–794.

[Fuj11] Fujino, O. On Kawamata's theorem. In *Classification of Algebraic Varieties*. EMS Series of Congress Reports, European Mathematical Society, Zürich, 2011, pp. 305–315.

[Ful98] Fulton, W. Intersection Theory, 2nd edn. Ergebnisse der Mathematik und ihrer Grenzgebiete. 3. Folge. A Series of Modern Surveys in Mathematics, Vol. 2. Berlin: Springer-Verlag, 1998.

[GHS03] Graber, T., Harris, J., and Starr, J. Families of rationally connected varieties. *J. Amer. Math. Soc.* **16** (2003) 57–67.

[GKKP11] Greb, D., Kebekus, S., Kovács, S. J., and Peternell, T. Differential forms on log canonical spaces, *Publ. Math. Inst. Hautes Études Sci.* **114** (2011) 87–169.

[Gra91] Grassi, A. On minimal models of elliptic threefolds. *Math. Ann.* **290**(2) (1991) 287–301.

[Gra93] Grassi, A. The singularities of the parameter surface of a minimal elliptic threefold. *Int. J. Math.* **4**(2) (1993) 203–230.

[Gra94] Grassi, A. On a question of J. Kollár. In *Classification of Algebraic Varieties (L'Aquila, 1992)*. Contemporary Mathematics, Vol. 162. Providence, RI: American Mathematical Society, 1994, pp. 209–214.

[Gri70] Griffiths, P. A. Periods of integrals on algebraic manifolds, III. Some global differential-geometric properties of the period mapping. *Inst. Hautes Études Sci. Publ. Math.* **38** (1970) 125–180.

[Gro62] Grothendieck, A. *Fondements de la géométrie algébrique. [Extraits du Séminaire Bourbaki, 1957–1962.*] Paris: Secrétariat mathématique, 1962.

[Gro94] Gross, M. A finiteness theorem for elliptic Calabi–Yau threefolds. *Duke Math. J.* **74**(2) (1994) 271–299.

[Gro97] Gross, M. Elliptic three-folds, II. Multiple fibres. *Trans. Amer. Math. Soc.* **349**(9) (1997) 3409–3468.

[Har77] Hartshorne, R. *Algebraic Geometry*. Graduate Texts in Mathematics, No. 52. New York: Springer-Verlag, 1977.

[Har10] Hartshorne, R. *Deformation Theory*. Graduate Texts in Mathematics, No. 257. New York: Springer-Verlag, 2010.

[HK10] Hacon, C. D. and Kovács, S. J. *Classification of Higher dimensional Algebraic Varieties*. Oberwolfach Seminars, Vol. 41. Basel: Birkhäuser, 2010.

[HK11] Hulek, K. and Kloosterman, R. Calculating the Mordell–Weil rank of elliptic threefolds and the cohomology of singular hypersurfaces. *Ann. Inst. Fourier (Grenoble)* **61**(3) (2011) 1133–1179.

[Hor76] Horikawa, E. On deformations of holomorphic maps, III. *Math. Ann.* **222**(3) (1976) 275–282.

[HX13] Hacon, C. D. and Xu, C. Existence of log canonical closures. *Invent. Math.* **192** (2013) 161–195.

[Kaw85a] Kawamata, Y. Pluricanonical systems on minimal algebraic varieties. *Invent. Math.* **79**(3) (1985) 567–588.

[Kaw85b] Kawamata, Y. Minimal models and the Kodaira dimension of algebraic fiber spaces. *J. Reine Angew. Math.* **363** (1985) 1–46.

[KL09] Kollár, J. and Larsen, M. *Quotients of Calabi–Yau Varieties*. Algebra, Arithmetic, and Geometry: In honor of Yu. I. Manin, Vol. II. Progress in Mathematics, Vol. 270. Boston: Birkhäuser, 2009, pp. 179–211.

[Klo13] Kloosterman, R. Cuspidal plane curves, syzygies and a bound on the MW-rank. *J. Algebra* **375** (2013) 216–234.

[KM98] Kollár, J. and Mori, S. *Birational Geometry of Algebraic Varieties*. Cambridge Tracts in Mathematics, Vol. 134. Cambridge: Cambridge University Press, 1998. With the collaboration of C. H. Clemens and A. Corti. Translated from the 1998 Japanese original.

[KMM92] Kollár, J. Miyaoka, Y. and Mori, S. Rationally connected varieties. *J. Alg. Geom.* **1**(3) (1992) 429–448.

[KN94] Kawamata, Y. and Namikawa, Y. Logarithmic deformations of normal crossing varieties and smoothing of degenerate Calabi–Yau varieties. *Invent. Math.* **118**(3) (1994) 395–409.

[Kod63] Kodaira, K. *On compact analytic surfaces, II, III*. *Ann. Math. (2)* **77** (1963) 563–626; **78** (1963) 1–40.

[Kol85] Kollár, J. Toward moduli of singular varieties. *Compos. Math.* **56**(3) (1985) 369–398.

[Kol86a] Kollár, J. Higher direct images of dualizing sheaves, I. *Ann. Math. (2)* **123** (1986) 11–42.

[Kol86b] Kollár, J. Higher direct images of dualizing sheaves, II. *Ann. Math. (2)* **124** (1986) 171–202.

[Kol92] Kollár, J. (ed.). Flips and abundance for algebraic threefolds. Société Mathématique de France, 1992. Papers from the Second Summer Seminar on Algebraic Geometry held at the University of Utah, Salt Lake City, UT, August 1991. Astérisque No. 211, 1992.

[Kol93] Kollár, J. Shafarevich maps and plurigenera of algebraic varieties. *Invent. Math.* **113** (1993) 177–215.

[Kol95] Kollár, J. Flatness criteria. *J. Algebra* **175**(2) (1995) 715–727.

[Kol96] Kollár, J. Rational Curves on Algebraic Varieties. Ergebnisse der Mathematik und ihrer Grenzgebiete. 3. Folge. Vol. 32. Berlin: Springer-Verlag, 1996.

[Kol10] Kollár, J. Exercises in the Birational Geometry of Algebraic Varieties. *Analytic and algebraic geometry*. IAS/Park City Mathematics Series, vol. 17. Providence, RI: American Mathematical Society, 2010, pp. 495–524.

[Kol13a] Kollár, J. Moduli of Varieties of General Type. Handbook of Moduli (G. Farkas and I. Morrison, eds), Advanced Lectures in Mathematics. Somerville, MA: International Press, 2013, pp. 131–158.

[Kol13b] Kollár, J. Singularities of the Minimal Model Program. Cambridge Tracts in Mathematics, Vol. 200, Cambridge: Cambridge University Press, 2013. With the collaboration of S. Kovács.

[KSC04] Kollár, J., Smith, K. E., and Corti, A. Rational and Nearly Rational Varieties. Cambridge Studies in Advanced Mathematics, Vol. 92. Cambridge: Cambridge University Press, 2004.

[Lai11] Lai, C.-J. Varieties fibered by good minimal models. *Math. Ann.* **350**(3) (2011) 533–547.

[Laz04] Lazarsfeld, R. Positivity in Algebraic Geometry, I–II. Ergebnisse der Mathematik und ihrer Grenzgebiete. 3. Folge. Vol. 48–49. Berlin: Springer-Verlag, 2004.

[Les12] Lesieutre, J. The diminished base locus is not always closed. ArXiv e-prints (2012).

[LM75] Lieberman, D. and Mumford, D. Matsusaka's big theorem. In *Algebraic Geometry* (Proceedings of Symposium on Pure Mathematics, Vol. 29, Humboldt State University, Arcata, CA, 1974), Providence, RI: American Mathematical Society, 1975, pp. 513–530.

[Mat68] Matsusaka, T. Algebraic deformations of polarized varieties. *Nagoya Math. J.* **31** (1968) 185–245. Corrections: **33** (1968) 137; **36** (1968) 119.

[Mat72] Matsusaka, T. Polarized varieties with a given Hilbert polynomial. *Amer. J. Math.* **94** (1972) 1027–1077.

[Miy88] Miyaoka, Y. On the Kodaira dimension of minimal threefolds. *Math. Ann.* **281**(2) (1988) 325–332.

[MM64] Matsusaka, T. and Mumford, D. Two fundamental theorems on deformations of polarized varieties. *Amer. J. Math.* **86** (1964) 668–684.

[MP97] Miyaoka, Y. and Peternell, T. Geometry of Higher-Dimensional Algebraic Varieties. DMV Seminar, Vol. 26, Basel: Birkhäuser, 1997.

[Nak88] Nakayama, N. The singularity of the canonical model of campact Kähler manifolds. *Math. Ann.* **280** (1988) 509–512.

[Nak91] Nakayama, N. Elliptic fibrations over surfaces, I. In *Algebraic Geometry and Analytic Geometry (Tokyo, 1990)*. ICM-90 Satellite Conference Proceedings. Tokyo: Springer-Verlag, 1991, pp. 126–137.

[Nak99] Nakayama, N. Projective algebraic varieties whose universal covering spaces are biholomorphic to $\mathbf{C}^n$. *J. Math. Soc. Japan* **51**(3) (1999) 643–654.

[Nak02a] Nakayama, N. Global structure of an elliptic fibration. *Publ. Res. Inst. Math. Sci.* **38**(3) (2002) 451–649.

[Nak02b] Nakayama, N. Local structure of an elliptic fibration. In *Higher Dimensional Birational Geometry (Kyoto, 1997)*. Advanced Studies in Pure Mathematics, Vol. 35. Mathematical Society of Japan, Tokyo, 2002, pp. 185–295.

[Ogu93] Oguiso, K. On algebraic fiber space structures on a Calabi–Yau 3-fold. *Int. J. Math.* **4**(3) (1993) 439–465. With an appendix by N. Nakayama.

[Rei83] Reid, M. Minimal models of canonical 3-folds. In *Algebraic Varieties and Analytic Varieties (Tokyo, 1981)*. Advanced Studies in Pure Mathematics, Vol. 1. Amsterdam: North-Holland, 1983, pp. 131–180.

[Ser01] Serre, J.-P. Espaces fibrés algèbriques. *Séminaire C. Chevalley (1958)*. Documents Mathématiques, Vol. 1. Paris: Société Mathématique de France, 2001.

[Ses63] Seshadri, C. S. Quotient space by an abelian variety. *Math. Ann.* **152** (1963) 185–194.

[Ste76] Steenbrink, J. Limits of Hodge structures. *Invent. Math.* **31**(3) (1975/76) 229–257.

[Vaf96] Vafa, C. Evidence for F-theory. *Nucl. Phys. B* **469** (1996) 403–415.

[Wil89] Wilson, P. M. H. Calabi–Yau manifolds with large Picard number. *Invent. Math.* **98** (1989) 139–155.

[Wil94] Wilson, P. M. H. The existence of elliptic fibre space structures on Calabi-Yau threefolds. *Math. Ann.* **300**(4) (1994) 693–703.

[Wil98] Wilson, P. M. H. The existence of elliptic fibre space structures on Calabi–Yau threefolds, II. *Math. Proc. Cambridge Philos. Soc.* **123**(2) (1998) 259–262.

15

Derived equivalence and non-vanishing loci II

L. Lombardi
Universität Bonn

M. Popa[a]
Northwestern University

Abstract

We prove a few cases of a conjecture on the invariance of cohomological support loci under derived equivalence by establishing a concrete connection with the related problem of the invariance of Hodge numbers. We use the main case in order to study the derived behavior of fibrations over curves.

Dedicated to Rob Lazarsfeld on the occasion of his 60th birthday, with warmth and gratitude

1 Introduction

This paper is concerned with the following conjecture made in [11] on the behavior of the non-vanishing loci for the cohomology of deformations of the canonical bundle under derived equivalence. We recall that given a smooth projective X these loci, more commonly called cohomological support loci, are the closed algebraic subsets of the Picard variety defined as

$$V^i(\omega_X) := \{\alpha \mid H^i(X, \omega_X \otimes \alpha) \neq 0\} \subseteq \mathrm{Pic}^0(X).$$

All varieties we consider are defined over the complex numbers. We denote by $\mathbf{D}(X)$ the bounded derived category of coherent sheaves $\mathbf{D}^b(\mathrm{Coh}(X))$.

Conjecture 1 ([11]) *Let X and Y be smooth projective varieties with $\mathbf{D}(X) \simeq \mathbf{D}(Y)$ as triangulated categories. Then*

$$V^i(\omega_X)_0 \simeq V^i(\omega_Y)_0 \quad \text{for all } i \geq 0,$$

[a] MP was partially supported by the NSF grant DMS-1101323.

where $V^i(\omega_X)_0$ denotes the union of the irreducible components of $V^i(\omega_X)$ passing through the origin, and similarly for Y.

We refer to [10] and [11] for a general discussion of this conjecture and its applications, and of the cases in which it has been known to hold (recovered below as well). The main point of this paper is to relate Conjecture 1 directly to part of the well-known problem of the invariance of Hodge numbers under derived equivalence; we state only the special case we need.

Conjecture 2 *Let X and Y be smooth projective varieties with $\mathbf{D}(X) \simeq \mathbf{D}(Y)$. Then*

$$h^{0,i}(X) = h^{0,i}(Y) \quad \text{for all } i \geq 0.$$

Our main result is the following:

Theorem 3 *Conjecture 2 implies Conjecture 1. More precisely, Conjecture 1 for a given i is implied by Conjecture 2 for $n - i$, where $n = \dim X$.*

This leads to a verification of Conjecture 1 in a few important cases, corresponding to the values of i for which Conjecture 2 is already known to hold.

Corollary 4 *Let X and Y be smooth projective varieties of dimension n, with $\mathbf{D}(X) \simeq \mathbf{D}(Y)$. Then*

$$V^i(\omega_X)_0 \simeq V^i(\omega_Y)_0 \quad \text{for} \quad i = 0, 1, n-1, n.$$

Proof According to Theorem 3, we need to know that derived equivalence implies the invariance of $h^{0,n}$, $h^{0,n-1}$, and $h^{0,1}$. The first two are well-known consequences of the invariance of Hochschild homology, while the last is the main result of [12]. □

A stronger result than Theorem 3 and Corollary 4, involving the dimension of cohomology groups related via the isomorphism, is in fact proved in Section 3 (see Conjecture 11 and Theorem 12). For $i = 0, 1$ this was proved in [10] by means of a twisted version of Hochschild homology. We note also that the corollary above recovers a result first proved in [10, Section 4], namely that Conjecture 1 holds for varieties of dimension up to 3. We can also conclude that it holds for an important class of irregular fourfolds.

Corollary 5 *Conjecture 1 holds in dimension up to three, and for fourfolds of maximal Albanese dimension.*

Proof The first part follows immediately from Corollary 4. For the second part, according to Corollary 4 and Theorem 3 it suffices to have $h^{0,2}(X) =$

$h^{0,2}(Y)$, which is proved for derived equivalent fourfolds of maximal Albanese dimension in [10, Corollary 1.8].[1] □

Overall, besides the unified approach, from the point of view of Conjecture 1 the key new result and applications here are in the case $i = n - 1$. By theorems of Beauville [2] and Green–Lazarsfeld [7], the cohomological support loci $V^{n-1}(\omega_X)$ are the most "geometric" among the V^i, corresponding in a quite precise way to fibrations of X over curves. This leads to the following structural application; note that while Fourier–Mukai equivalences between smooth projective surfaces are completely classified [4, 9], in higher dimension few results toward classification are available (see, e.g., [16]).

Theorem 6 *Let X and Y be smooth projective varieties with* $\mathbf{D}(X) \simeq \mathbf{D}(Y)$, *such that X admits a surjective morphism to a smooth projective curve C of genus* $g \geq 2$. *Then:*

(i) Y admits a surjective morphism to a curve of genus $\geq g$.
(ii) If X has a Fano fibration structure over C, then so does Y, and X and Y are K-equivalent.[2] *In particular, if X is a Mori fiber space over C, then X and Y are isomorphic.*

A slightly stronger statement is given in Theorem 14. We remark that it is known from results of Beauville and Siu that X admits a surjective morphism to a curve of genus $\geq g$ if and only if $\pi_1(X)$ has a surjective homomorphism onto Γ_g, the fundamental group of a Riemann surface of genus g (see the Appendix to [6]). On the other hand, it is also known that derived equivalent varieties do not necessarily have isomorphic fundamental groups (see [1, 14]), so this would not suffice in order to deduce Theorem 6 (*i*). A more precise version of (*i*) can be found in the Remark on p. 302; see also Question 13. The refinement we give in (*ii*) in the case of Fano fibrations answers a question posed to us by Y. Kawamata; for this, the method of proof is completely independent of the study of $V^i(\omega_X)$, relying instead on Kawamata's kernel technique [9] and on the structure of the Albanese map for varieties with nef anticanonical bundle [17]. The result, however, fits naturally in the present context.

Going back to the main results, the isomorphism between the V^i_0 is realized, as in [10], via the Rouquier isomorphism associated to a Fourier–Mukai equivalence (see Section 2). To relate this to the behavior of Hodge numbers of type $h^{0,i}$ as in Theorem 3, the main new ingredients are Simpson's result describing

[1] Note that the same holds for fourfolds of non-negative Kodaira dimension whose Albanese image has dimension 3, and for those with non-affine $\mathrm{Aut}^0(X)$.

[2] Recall that this means there exist a smooth projective Z and birational morphisms $f: Z \to X$ and $g: Z \to Y$ such that $f^*\omega_X \simeq g^*\omega_Y$.

the components of all $V^i(\omega_X)$ as torsion translates of abelian subvarieties of $\mathrm{Pic}^0(X)$, used via a density argument involving torsion points of special prime order, and the comparison of the derived categories of cyclic covers associated to torsion line bundles mapped to each other via the Rouquier isomorphism, modeled after and slightly extending results of Bridgeland–Maciocia [3] on equivalences of canonical covers.

2 Derived equivalences of cyclic covers

Cyclic covers. Let X be a complex smooth projective variety and α be a d-torsion element of $\mathrm{Pic}^0(X)$. We denote by

$$\pi_\alpha : X_\alpha \to X$$

the étale cyclic cover of order d associated to α (see, e.g., [8, Section 7.3]). Then

$$\pi_{\alpha*}\mathcal{O}_{X_\alpha} \simeq \bigoplus_{i=0}^{d-1} \alpha^{-i} \tag{1}$$

and there is a free action of the group $G := \mathbf{Z}/d\mathbf{Z}$ on X_α such that $X_\alpha/G \simeq X$. The following lemma is analogous to [3, Proposition 2.5(b)]. We include a proof for completeness, entirely inspired by the approach in [3, Proposition 2.5(a)].

Lemma 7 *Let E be an object of $\mathbf{D}(X)$. There is an object E_α in $\mathbf{D}(X_\alpha)$ such that $\pi_{\alpha*}E_\alpha \simeq E$ if and only if $E \otimes \alpha \simeq E$.*

Proof For the nontrivial implication, let

$$s : E \xrightarrow{\simeq} E \otimes \alpha$$

be an isomorphism. We proceed by induction on the number r of nonzero cohomology sheaves of E. If E is a sheaf concentrated in degree zero, then the lemma is a standard fact. Indeed, it is well known that

$$\pi_{\alpha*} : \mathrm{Coh}(X_\alpha) \to \mathrm{Coh}(\mathcal{A})$$

is an equivalence between the category of coherent $\mathcal{O}_{X_\alpha}$-modules and the category of coherent $\mathcal{A} := (\bigoplus_{i=0}^{d-1} \alpha^i)$-algebras, while a coherent sheaf E on X belongs to $\mathrm{Coh}(\mathcal{A})$ if and only if $E \otimes \alpha \simeq E$.

Suppose now that the lemma is true for all objects having at most $r-1$ nonzero cohomology sheaves, and consider an object E with r nonzero cohomology sheaves. By shifting E, we can assume that $\mathcal{H}^i(E) = 0$ for

$i \notin [-(r-1), 0]$. Since $E \otimes \alpha \simeq E$, we also have $\mathcal{H}^0(E) \otimes \alpha \simeq \mathcal{H}^0(E)$. Therefore, by the above, there exists a coherent sheaf M_α on X_α such that $\pi_{\alpha *} M_\alpha \simeq \mathcal{H}^0(E)$. Now the natural morphism $E \xrightarrow{j} \mathcal{H}^0(E)$ induces a distinguished triangle

$$E \xrightarrow{j} \mathcal{H}^0(E) \xrightarrow{f} F \to E[1]$$

such that the object F has $r-1$ nonzero cohomology sheaves. By the commutativity of the following diagram:

$$\begin{array}{ccccccc} E & \xrightarrow{j} & \mathcal{H}^0(E) & \xrightarrow{f} & F & \longrightarrow & E[1] \\ \downarrow & & \downarrow & & & & \downarrow \\ E \otimes \alpha & \xrightarrow{j \otimes \alpha} & \mathcal{H}^0(E) \otimes \alpha & \xrightarrow{f \otimes \alpha} & F \otimes \alpha & \longrightarrow & (E \otimes \alpha)[1] \end{array}$$

we obtain an isomorphism $F \simeq F \otimes \alpha$, and therefore by induction an object F_α in $\mathbf{D}(X_\alpha)$ such that $\pi_{\alpha *} F_\alpha \simeq F$.

To show the existence of an object E_α in $\mathbf{D}(X_\alpha)$ such that $\pi_{\alpha *} E_\alpha \simeq E$, we assume for a moment that there exists a morphism $f_\alpha \colon M_\alpha \to F_\alpha$ such that $\pi_{\alpha *} f_\alpha = f$. This is enough to conclude, since by completing f_α to a distinguished triangle

$$M_\alpha \xrightarrow{f_\alpha} F_\alpha \to E_\alpha[1] \to M_\alpha[1]$$

and applying $\pi_{\alpha *}$, we obtain $\pi_{\alpha *} E_\alpha \simeq E$.

We are left with showing the existence of f_α. Let $\lambda_\alpha \colon \pi_{\alpha *} M_\alpha \to \pi_{\alpha *} M_\alpha \otimes \alpha$ and $\mu_\alpha \colon \pi_{\alpha *} F_\alpha \to \pi_{\alpha *} F_\alpha \otimes \alpha$ be the isomorphisms determined by the diagram above. Note that

$$\mu_\alpha \circ f = (f \otimes \alpha) \circ \lambda_\alpha \quad \text{in} \quad \mathbf{D}(X). \tag{2}$$

We can replace F_α by an injective resolution

$$\cdots \to \mathcal{I}_\alpha^{-1} \xrightarrow{d^{-1}} \mathcal{I}_\alpha^{0} \xrightarrow{d^{0}} \mathcal{I}_\alpha^{1} \xrightarrow{d^{1}} \cdots,$$

so that f is represented (up to homotopy) by a morphism of $\mathcal{O}_X$-modules

$$u \colon \pi_{\alpha *} M_\alpha \to \pi_{\alpha *} \mathcal{I}_\alpha^0.$$

Let V be the image of the map

$$\mathrm{Hom}(\pi_{\alpha *} M_\alpha, -) \colon \mathrm{Hom}(\pi_{\alpha *} M_\alpha, \pi_{\alpha *} \mathcal{I}_\alpha^{-1}) \to \mathrm{Hom}(\pi_{\alpha *} M_\alpha, \pi_{\alpha *} \mathcal{I}_\alpha^{0}).$$

By (2), we have isomorphisms of $\mathcal{O}_X$-modules $a_1 \colon \pi_{\alpha *} M_\alpha \to \pi_{\alpha *} M_\alpha \otimes \alpha$ and $b_1 \colon \pi_{\alpha *} \mathcal{I}_\alpha^0 \to \pi_{\alpha *} \mathcal{I}_\alpha^0 \otimes \alpha$ such that

$$b_1 \circ u = (u \otimes \alpha) \circ a_1 \qquad \text{up to homotopy.} \tag{3}$$

By setting $a_i := (a_1 \otimes \alpha^{i-1}) \circ \cdots \circ (a_1 \otimes \alpha) \circ a_1$ (for $i \geq 2$) and similarly for b_i, we define an action of $G := \mathbf{Z}/d\mathbf{Z}$ on V as

$$g^i \cdot (-) := b_i^{-1} \circ (- \otimes \alpha^i) \circ a_i,$$

where g is a generator of G. Moreover, we define operators A and B on V as

$$A := \sum_{i=0}^{d-1} g^i \cdot (-), \qquad B := 1 - g \cdot (-).$$

Since $AB = 0$, we note that $\operatorname{Ker} A = \operatorname{Im} B$.

By (3), we have that $B(u) = u - b_1^{-1} \circ (u \otimes \alpha) \circ a_1$ is null-homotopic, and therefore $B(u) \in V$. Since $\operatorname{Ker} A = \operatorname{Im} B$, there exists a morphism $\eta \in V$ such that $B(\eta) = B(u)$. Consider the morphism $t := u - \eta \in \operatorname{Hom}(\pi_{\alpha *} M_\alpha, \pi_{\alpha *} \mathcal{I}_\alpha^0)$. It is easy to check that t is homotopic to u and therefore it represents f as well. But now $B(t) = 0$, so $t = \pi_{\alpha *}(v)$ for some morphism $v \colon M_\alpha \to \mathcal{I}_\alpha^0$, which concludes the proof. □

Rouquier isomorphism. It is well known by Orlov's criterion that every equivalence $\Phi \colon \mathbf{D}(X) \to \mathbf{D}(Y)$ is of Fourier–Mukai type, i.e., induced by an object $\mathcal{E} \in \mathbf{D}(X \times Y)$, unique up to isomorphism, via

$$\Phi = \Phi_{\mathcal{E}} \colon \mathbf{D}(X) \to \mathbf{D}(Y), \quad \Phi_{\mathcal{E}}(-) = \mathbf{R}p_{Y*}(p_X^*(-) \overset{\mathbf{L}}{\otimes} \mathcal{E}).$$

For every such equivalence, Rouquier [13, Théorème 4.18] showed that there is an induced isomorphism of algebraic groups

$$F_{\mathcal{E}} \colon \operatorname{Aut}^0(X) \times \operatorname{Pic}^0(X) \to \operatorname{Aut}^0(Y) \times \operatorname{Pic}^0(Y)$$

which usually mixes the two factors. A concrete formula for $F_{\mathcal{E}}$ was worked out in [12, Lemma 3.1], namely

$$F_{\mathcal{E}}(\varphi, \alpha) = (\psi, \beta) \iff p_X^* \alpha \otimes (\varphi \times \mathrm{id}_Y)^* \mathcal{E} \simeq p_Y^* \beta \otimes (\mathrm{id}_X \times \psi)_* \mathcal{E}. \tag{4}$$

Derived equivalences of cyclic covers. Before stating the main theorem of this section, we recall two definitions from [3] (see also [8, Section 7.3]). Let $\widetilde{X}$ and $\widetilde{Y}$ be two smooth projective varieties on which the group $G := \mathbf{Z}/d\mathbf{Z}$ acts freely. Denote by $\pi_X \colon \widetilde{X} \to X$ and $\pi_Y \colon \widetilde{Y} \to Y$ the quotient maps of $\widetilde{X}$ and $\widetilde{Y}$ respectively.

Definition 8 A functor $\widetilde{\Phi} \colon \mathbf{D}(\widetilde{X}) \to \mathbf{D}(\widetilde{Y})$ is *equivariant* if there exist an automorphism μ of G and isomorphisms of functors

$$g^* \circ \widetilde{\Phi} \simeq \widetilde{\Phi} \circ \mu(g)^* \quad \text{for all} \quad g \in G.$$

Definition 9 Let $\Phi\colon \mathbf{D}(X) \to \mathbf{D}(Y)$ be a functor. A *lift* of Φ is a functor $\widetilde{\Phi}\colon \mathbf{D}(\widetilde{X}) \to \mathbf{D}(\widetilde{Y})$ inducing isomorphisms

$$\pi_{Y*} \circ \widetilde{\Phi} \simeq \Phi \circ \pi_{X*}, \tag{5}$$

$$\pi_Y^* \circ \Phi \simeq \widetilde{\Phi} \circ \pi_X^*. \tag{6}$$

Remark If $\Phi\colon \mathbf{D}(X) \to \mathbf{D}(Y)$ and $\widetilde{\Phi}\colon \mathbf{D}(\widetilde{X}) \to \mathbf{D}(\widetilde{Y})$ are equivalences, then by taking the adjoints (5) holds if and only if (6) holds.

Now we are ready to prove the main result of this section. It is a slight extension of the result of [3] on canonical covers, whose proof almost entirely follows the one given there, and which serves as a technical tool for our main theorem.

Theorem 10 *Let X and Y be smooth projective varieties, and $\alpha \in \mathrm{Pic}^0(X)$ and $\beta \in \mathrm{Pic}^0(Y)$ d-torsion elements. Denote by $\pi_\alpha\colon X_\alpha \to X$ and $\pi_\beta\colon Y_\beta \to Y$ the cyclic covers associated to α and β respectively.*

(i) Suppose that $\Phi_{\mathcal{E}}\colon \mathbf{D}(X) \to \mathbf{D}(Y)$ is an equivalence, and that $F_{\mathcal{E}}(\mathrm{id}_X, \alpha) = (\mathrm{id}_Y, \beta)$. Then there exists an equivariant equivalence $\Phi_{\widetilde{\mathcal{E}}}\colon \mathbf{D}(X_\alpha) \to \mathbf{D}(Y_\beta)$ lifting $\Phi_{\mathcal{E}}$.

(ii) Suppose that $\Phi_{\widetilde{\mathcal{F}}}\colon \mathbf{D}(X_\alpha) \to \mathbf{D}(Y_\beta)$ is an equivariant equivalence. Then $\Phi_{\widetilde{\mathcal{F}}}$ is the lift of some equivalence $\Phi_{\mathcal{F}}\colon \mathbf{D}(X) \to \mathbf{D}(Y)$.

Proof To see (*i*), consider the following commutative diagram, where p_1, p_2, r_1 and r_2 are projection maps:

$$\begin{array}{ccccc}
X_\alpha & \xleftarrow{\ r_1\ } & X_\alpha \times Y & \xrightarrow{\ r_2\ } & Y \\
\downarrow{\scriptstyle \pi_\alpha} & & \downarrow{\scriptstyle \pi_\alpha \times \mathrm{id}_Y} & & \| \\
X & \xleftarrow{\ p_1\ } & X \times Y & \xrightarrow{\ p_2\ } & Y
\end{array}$$

By (4), the condition $F_{\mathcal{E}}(\mathrm{id}_X, \alpha) = (\mathrm{id}_Y, \beta)$ is equivalent to the isomorphism in $\mathbf{D}(X \times Y)$:

$$p_1^*\alpha \otimes \mathcal{E} \simeq p_2^*\beta \otimes \mathcal{E}. \tag{7}$$

Pulling (7) back via the map $(\pi_\alpha \times \mathrm{id}_Y)$, we get an isomorphism

$$(\pi_\alpha \times \mathrm{id}_Y)^*\mathcal{E} \simeq r_2^*\beta \otimes (\pi_\alpha \times \mathrm{id}_Y)^*\mathcal{E}$$

as $\pi_\alpha^*\alpha \simeq \mathcal{O}_{X_\alpha}$. As the map $(\mathrm{id}_{X_\alpha} \times \pi_\beta)\colon X_\alpha \times Y_\beta \to X_\alpha \times Y$ is the étale cyclic cover associated to the line bundle $r_2^*\beta$, by Lemma 7 there exists an object $\widetilde{\mathcal{E}}$ such that

$$(\mathrm{id}_{X_\alpha} \times \pi_\beta)_*\widetilde{\mathcal{E}} \simeq (\pi_\alpha \times \mathrm{id}_Y)^*\mathcal{E}.$$

By [3, Lemma 4.4], there is an isomorphism

$$\pi_{\beta*} \circ \Phi_{\widetilde{\mathcal{E}}} \simeq \Phi_{\mathcal{E}} \circ \pi_{\alpha*}. \tag{8}$$

We now show that $\Phi_{\widetilde{\mathcal{E}}}$ is an equivalence. Let $\Psi_{\mathcal{E}'} : \mathbf{D}(Y) \to \mathbf{D}(X)$ be a quasi-inverse of $\Phi_{\mathcal{E}}$. Since $F_{\mathcal{E}'} = F_{\mathcal{E}}^{-1}$, we have that $F_{\mathcal{E}'}(\mathrm{id}_X, \beta) = (\mathrm{id}_Y, \alpha)$. By repeating the previous argument, one then sees that there exists an object $\widetilde{\mathcal{E}'}$ such that

$$(\pi_\alpha \times \mathrm{id}_{Y_\beta})_* \widetilde{\mathcal{E}'} \simeq (\mathrm{id}_X \times \pi_\beta)^* \mathcal{E}'$$

and an isomorphism of functors

$$\pi_{\alpha*} \circ \Psi_{\widetilde{\mathcal{E}'}} \simeq \Psi_{\mathcal{E}'} \circ \pi_{\beta*}. \tag{9}$$

Since $\Psi_{\mathcal{E}'} \circ \Phi_{\mathcal{E}} \simeq \mathrm{id}_{\mathbf{D}(X)}$, using (8) and (9) we get an isomorphism

$$\pi_{\alpha*} \circ \Psi_{\widetilde{\mathcal{E}'}} \circ \Phi_{\widetilde{\mathcal{E}}} \simeq \Psi_{\mathcal{E}'} \circ \pi_{\beta*} \circ \Phi_{\widetilde{\mathcal{E}}} \simeq \Psi_{\mathcal{E}'} \circ \Phi_{\mathcal{E}} \circ \pi_{\alpha*} \simeq \pi_{\alpha*}. \tag{10}$$

Hence, following the proof of [3, Lemma 4.3], we have that $\Psi_{\widetilde{\mathcal{E}'}} \circ \Phi_{\widetilde{\mathcal{E}}} \simeq g_*(L \otimes -)$ for some $g \in G$ and $L \in \mathrm{Pic}(X_\alpha)$. By taking left adjoints in (10), we obtain on the other hand that

$$(L^{-1} \otimes -) \circ g^* \circ \pi_\alpha^* \simeq \pi_\alpha^*,$$

which applied to $\mathcal{O}_X$ yields $L \simeq \mathcal{O}_{X_\alpha}$. This gives $\Psi_{\widetilde{\mathcal{E}'}} \circ \Phi_{\widetilde{\mathcal{E}}} \simeq g_*$. Similarly, we can show that $\Phi_{\widetilde{\mathcal{E}}} \circ \Psi_{\widetilde{\mathcal{E}'}} \simeq h_*$ for some $h \in G$, and hence that $g^* \circ \Psi_{\widetilde{\mathcal{E}'}}$, or equivalently $\Psi_{\widetilde{\mathcal{E}'}} \circ h^*$, is a quasi-inverse of $\Phi_{\widetilde{\mathcal{E}}}$. Finally, Remark 2 implies that $\Phi_{\widetilde{\mathcal{E}}}$ is a lift of $\Phi_{\mathcal{E}}$.

The proofs of the fact that $\Phi_{\widetilde{\mathcal{E}}}$ is equivariant and of *(ii)* are now completely analogous to those of the corresponding statements in [3, Theorem 4.5]. □

3 Comparison of cohomological support loci

A more precise statement than that of Conjecture 1 naturally involves the dimension of the cohomology groups of the line bundles mapped to each other via the Rouquier isomorphism. Such a statement was proved in [10] when $i = 0, 1$. The concrete statement, which also contains Conjecture 2 by specializing at the origin, is the following:

Conjecture 11 *Let X and Y be smooth projective varieties of dimension n, and let $\Phi_{\mathcal{E}}$ be a Fourier–Mukai equivalence between $\mathbf{D}(X)$ and $\mathbf{D}(Y)$. If $F = F_{\mathcal{E}}$ is the induced Rouquier isomorphism, then*

$$F\left(\mathrm{id}_X, V^i(\omega_X)_0\right) = \left(\mathrm{id}_Y, V^i(\omega_Y)_0\right)$$

for all i, so that $V^i(\omega_X)_0 \simeq V^i(\omega_Y)_0$. *Moreover if* $\alpha \in V^i(\omega_X)_0$ *and* $F(\mathrm{id}_X, \alpha) = (\mathrm{id}_Y, \beta)$, *then*

$$h^i(X, \omega_X \otimes \alpha) = h^i(Y, \omega_Y \otimes \beta).$$

In order to address this statement, we consider for each $m \geq 1$ the more refined cohomological support loci

$$V^i_m(\omega_X) := \{\alpha \in \mathrm{Pic}^0(X) \mid h^i(X, \omega_X \otimes \alpha) \geq m\}.$$

In this notation, $V^i(\omega_X)$ becomes $V^i_1(\omega_X)$. The following result is a strengthening of Theorem 3 in the Introduction.[3]

Theorem 12 *Conjecture 2 is equivalent to Conjecture 11. Specifically, if* $n = \dim X$, *Conjecture 2 for* $n - i$ *implies*

$$F\left(\mathrm{id}_X, V^i_m(\omega_X)_0\right) = \left(\mathrm{id}_Y, V^i_m(\omega_Y)_0\right)$$

for all $m \geq 1$.

Proof Note first that F induces an isomorphism on the locus of line bundles $\alpha \in \mathrm{Pic}^0(X)$ with the property that $F(\mathrm{id}_X, \alpha) = (\mathrm{id}_Y, \beta)$ for some $\beta \in \mathrm{Pic}^0(Y)$, so indeed the first assertion in the statement follows from the second one, which we prove in a few steps.

Step 1. We first show that if $\alpha \in V^i(\omega_X)_0$, then it does satisfy the property above, namely there exists $\beta \in \mathrm{Pic}^0(Y)$ such that

$$F(\mathrm{id}_X, \alpha) = (\mathrm{id}_Y, \beta).$$

A more general statement has already been proved in [10, Theorem 3.2]. We extract the argument we need here in order to keep the proof self-contained, following [12, Section 3] as well. The Rouquier isomorphism F induces a morphism

$$\pi\colon \mathrm{Pic}^0(Y) \to \mathrm{Aut}^0(X), \quad \pi(\beta) = p_1(F^{-1}(\mathrm{id}_Y, \beta)),$$

whose image is an abelian variety A and where p_1 is the projection from $\mathrm{Aut}^0(X) \times \mathrm{Pic}^0(X)$ onto the first factor. If A is trivial there is nothing to prove, so we can assume that A is positive-dimensional.

As A is an abelian variety of automorphisms of X, according to [5, Section 3] there exists a finite subgroup $H \subset A$ and an étale locally trivial fibration

[3] We are grateful to the referee, who suggested that our original argument for $m = 1$ applies in fact to all m. This was crucial for deducing this more refined statement, as opposed to just that of Conjecture 1, from Conjecture 2.

$p\colon X \to A/H$ which is trivialized by base change to A. In other words, there is a cartesian diagram

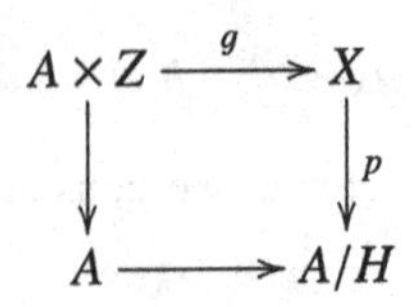

where $Z = p^{-1}(0)$. Restricting g to the fiber $A \times \{z_0\}$, where z_0 is an arbitrary point in Z, we obtain a morphism $f\colon A \to X$, which is in fact an orbit of the action of A on X. It is shown in the proof of [12, Theorem A] that $\mathrm{Ker}(f^*)^0 \simeq \mathrm{Ker}(\pi)^0$, where $(\cdot)^0$ denotes the connected component of the identity; this is based on a theorem of Matsumura–Nishi, essentially saying that the induced $f^*\colon \mathrm{Pic}^0(X) \to \mathrm{Pic}^0(A)$ is surjective. Consequently, we only need to show that $\alpha \in \mathrm{Ker}(f^*)$; it will then automatically be in $\mathrm{Ker}(f^*)^0$, since it lives in $V^i(\omega_X)_0$, which is a union of abelian subvarieties.

To this end, note that $\alpha \in V^i(\omega_X)_0$ implies that

$$H^i\left(A \times Z, g^*(\omega_X \otimes \alpha)\right) \simeq H^i\left(A \times Z, f^*\alpha \boxtimes (\omega_Z \otimes \alpha_{|Z})\right) \neq 0.$$

Applying the Künneth formula, we conclude that we must have

$$H^k(A, f^*\alpha) \neq 0 \quad \text{for some } 0 \leq k \leq i,$$

which implies that $f^*\alpha \simeq \mathcal{O}_A$.

Step 2. Since one can repeat the argument in Step 1 for F^{-1}, it is enough to show that if $\alpha \in V^i_m(\omega_X)_0$ and $F(\mathrm{id}_X, \alpha) = (\mathrm{id}_Y, \beta)$, then $\beta \in V^i_m(\omega_Y)_0$ as well. In this step we show that it is enough to prove this assertion in the case when $\alpha \in \mathrm{Pic}^0(X)$ is a torsion point of (special) prime order. First, since F is a group isomorphism, α is torsion of some order if and only if β is torsion of the same order.

According to a well-known theorem of Simpson [15], every irreducible component Z of $V^i_m(\omega_Y)$ is a torsion translate $\tau_Z + A_Z$ of an abelian subvariety of $\mathrm{Pic}^0(Y)$. We consider the set P_i of all prime numbers that do not divide $\mathrm{ord}(\tau_Z)$ for any such component Z. As $V^i_m(\omega_Y)$ is an algebraic set by the semicontinuity theorem, we are only throwing away a finite set of primes. We will show that it is enough to prove the assertion above when α is torsion with order in P_i. First note that it is a standard fact that torsion points of prime order are Zariski dense in a complex abelian variety.[4] Consequently, torsion points with order in the set P_i are dense as well.

[4] This follows for instance from the fact that real numbers can be approximated by rational numbers with prime denominators.

Now let W be a component of $V^i_m(\omega_X)_0$. It suffices to show that

$$Z := p_2\,(F(\mathrm{id}_X, W)) \subset V^i_m(\omega_Y)_0,$$

where p_2 is the projection onto the second component of $\mathrm{Aut}^0(Y) \times \mathrm{Pic}^0(Y)$. Indeed, since one can repeat the same argument for the inverse homomorphism F^{-1}, this implies that Z has to be a component of $V^i_m(\omega_Y)_0$, isomorphic to W via F. Now Z is an abelian variety, and therefore by the discussion above torsion points β of order in P_i are dense in Z. By semicontinuity, it suffices to show that $\beta \in V^i_m(\omega_Y)_0$. These βs are precisely the images of $\alpha \in W$ of order in P_i, which concludes our reduction step.

Step 3. Now let $\alpha \in V^i_m(\omega_X)_0$ be a torsion point of order belonging to the set P_i, and $F(\mathrm{id}_X, \alpha) = (\mathrm{id}_Y, \beta)$. Denote

$$p = \mathrm{ord}(\alpha) = \mathrm{ord}(\beta).$$

Consider the cyclic covers $\pi_\alpha \colon X_\alpha \to X$ and $\pi_\beta \colon Y_\beta \to Y$ associated to α and β respectively. We can apply Theorem 10 to conclude that there exists a Fourier–Mukai equivalence

$$\Phi_{\widetilde{\mathcal{E}}} \colon \mathbf{D}(X_\alpha) \to \mathbf{D}(Y_\beta)$$

lifting $\Phi_{\mathcal{E}}$. Assuming Conjecture 2, we have in particular that

$$h^{0,n-i}(X) = h^{0,n-i}(Y) \quad \text{and} \quad h^{0,n-i}(X_\alpha) = h^{0,n-i}(Y_\beta).$$

On the other hand, using (1), we have

$$H^{n-i}(X_\alpha, \mathcal{O}_{X_\alpha}) \simeq \bigoplus_{j=0}^{p-1} H^{n-i}(X, \alpha^{-j}) \quad \text{and}$$

$$H^{n-i}(Y_\beta, \mathcal{O}_{Y_\beta}) \simeq \bigoplus_{j=0}^{p-1} H^{n-i}(Y, \beta^{-j}).$$

The terms on the left hand side and the terms corresponding to $j = 0$ on the right hand side have the same dimension. On the other hand, since every component of $V^i_m(\omega_X)_0$ is an abelian subvariety of $\mathrm{Pic}^0(X)$, we have that $\alpha^j \in V^i_m(\omega_X)_0$ for all j, so

$$h^{n-i}(X, \alpha^{-j}) \geq m \quad \text{for all } j.$$

We conclude that

$$h^{n-i}(Y, \beta^{-k}) \geq m \quad \text{for some} \quad 1 \leq k \leq p-1.$$

This says that $\beta^k \in V^i_m(\omega_Y)$. We claim that in fact $\beta^k \in V^i_m(\omega_Y)_0$. Assuming that this is the case, we can conclude the argument. Indeed, pick a component

$T \subset V^i_m(\omega_Y)_0$ such that $\beta^k \in T$. But β^k generates the cyclic group of prime order $\{1, \beta, \ldots, \beta^{p-1}\}$, so $\beta \in T$ as well, since T is an abelian variety.

We are left with proving that $\beta^k \in V^i_m(\omega_Y)_0$. Pick any component S in $V^i_m(\omega_Y)$ containing β^k. By the Simpson theorem mentioned above, we have that $S = \tau + B$, where τ is a torsion point and B is an abelian subvariety of $\mathrm{Pic}^0(Y)$. We claim that we must have $\tau \in B$, so that $S = B$, confirming our statement.[5] To this end, switching abusively to additive notation, say $k\beta = \tau + b$ with $b \in B$, and denote the torsion order of τ by r. Since the order p of β is assumed to be in the set P_i, we have that r and p are coprime. Now on the one hand $r\tau = 0 \in B$, while on the other hand $p\tau + pb = kp\beta = 0$, so $p\tau \in B$ as well. Since r and p are coprime, one easily concludes that $\tau \in B$. □

4 Fibrations over curves

Fibration structure via derived equivalence. We now apply the derived invariance of $V^{n-1}(\omega_X)_0$ to deduce Theorem 6 (*i*) in the Introduction.

Proof of Theorem 6 (i) Let $f : X \to C$ be a surjective morphism onto a smooth projective curve of genus $g \geq 2$. Using Stein factorization, we can assume that f has connected fibers. We have that $f^* \mathrm{Pic}^0(C) \subset V^{n-1}(\omega_X)_0$. Since by Corollary 4 we have $V^{n-1}(\omega_X)_0 \simeq V^{n-1}(\omega_Y)_0$, there exists a component T of $V^{n-1}(\omega_Y)_0$ of dimension at least g. By [2, Corollaire 2.3], there exists a smooth projective curve D and a surjective morphism with connected fibers $g : Y \to D$ such that $T = g^* \mathrm{Pic}^0(D)$. Note that $g(D) = \dim T \geq g$. □

Remark The discussion above shows in fact the following more refined statement. For a smooth projective variety Z, define

$$A_Z := \{g \in \mathbf{N} \mid g = \dim T \text{ for some irreducible component } T \subset V^{n-1}(\omega_Z)_0\}.$$

Then if $\mathbf{D}(X) \simeq \mathbf{D}(Y)$, we have $A_X = A_Y$. Denoting this set by A, for each $g \in A$ both X and Y have surjective maps onto curves of genus g. The maximal genus of a curve admitting a surjective map from X (or Y) is $\max(A)$.

Question 13 If $\mathbf{D}(X) \simeq \mathbf{D}(Y)$, is the set of curves of genus at least 2 admitting non-constant maps from X the same as that for Y? Or at least the set of curves corresponding to irreducible components of $V^{n-1}(\omega_X)_0$?

Fano fibrations. The following is a slightly more precise version of Theorem 6 (*ii*) in the Introduction.

[5] Note that in fact we are proving something stronger: β^k belongs *only* to components of $V^i_m(\omega_Y)$ passing through the origin.

Theorem 14 *Let X and Y be smooth projective complex varieties such that $\mathbf{D}(X) \simeq \mathbf{D}(Y)$. Assume that there is an algebraic fiber space $f : X \to C$ such that C is a smooth projective curve of genus at least 2 and the general fiber of f is Fano. Then:*

(i) X and Y are K-equivalent.
(ii) There is an algebraic fiber space $g : Y \to C$ such that for $c \in C$ where the fibers X_c and Y_c are smooth, with X_c Fano, one has $Y_c \simeq X_c$.
(iii) If ω_X^{-1} is f-ample (e.g., if f is a Mori fiber space), then $X \simeq Y$.

Proof Let p and q be the projections of $X \times Y$ onto the first and second factor respectively. Consider the unique-up-to-isomorphism $\mathcal{E} \in \mathbf{D}(X \times Y)$ such that the given equivalence is the Fourier–Mukai functor $\Phi_{\mathcal{E}}$. Then by [8, Corollary 6.5], there exists a component Z of $\mathrm{Supp}(\mathcal{E})$ such that $p_{|Z} : Z \to X$ is surjective. We first claim that $\dim Z = \dim X$.

Assuming by contradiction that $\dim Z > \dim X$, we show that ω_X^{-1} is nef. We denote by F the general fiber of f, which is Fano. We also define $Z_F := p_Z^{-1}(F) \subset Z$, while $q_F : Z_F \to Y$ is the projection obtained by restricting q to Z_F. Since ω_F^{-1} is ample, we obtain that q_F is finite onto its image; see [8, Corollary 6.8]. On the other hand, the assumption that $\dim Z > \dim X$ implies that $\dim Z_F \geq \dim X = \dim Y$, so q_F must be surjective (and consequently $\dim Z_F = \dim X$).

By passing to its normalization if necessary, we can assume without loss of generality that Z_F is normal. Denoting by p_F the projection of Z_F to X, by [8, Corollary 6.9] we have that there exists $r > 0$ such that

$$p_F^* \, \omega_X^{-r} \simeq q_F^* \, \omega_Y^{-r}.$$

Now since p_F factors through F and ω_F^{-1} is ample, we have that $p_F^* \omega_X^{-1}$ is nef, hence by the isomorphism above so is $q_F^* \, \omega_Y^{-1}$. Finally, since q_F is finite and surjective, we obtain that ω_Y^{-1} is nef, so by [9, Theorem 1.4], ω_X^{-1} is nef as well.

We can now conclude the proof of the claim using the main result of Zhang [17] (part of a conjecture of Demailly–Peternell–Schneider), saying that a smooth projective variety with nef anticanonical bundle has surjective Albanese map. In our case, since the general fiber of f is Fano, the Albanese map of X is obtained by composing f with the Abel–Jacobi embedding of C. But this implies that C has genus at most 1, a contradiction. The claim is proved, so

$$\dim Z = \dim X = \dim Y.$$

At this stage, the K-equivalence statement follows from Lemma 15 below.

For statements (*ii*) and (*iii*) we emphasize that, once we know that X and Y are K-equivalent, the argument is standard and independent of derived equivalence.[6] Note first that smooth birational varieties have the same Albanese variety and Albanese image. Since f is the Albanese map of X, it follows that the Albanese map of Y is a surjective morphism $g\colon Y \to C$. Furthermore, C is the Albanese image of any other birational model as well, hence any smooth model Z inducing a K-equivalence between X and Y sits in a commutative diagram

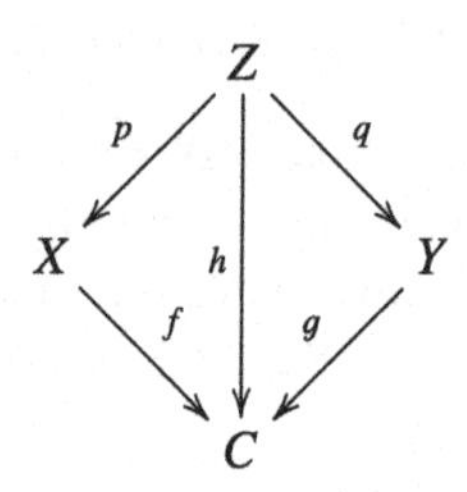

Note that in particular g has connected fibers since f does.

For a point $c \in C$, denote by X_c, Y_c, Z_c the fibers of f, g, h over c. By adjunction, Z_c realizes a K-equivalence between X_c and Y_c. First, assuming that c is chosen such that X_c and Y_c are smooth, with X_c Fano, we show that $X_c \simeq Y_c$.

To this end, if we assume that the induced rational map $\varphi_c\colon Y_c \to X_c$ is not a morphism, there must be a curve $B \subset Z_c$ which is contracted by q_c but not by p_c. Then $q_c^*\omega_{Y_c} \cdot B = 0$, and so $p_c^*\omega_{X_c} \cdot B = 0$ as well. On the other hand $\omega_{X_c}^{-1} \cdot p_c(B) < 0$, which is a contradiction. Therefore we obtain that φ_c is a birational morphism with the property that $\varphi_c^*\omega_{X_c} \simeq \omega_{Y_c}$, which implies that φ_c is an isomorphism.

If in fact ω_X^{-1} is f-ample, this argument can be globalized: indeed, assuming that the rational map $\varphi\colon Y \to X$ is not a morphism, there exists a curve $B \subset Z$ which is contracted by q and hence h, but not by p. Since B lives in a fiber of f (by the commutativity of the diagram), we again obtain a contradiction. Once we know that φ is a morphism, the same argument as above implies that it is an isomorphism. □

The following lemma used in the proof above is due to Kawamata, and can be extracted from his argument leading to the fact that derived equivalent varieties of general type are K-equivalent [9]; we sketch the argument for convenience.

[6] We thank Alessio Corti for pointing this out to us.

Lemma 15 *Let $\Phi_{\mathcal{E}}\colon \mathbf{D}(X) \to \mathbf{D}(Y)$ be a derived equivalence, and assume that there exists a component Z of the support of $\mathcal{E}$ such that* $\dim Z = \dim X$ *and Z dominates X. Then X and Y are K-equivalent.*

Proof Denote by p and q the projections of Z to X and Y. Since p is surjective, [8, Corollary 6.12] tells us that p is birational, and Z is the unique component of $\mathrm{Supp}(\mathcal{E})$ dominating X. We claim that q is also surjective, in which case by the same reasoning q is birational as well. Since (on the normalization of Z) we have $p^*\omega_X^r \simeq q^*\omega_Y^r$ for some $r \geq 1$, this suffices to conclude that X and Y are K-equivalent as in [9, Theorem 2.3] (see also [8, p. 149]).

Assuming that q is not surjective, we can find general points x_1 and x_2 in X such that $p^{-1}(x_1)$ and $p^{-1}(x_2)$ consist of one point, and $q(p^{-1}(x_1)) = q(p^{-1}(x_2)) = y$ for some $y \in Y$. One then sees that

$$\mathrm{Supp}\ \Phi_{\mathcal{E}}(\mathcal{O}_{x_1}) = \mathrm{Supp}\ \Phi_{\mathcal{E}}(\mathcal{O}_{x_2}) = \{y\}.$$

This implies in standard fashion that

$$\mathrm{Hom}^{\bullet}_{\mathbf{D}(X)}(\mathcal{O}_{x_1}, \mathcal{O}_{x_2}) \simeq \mathrm{Hom}^{\bullet}_{\mathbf{D}(Y)}(\Phi_{\mathcal{E}}(\mathcal{O}_{x_1}), \Phi_{\mathcal{E}}(\mathcal{O}_{x_2})) \neq 0,$$

a contradiction. □

Acknowledgments We thank Yujiro Kawamata and Anatoly Libgober for discussions that motivated this work, and Caucher Birkar, Alessio Corti, Lawrence Ein, Anne-Sophie Kaloghiros, Artie Prendergast-Smith, and Christian Schnell for answering numerous questions. We are grateful to the referee, who suggested how to improve the statement of Theorem 3 in the form of Theorem 12. Special thanks go to Rob Lazarsfeld, our first and second-generation teacher, whose fundamental results in the study of cohomological support loci have provided the original inspiration for this line of research.

References

[1] Bak, A. 2009. The spectral construction for a (1, 8)-polarized family of abelian varieties. arXiv:0903.5488.

[2] Beauville, A. 1992. Annulation du H[1] pour les fibrés en droites plats. In *Complex Algebraic Varieties (Bayreuth, 1990)*, pp. 1–15. Lectures Notes in Mathematics, Vol. 1507. Berlin: Springer-Verlag.

[3] Bridgeland, T. and Maciocia, A. 1998. Fourier–Mukai transforms for quotient varieties. arXiv:9811101.

[4] Bridgeland, T. and Maciocia, A. 2001. Complex surfaces with equivalent derived categories. *Math. Z.*, **236**(4), 677–697.

[5] Brion, M. 2009. Some basic results on actions of non-affine algebraic groups. In *Symmetry and Spaces: In honor of Gerry Schwarz*, pp. 1–20. Progress in Mathematics, Vol. 278, Boston: Birkhäuser.

[6] Catanese, F. 1991. Moduli and classification of irregular Kaehler manifolds (and algebraic varieties) with Albanese general type fibrations. *Invent. Math.*, **104**(2), 389–407.

[7] Green, M. and Lazarsfeld, R. 1991. Higher obstructions to deforming cohomology groups of line bundles. *J. Amer. Math. Soc.*, **1**(4), 87–103.

[8] Huybrechts, H. 2006. *Fourier–Mukai Transforms in Algebraic Geometry*. Oxford Mathematical Monographs. Oxford: The Clarendon Press.

[9] Kawamata, Y. 2002. D-equivalence and K-equivalence. *J. Diff. Geom.*, **61**(1), 147–171.

[10] Lombardi, L. 2014. Derived invariants of irregular varieties and Hochschild homology. *Algebra and Number Theory*, **8**(3), 513–542.

[11] Popa, M. 2013. Derived equivalence and non-vanishing loci. "A celebration of Algebraic Geometry", *Clay Math. Proc.* 18, 567–575, Amer. Math. Soc., Providence, RI.

[12] Popa, M. and Schnell, C. 2011. Derived invariance of the number of holomorphic 1-forms and vector fields. *Ann. Sci. ENS*, **44**(3), 527–536.

[13] Rouquier, R. 2011. Automorphismes, graduations et catégories triangulées. *J. Inst. Math. Jussieu*, **10**(3), 713–751.

[14] Schnell, C. 2012. The fundamental group is not a derived invariant. In *Derived Categories in Algebraic Geometry*, pp. 279–285. EMS Series of Congress Reports. Zürich: European Mathematical Society.

[15] Simpson, C. 1993. Subspaces of moduli spaces of rank one local systems. *Ann. Sci. ENS*, **26**(3), 361–401.

[16] Toda, T. 2006. Fourier-Mukai transforms and canonical divisors. *Compos. Math.*, **142**(4), 962–982.

[17] Zhang, Q. 1996. On projective manifolds with nef anticanonical bundles. *J. Reine Angew. Math.*, **478**, 57–60.

16

The automorphism groups of Enriques surfaces covered by symmetric quartic surfaces

S. Mukai[a]
Research Institute for Mathematical Sciences, Kyoto University

H. Ohashi[a]
Tokyo University of Science

Abstract

Let S be the (minimal) Enriques surface obtained from the symmetric quartic surface $(\sum_{i<j} x_i x_j)^2 = kx_1x_2x_3x_4$ in $\mathbb{P}^3$ with $k \neq 0, 4, 36$ by taking a quotient of the Cremona action $(x_i) \mapsto (1/x_i)$. The automorphism group of S is a semi-direct product of a free product $\mathcal{F}$ of four involutions and the symmetric group $\mathfrak{S}_4$. Up to action of $\mathcal{F}$, there are exactly 29 elliptic pencils on S.

Dedicated to Prof. Robert Lazarsfeld on his 60th birthday

The automorphism groups of very general Enriques surfaces, namely those corresponding to very general points in moduli, were computed by Barth and Peters [1] as an explicitly described infinite arithmetic group. Also, many authors [1, 3, 5, 9] have studied Enriques surfaces with only finitely many automorphisms. The article [1] also includes an example whose automorphism group is infinite but still virtually abelian. In this paper we give a concrete example of an Enriques surface whose automorphism group is not virtually abelian. Moreover, the automorphism group is described explicitly in terms of generators and relations. See also Remark 5.

We work over any algebraically closed field whose characteristic is not 2. Let us introduce the quartic surface with parameters k and l,

$$\overline{X}\colon \{s_2^2 = ks_4 + ls_1s_3\} \subset \mathbb{P}^3, \tag{1}$$

[a] Supported in part by the JSPS Grant-in-Aid for Scientific Research (B) 22340007, (S) 19104001, (S) 22224001, (S) 25220701, (A) 22244003, for Exploratory Research 20654004 and for Young Scientists (B) 23740010.

where s_d are the fundamental symmetric polynomials of degree d in the homogeneous coordinates $x_1, \dots, x_4$. It is singular at the four coordinate points $(1:0:0:0), \dots, (0:0:0:1)$ and has an action of the symmetric group $\mathfrak{S}_4$. It also admits the action of the standard Cremona transformation

$$\varepsilon\colon (x_1 : \cdots : x_4) \mapsto \left(\frac{1}{x_1} : \cdots : \frac{1}{x_4}\right)$$

which commutes with $\mathfrak{S}_4$. After taking the minimal resolution X, the quotient surface $S = X/\varepsilon$ becomes an Enriques surface, whenever $\overline{X}$ avoids the eight fixed points $(\pm 1 : \pm 1 : \pm 1 : 1)$ of ε. This condition is equivalent to $k + 16l \neq 36$, $k \neq 4$, and $4l + k \neq 0$.

The projection from one of four coordinate points exhibits X as a double cover of the projective plane $\mathbb{P}^2$. The associated covering involution commutes with ε and defines an involution of the Enriques surface S. In this way we obtain four involutions σ_i $(i = 1, \dots, 4)$. The action of $\mathfrak{S}_4$ also descends to S. Therefore, by mapping the generators of C_2^{*4} to σ_i, we obtain a group homomorphism

$$\mathfrak{S}_4 \ltimes (C_2^{*4}) \to \mathrm{Aut}(S), \tag{2}$$

where $\mathfrak{S}_4$ acts on the free product as a permutation of the four factors.

In this paper we study the automorphism group and elliptic fibrations of S in the case $l = 0$. Our main result is as follows:

Theorem 1 (= Theorem 3.7) *In equation* (1), *let* $l = 0$ *and* $k \neq 0, 4, 36$. *Then* (2) *is an isomorphism. Namely,* $\mathrm{Aut}(S)$ *is isomorphic to the semi-direct product of the free product* $\mathcal{F}$ *of four involutions* σ_i $(i = 1, \dots, 4)$ *and the symmetric group* $\mathfrak{S}_4$.

In the proof of this theorem, we also obtain the following results on elliptic pencils and smooth rational curves. Let S be as in Theorem 1.

Theorem 2 (= Theorem 3.4) *Up to the action of the free product* $\mathcal{F} \simeq C_2^{*4}$, *there are exactly* 29 *elliptic pencils on* S. *They are classified into five types and the main properties are as follows:*

	Singular fibers	*Mordell–Weil rank*	*Number*
(1)	$\tilde{E}_7 + \tilde{A}_1$	0	12
(2)	$\tilde{E}_6 + \tilde{A}_2$	0	4
(3)	$\tilde{D}_6 + \tilde{A}_1$	1	6
(4)	$\tilde{A}_7 + \tilde{A}_1$	0	3
(5)	$2\tilde{A}_5 + \tilde{A}_2 + \tilde{A}_1$	0	4

Here $2\tilde{A}_n$ denotes the multiple singular fiber of type $\tilde{A}_n$ and the Mordell–Weil rank stands for that of its Jacobian fibration.

Theorem 3 (= Theorem 3.3) *Up to the action of the free product $\mathcal{F} \simeq C_2^{*4}$, there are exactly 16 smooth rational curves on S. They are represented by the curves in the configuration $10A + 6B$ (see below).*

The proof of Theorem 1 uses some 16 smooth rational curves on S and the fact that four involutions $\sigma_1, \ldots, \sigma_4$ are numerically reflective. First using the four singularities of type D_4 and four tropes on $\overline{X}$, we find 10 smooth rational curves on S with the dual graph as in Figure 1 (Section 1). We call this the $10A$ configuration.

Also, by looking at some other plane sections, we find six further smooth rational curves on S with the dual graph as in Figure 2. This is called the $6B$ configuration.

We denote by $NS(S)_f$ the Néron–Severi lattice of S modulo torsion. The action of involutions $\sigma_1, \ldots, \sigma_4$ on $NS(S)_f$ is the reflection in (-2) classes $G_1, \ldots, G_4 \in NS(S)_f$, respectively. For instance, the class G_1 is $E_2 + E_{23} + E_3 + E_{34} + E_4 + E_{24} - E_1$ in terms of Figure 1 (Proposition 2.1). The dual graph of these four (-2) classes is the complete graph in four vertices with doubled edges. This is called the $4C$ configuration.

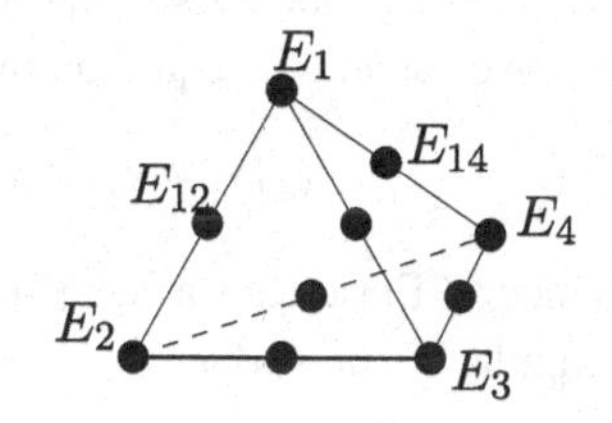

Figure 1 The $10A$ configuration.

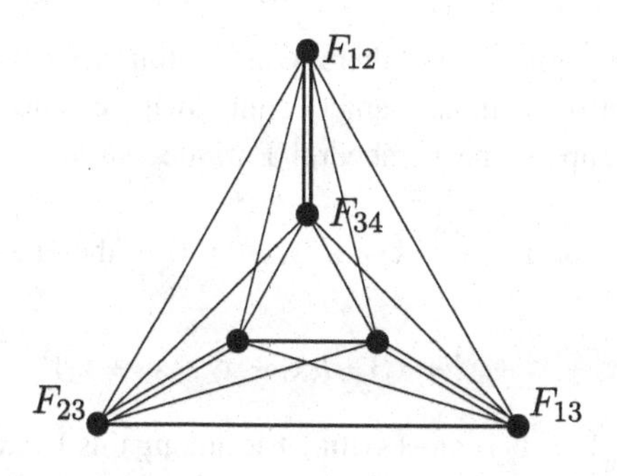

Figure 2 The $6B$ configuration.

We can check that the 20 (-2) classes E_i, E_{ij}, F_{ij}, G_i define a convex polyhedron whose Coxeter diagram satisfies Vinberg's condition [12]. Namely, the subgroup $W(10A + 6B + 4C)$ generated by reflections in these 20 classes has finite index in the orthogonal group $O(NS(S)_f)$. In fact, the limit of our Enriques surfaces as $k \to \infty$ is of type V in Kondo [5] (see Remark 2.3), and our diagram coincides with Kondo's. Although in his case the classes $G_1, \ldots, G_4$ were also represented by smooth rational curves, in our case they appear just as the *center* of the reflective involutions $\sigma_1, \ldots, \sigma_4$ and are not effective (Corollary 2.2).

To prove our Theorem 1, we divide the generators of $W(10A + 6B + 4C)$ into two parts, those coming from $10A + 6B$ and those from $4C$. By a lemma of Vinberg [11], $W(10A+6B+4C)$ is the semi-direct product $W(4C) \ltimes \overline{N}(W(10A+6B))$, where $\overline{N}$ denotes the normal closure. Since the whole $10A + 6B + 4C$ configuration has only $\mathfrak{S}_4$-symmetry, we obtain our Theorem 1 and the others (Section 3).

Remark 4 There are some interesting cases in $l \neq 0$ too.

(1) When $(k - 4)(l - 4) = 16$, the surface $\overline{X}$ is Kummer's quartic surface $\mathrm{Km}(J(C))$ written in Hutchinson's form. It has 16 nodes. Our four involutions σ_i are called *projections*. As is shown in [7], S is an Enriques surface of Hutchinson–Göpel type and the four involutions are numerically reflective. Especially in the case $(k, l) = (-4, 2)$, the hyperelliptic curve C branches over the vertices of a regular octahedron and the equation of $\overline{X}$ becomes

$$(x_1^2x_2^2 + x_3^2x_4^2) + (x_1^2x_3^2 + x_2^2x_4^2) + (x_1^2x_4^2 + x_2^2x_3^2) + 2x_1x_2x_3x_4 = 0.$$

This is the case of the octahedral Enriques surface [8] and S is isomorphic to the normalization of the singular sextic surface

$$x_1^2 + x_2^2 + x_3^2 + x_4^2 + \sqrt{-1}\left(\frac{1}{x_1^2} + \frac{1}{x_2^2} + \frac{1}{x_3^2} + \frac{1}{x_4^2}\right)x_1x_2x_3x_4 = 0.$$

In these cases we know that there exist automorphisms on S induced from X other than projections, namely some switches and correlations. The automorphism group of the octahedral Enriques surface will be discussed elsewhere.

(2) The quartic surface $\overline{X} : ks_4 + ls_1s_3 = 0$ is the Hessian of the cubic surface

$$k(x_1^3 + x_2^3 + x_3^3 + x_4^3) + l(x_1 + x_2 + x_3 + x_4)^3 = 0.$$

The case $(k : l) = (1 : -1)$ is most symmetric among this 1-parameter family. In this special case, the Enriques surface $S = X/\varepsilon$ is of type VI in Kondo [5] and the automorphism group is isomorphic to $\mathfrak{S}_5$. In particular, the homomorphism (2) is neither injective nor surjective.

Remark 5 In terms of virtual cohomological dimensions of discrete groups [10], our example can be located in the following way. The virtual cohomological dimension is equal to 0 for finite groups. At the other extreme, the discrete group Aut(S) for very general Enriques surfaces S has the virtual cohomological dimension 8. See [2]. In our case, the automorphism group has virtual cohomological dimension 1.

1 Smooth rational curves

Under the condition $l = 0$, equation (1) becomes

$$\overline{X}\colon (x_1x_2 + x_1x_3 + x_1x_4 + x_2x_3 + x_2x_4 + x_3x_4)^2 = kx_1x_2x_3x_4. \tag{3}$$

This surface has four rational double points of type D_4 at the four coordinate points $(1 : 0 : 0 : 0), \ldots, (0 : 0 : 0 : 1)$ and by taking the quotient of the minimal resolution X by the standard Cremona involution

$$\varepsilon\colon (x_1 : \cdots : x_4) \mapsto \left(\frac{1}{x_1} : \cdots : \frac{1}{x_4}\right),$$

we obtain an Enriques surface $S = X/\varepsilon$. We begin with a study of the configuration of smooth rational curves on the surfaces.

The desingularization X has 16 smooth rational curves as the exceptional curves of the four D_4 singularities. Also, each coordinate plane cuts the quartic doubly along a conic, which defines a smooth rational curve on X. These are called *tropes*. The configuration of these 20 curves is shown in Figure 3, which depicts the dual graph. Black vertices come from the singularities and white ones are tropes.

The standard Cremona involution ε acts on Figure 3 by point symmetry. Therefore, the Enriques surface S has 10 smooth rational curves whose dual

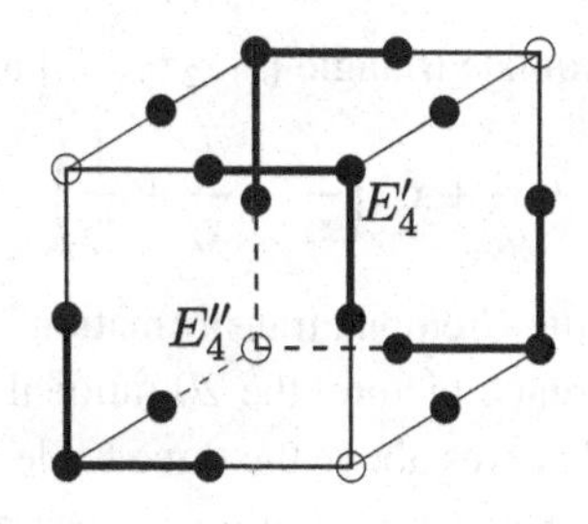

Figure 3 The quartic surface with four D_4 singularities.

graph is that in Figure 1. In what follows, we call these 10 curves on S the $10A$ configuration. The indexing is given as follows. Since a vertex of the tetrahedron corresponds to two curves on X, namely the trope $\{x_i = 0\}$ and the central component of the exceptional curves at $(0 : \cdots : 1 : \cdots : 0)$ (the ith coordinate is 1), we denote the curve at the vertex by E_i $(i = 1, \ldots, 4)$. Also, if a vertex at the middle of an edge is connected to two vertices, say E_i and E_j, then we denote the curve by E_{ij}. This is the first configuration of smooth rational curves on S of interest. It is convenient to note that the 10 curves $\{E_i, E_{ij}\}$ generate $NS(S)_f$ over the rationals; the Gram matrix of these curves has determinant -64.

Next let us consider the six plane sections $\{x_i + x_j = 0\}$ $(i = 1, \ldots, 4)$. In equation (3), we see that each plane section decomposes into two conics which are disjoint on X and exchanged by ε. Thus we obtain six further smooth rational curves on S, naturally indexed as F_{ij}. The intersection relation between these curves is shown in Figure 2. We call it the $6B$ configuration. Moreover, we can clarify the intersection relations between the configurations as follows:

$$(E_k, F_{ij}) = 0; \qquad (E_{kl}, F_{ij}) = \begin{cases} 2 & \text{if } \{k, l\} = \{i, j\}, \\ 0 & \text{otherwise.} \end{cases}$$

The configuration of 16 curves thus obtained is denoted by $10A + 6B$.

2 Numerically reflective involutions

The quartic surface (3) can be exhibited as a double cover of $\mathbb{P}^2$ by the projection from one of the coordinate points, say $(0 : 0 : 0 : 1)$. The branch $B \subset \mathbb{P}^2$ is the sextic plane curve defined by

$$x_1x_2x_3\left\{4(x_1 + x_2 + x_3)\left(\frac{1}{x_1} + \frac{1}{x_2} + \frac{1}{x_3}\right)x_1x_2x_3 - kx_1x_2x_3\right\} = 0. \tag{4}$$

It is the union of the coordinate triangle $\{x_1x_2x_3 = 0\}$ and the cubic curve

$$C\colon 4(x_1 + x_2 + x_3)\left(\frac{1}{x_1} + \frac{1}{x_2} + \frac{1}{x_3}\right) - k = 0, \tag{5}$$

which is invariant under the Cremona transformation $(x_i) \mapsto (1/x_i)$ of $\mathbb{P}^2$. See Figure 4. In this double-plane picture, the 20 rational curves in Figure 3 can be seen as the 12 rational curves above the three triple points of B, three rational curves above the three nodes of B, three tropes as the inverse image of the coordinate triangle, and some components of inverse images of the curves

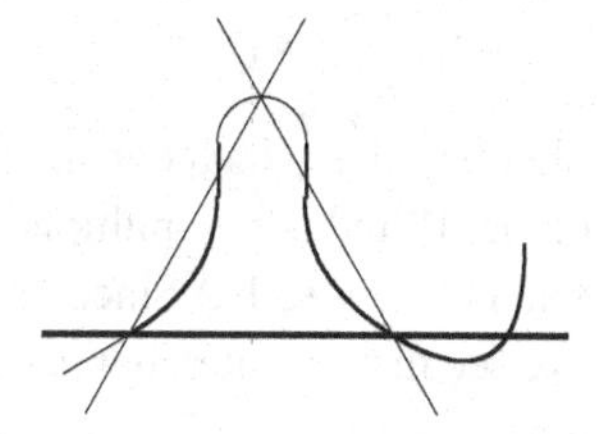

Figure 4 The branch sextic B.

L: $\{x_1 + x_2 + x_3 = 0\}$ and Q: $\{\frac{1}{x_1} + \frac{1}{x_2} + \frac{1}{x_3} = 0\}$. (We note that the line L must pass through the three simple intersection points of C with the triangle in Figure 4, although it is not visible.)

The covering transformation of this double cover $X \to \mathbb{P}^2$ is called the *projection*. It is an antisymplectic involution acting on X. It stabilizes all the curves above the branch curve B (including those above the singularities of B). In particular, in Figure 3, if E_4'' is the trope $\{x_4 = 0\}$, then the projection stabilizes all the curves except for E_4'' and its antipodal E_4' (coming from the singularity at $(0 : 0 : 0 : 1)$). It is easy to determine the fixed curves of the projection, consisting of six rational curves (vertices of the cube except for E_4' and E_4'') and the inverse image of the elliptic curve C.

Since the projection commutes with the Cremona involution of $\mathbb{P}^3$, we obtain an involution of the Enriques surface S. It is denoted by σ_4, where the index is in accordance with the center of the projection $(0 : 0 : 0 : 1)$.

Proposition 2.1 *The involution $\sigma_4 \in \mathrm{Aut}(S)$ is numerically reflective. Moreover, its action on the Néron–Severi lattice $NS(S)_f$ is the reflection in the divisor $G_4 = E_1 + E_{12} + E_2 + E_{23} + E_3 + E_{13} - E_4$ of self-intersection (-2). In Figure 1, the six positive components in G_4 are just the cycle of curves disjoint from E_4.*

Proof From our description of fixed curves of the projection as above, we see that σ_4 preserves all the curves E_i and E_{ij} except for E_4. Compare Figures 1 and 3.

Consider the elliptic fibration $f: S \to \mathbb{P}^1$ defined by the divisor $2D_f = 2(E_1 + E_{12} + E_2 + E_{23} + E_3 + E_{13})$. It gives the multiple fiber of type ${}_2\mathrm{I}_6$ in Kodaira's notation. From Figure 1, we see that the curve E_4 sits inside a reducible fiber which we denote by D'. In comparison with Figure 4, f corresponds to the pencil $\mathcal{L}$ of cubics on $\mathbb{P}^2$ spanned by the triangle $\{x_1x_2x_3 = 0\}$ and the cubic curve C of (5). Thus we see that the multiple fibers of f are exactly the transform of the triangle, which is nothing but the divisor $2D_f$ of type ${}_2\mathrm{I}_6$, and the transform of C, namely some irreducible fiber of type ${}_2\mathrm{I}_0$. On the other hand, the cubic

$$C_\infty := L + Q \in \mathcal{L} \tag{6}$$

corresponds to the reducible fiber of f which contains E_4. Therefore, the fiber $D' = E_4 + B$ is of Dynkin type $\tilde{A}_1$ and is not multiple. (More precisely, it is of type III in characteristic 3 and otherwise I_2.) Since the Cremona involution of $\mathbb{P}^2$ interchanges L and Q, we see that σ_4 interchanges E_4 and B.

From the linear equivalence $E_4 + B \sim 2(E_1 + E_{12} + E_2 + E_{23} + E_3 + E_{13})$, we see that the action is

$$\sigma_4 \colon E_4 \mapsto B = 2(E_1 + E_{12} + E_2 + E_{23} + E_3 + E_{13}) - E_4.$$

By taking the first paragraph into account, we see that σ_4 is numerically reflective and acts on $NS(S)_f$ by the reflection in the divisor

$$G_4 = E_1 + E_{12} + E_2 + E_{23} + E_3 + E_{13} - E_4.$$

□

By symmetry, we obtain divisors G_i $(i = 1, \ldots, 4)$ which describe the numerically reflective involutions σ_i in a similar manner. We see that $(G_i, G_j) = 2$ for $i \neq j$, so that the intersection diagram associated with divisors $G_1, \ldots, G_4$ is the complete graph in four vertices with all edges doubled. In what follows we denote this configuration by $4C$.

We note that the automorphism σ_i sends G_i to its negative. It implies the following corollary:

Corollary 2.2 *The numerical classes of G_i are not effective.*

We can compute the intersections of G_i and the $10A + 6B$ configuration. We have the following:

$$(G_i, E_j) = \begin{cases} 2 & \text{if } i = j, \\ 0 & \text{otherwise,} \end{cases} \quad (G_i, E_{kl}) = 0, \quad (G_i, F_{kl}) = \begin{cases} 2 & \text{if } i \notin \{k, l\}, \\ 0 & \text{if } i \in \{k, l\}. \end{cases}$$

Remark 2.3 The limit of our quartic surface $\overline{X}$ in (3) as $k \to \infty$ is the double $\mathbb{P}^2$ with branch the union of the coordinate triangle and the reducible cubic C_∞ in (6). Hence the limit of our Enriques surfaces is of type V in Kondo [5]. (See [6] also.) In this limit our divisor class G_4 becomes effective and corresponds to the new singular point coming from the intersection $L \cap Q$ in (6). Thus G_4 can be regarded as the vanishing cycle of this specialization $k \to \infty$. Furthermore, the numerically reflective involution σ_4 becomes numerically trivial in this limit. More precisely, the limit of its graph as $k \to \infty$ is the union of that of the limit involution and the product $C_4 \times C_4$, where C_4 is the unique (-2) curve representing G_4 in the limiting Enriques surface.

3 Proof of the theorems

In the previous two sections, we obtained 16 smooth rational curves with the configuration $10A + 6B$ and four numerically reflective involutions σ_i whose centers G_i have the configuration $4C$. We begin with consideration of the natural representation $r\colon \mathrm{Aut}(S) \to O(NS(S)_f)$.

Proposition 3.1 *The homomorphism r is injective, namely there are no nontrivial numerically trivial automorphisms on S.*

Proof Let g be a numerically trivial automorphism of S, which is tame by virtue of Dolgachev [4]. It preserves each (-2) curve, in particular in the $10A+6B$ configuration. The curves $E_1, \ldots, E_4$ in Figure 1 must be pointwise fixed, since $\mathrm{Aut}(\mathbb{P}^1)$ is sharply triply transitive and since each E_i has three distinct intersections with its neighbors.

We again focus on the elliptic fibration $f\colon S \to \mathbb{P}^1$ defined by $D_f = E_1 + E_{12} + E_2 + E_{23} + E_3 + E_{13}$ as in Proposition 2.1. We saw that $E_4 + B = E_4 + \sigma_4(E_4)$ is a non-multiple fiber of f. Therefore the bisections E_{14}, E_{24}, E_{34} of f must intersect B. By a suitable choice of a bisection $C_f \in \{E_{14}, E_{24}, E_{34}\}$, we can assume that C_f does not pass through the intersection $E_4 \cap B$. Then, since g preserves all (-2) curves, the curve C_f has three distinct fixed points $E_4 \cap C_f$, $B \cap C_f$, and $E_i \cap C_f$, where E_i is another vertex of the edge containing C_f in Figure 1. It follows that g fixes C_f pointwise, hence the singular curve $E_4 + C_f$ too. It follows that $g = \mathrm{id}_S$, since g is tame and of finite order. □

In what follows we denote the hyperbolic lattice $NS(S)_f$ by L. Let us denote by $O'(L)$ the group of integral isometries whose $\mathbb{R}$-extensions preserve the positive cone of $L \otimes \mathbb{R}$. We denote by Λ the 9-dimensional Lobachevsky space associated with the positive cone. Then $O'(L)$ acts on Λ as a discrete group of motions. We refer the reader to [12] for the theory of discrete groups generated by reflections acting on Lobachevsky spaces.

We let

$$P^c = \{\mathbb{R}_+ x \in PS(L) \mid (x, E) \geq 0 \text{ for all } E \in \{E_i, E_{ij}, F_{ij}, G_i\}\}$$

be the convex polyhedron defined by the 20 roots from the $10A + 6B + 4C$ configuration in the projective sphere $PS(L) = (L - \{0\})/\mathbb{R}_+$ (see [12, Section 2]). We have seen that every intersection number of two distinct divisors in $10A + 6B + 4C$ is between 0 and 2, hence the Coxeter diagram associated with these 20 roots has no dotted lines or Lanner's subdiagrams. Also, by an easy check of the $10A + 6B + 4C$ configuration, we have the following:

Lemma 3.2 *The Coxeter diagram of the polyhedron $P = P^c \cap \Lambda$ has exactly 29 parabolic subdiagrams of maximal rank 8. They are as follows:*

	Subdiagram	*Number*	$10A$	$6B$	$4C$
(1)	$\tilde{E}_7 + \tilde{A}_1$	12	8	1	1
(2)	$\tilde{E}_6 + \tilde{A}_2$	4	7	3	0
(3)	$\tilde{D}_6 + \tilde{A}_1 + \tilde{A}_1$	6	8	1	2
(4)	$\tilde{A}_7 + \tilde{A}_1$	3	8	2	0
(5)	$\tilde{A}_5 + \tilde{A}_2 + \tilde{A}_1$	4	7	3	1

Here, each column $10A, 6B, 4C$ *shows the number of vertices used from the configuration.*

It is easy to check that every connected parabolic subdiagram is a connected component of some parabolic subdiagram of rank 8, using the previous table. By Theorem 2.6 of [12], we see that P has finite volume and we obtain $P^c \subset \overline{\Lambda}$. This polyhedron gives the fundamental domain of the associated discrete reflection group generated by 20 reflections in the 20 roots $\{E_i, E_{ij}, F_{ij}, G_i\}$, which we denote by $W = W(10A + 6B + 4C)$. Algebro-geometrically, 16 of the generators are the Picard–Lefschetz transformations in (-2) curves in the $10A + 6B$ configuration and the remaining four are the involutions σ_i $(i = 1, \ldots, 4)$ corresponding to $4C$. As an abstract group, we see that W has the structure of a Coxeter group whose fundamental relations are given by the Coxeter diagram (see [12]) of P. We note that the quasi-polarization (namely a nef and big divisor)

$$H = \sum_i E_i + \sum_{i<j} E_{ij}$$

defines an element $\mathbb{R}_+ H$ in P.

Now let $W(4C)$ be the subgroup of W generated by four reflections in G_i. Via the homomorphism $r\colon \mathrm{Aut}(S) \to O(NS(S)_f)$, the subgroup $\mathcal{F} \subset \mathrm{Aut}(S)$ generated by the four numerically reflective involutions σ_i is mapped onto this Coxeter subgroup $W(4C) \simeq C_2^{*4}$. It follows that $\mathcal{F} \simeq W(4C)$. Let $W(10A + 6B)$ be the subgroup generated by 16 reflections in E_i, E_{ij} and F_{ij}, and let $\overline{N}(W(10A + 6B))$ be the minimal normal subgroup of W which contains $W(10A + 6B)$. Since the intersection numbers between elements of $4C$ and $10A + 6B$ are all even, by [11, Proposition] we have the exact sequence

$$1 \longrightarrow \overline{N}(W(10A + 6B)) \longrightarrow W \longrightarrow W(4C) \longrightarrow 1. \tag{7}$$

The kernel is exactly the subgroup generated by the conjugates

$$\{\sigma g \sigma^{-1} \mid \sigma \in W(4C), g \text{ a generator of } W(10A + 6B))\}.$$

We have the corresponding geometric consequence as follows:

Theorem 3.3 *There are exactly 16 smooth rational curves on S up to the action of $\mathcal{F}$:*

Proof Let E be a smooth rational curve on S. We consider the orbit $\mathcal{F}.E$. Since the divisor H above is nef, we can choose $E_0 \in \mathcal{F}.E$ such that the degree (E_0, H) is minimal. We show that E_0 is one of 16 curves in $10A + 6B$.

In fact, by the automorphism σ_i, we have

$$(E_0, H) \leq (\sigma_i(E_0), H) = (E_0, H) + (E_0, G_i)(G_i, H),$$

and $0 \leq (E_0, G_i)$ for all i. Suppose that E_0 intersects non-negatively all 16 curves in $10A + 6B$. Then we have $\mathbb{R}_+ E_0 \in P^c$. But from $P^c \subset \overline{\Lambda}$ we obtain $(E_0^2) \geq 0$, which is a contradiction. Hence E_0 is negative on some curve in $10A + 6B$ and we see that E_0 is one of them.

Next let us show that two distinct curves E, E' in the $10A + 6B$ configuration are inequivalent under $\mathcal{F}$. For the six curves E_{ij} from $10A$, we have $(E_{ij}, G_k) = 0$ for all $i < j$ and k. Therefore, by an easy induction, we see that the sextuple $((E_{ij}, E))_{1 \leq i < j \leq 4}$ consisting of intersection numbers is an invariant of the orbit $\mathcal{F}.E$. Suppose that $(E_{ij}, E) = (E_{ij}, E')$ for all $i < j$. Since E and E' are both in the $10A + 6B$ configuration, we see easily that $E = E'$. This shows that the orbits $\mathcal{F}.E$ and $\mathcal{F}.E'$ are disjoint. □

In other words, the group $\overline{N}(W(10A + 6B))$ is nothing but the Weyl group of S generated by Picard–Lefschetz reflections in all (-2) curves. We can proceed to elliptic pencils.

Theorem 3.4 *There are exactly 29 elliptic pencils on S up to the action of $\mathcal{F}$. Their properties are as in the table of Theorem 2.*

Proof Let $2f$ be a fiber class of an elliptic pencil on S. As before, we choose an element $f_0 \in \mathcal{F}.f$ such that the degree (f_0, H) is minimal. We have

$$(f_0, H) \leq (\sigma_i(f_0), H) = (f_0, H) + (f_0, G_i)(G_i, H),$$

hence $(f_0, G_i) \geq 0$ for all i. Moreover, since f_0 is nef we have $(f_0, E) \geq 0$ for all E in the $10A + 6B$ configuration. Therefore, $f_0 \in P^c$. This shows that f_0 corresponds to one of the maximal parabolic subdiagrams classified in Lemma 3.2.

Conversely, we can construct 29 elliptic pencils from the 29 subdiagrams in Lemma 3.2 as follows. The two types (2) and (4) in the lemma are easiest since they do not contain a class in $4C$. The elliptic pencils of types (2) and (4) have singular fibers of type $\tilde{E}_6 + \tilde{A}_2$ and $\tilde{A}_7 + \tilde{A}_1$, respectively as in the case of [5, Table 2].

In the case of type (1) (resp. (5)), one component of the parabolic subdiagram is $\tilde{A}_1$ consisting of a (-2) curve E in $6B$ (resp. $10A$) and (-2) class G in $4C$. Moreover, the sum $E + G$ is half of $\tilde{E}_7$ (resp. $\tilde{A}_2$). Hence $E + \sigma(E)$ is a non-multiple fiber of type $\tilde{A}_1$ since it is linearly equivalent to $2(E + G)$, where σ is the reflection in G.

In the case of type (3), one component is $\tilde{A}_1$ consisting of two classes G and G' in $4C$. But the other two components consist of (-2) curves. Therefore, the Mordell–Weil group is of rank 1 since neither G nor G' is effective. (The composite $\sigma\sigma'$ of two reflections in G and G' is the translation by a generator of the Mordell–Weil group.)

That these 29 pencils are inequivalent under $\mathcal{F}$ follows from the previous result for rational curves. □

To study the image of the representation $r\colon \mathrm{Aut}(S) \to O(NS(S)_f)$, we need some lemma. We denote by $4A'$ the set of four roots $\{E_i\}$ and by $6A''$ the set $\{E_{ij}\}$. Recall that by Theorem 3.3, all (-2) curves on S are in the $\mathcal{F}$-orbit of the three sets $4A', 6A''$, and $6B$.

Lemma 3.5 *Let τ be any automorphism of S. Then τ preserves each of the three orbits of rational curves $\mathcal{F}.(4A')$, $\mathcal{F}.(6A'')$, and $\mathcal{F}.(6B)$.*

Proof Any automorphism τ permutes smooth rational curves on S, and hence induces a symmetry of the dual graph of the set of rational curves. Thus, for the proof, it suffices to give a characterization of each orbit in terms of this infinite graph. We use the (full) subgraphs which are isomorphic to the dual graph of reducible fibers of elliptic fibrations.

Consider a vertex v in $\mathcal{F}.(6B)$. Then there exists a subgraph of fiber type I_3 passing through v. Conversely, if for a vertex v there is a subgraph of fiber type I_3, by Theorem 3.4, it is equivalent to a vertex in $6B$ under $\mathcal{F}$. Thus the vertices in $\mathcal{F}.(6B)$ are characterized by the property that there exists a subgraph of fiber type I_3 passing through them.

Similarly, vertices in $\mathcal{F}.(10A)$ are characterized by subgraphs of type I_8. Moreover, the vertices v in $\mathcal{F}.(4A')$ are characterized by the property that there exists a subgraph of type IV^* which has v as its end. In the opposite way, vertices in $\mathcal{F}.(6A'')$ are those which do not have such IV^* subgraphs. Thus the three orbits are all characterized and τ preserves these orbits. □

Corollary 3.6 *The set of six curves* $\{E_{ij}\}$ *is preserved under any automorphism.*

Proof In fact, for any E_{kl} and any σ_i we have $\sigma_i(E_{kl}) = E_{kl}$. Hence $\mathcal{F}.(6A'') = \{E_{ij}\}$. □

Recall that S has action by $\mathfrak{S}_4$ from the symmetry of the defining equation of $\overline{X}$. Explicitly, it acts on the curves in the $10A + 6B$ configuration by the permutation of indices. For involutions σ_i, the same holds true if we regard the action as taking conjugates. It is easy to see that this group $\mathfrak{S}_4$ can be identified with the symmetry group Sym(P) of the polyhedron $P \subset \Lambda$ via r. We can also regard this group as acting on the reflection group W and the exact sequence (7) is preserved under this action. In particular, $W(4C)$ and Sym(P) generate a group isomorphic to $\mathfrak{S}_4 \ltimes C_2^{*4}$.

Theorem 3.7 *The representation r induces an isomorphism of* Aut(S) *onto the group generated by* $W(4C)$ *and* Sym(P), *hence we obtain* $\mathrm{Aut}(S) \simeq \mathfrak{S}_4 \ltimes C_2^{*4}$.

Proof Since r maps $\mathcal{F}$ onto $W(4C)$, the image of r includes the groups $W(4C)$ and Sym(P).

Conversely, let us pick up an arbitrary automorphism τ. We consider the image $\tau(H)$ of H. We use the elliptic fibration defined by the divisor $f = H - E_{12} - E_{34}$ of type I_8. By Theorem 3.4, the image $\tau(f)$ is equivalent to one of three elliptic pencils described in item (4) under $\mathcal{F}$. Moreover, since $\mathfrak{S}_4$ acts transitively on these three pencils, we can assume that $\tau(f) = f$ by composing τ with some elements of $\mathcal{F}$ and $\mathfrak{S}_4$. Thus we have $\tau(H) = f + \tau(E_{12}) + \tau(E_{34})$. By the previous corollary, $\tau(E_{12})$ and $\tau(E_{34})$ are in the set $\{E_{ij}\}$. By an easy check of intersection numbers, we see that $\tau(E_{12}) + \tau(E_{34}) = E_{12} + E_{34}$. In particular we obtain $\tau(H) = H$ as divisors. Since any permutation of the $10A$ configuration can be induced from the automorphism group $\mathfrak{S}_4$, this shows that the image of r is contained in the group generated by $W(4C)$ and Sym(P). □

References

[1] Barth, W. and Peters, C. Automorphisms of Enriques surfaces. *Invent. Math.* **73** (1983) 383–411.

[2] Borel, A. and Serre, J. P. Corners and arithmetic groups. *Comment. Math. Helv.* **48** (1973) 436–491.

[3] Dolgachev, I. On automorphisms of Enriques surfaces. *Invent. Math.* **76** (1984) 163–177.

[4] Dolgachev, I. Numerical trivial automorphisms of Enriques surfaces in arbitrary characteristic. In *Arithmetic and Geometry of K3 Surfaces and Calabi–Yau Threefolds*. Fields Institute Communication No. 67, 2013, pp. 267–283.

[5] Kondo, S. Enriques surfaces with finite automorphism groups. *Japan. J. Math.* **12** (1986) 191–282.

[6] Mukai, S. Numerically trivial involutions of Kummer type of an Enriques surface. *Kyoto J. Math.* **50** (2010) 889–902.

[7] Mukai, S. Kummer's quartics and numerically reflective involutions of Enriques surfaces. *J. Math. Soc. Japan* **64** (2012) 231–246.

[8] Mukai, S. and Ohashi, H. Enriques surfaces of Hutchinson–Göpel type and Mathieu automorphisms. In *Arithmetic and Geometry of K3 Surfaces and Calabi–Yau Threefolds*. Fields Institute Communication No. 67, 2013, pp. 429–454.

[9] Nikulin, V. V. On a description of the automorphism groups of Enriques surfaces. *Soviet Math. Dokl.* **30** (1984) 282–285.

[10] Serre, J. P. Cohomologie des groupes discrets. In *Prospects in Mathematics* (Proceedings of Symposium, Princeton University, Princeton, NJ, 1970), pp. 77–169. Annals of Mathematical Studies, No. 70. Princeton, NJ: Princeton University Press, 1971.

[11] Vinberg, E. B. The two most algebraic K3 surfaces. *Math. Ann.* **265** (1983) 1–21.

[12] Vinberg, E. B. Some arithmetical discrete groups in Lobaĉevskiĭ spaces. In *Discrete Subgroups of Lie Groups and Applications to Moduli (Bombay 1973)*. Oxford: Oxford University Press, 1975, pp. 323–348.

17

Lower-order asymptotics for Szegö and Toeplitz kernels under Hamiltonian circle actions

R. Paoletti
Università degli Studi di Milano Bicocca

Abstract

We consider a natural variant of Berezin–Toeplitz quantization of compact Kähler manifolds, in the presence of a Hamiltonian circle action lifting to the quantizing line bundle. Assuming that the moment map is positive, we study the diagonal asymptotics of the associated Szegö and Toeplitz operators, and specifically their relation to the moment map and to the geometry of a certain symplectic quotient. When the underlying action is trivial and the moment map is taken to be identically equal to one, this scheme coincides with the usual Berezin–Toeplitz quantization. This continues previous work on near-diagonal scaling asymptotics of equivariant Szegö kernels in the presence of Hamiltonian torus actions.

Dedicated to Rob Lazarsfeld on the occasion of his 60th birthday

1 Introduction

The object of this paper are the asymptotics of Szegö and Toeplitz operators in a non-standard version of the Berezin–Toeplitz quantization of a complex projective Kähler manifold (M, J, ω).

In Berezin–Toeplitz quantization, one typically adopts as "quantum spaces" the Hermitian spaces $H^0\left(M, A^{\otimes k}\right)$ of global holomorphic sections of higher powers of the polarizing line bundle (A, h); here (A, h) is a positive, hence ample, Hermitian holomorphic line bundle on M. Quantum observables, on the contrary, correspond to Toeplitz operators associated with real C^∞ functions on M.

From *Recent Advances in Algebraic Geometry*, edited by Christopher Hacon, Mircea Mustaţă and Mihnea Popa

Here we assume given a Hamiltonian action μ^M of the circle group $U(1) = \mathbf{T}^1$ on M, with positive moment map Φ, and admitting a metric-preserving linearization to A. It is then natural to replace the spaces $H^0\left(M, A^{\otimes k}\right)$ with certain new "quantum spaces" which arise by decomposing the Hardy space associated with A into isotypes for the action; these are generally not spaces of sections of powers of A. One is thus led to consider analogues of the usual constructs of Berezin–Toeplitz quantization. In particular, it is interesting to investigate how the symplectic geometry of the underlying action, encapsulated in Φ, influences the semiclassical asymptotics in this quantization scheme.

This picture generalizes the usual Berezin–Toeplitz quantization of (M, J, ω), for one falls back on the standard case by considering the trivial action of $\mathbf{T}^1$ on M with moment map $\Phi = 1$. Then the lifted action is essentially fiberwise scalar multiplication, and the corresponding equivariant spaces are the usual spaces of global holomorphic sections.

This theme was considered in [28] for general Hamiltonian torus actions; the focus there was on near-diagonal scaling asymptotics of the associated equivariant Szegö kernels. Here we shall restrict our analysis to circle actions, and investigate the lower-order terms of these asymptotic expansions, as well as their analogues for Toeplitz operators.

In the usual standard setting of Berezin–Toeplitz quantization, a huge amount of work has been devoted to these themes, involving a variety of approaches and techniques; see for example (obviously with no pretense of being exhaustive) [1, 2, 4, 5, 7, 8, 10–12, 14, 22, 25, 26, 29–31, 34, 35] and references cited therein.

In the present paper, we follow the general approach to quantization based on the microlocal analysis of the Szegö kernel on the circle bundle X of $A^\vee$; this train of thought was first introduced in the grounding work [5], and afterwards explored by many authors. We shall also specifically build on ideas and results from [13, 17, 19]; in fact, the present paper was inspired considerably by the concise approach in [19] to the derivation of the lower-order terms in the TYZ expansion for real-analytic metrics.

Now let us make our discussion more precise. Let (M, J) be a connected complex d-dimensional projective manifold, and let A be an ample holomorphic line bundle on M, with dual line bundle $A^\vee$ and projection $\widehat{\pi}\colon A^\vee \to M$.

There is an Hermitian metric ℓ_A on A such that the unique covariant derivative ∇_A on A compatible with ℓ_A and the holomorphic structure has curvature $\Theta_A = -2i\,\omega$, where ω is a Kähler form on M. Then $dV_M =: \omega^{\wedge d}/d!$ is a volume form on M.

Let $X \subseteq A^\vee$ be the unit circle bundle, with projection $\pi = \widehat{\pi}|_X : X \to M$. Then ∇ corresponds to a connection contact form $\alpha \in \Omega^1(X)$, such that $d\alpha = 2\pi^*(\omega)$ and $dV_X =: (1/2\pi)\,\alpha \wedge \pi^*(dV_M)$ is a volume form on X. Let $L^2(X) =: L^2(X, dV_X)$, and identify functions with densities and half-densities on X. Also, let $H(X) =: \ker(\overline{\partial}_b) \cap L^2(X)$ be the Hardy space of X, where $\overline{\partial}_b$ is the Cauchy–Riemann operator on X.

Suppose that the action $\mu^M : \mathbf{T}^1 \times M \to M$ is holomorphic with respect to J and Hamiltonian with respect to $2\,\omega$, with moment map $\Phi : M \to \mathbb{R}$; suppose furthermore that (μ^M, Φ) can be linearized to a holomorphic action μ^A on A leaving ℓ_A invariant. Then $\mathbf{T}^1$ acts on X as a group of contactomorphisms under the naturally induced action $\mu^X : \mathbf{T}^1 \times X \to X$ lifting μ^M.

Infinitesimally, the relation between μ^M and μ^X is as follows. Let $\partial/\partial\theta \in \mathfrak{X}(X)$ be the infinitesimal generator of the action of $\mathbf{T}^1$ on X given by fiberwise scalar multiplication, mult: $(e^{i\theta}, x) \mapsto e^{i\theta} \cdot x$; also, let $\xi_M \in \mathfrak{X}(M)$ be the infinitesimal generator of μ^M, with horizontal lift $\xi_M^\sharp$ to X. Then the infinitesimal generator $\xi_X \in \mathfrak{X}(X)$ of μ^X is given by

$$\xi_X = \xi_M^\sharp - \Phi\,\frac{\partial}{\partial\theta}, \tag{1}$$

where we write Φ for $\Phi \circ \pi$. Thus μ^X depends crucially on the choice of Φ; for example, when μ^M is trivial, choosing $\Phi = 0$ yields the trivial action on X, while if $\Phi = 1$ we obtain the action

$$\nu^X : \mathbf{T}^1 \times X \to X, \ (e^{i\theta}, x) \mapsto e^{-i\theta} \cdot x. \tag{2}$$

Since μ^X preserves α and is a lifting of the holomorphic action μ^M, it leaves $H(X)$ invariant; therefore, it determines a unitary action of $\mathbf{T}^1$ on $H(X)$. Thus $H(X)$ equivariantly and unitarily decomposes into the Hilbert direct sum of its isotypes,

$$H_k^\mu(X) =: \left\{ f \in H(X) \, : \, f\left(\mu^X_{g^{-1}}(x)\right) = g^k f(x) \ \forall\,(g,x) \in \mathbf{T}^1 \times X \right\}, \tag{3}$$

where $k \in \mathbb{Z}$. If μ^M is trivial and $\Phi = 1$, (3) is the standard kth equivariant Szegö kernel $H_k(X)$, which is unitarily isomorphic to $H^0\left(M, A^{\otimes k}\right)$ in a natural manner. However, in general $H_k^\mu(X)$ is not a space of sections of powers of A, and may even be infinite-dimensional. For example, if μ^M is trivial and $\Phi = 0$ then $H_0^\mu(X) = H(X)$, while $H_k^\mu(X)$ is the null space for $k \neq 0$.

Nonetheless, if $\Phi > 0$ then $H_k^\mu(X)$ is finite-dimensional for any $k \in \mathbb{Z}$, and is the null space if $k < 0$ [28]; in particular, the orthogonal projector

$\Pi_k^\mu : L^2(X) \to H_k^\mu(X)$ is a smoothing operator, with Schwartz kernel $\Pi_k^\mu(\cdot,\cdot) \in C^\infty(X \times X)$ given by

$$\Pi_k^\mu(x,y) = \sum_j s_j^{(k)}(x) \cdot \overline{s_j^{(k)}(y)} \quad (x, y \in X) \tag{4}$$

for any choice of an orthonormal basis $\left(s_j^{(k)}\right)$ of $H_k^\mu(X)$. The diagonal restriction $x \mapsto \Pi_k^\mu(x,x)$ descends to a well-defined C^∞ function on M.

Also, if $\Phi > 0$ then $\xi_X(x) \neq 0$ for every $x \in X$ by (1); hence μ^X is locally free, and every $x \in X$ has finite stabilizer $T_x \subseteq \mathbf{T}^1$. As μ^X commutes with scalar multiplication, T_x only depends on $m = \pi(x) \in M$; we shall emphasize this by denoting T_x by T_m. For instance, $T_m = \{1\}$ for every $m \in M$ if μ^M is trivial and $\Phi = 1$. While T_m is generally not constant on M, it equals a fixed finite subgroup $T_{\mathrm{gen}} \subseteq \mathbf{T}^1$ on a dense open subset $M' \subseteq M$. Then T_{gen} stabilizes every $x \in X$; after passing to the quotient, we may reduce to the case $T_{\mathrm{gen}} = \{1\}$. By Corollary 1.1 of [28], at a point $x \in X$ where T_m is trivial, $\Pi_k^\mu(x,x)$ satisfies an asymptotic expansion as $k \to +\infty$ of the form

$$\Pi_k^\mu(x,x) \sim \left(\frac{k}{\pi}\right)^d \sum_{j \geq 0} k^{-j} S_j^\mu(m), \tag{5}$$

where $S_0^\mu(m) = \Phi(m)^{-(d+1)}$. Here we shall focus on the lower-order terms S_j^μ.

More generally, given a real $f \in C^\infty(M)$, one can consider the associated Toeplitz operators $T_k^\mu[f] =: \Pi_k^\mu \circ M_f \circ \Pi_k^\mu$, viewed as self-adjoint endomorphisms of $H_k^\mu(X)$; here $M_f : L^2(X) \to L^2(X)$ is multiplication by $f \circ \pi$. Assuming $\Phi > 0$, this is also a smoothing operator, whose distributional kernel may be expressed as

$$\begin{aligned} T_k^\mu[f](x,x') &= \int_X \Pi_k^\mu(x,y)\, f(y)\, \Pi_k^\mu(y,x')\, dV_X(y) \\ &= \sum_j T_k^\mu[f](s_j^{(k)})(x) \cdot \overline{s_j^{(k)}(y)} \qquad (x, y \in X), \end{aligned} \tag{6}$$

where we write $f(y)$ for $f(\pi(y))$. The diagonal restriction $x \mapsto T_k^\mu[f](x,x)$ also descends to M. We shall see that $T_k^\mu[f](x,x')$ has near diagonal scaling asymptotics (that is, for $x \to x'$) analogous to those of Π_k^μ in Theorem 1 of [28], and investigate the lower-order terms in the asymptotics of the diagonal restriction $T_k^\mu[f](x,x)$. We shall then derive from this an asymptotic expansion for an "equivariant Berezin transform," and consider the relation between commutators of Toeplitz operators and Poisson brackets of the corresponding Hamiltonians. Before describing our results in detail, we need to specify the geometric setting somewhat.

We shall assume without loss that T_{gen} is trivial; then μ^X is free on a dense $\nu^X \times \mu^X$-invariant open subset $X' \subseteq X$ (since ν^X – given by (2) – and μ^X commute, we may consider the product action). Thus $M' =: \pi(X') \subseteq M$ is also open and dense.

The quotient $N = X/\mathbf{T}^1$ is an orbifold, and the dense open subset $N' =: X'/\mathbf{T}^1 \subseteq N$ is a manifold; the restricted projection $\kappa \colon X' \to N'$ is a circle bundle, and passing from π to κ the roles of μ^X and ν^X get interchanged.

More precisely, $\beta =: \alpha/\Phi$ is a connection 1-form for κ, defining the same horizontal distribution as α, and there is on N' a naturally induced Kähler structure (N', I, η) with $d\beta = 2\,\kappa^*(\eta)$, and if ω is real-analytic then so is η. Furthermore, ν^X descends to an action $\nu^N \colon \mathbf{T}^1 \times N' \to N'$, which turns out to be holomorphic with respect to I and Hamiltonian with respect to $2\,\eta$. If as generating Hamiltonian for ν^N we choose Φ^{-1}, descended to a function on N', ν^X is the corresponding contact lift of ν^N to (X', β) in the sense of (1).

Every μ^M-invariant C^∞ function $f = f(m)$ on M lifts to a $\nu^X \times \mu^X$-invariant function $f = f(x)$ on X, and then descends to a ν^N-invariant C^∞ function $f = f(n)$ on N'. In the reverse direction, a C^∞ ν^N-invariant function $f = f(n)$ on N' yields a μ^M-invariant C^∞ function $f = f(m)$ on M'. We thus have a natural algebraic isomorphism between spaces of invariant smooth functions:

$$C^\infty(M')^\mu \cong C^\infty(N')^\nu.$$

If ω is real-analytic, this restricts to an isomorphism between the corresponding subspaces of invariant real-analytic functions:

$$C^\varpi(M')^\mu \cong C^\varpi(N')^\nu.$$

With this understanding, we shall think of Φ as being defined on M, X, or N according to the context, and drop the symbols of pull-back or push-forward. Similarly, let ϱ_N be the scalar curvature of the Kähler structure $(N', I, 2\,\eta)$; then ϱ_N is ν^N-invariant, and may be viewed as a μ^M-invariant function on M'. By the same principle, the Laplace–Beltrami operator Δ_N of $(N', I, 2\,\eta)$ acts on μ^M-invariant functions on M' (see Section 2.1 for precise definitions).

An important ingredient of the present analysis is the study by Engliš of the asymptotics of Laplace integrals on a real-analytic Kähler manifold. Namely, let $(g_{k\bar{l}})$ be a real-analytic Kähler metric on an open subset $U \subseteq \mathbf{C}^d$, and suppose that Ξ is a Kähler potential for $(g_{k\bar{l}})$ on U. Let $\widetilde{\Xi}$ be a sesquiholomorphic extension of Ξ to some open neighborhood $\widehat{U} \subseteq U \times U$ of the diagonal. Calabi's *diastasis function* is given by

$$\mathcal{D}(z, w) =: \Xi(z) + \Xi(w) - \widetilde{\Xi}(z, w) - \widetilde{\Xi}(w, z) \qquad \left((z, w) \in \widehat{U}\right); \tag{7}$$

it is an intrinsic attribute of $(g_{k\bar{l}})$, that is, it does not depend on the choice of Ξ, and it satisfies $\mathcal{D}(z,z) = 0$ and $\mathcal{D}(z,w) > 0$ if $z \neq w$ [9] (see also the discussions in [8] and [20]).

In [13], Engliš considers the asymptotics as $\lambda \to +\infty$ of integrals of the form

$$I(\lambda, y) =: \int_U e^{-\lambda \mathcal{D}(x,y)} f(x)\, g(x)\, dx, \tag{8}$$

where $g =: \det[g_{k\bar{l}}]$ and dx denotes the Lebesgue measure on $\mathbb{C}^d$. By Theorem 3 of [13], there is an asymptotic expansion of the form

$$I(\lambda, y) \sim \left(\frac{\pi}{\lambda}\right)^d \sum_{j\geq 0} \lambda^{-j}\, R^U_j(f)\Big|_y, \tag{9}$$

where the R^U_js are covariant differential operators, that may be expressed in a universal manner in terms of the metric, the curvature tensor, and their covariant derivatives; in particular, $R_0 = \mathrm{id}$ and $R_1 = \Delta_N - \varrho_N/2$ (the opposite sign convention is used in [13] for the curvature tensor and for ϱ_N). Engliš also provided an explicit description of R^U_j for $j \leq 3$; the higher R^U_js and their differential geometric significance were investigated further in [20], and a graph-theoretic formula for them was given in [33]. Because $\mathcal{D}$ and the R^U_ls are intrinsically defined, the expansion (9) holds globally on any real-analytic Kähler manifold (S, g), in which case we shall denote the covariant operators by R^S_j.

Theorem 1 *With the notation above, suppose that ω is real-analytic, $\Phi > 0$, and T_{gen} is trivial. Then the invariant functions $S^\mu_j : M' \to \mathbb{R}$ in (5) are determined as follows. First, $S^\mu_0 = \Phi^{-(d+1)}$. Next, for some $j \geq 0$ suppose inductively that*

$$S^\mu_0, \ldots, S^\mu_j \in C^\varpi(M')^\mu \cong C^\varpi(N')^\nu$$

have been constructed, and let $\widetilde{S^\mu_0}, \ldots, \widetilde{S^\mu_j}$ be their respective sesquiholomorphic extensions as elements of $C^\varpi(N')^\nu$. Define

$$Z_j(n_0, n) =: \Phi(n)^{d+1} \sum_{a+b=j} \widetilde{S^\mu_a}(n_0, n)\, \widetilde{S^\mu_b}(n, n_0). \tag{10}$$

Then, thinking of the R^N_rs as acting on the variable n and of n_0 as a parameter,

$$\begin{aligned} S^\mu_{j+1}(n_0) = &-\Phi(n_0)^{d+1} \sum_{l=1}^{j} S^\mu_l(n_0)\, S^\mu_{j+1-l}(n_0) \\ &- \sum_{r=1}^{j+1} R^N_r\big(Z_{j+1-r}(n_0, \cdot)\big)\big|_{n=n_0}. \end{aligned} \tag{11}$$

Since the R_r^Ns are universal intrinsic attributes of the Kähler manifold (N, K, η), (11) expresses the S_j^μs as a universal intrinsic attribute of the Hamiltonian action, through the geometry of its quotient. As mentioned, the R_r^Ns were computed in Section 4 of [13], in [20] and [33]; thus, in principle, (11) determines S_l^μ explicitly in terms of the geometry of the quotient N'. Let us consider S_1^μ:

Corollary 2 *Under the assumptions of Theorem 1, we have*

$$S_1^\mu = \frac{1}{2}\varrho_N \,\Phi^{-(d+1)}$$
$$+(d+1)\,\Phi(n_0)^{-(d+2)}\left[\frac{1}{2\,\Phi}\left\|\mathrm{grad}_N(\Phi)\right\|^2 - \Delta_N(\Phi)\right].$$

Here ϱ_N, the gradient $\mathrm{grad}_N\Phi$ of Φ as a function on N', and the Laplacian $\Delta_N(\Phi)$ are taken with respect to the Kähler structure $(N, I, 2\eta)$, and $\|\cdot\|_N$ is the norm in the same metric. Their relation to the corresponding objects on M is explained in Sections 2.2 and 2.8 (see (34) and (71)). If $\Phi = 1$, we recover Lu's subprincipal term [22].

Remark 3 A notational remark is in order. If, working in a system of local holomorphic coordinates, $\gamma_{a\bar{b}}$ is a Kähler form, the corresponding Kähler metric here is $\rho_{a\bar{b}} = -i\,\gamma_{a\bar{b}}$ (see the discussion in Section 2.1 and (17)). In the literature, often a factor $1/2$ (or $1/(2\pi)$) is included on the LHS of the previous relation; with this convention, the previous invariants would be associated with (N, I, η) [22, 32].

Next let us dwell on the local asymptotics of the Toeplitz kernels $T_k^\mu[f](\cdot,\cdot)$. Firstly, by Theorem 1 of [28] we have $\Pi_k^\mu(x', x'') = O(k^{-\infty})$ uniformly for $\mathrm{dist}_X\left(\mathbf{T}^1 \cdot x', x''\right) \geq C\,k^{\epsilon-1/2}$, for any given $\epsilon > 0$. In view of (6), the same holds of $T_k^\mu[f]$. We can then focus on the local asymptotics of $T_k^\mu[f](x', x'')$ for $x'' \to \mathbf{T}^1 \cdot x'$. In view of (3) and (4), for any $e^{i\vartheta} \in \mathbf{T}^1$ we have

$$T_k^\mu[f]\left(\mu^X_{e^{-i\vartheta}}(x'), x''\right) = e^{ik\vartheta}\,T_k^\mu[f]\,(x', x'') = T_k^\mu[f]\left(x', \mu^X_{e^{i\vartheta}}(x'')\right). \tag{12}$$

Therefore, we need only consider the asymptotics of $T_k^\mu[f](x', x'')$ for $x'' \to x'$. Predictably, these exhibit the same kind of scaling behavior as the asymptotics of $\Pi_k^\mu(x', x'')$ for $x' \to x''$ (Theorem 2 of [28]).

This is best expressed in terms of Heisenberg local coordinates (in the following: HLC for short) $x + (\theta, \mathbf{v})$ centered at a given $x \in X$; here $(\theta, \mathbf{v}) \in (-\pi, \pi) \times B_{2d}(\mathbf{0}, \delta)$, where $B_{2d}(\mathbf{0}, \delta) \subseteq \mathbb{C}^d$ is the open ball centered at the origin and of radius $\delta > 0$. It is in these coordinates that the near-diagonal scaling asymptotics of the standard equivariant Szegö kernels Π_k exhibit their universal nature [3, 30], and by [28] the same holds of the Π_k^μs. While we refer to [30] for a precise definition, let us recall that Heisenberg local coordinates enjoy the following properties.

Firstly, the parameterized submanifold $\gamma_x\colon \mathbf{v} \mapsto x + (0, \mathbf{v})$ is horizontal, that is, tangent to $\ker(\alpha) \subseteq TX$, at $\mathbf{v} = \mathbf{0}$. In view of (1), and given that $\Phi > 0$, γ_x is transverse to the μ^X-orbit $\mathbf{T}^1 \cdot x$; hence for $\mathbf{v} \sim \mathbf{0}$ we have

$$D_2\,\|\mathbf{v}\| \geq \mathrm{dist}_X\left(\mathbf{T}^1 \cdot x, x + \mathbf{v}\right) \geq D_1\,\|\mathbf{v}\|, \tag{13}$$

for some fixed D_1, $D_2 > 0$.

Since HLC centered at $x \in X$ come with a built-in unitary isomorphism $T_mM \cong \mathbb{C}^d$, where $m = \pi(x) \in X$, we may use the expression $x + (\theta, \mathbf{v})$ when $\mathbf{v} \in T_mM$ has sufficiently small norm.

Finally, scalar multiplication by $e^{i\vartheta} \in \mathbf{T}^1$ is expressed in HLC by a translation in the angular coordinate: where both terms are defined, we have

$$e^{i\vartheta} \cdot (x + (\theta, \mathbf{v})) = x + (\vartheta + \theta, \mathbf{v}). \tag{14}$$

We shall set $x + \mathbf{v} =: x + (0, \mathbf{v})$.

Given (12) and the previous transversality argument, we need only consider the asymptotics of $T^\mu_k[f](x + \mathbf{v}, x + \mathbf{w})$ for $\mathbf{v}, \mathbf{w} \to 0$. Following [30], let us define, for $\mathbf{v}, \mathbf{w} \in T_mM$,

$$\psi_2(\mathbf{v}, \mathbf{w}) =: -i\,\omega_m(\mathbf{v}, \mathbf{w}) - \frac{1}{2}\,\|\mathbf{v} - \mathbf{w}\|^2_m, \tag{15}$$

where $\|\cdot\|_m$ is the Euclidean norm on the unitary vector space (T_mM, ω_m, J_m).

Theorem 4 *Assume as above that* $\mathbf{\Phi} > 0$. *Then for any* $f \in C^\infty(M)^\mu$ *we have*

(1) $T^\mu_k[f] = 0$ *for any* $k \leq 0$.
(2) For any $C, \epsilon > 0$, *we have* $T^\mu_k[f](x', x'') = O(k^{-\infty})$ *as* $k \to +\infty$, *uniformly for* $\mathrm{dist}_X\left(\mathbf{T}^1 \cdot x', x''\right) \geq C\,k^{\epsilon - 1/2}$.
(3) Suppose $x \in \mathbf{X}$ *and fix a system of HLC on X centered at x. Set* $m =: \pi(x)$. *Then uniformly for* $\mathbf{v}, \mathbf{w} \in T_mM$ *with* $\|\mathbf{v}\|, \|\mathbf{w}\| \leq C\,k^{1/9}$, *as* $k \to +\infty$ *we have an asymptotic expansion of the form*

$$T^\mu_k[f]\left(x + \frac{\mathbf{v}}{\sqrt{k}}, x + \frac{\mathbf{w}}{\sqrt{k}}\right) = \left(\frac{k}{\pi}\right)^d \sum_{t \in T_m} t^k\, e^{\psi_2\left(d_m\mu^M_{t^{-1}}(\mathbf{v}), \mathbf{w}\right)/\Phi(m)} \cdot A_t(m, \mathbf{v}, \mathbf{w}),$$

with

$$A_t(m, \mathbf{v}, \mathbf{w}, f) \sim \sum_{j \geq 0} k^{-j/2}\, R_j\left(m, d_m\mu^M_{t^{-1}}(\mathbf{v}), \mathbf{w}, f\right),$$

where the $R_j(\cdot,\cdot,\cdot,\cdot)$'s are polynomial in $\mathbf{v}$ *and* $\mathbf{w}$ *and differential operators in* f. *In particular,*

$$R_0\left(m, d_m\mu_{t^{-1}}^M(\mathbf{v}), \mathbf{w}, f\right) = \Phi(m)^{-(d+1)} f(m).$$

(4) The previous asymptotic expansion goes down by integer steps when $\mathbf{v} = \mathbf{w} = 0$ *(that is, only powers of* k^{-1} *appear in the diagonal asymptotics).*

Theorem 4 might be proven by a microlocal argument along the lines of that used for Theorem 1 of [28]; to avoid introducing too much machinery, we shall instead deduce it as a consequence of Theorem 1 of [28], by inserting in (6) the near-diagonal scaling asymptotics for Π_k^μ.

Corollary 5 *In the situation of Theorem 4, suppose in addition that* T_{gen} *is trivial. If* $x \in X'$, *then as* $k \to +\infty$ *there is an asymptotic expansion*

$$T_k^\mu[f](x,x) \sim \left(\frac{k}{\pi}\right)^d \sum_{j\geq 0} k^{-j} S_j^\mu[f](m),$$

where $m = \pi(x)$ *and every* $S_j^\mu[f] \in C^\infty(M')^\mu$. *In particular, we have*

$$S_0^\mu[f] = \Phi^{-(d+1)} \cdot f.$$

When $\Phi = 1$, corresponding results were obtained in Lemma 4.6 of [24] and Lemma 7.2.4 of [23], covering the case of symplectic manifolds in the presence of a twisting vector bundle.

Let us consider the lower-order $S_j^\mu[f]$s.

Theorem 6 *Under the assumptions of Corollary 5, assume also that* ω *is real-analytic. Then for every* $j = 0, 1, 2, \ldots$ *we have* $S_j^\mu[f] = P_j^\mu(f)$, *where each* P_j^μ *is a differential operator of degree* $\leq 2j$. *More precisely, viewed as a* ν^N*-invariant function on* N, $S_j^\mu[f]$ *is given by*

$$S_j^\mu[f](n_0) = P_j^\mu(f)(n_0) = \sum_{r+s=j} R_r^N(f(\cdot)\, Z_s(n_0,\cdot))\Big|_{n=n_0}.$$

Remark 7 Clearly, $S_j^\mu = S_j^\mu[1]$ for every $j \geq 0$.

Corollary 8 *In the situation of Theorem 6,*

$$S_1^\mu[f] = \Phi^{-(d+1)}\, \Delta_N(f) + S_1^\mu \cdot f.$$

For $\Phi = 1$, the corresponding result to Corollary 8 was obtained in (0.13) of [26].

For a general discussion of the Berezin transform in the Kähler context, we refer, say, to [1, 7, 13, 21, 29]. Here we adopt the following natural adjustment:

Definition 9 If $f \in C^\infty(M)$ and $k = 0, 1, 2, \ldots$, let the kth *μ-equivariant Berezin transform* of f be given by

$$\mathrm{Ber}^\mu_k[f](m) =: \frac{T^\mu_k[f](x,x)}{\Pi^\mu_k(x,x)} \qquad (m \in M)$$

for any choice of $x \in \pi^{-1}(m)$.

Corollary 10 *Assume that ω is real-analytic, $\Phi > 0$, and $T_{\mathrm{gen}} = \{1\}$. If $f \in C^\infty(M)^\mu$, then as $k \to +\infty$ on M', uniformly on compact subsets of M', there is an asymptotic expansion of the form*

$$\mathrm{Ber}^\mu_k[f] \sim \sum_{j \geq 0} k^{-j} B^\mu_j(f),$$

where every B^μ_j is a differential operator of degree $2j$. In particular, $B^\mu_0 = \mathrm{id}$ and $B^\mu_1 = \Delta_N$.

A corresponding result for $\Phi = 1$ was given in [13].

The following analogue of the Heisenberg correspondence relates the commutator of two equivariant Toeplitz operators to the Poisson brackets of the corresponding Hamiltonians. Let $\{\cdot,\cdot\}_M$ and $\{\cdot,\cdot\}_N$ denote, respectively, Poisson brackets on $(M, 2\,\omega)$ and $(N', 2\,\eta)$. By restriction, they yield maps

$$\{\cdot,\cdot\}_M, \{\cdot,\cdot\}_N : C^\infty(M')^\mu \times C^\infty(M')^\mu \to C^\infty(M')^\mu.$$

Theorem 11 *Assume that ω is real-analytic, $\Phi > 0$, and $T_{\mathrm{gen}} = \{1\}$. Let $f, g \in C^\infty(M)^\mu$ be real-valued, and denote by $E^\mu_k[f,g](\cdot,\cdot) \in C^\infty(X \times X)$ the Schwartz kernel of the composition $T^\mu_k[f] \circ T^\mu_k[g]$. Then uniformly on compact subsets of M' as $k \to +\infty$ we have*

$$\begin{aligned} &E^\mu_k[f,g](x,x) - E^\mu_k[f,g](x,x) \\ &= \left(\frac{k}{\pi}\right)^d \left[-\frac{i}{k}\,\Phi(m)^{-(d+1)}\,\{f,g\}_N(m) + O\left(k^{-2}\right)\right] \\ &= \left(\frac{k}{\pi}\right)^d \left[-\frac{i}{k}\,\Phi(m)^{-d}\,\{f,g\}_M(m) + O\left(k^{-2}\right)\right], \end{aligned}$$

for any $x \in \pi^{-1}(m)$.

In the course of the proof, one actually establishes an asymptotic expansion for $E^\mu_k[f,g](x,x)$ (see (118)):

$$E^\mu_k[f,g](x,x) \sim \left(\frac{k}{\pi}\right)^d \sum_j k^{-j} A_j[f,g](x), \tag{16}$$

where $A_0[f,g] = \Phi^{-(d+1)} \cdot f\,g$ and

$$A_1[f,g] = \Phi^{-(d+1)} \left[f\,\Delta_N g + g\,\Delta_N f + \left\langle \mathrm{grad}_N(f)^{(0,1)}, \mathrm{grad}_N(g)^{(1,0)} \right\rangle \right] + S_1^{\mu} \cdot f\,g$$

(we leave the explicit computation to the reader). When $\Phi = 1$, the formula for $A_1[f,g]$ was obtained in (0.16) of [26].

As explained in the references above for the standard case, this expansion can be used to define in a natural manner a $*$-product on $C^\infty(M')^\mu$ (depending on Φ), but we won't discuss this here.

2 Preliminaries

2.1 Some notation and recalls from Kähler geometry

Let (P,K) be a d-dimensional complex manifold and let (P,K,γ) be a Kähler structure on it, with associated covariant metric tensor $\rho(\cdot,\cdot) =: \gamma(\cdot, K(\cdot))$; also, let $\ell =: \rho - i\gamma$ be the associated Hermitian metric. Given holomorphic local coordinates (z_a) on P, we shall let $\partial_a =: \partial/\partial z_a$ and $\partial_{\overline{a}} =: \partial/\partial \overline{z}_a$, $\rho_{a\overline{b}} =: \rho(\partial_a, \partial_{\overline{b}})$, $\gamma_{a\overline{b}} =: \gamma(\partial_a, \partial_{\overline{b}})$. $\ell_{a\overline{b}} =: \ell(\partial_a, \partial_{\overline{b}})$. Then locally

$$\gamma = \sum_{a,b} \gamma_{a\overline{b}}\, dz_a \wedge d\overline{z}_b = i \sum_{a,b} \rho_{a\overline{b}}\, dz_a \wedge d\overline{z}_b = \frac{i}{2} \sum_{a,b} \ell_{a\overline{b}}\, dz_a \wedge d\overline{z}_b. \tag{17}$$

Consider the real local frame $\mathcal{B} = (\partial/\partial x_1, \ldots, \partial/\partial x_d, \partial/\partial y_1, \ldots, \partial/\partial y_d)$, where $z_j = x_j + i\,y_j$ is the decomposition in real and imaginary parts, and denote by $M_{\mathcal{B}}(\rho)$ the matrix representing ρ in this frame. Then

$$\det M_{\mathcal{B}}(\rho) = 4^d \det\left([\rho_{a\overline{b}}]\right)^2.$$

Therefore, the Riemannian volume form of (P,ρ) is

$$\begin{aligned} dV_P = \frac{1}{d!}\gamma^{\wedge d} &= \sqrt{\det\left(M_{\mathcal{B}}(\rho)\right)} \cdot dx_1 \wedge \cdots dx_d \wedge dy_1 \cdots \wedge dy_d \\ &= 2^d \det\left([\rho_{k\overline{l}}]\right) \cdot dx_1 \wedge \cdots dx_d \wedge dy_1 \cdots \wedge dy_d \\ &= \det\left([2\,\rho_{k\overline{l}}]\right) \cdot dx_1 \wedge \cdots dx_d \wedge dy_1 \cdots \wedge dy_d. \end{aligned} \tag{18}$$

Let R be the covariant curvature tensor of the Riemannian manifold (P,ρ), with components $R_{a\overline{b}c\overline{d}} = R(\partial_a, \partial_{\overline{b}}, \partial_c, \partial_{\overline{d}})$ [32].

We shall set (leaving the metric understood and adopting Einstein notation)

$$\varrho_P =: \rho^{\overline{b}a}\,\rho^{\overline{d}c}\,R_{a\overline{b}c\overline{d}}; \tag{19}$$

this is 1/2 of the ordinary Riemannian scalar curvature scal_P.

Similarly, for $f \in C^\infty$, we shall let

$$\Delta_P(f) =: \rho^{\overline{b}a}\, \partial_a \partial_{\overline{b}} f, \tag{20}$$

which is $1/2$ times the ordinary Riemannian Laplace–Beltrami operator.

The gradient of f is locally given by

$$\operatorname{grad}_P(f) = \rho^{\overline{b}a}\, (\partial_{\overline{b}} f)\, \partial_a + \rho^{\overline{b}a}\, (\partial_a f)\, \partial_{\overline{b}}, \tag{21}$$

and its square norm is given by

$$\left\|\operatorname{grad}_P(f)\right\|^2 = 2\,\rho^{\overline{b}a}\, (\partial_a f)\, (\partial_{\overline{b}} f). \tag{22}$$

Since Δ_P here is $1/2$ times the ordinary Laplace–Beltrami operator, we have for any $f_1, f_2 \in C^\infty(P)$:

$$\Delta_P(f_1 \cdot f_2) = f_1\, \Delta_P(f_2) + \rho(\operatorname{grad}_P(f_1), \operatorname{grad}_P(f_2)) + f_2\, \Delta_P(f_1).$$

It follows inductively that for any $f \in C^\infty(P)$ and $l \geq 0$, we have

$$\Delta_P\left(f^l\right) = l\, f^{l-1}\, \Delta_P(f) + \frac{(l-1)\,l}{2}\, f^{l-2}\, \left\|\operatorname{grad}_P(f)\right\|^2. \tag{23}$$

Let us now consider the Poisson brackets $\{f, g\}_P = \gamma(H_f, H_g)$ of two real functions $f, g \in C^\infty(P)$ in the symplectic structure (P, γ); here H_f is the Hamiltonian vector field of f with respect to γ. We have $H_f = -K(\operatorname{grad}_P(f))$, hence given (21)

$$\begin{aligned}
\{f, g\}_P &= \gamma\big(K(\operatorname{grad}_P(f)), K(\operatorname{grad}_P(g))\big) = \gamma(\operatorname{grad}_P(f), \operatorname{grad}_P(g)) \\
&= \rho\big(K(\operatorname{grad}_P(f)), \operatorname{grad}_P(g)\big) \\
&= i\,\rho\big(\rho^{\overline{b}a}\, (\partial_{\overline{b}} f)\, \partial_a - \rho^{\overline{b}a}\, (\partial_a f)\, \partial_{\overline{b}}, \rho^{\overline{d}c}\, (\partial_{\overline{d}} g)\, \partial_c + \rho^{\overline{d}c}\, (\partial_c g)\, \partial_{\overline{d}}\big) \\
&= \frac{1}{i}\, \rho^{\overline{d}c}\, \big[(\partial_c f)\, (\partial_{\overline{d}} g) - (\partial_c g)\, (\partial_{\overline{d}} f)\big].
\end{aligned} \tag{24}$$

Lemma 12 *Let (P, K, γ) be a Kähler manifold, with γ real-analytic. Let $\Phi \colon P \to \mathbb{R}$ be a real $C^\infty(M)$ function whose Hamiltonian flow with respect to γ is holomorphic with respect to K. Then Φ is real-analytic.*

Proof Let $T^c P = TP \otimes \mathbb{C}$ be the complexified tangent bundle of P, and $T^c P = T'P \oplus T''P$ its decomposition into $\pm i$-eigenbundles of K. Let $\upsilon_\Phi \in \mathfrak{X}(P)$ be the Hamiltonian vector field of Φ with respect to γ. If $\upsilon_\Phi = \upsilon'_\Phi \oplus \upsilon''_\Phi$, with $\upsilon'_\Phi \in T'P$ and $\upsilon''_\Phi = \overline{\upsilon'_\Phi} \in T''P$, then υ'_Φ is holomorphic, whence real-analytic. Then clearly υ_Φ is real-analytic as well, and therefore so is its differential

$dv_\Phi = \iota(v_\Phi)\gamma$. This forces Φ itself to be real-analytic (say by Proposition 2.2.10 of [18]). □

The Laplacian and sesquiholomorphic extensions

We give here a couple of technical lemmas that will be handy in the proof of Corollary 2.

Lemma 13 *Let (P, K, γ) be a Kähler manifold, and consider $f \in C^\varpi(P)$ with $f > 0$. Let $\widetilde{f}(\cdot,\cdot)$ be the sesquiholomorphic extension of f to an open neighborhood $\widetilde{P} \subseteq P \times P$ of the diagonal (thus $\widetilde{f}(\cdot,\cdot)$ is holomorphic in the first entry and antiholomorphic in the second, and $\widetilde{f}(p,p) = f(p)$ for all $p \in P$). Given $p_0 \in P$, let $P' \subseteq P$ be an open neighborhood of p_0 so small that $P' \times P' \subseteq \widetilde{P}$ and $\widetilde{f}(p_0, p) \neq 0$ for all $p \in P'$. Define f_1, f_2, $F_f \in C^\varpi(P')$ by setting for $p \in P'$:*

$$f_1(p) =: f(p_0, p), \quad f_2(p) =: f(p, p_0) = \overline{f_1(p)}, \quad F_f(p) = \frac{f(p)}{f_1(p)\, f_2(p)}.$$

Thus f_1 is antiholomorphic, f_2 is holomorphic, and $F_f > 0$. Then

$$\Delta_P(F_f)(p_0) = \frac{1}{f(p_0)^2}\left[\Delta_P(f)(p_0) - \frac{1}{2\, f(p_0)}\left\|\mathrm{grad}_P(f)(p_0)\right\|^2\right], \tag{25}$$

where the terms involved are given by (20) and (22).

Remark 14 To be precise, we should really write F_{f,p_0} for F_f, since the latter also depends on the reference point.

Proof As above, let ρ be the metric tensor. In a local holomorphic chart (z_a) for P centered at p_0, given that $\partial_a \partial_{\overline{b}} f_j = 0$ we have

$$\begin{aligned}
\Delta_P(F) &= \rho^{\overline{b}a}\, \partial_a \partial_{\overline{b}} \left(\frac{f}{f_1\, f_2}\right) \\
&= \rho^{\overline{b}a}\, \partial_a \left(\frac{1}{f_1\, f_2}\, \partial_{\overline{b}} f - \frac{f}{f_1^2\, f_2}\, \partial_{\overline{b}} f_1\right) \\
&= \rho^{\overline{b}a} \left(-\frac{1}{f_1\, f_2^2}\, \partial_a f_2\, \partial_{\overline{b}} f + \frac{1}{f_1\, f_2}\, \partial_a\, \partial_{\overline{b}} f - \frac{1}{f_1^2\, f_2}\, \partial_a f\, \partial_{\overline{b}} f_1 \right.\\
&\quad \left. + \frac{f}{f_1^2\, f_2^2}\, \partial_a f_2\, \partial_{\overline{b}} f_1\right).
\end{aligned} \tag{26}$$

At p_0, $\partial_a f_2(p_0) = \partial_a f(p_0)$, $\partial_{\overline{b}} f_1(p_0) = \partial_{\overline{b}} f(p_0)$, and $f_1(p_0) = f_2(p_0) = f(p_0)$.

Thus, (26) yields

$$\begin{aligned}\Delta_P(F)(p_0) &= \rho^{\overline{b}a}(p_0)\left(-\frac{1}{f(p_0)^3}\,\partial_a f(p_0)\,\partial_{\overline{b}} f(p_0) + \frac{1}{f(p_0)^2}\,\partial_a\,\partial_{\overline{b}} f(p_0)\right.\\ &\quad\left.-\frac{1}{f(p_0)^3}\,\partial_a f(p_0)\,\partial_{\overline{b}} f(p_0) + \frac{1}{f(p_0)^3}\,\partial_a f(p_0)\,\partial_{\overline{b}} f(p_0)\right)\\ &= \frac{1}{f(p_0)^2}\left[\rho^{\overline{b}a}(p_0)\,\partial_a\,\partial_{\overline{b}} f(p_0) - \frac{1}{f(p_0)}\,\rho^{\overline{b}a}(p_0)\,\partial_a f(p_0)\,\partial_{\overline{b}} f(p_0)\right]\\ &= \frac{1}{f(p_0)^2}\left[\Delta_P(f)(p_0) - \frac{1}{2\,f(p_0)}\,\left\|\mathrm{grad}_P(f)(p_0)\right\|_P^2\right].\end{aligned}$$

□

Lemma 15 *With the hypothesis and notation of Lemma 13, we have*

$$\mathrm{grad}_P(F_f)(p_0) = 0.$$

Proof Let again (z_a) be a local holomorphic coordinate chart for P centered at p_0. Then for every a we have

$$\begin{aligned}\partial_a(F_f)(p_0) &= \frac{1}{f_1(p_0)}\,\partial_a\left(\frac{f}{f_2}\right)(p_0)\\ &= \frac{f_2(p_0)\,\partial_a f(p_0) - f(p_0)\,\partial_a f_2(p_0)}{f_1(p_0)\,f_2(p_0)^2} = \frac{f(p_0)\,\partial_a f(p_0) - f(p_0)\,\partial_a f(p_0)}{f(p_0)^3} = 0.\end{aligned}$$

Similarly, $\partial_{\overline{a}} F_f(p_0) = 0$ for every a. □

2.2 The Kähler structure on N'

We are assuming $\Phi > 0$ and T_{gen} trivial. Then the two projections

$$M' \overset{\pi}{\longleftarrow} X' \overset{\kappa}{\longrightarrow} N'$$

are circle bundle structures; the fibers of π are the orbits in X' of ν^X and those of κ are the orbits in X' of μ^X.

Let $\mathcal{H} = \ker(\alpha) \subseteq TX$ be the horizontal distribution for π. Since α is μ^X-invariant, so is $\mathcal{H}$. In addition, by (1) $\mathcal{H}$ is transverse to every μ^X-orbit. Therefore, it may be viewed as an invariant horizontal distribution for κ as well.

Let $J_{\mathcal{H}}$ be the complex structure that $\mathcal{H}$ inherits from J by the isomorphism $d\pi|_{\mathcal{H}} : \mathcal{H} \cong \pi^*(TM)$. Since μ^M is holomorphic, $J_{\mathcal{H}}$ is μ^X-invariant. Therefore, given the isomorphism $d\kappa|_{\mathcal{H}} : \mathcal{H} \cong \kappa^*(TN')$, it descends to an almost complex structure I on N'.

Proposition 16 *I is a complex structure.*

Proof Let $\mathcal{J}$ be the complex distribution on M associated with J. Thus

$$\mathcal{J} = \{\mathbf{v} - i\,J(\mathbf{v}) \,:\, \mathbf{v} \in TM\} = \ker(J - i\,\mathrm{id}) \subseteq TM \otimes \mathbb{C}.$$

As J is integrable, $\mathcal{J}$ is involutive.

Similarly, let

$$\mathcal{J}_{\mathcal{H}} = \{\mathbf{h} - i\,J_{\mathcal{H}}(\mathbf{h}) \,:\, \mathbf{h} \in \mathcal{H}\} = \ker(J_{\mathcal{H}} - i\,\mathrm{id}) \subseteq \mathcal{H} \otimes \mathbb{C}.$$

Evidently, $\mathcal{J}_{\mathcal{H}}$ is the horizontal lift of $\mathcal{J}_M$.

Lemma 17 *$\mathcal{J}_{\mathcal{H}}$ is involutive.*

Proof If $V \in \mathfrak{X}(M)$ is a real vector field on M, then $U =: V - i\,J(V)$ is a complex vector field on X tangent to $J_{\mathcal{H}}$, and its horizontal lift

$$U^\sharp = V^\sharp - i\,J(V)^\sharp = V^\sharp - i\,J_{\mathcal{H}}\left(V^\sharp\right)$$

is a complex vector field on X tangent to $\mathcal{J}_{\mathcal{H}}$. It is clear that $\mathcal{J}_{\mathcal{H}}$ is locally spanned by vector fields of this form, so it suffices to show that $\left[U_1^\sharp, U_2^\sharp\right]$ is tangent to $\mathcal{J}_{\mathcal{H}}$, for any pair of complex vector fields U_1, U_2 on M tangent to $\mathcal{J}_M$.

Since $\mathcal{J}_M$ is involutive, $[U_1, U_2]$ is tangent to $\mathcal{J}_M$. Given that $\left[U_1^\sharp, U_2^\sharp\right]$ is π-correlated to $[U_1, U_2]$, to show that $\left[U_1^\sharp, U_2^\sharp\right]$ is tangent to $\mathcal{J}_{\mathcal{H}}$ it suffices to show that it is horizontal.

On the one hand, by compatibility of ω and J and because by construction $J(U_l) = i\,U_l$, we have

$$\omega(U_1, U_2) = \omega(J(U_1), J(U_2)) = i^2\,\omega(U_1, U_2),$$

so that $\omega(U_1, U_2) = 0$. On the other hand, since $U_l^\sharp$ is horizontal we have $\alpha\left(U_l^\sharp\right) = 0$; therefore, given that $d\alpha = 2\,\pi^*(\omega)$, we get

$$\begin{aligned} 0 = 2\,\omega(U_1, U_2) &= 2\,\pi^*(\omega)\left(U_1^\sharp, U_2^\sharp\right) = d\alpha\left(U_1^\sharp, U_2^\sharp\right) \\ &= U_1^\sharp \cdot \alpha\left(U_2^\sharp\right) - U_2^\sharp \cdot \alpha\left(U_1^\sharp\right) - \alpha\left(\left[U_1^\sharp, U_2^\sharp\right]\right) \\ &= -\alpha\left(\left[U_1^\sharp, U_2^\sharp\right]\right). \end{aligned}$$

□

Finally, let us set

$$\mathcal{I} = \{\mathbf{s} - i\,I(\mathbf{s}) \,:\, \mathbf{s} \in TN'\} = \ker(I - i\,\mathrm{id}) \subseteq TN' \otimes \mathbb{C}.$$

We need to prove that $\mathcal{I}$ is an involutive complex distribution. Let $S_1, S_2 \in \mathfrak{X}(N') \otimes \mathbb{C}$ be complex vector fields on N' tangent to $\mathcal{I}$, and let $\widehat{S}_1$, $\widehat{S}_2$ be their horizontal lifts to X. By definition of I, it follows that the restriction of $\mathcal{J}_{\mathcal{H}}$ to X' is the horizontal lift of $\mathcal{I}$ under κ. Therefore, $\widehat{S}_l$ is tangent to $\mathcal{J}_{\mathcal{H}}$ and μ^X-invariant. Then the same holds of their commutator $\left[\widehat{S}_1, \widehat{S}_2\right]$ because $\mathcal{J}_{\mathcal{H}}$ is involutive and μ^X-invariant. Since $\left[\widehat{S}_1, \widehat{S}_2\right]$ is κ-correlated to $[S_1, S_2]$, we conclude that $[S_1, S_2]$ is tangent to $\mathcal{I}$. □

Let us define

$$\beta =: \frac{1}{\Phi}\,\alpha. \tag{27}$$

Lemma 18 *β is a connection form for $\kappa\colon X' \to N'$, with respect to which the horizontal tangent bundle is $\mathcal{H}$ (the horizontal tangent bundle of π).*

Proof Since μ^X preserves α and lifts μ^M, which is a Hamiltonian action with moment map Φ, β is μ^X-invariant. Furthermore, we see from (1) and (27) that $\beta(\xi_X) = -1$. □

Thus $\mathcal{H} \subseteq TX'$ is the horizontal tangent space for both π and κ. If V is a vector field on M, we shall denote by $V^\sharp$ its horizontal lift to X under π; it is a ν^X-invariant section of $\mathcal{H}$ on X. Similarly, if U is a vector field on N', we shall denote by $\widehat{U}$ its horizontal lift to X' under κ; it is a μ^X-invariant section of $\mathcal{H}$ on X'. Clearly, vector fields on M are the same as ν^X-invariant sections of $\mathcal{H}$ on X, and vector fields on N' are the same as μ^X-invariant sections of $\mathcal{H}$ on X.

Lemma 19 *There exists a unique Kähler form η on N' such that $d\beta = 2\,\kappa^*(\eta)$.*

Proof We have

$$d\beta = \frac{1}{\Phi}\,d\alpha - \frac{1}{\Phi^2}\,d\Phi \wedge \alpha = \frac{2}{\Phi}\,\pi^*(\omega) - \frac{1}{\Phi^2}\,d\Phi \wedge \alpha, \tag{28}$$

and direct inspection using (1) shows that $\iota(\xi_X^\sharp)d\beta = 0$. Since $d\beta$ is μ^X-invariant, it follows that there exists a necessarily unique 2-form η on N' such that $d\beta = \kappa^*(2\,\eta)$.

Thus, η is a closed 2-form on N'. To see that it is in fact a Kähler form, we need to check that it is compatible with the complex structure and non-degenerate. To this end, we fix an arbitrary $n \in M'$, choose an arbitrary $x \in \kappa^{-1}(n)$, and set $m = \pi(x)$. Our construction then yields natural complex-linear isomorphisms $(T_mM, J_m) \cong (\mathcal{H}_x, J_{\mathcal{H}_x}) \cong (T_nN', I_n)$. To see that η_n is non-degenerate on T_nN' and compatible with I_n, it then suffices to see that the restriction of $d\beta$ is non-degenerate on $\mathcal{H}_x$, and compatible with $J_{\mathcal{H}_x}$.

By (28), under the complex-linear isomorphism $(T_mM, J_m) \cong (\mathcal{H}_x, J_{\mathcal{H}_x})$, the restriction of $d\beta$ on $\mathcal{H}_x$ may be identified with $2\omega_m/\Phi(m)$ on T_mM. Since ω is Kähler on (M, J), it is non-degenerate on T_mM and compatible with J_m, and this completes the proof. □

Suppose $f \in C^\infty(M')^\mu \cong C^\infty(N')^\nu$, and let H_f be its Hamiltonian vector field on $(M', 2\,\omega)$. Since f is μ^M-invariant, so is H_f. Let $H_f^\sharp$ be the horizontal lift of H_f to X'. Then $H_f^\sharp$ is a $\mu^X \times \nu^X$-invariant horizontal vector field on X', and therefore it descends to a ν^N-invariant vector field $\overline{H}_f$, respectively.

Lemma 20 *Let K_f be the Hamiltonian vector field of $f \in C^\infty(M')^\mu \cong C^\infty(N')^\nu$ on $(N, 2\,\eta)$. Then $K_f = \Phi\,\overline{H}_f$.*

Proof We need to show that for any $n \in N'$ and $\mathbf{u} \in T_n N'$ we have

$$2\,\Phi(n) \cdot \eta_n(\overline{H}_f(n), \mathbf{u}) = d_n^N f(\mathbf{u}), \tag{29}$$

where $d^N f$ is the differential of f when f is viewed as a function on N.

Choose as before $x \in \kappa^{-1}(n)$ and let $m =: \pi(x) \in M'$. Let $\widehat{\mathbf{u}} \in \mathcal{H}_x$ be the horizontal lift of $\mathbf{u}$ under κ, and set $\mathbf{v} = d_x\pi(\widehat{\mathbf{u}})$. Thus $\widehat{\mathbf{u}} = \mathbf{v}^\sharp$. Since f is invariant, on X we have $f \circ \pi = f \circ \kappa$; thus,

$$d_m^M f(\mathbf{v}) = d_x^X f\left(\mathbf{v}^\sharp\right) = d_x^X f\left(\widehat{\mathbf{u}}\right) = d_n^N f(\mathbf{u}). \tag{30}$$

On the contrary, since $H_f^\sharp = \widehat{\overline{H}_f}$, we have

$$\begin{aligned} &2\,\Phi(n) \cdot \eta_n\left(\overline{H}_f(n), \mathbf{u}\right) \\ &= \Phi(m) \cdot d_x\beta\left(H_f(m)^\sharp, \mathbf{v}^\sharp\right) = \Phi(m) \cdot \frac{1}{\Phi(m)}\, d_x\alpha\left(H_f(m)^\sharp, \mathbf{v}^\sharp\right) \\ &= 2\,\omega_m(H_f(m), \mathbf{v}) = d_m^M f(\mathbf{v}). \end{aligned} \tag{31}$$

(29) follows from (30) and (31). □

Suppose $f, g \in C^\infty(M)^\mu$. Since $C^\infty(M')^\mu \cong C^\infty(N')^\nu$, we have Poisson brackets $\{f, g\}_M \in C^\infty(M')^\mu$ and $\{f, g\}_N \in C^\infty(N')^\nu$ on $(M', 2\,\omega)$ and $(N', 2\,\eta)$, respectively. The relation between them under the previous isomorphism is as follows:

Corollary 21 *For any $f, g \in C^\infty(M')^\mu \cong C^\infty(N')^\nu$, we have $\{f, g\}_N = \Phi\,\{f, g\}_M$.*

Proof We have, omitting symbols of pull-back,

$$\begin{aligned} \{f, g\}_N &= 2\,\eta(K_f, K_g) = \Phi^2\, d\beta\left(H_f^\sharp . H_g^\sharp\right) \\ &= \Phi^2\, \frac{1}{\Phi}\, d\alpha\left(H_f^\sharp . H_g^\sharp\right) = \Phi \cdot 2\,\omega(H_f, H_g) \\ &= \Phi \cdot \{f, g\}_M. \end{aligned} \tag{32}$$

□

We can similarly relate the gradients $\mathrm{grad}_M(f)$ and $\mathrm{grad}_N(f)$ of an invariant f on $(M', 2g)$ and $(N', 2h)$, where $g(\cdot,\cdot) = \omega(\cdot, J(\cdot))$ and $h = \eta(\cdot, I(\cdot))$ are the Riemannian metrics on M and N, respectively. We have

$$\mathrm{grad}_N(f) = I(K_f) = \Phi\, I\left(\overline{H}_f\right) = \Phi\, \overline{J(H_f)} = \Phi\, \overline{\mathrm{grad}_M(f)}. \tag{33}$$

Passing to square norms, we get

$$\begin{aligned}
\|\mathrm{grad}_N(f)\|_N^2 &= 2\, h(\mathrm{grad}_N(f), \mathrm{grad}_N(f)) = 2\,\eta\big(\mathrm{grad}_N(f), I(\mathrm{grad}_N(f))\big) \\
&= 2\,\Phi^2\, \eta\left(\overline{\mathrm{grad}_M(f)}, \overline{J(\mathrm{grad}_M(f))}\right) \\
&= 2\,\Phi^2\, d\beta\left(\mathrm{grad}_M(f)^\sharp, J(\mathrm{grad}_M(f)^\sharp)\right) \\
&= 2\,\Phi\, d\alpha\left(\mathrm{grad}_M(f)^\sharp, J(\mathrm{grad}_M(f)^\sharp)\right) \\
&= \Phi \cdot 2\,\omega(\mathrm{grad}_M(f), J(\mathrm{grad}_M(f)) \\
&= \Phi\, \|\mathrm{grad}_M(f)\|_M^2.
\end{aligned} \tag{34}$$

2.3 The descended action on N

Let us dwell on the Hamiltonian nature of the descended action ν^N. Recall that the action ν^X given by (2), that is, scalar multiplication composed with inversion, commutes with μ^X, hence it descends to an action $\nu^N : \mathbf{T}^1 \times N \to N$.

Lemma 22 *ν^N is an holomorphic action on (N', I).*

Proof Choose $n \in N'$ and $x \in \kappa^{-1}(n)$, and let $m =: \pi(x)$. Fix $t = e^{i\theta} \in \mathbf{T}^1$. By construction, we have complex-linear isomorphisms $T_nN' \cong \mathcal{H}_x \cong T_mM$ that inter-twine $d_n\nu_t^N : T_nN' \to T_{\nu_t^N(n)}N'$ with $d_x\nu_t^X : \mathcal{H}_x \to \mathcal{H}_{\nu_t^X(x)} = \mathcal{H}_{e^{-i\theta}\cdot x}$, hence with the identity map of T_mM. The statement follows. □

Lemma 23 *ν^N is a symplectic action on (N', η).*

Proof This follows as for Lemma 22, since in view of (28) under the previous isomorphism η_n corresponds to $\omega_m/\Phi(m)$. □

Thus ν_t^N is an automorphism of the Kähler manifold (N', I, η), for each $t \in \mathbf{T}^1$.

Lemma 24 *ν^N is a Hamiltonian action on $(N', 2\,\eta)$, with moment map $1/\Phi$ (viewed as a function on N).*

Proof The vector field $-\partial/\partial\theta$ on X is μ^X-invariant, hence it descends to a vector field υ on N, which is the infinitesimal generator of ν^X. We need to show that $2\,\iota(\upsilon)\,\eta = d^N(1/\Phi)$, that is, for any $n \in N'$ and any $\mathbf{u} \in T_nN'$ we have

$$2\,\eta_n(\upsilon, \mathbf{u}) = -\Phi(n)^{-2}\, d_n^N\Phi(\mathbf{u}). \tag{35}$$

As before, let $\widehat{\mathbf{u}}$ be the horizontal lift of $\mathbf{u}$ with respect to κ, and set $\mathbf{v} =: d_x\pi(\widehat{\mathbf{u}})$, so that $\widehat{\mathbf{u}} = \mathbf{v}^\sharp$. Thus

$$
\begin{aligned}
-\Phi(n)^{-2}\, d_n^N\Phi(\mathbf{u}) &= -\Phi(x)^{-2}\, d_x^X\Phi(\widehat{\mathbf{u}}) \\
&= -\Phi(x)^{-2}\, d_x^X\Phi\left(\mathbf{v}^\sharp\right) = -\Phi(m)^{-2}\, d_m^M\Phi(\mathbf{v}).
\end{aligned} \tag{36}
$$

On the contrary, since $\kappa^*(2\eta) = d\beta$, we have

$$
d_x\beta\left(-\frac{\partial}{\partial\theta}, \mathbf{v}^\sharp\right) = d_x\beta\left(-\frac{\partial}{\partial\theta}, \widehat{\mathbf{u}}\right) = 2\,\eta_n(\upsilon, \mathbf{u}). \tag{37}
$$

Then (35) is equivalent to the equality

$$
d_x\beta\left(\frac{\partial}{\partial\theta}, \mathbf{v}^\sharp\right) = \frac{1}{\Phi(m)^2}\, d_m^M\Phi(\mathbf{v}), \tag{38}
$$

for any $m \in M$, $\mathbf{v} \in T_mM$, and $x \in \pi^{-1}(x)$. The latter is an immediate consequence of (28). □

Now β is a connection 1-form for the circle bundle $\kappa\colon X' \to N'$ and is preserved by ν^X; therefore, for an appropriate constant c, ν^X is a contact lift to (X', β) of ν^N, with respect to the Hamiltonian $c + 1/\Phi$.

Lemma 25 *The correct choice is $c = 0$. Furthermore, the horizontal lift of υ with respect to κ is*

$$
\widehat{\upsilon} = -\frac{1}{\Phi}\,\xi_M^\sharp.
$$

Proof We want to give a decomposition of $-\partial/\partial\theta$ analogous to (1), but referred to the circle bundle structure $\kappa\colon X' \to N'$. To this end, let β be a locally defined angular coordinate on X' referred to κ, so that $\xi_X = -\partial/\partial\beta$. Since the horizontal component of $-\partial/\partial\theta$ with respect to κ is $\widehat{\upsilon}$, the analogue of (1) is

$$
\begin{aligned}
-\frac{\partial}{\partial\theta} &= \widehat{\upsilon} - \left(c + \frac{1}{\Phi}\right)\frac{\partial}{\partial\beta} \\
&= \widehat{\upsilon} + \left(c + \frac{1}{\Phi}\right)\left(\xi_M^\sharp - \Phi\,\frac{\partial}{\partial\theta}\right) \\
&= \left[\widehat{\upsilon} + \left(c + \frac{1}{\Phi}\right)\xi_M^\sharp\right] - (1 + c\,\Phi)\,\frac{\partial}{\partial\theta},
\end{aligned} \tag{39}
$$

where the latter is a decomposition into horizontal and vertical components with respect to π. The latter equality is equivalent to the claimed statement. □

2.4 The complexified action on $A_0^\vee$

The action $\mu^M : \mathbf{T}^1 \times M \to M$ complexifies to a holomorphic action $\widetilde{\mu}^M : \mathbb{T}^1 \times M \to M$, where $\mathbb{T}^1 = \mathrm{GL}(1, \mathbb{C}) \cong \mathbb{C}^*$ (see, for instance, the discussion in Section 4 of [16]). Let (ρ, ϑ) be polar coordinates on $\mathbf{C}^*$, and let $\xi =: \partial/\partial\vartheta$, $\eta =: -\rho\,\partial/\partial\rho$; then $\eta = J_0(\xi)$ (J_0 being the complex structure on $\mathbf{C}^*$). By holomorphicity, if ξ_M and η_M are the induced vector fields on M, then $\eta_M = J_M(\xi_M)$.

On the contrary, the contact lift $\mu^X : \mathbf{T}^1 \times X \to X$ of μ^M extends to a linearized action $\mu^{A^\vee} : \mathbf{T}^1 \times A_0^\vee \to A_0^\vee$. There is a natural diffeomorphism $X \times \mathbb{R}_+ \cong A_0^\vee$, given by $(x, r) \mapsto r \cdot x$; as a function on $A_0^\vee$, r is simply the norm for the given Hermitian structure. If θ is a locally defined angular coordinate on X, depending on the choice of a local unitary frame of $A^\vee$, then (r, θ) restrict to polar coordinates along the fibers of $A_0^\vee$. Thus, if $J_{A^\vee}$ is the complex structure of $A^\vee$, then the globally defined vertical vector fields $\partial/\partial\theta$ and $\partial/\partial r$ on $A_0^\vee$ are related by $J_{A^\vee}(\partial/\partial\theta) = -r\,\partial/\partial r$. By (1), the infinitesimal generator of $\mu^{A^\vee}$ is

$$\xi_{A^\vee} = \xi_M^\sharp - \Phi\,\frac{\partial}{\partial\theta}, \tag{40}$$

where the horizontal lift is now taken in the tangent bundle of $A^\vee$, with respect to the extended connection.

The action $\mu^{A^\vee} : \mathbf{T}^1 \times A_0^\vee \to A_0^\vee$ again extends to a holomorphic action $\widetilde{\mu}^{A^\vee} : \mathbb{T}^1 \times A_0^\vee \to A_0^\vee$, which is of course a linearization of $\widetilde{\mu}^M$ (see the discussion in Section 5 of [16]). By holomorphicity, the induced vector fields $\xi_{A^\vee}$ and $\eta_{A^\vee}$, with $\xi_{A^\vee}$ given by (40), satisfy

$$\eta_{A^\vee} = J_{A^\vee}(\xi_{A^\vee}) = \eta_M^\sharp + \Phi\, r\,\frac{\partial}{\partial r}. \tag{41}$$

Let $\mathcal{N}_A : A_0^\vee \to \mathbb{R}$ be the square norm function; thus $\mathcal{N}_A = r^2$ under the previous diffeomorphism $A_0^\vee \cong X \times \mathbb{R}_+$. Then

$$\xi_{A^\vee}(\mathcal{N}_A) = 0, \qquad \eta_{A^\vee}(\mathcal{N}_A) = 2\,\Phi\,\mathcal{N}_A > 0. \tag{42}$$

Lemma 26 *Let $a =: \min|\Phi|$, $A =: \max|\Phi|$. Then, for every $\lambda \in A_0^\vee$, we have*

$$e^{2at}\,\mathcal{N}_A(\lambda) \le \mathcal{N}_A\left(\widetilde{\mu}^{A^\vee}_{e^{-t}}(\lambda)\right) \le e^{2At}\,\mathcal{N}_A(\lambda)$$

if $t \ge 0$, and

$$e^{2At}\,\mathcal{N}_A(\lambda) \le \mathcal{N}_A\left(\widetilde{\mu}^{A^\vee}_{e^{-t}}(\lambda)\right) \le e^{2at}\,\mathcal{N}_A(\lambda)$$

if $t < 0$.

Proof The invariant vector field $\eta = -\rho\,\partial/\partial\rho$ on $\mathbb{C}^*$ is associated with the 1-parameter subgroup $t \mapsto e^{-t}$. Therefore, if given $\lambda \in A_0^\vee$ we define

$$\mathcal{N}_A^\lambda(t) =: \mathcal{N}_A\left(\widetilde{\mu}^{A^\vee}\left(e^{-t}, \lambda\right)\right) \qquad (t \in \mathbb{R})$$

then by (42)

$$\frac{d}{dt}\mathcal{N}_A^\lambda(t) = \eta_{A^\vee}(\mathcal{N}_A)\left(\widetilde{\mu}^{A^\vee}\left(e^{-t}, \lambda\right)\right) = 2\,\Phi\left(\widetilde{\mu}^{A^\vee}_{e^{-t}}(\lambda)\right)\mathcal{N}_A^\lambda(t),$$

which can be rewritten

$$\frac{d}{dt}\ln(\mathcal{N}_A^\lambda)\bigg|_{t=t_0} = 2\,\Phi\left(\widetilde{\mu}^{A^\vee}_{e^{-t_0}}(\lambda)\right) \tag{43}$$

for any $t_0 \in \mathbb{R}$. We deduce from (43) that

$$2\,a \leq \frac{d}{dt}\ln(\mathcal{N}_A^\lambda) \leq 2\,A, \tag{44}$$

which easily implies the claim. □

Let us set, for $(z, \lambda) \in \mathbb{C}^* \times A_0^\vee$:

$$z \bullet \lambda =: \widetilde{\mu}^{A^\vee}\left(z^{-1}, \lambda\right) = \widetilde{\mu}^{A^\vee}_{z^{-1}}(\lambda). \tag{45}$$

Corollary 27 *If* $|z| \geq 1$, *then*

$$|z|^{2a}\,\mathcal{N}_A(\lambda) \leq \mathcal{N}_A(z \bullet \lambda) \leq |z|^{2A}\,\mathcal{N}_A(\lambda).$$

If $0 < |z| < 1$, *then*

$$|z|^{2A}\,\mathcal{N}_A(\lambda) \leq \mathcal{N}_A(z \bullet \lambda) \leq |z|^{2a}\,\mathcal{N}_A(\lambda).$$

Proof If $z = e^{t+is}$, with $t, s \in \mathbb{R}$ and $|z| = e^t$, then because the action of $e^{is} \in \mathbf{T}^1$ is metric-preserving we have

$$\mathcal{N}_A(z \bullet \lambda) = \mathcal{N}_A\left(e^t \bullet \lambda\right) = \mathcal{N}_A\left(\widetilde{\mu}^{A^\vee}_{e^{-t}}(\lambda)\right).$$

Thus the corollary is just a restatement of Lemma 26. □

Corollary 28 *The* $\mathcal{C}^\infty$ *map* $\Upsilon: \mathbb{R}_+ \times X' \to A_0^{\vee\prime}$ *given by* $(t, x) \mapsto t \bullet x$ *is a bijection.*

Proof By Corollary 27, $\mathcal{N}_A(t \bullet \lambda) \to +\infty$ as $t \to +\infty$, and $\mathcal{N}_A(t \bullet \lambda) \to 0^+$ as $t \to 0^+$; thus for any $\lambda \in A_0^{\vee\prime}$ there exists $t_\lambda \in \mathbb{R}_+$ such that $t_\lambda^{-1} \bullet \lambda \in X'$. Corollary 27 also implies that $\mathcal{N}_A(t \bullet x) > \mathcal{N}_A(x)$ for any $t > 1$ and $x \in X'$. Therefore, $t \mapsto \mathcal{N}_A(t \bullet \lambda)$ is a strictly increasing function, since if $t_1 < t_2$ then

$$\mathcal{N}_A(t_2 \bullet \lambda) = \mathcal{N}_A\left(\left(\frac{t_2}{t_1}t_1\right)\bullet\lambda\right)$$
$$= \mathcal{N}_A\left(\left(\frac{t_2}{t_1}\right)\bullet(t_1 \bullet \lambda)\right) > \mathcal{N}_A(t_1 \bullet \lambda). \tag{46}$$

Hence t_λ is in fact unique. □

Remark 29 More is true. Since $X' \subseteq A_0^\vee$ is a real-analytic submanifold (see Corollary 33), $\mathbb{R}_+ \times X'$ is a real-analytic submanifold of $\mathbb{C}^* \times A_0^\vee$. Being the restriction of the holomorphic map (45), Υ is then a real-analytic bijection of real-analytic manifolds. It is in fact also a local diffeomorphism, for its differential has everywhere maximal rank; by the real-analytic inverse function theorem (Theorem 2.5.1 of [18]), Υ is a real-analytic equivalence between $\mathbb{R}_+\times X'$ and $A_0^{\vee\prime}$

Recall that a Lie group action on a manifold P is called *proper* if the associated *action map* $G \times P \to P \times P$, $(g, p) \mapsto (g \cdot p, p)$, is proper (Definition B2 of [15]).

Let $A_0^{\vee\prime} \subseteq A_0^\vee$ be the inverse image of M'; in other words, in terms of the diffeomorphism $A_0^\vee \cong X \times \mathbb{R}_+$, we have $A_0^{\vee\prime} \cong X' \times \mathbb{R}_+$.

Corollary 30 *The complexified action $\widetilde{\mu}^{A^\vee} : \mathbb{T}^1 \times A_0^\vee \to A_0^\vee$ is proper. In addition, its restriction to $A_0^{\vee\prime}$ is free.*

Proof Let $\Upsilon : \mathbb{T}^1 \times A_0^\vee \to A_0^\vee \times A_0^\vee$ be the action map of $\widetilde{\mu}^{A^\vee}$, and let $R \subset A_0^\vee \times A_0^\vee$ be a compact subset. If $\pi_j : A_0^\vee \times A_0^\vee \to A_0^\vee$ is the projection onto the jth factor, let $R_j =: \pi_j(R)$. Then R_j is compact and $R \subset R_1 \times R_2$. Therefore, to prove that Υ is proper, it suffices to show that $\Upsilon^{-1}(R_1 \times R_2)$ is compact, for any pair of compact subsets $R_1, R_2 \subset A_0^\vee$. Clearly, $\Upsilon^{-1}(R_1 \times R_2) \subseteq \mathbb{T}^1 \times R_2$. For $j = 1, 2$ let $\ell_j =: \min_{R_j} \mathcal{N}_A$ and $L_j =: \max_{R_j} \mathcal{N}_A$.

Suppose $(w, \lambda) \in \Upsilon^{-1}(R_1 \times R_2)$, and set $z = w^{-1}$, so that $\widetilde{\mu}^{A^\vee}(w, \lambda) = z \bullet \lambda$. If $|w| \leq 1$, then $|z| \geq 1$ and so by Corollary 27

$$|z|^{2a}\, \mathcal{N}_A(\lambda) \leq \mathcal{N}_A(z \bullet \lambda) \leq |z|^{2A}\, \mathcal{N}_A(\lambda).$$

Since $\lambda \in R_2$, we have $l_2 \leq \mathcal{N}_A(\lambda)$, and since $z\bullet\lambda \in R_1$, we have $\mathcal{N}_A(z\bullet\lambda) \leq L_1$. Therefore, if $|z| \geq 1$ then

$$\ell_2\, |z|^{2a} \leq L_1 \quad\Longrightarrow\quad |z| \leq \left(\frac{L_1}{l_2}\right)^{1/2a}.$$

In other words, if $|w| \leq 1$ then $(\ell_2/L_1)^{1/2a} \leq |w|$. Similarly, one sees that if $|w| \geq 1$ then $|w| \leq (L_2/\ell_1)^{1/2a}$.

Therefore, if $p\colon \mathbb{T}^1 \times A_0^\vee \to A_0^\vee$ is the projection onto the first factor, then $S\left(\Upsilon^{-1}(R_1 \times R_2)\right)$ is compact. Hence, $\Upsilon^{-1}(R_1 \times R_2) \subseteq S \times R_2$ is also compact, and this completes the proof that the action is proper.

The statement about the freeness of the action follows immediately from (44) and the definition of X'. □

2.5 The circle bundle structures

Let us view $\kappa\colon X' \to N'$ as a circle bundle over N', with the action of $\mathbf{T}^1$ on X' given by $\left(e^{i\theta}, x\right) \mapsto e^{i\theta} \bullet x$; the latter is defined in (45). On N', associated with the Kähler structure we have the volume form $dV_N = \eta^{\wedge d}/d!$; thus on X', viewed as a circle bundle over N', we have the natural choice of a volume form $dW_X = (1/2\pi)\beta \wedge \kappa^*(dV_N)$. Algebraically, $L^2(X, dV_X) = L^2(X', dV_X) = L^2(X', dW_X)$, although the metrics are different. Explicitly,

$$\begin{aligned} dW_X &= \frac{1}{2\pi}\,\beta \wedge \kappa^*(dV_N) \\ &= \frac{1}{2\pi}\,\frac{\alpha}{\Phi} \wedge \frac{1}{d!}\left(\frac{1}{\Phi}\,\pi^*(\omega) - \frac{1}{2\Phi^2}\,d\Phi \wedge \alpha\right)^{\wedge d} \\ &= \Phi^{-(d+1)}\left[\frac{1}{2\pi}\,\alpha \wedge \pi^*(\omega)^{\wedge d}\right] \\ &= \Phi^{-(d+1)}\,dV_X. \end{aligned} \tag{47}$$

Furthermore, the two circle bundles π and κ have different CR structures, because they do not have the same vertical tangent bundle. However, by construction they share the same horizontal distribution, and the same horizontal complex structure $J_{\mathcal{H}}$. Let $\mathcal{H}^{(0,1)} \subseteq \mathcal{H} \otimes \mathbb{C}$ be the $-i$-eigenbundle of $J_{\mathcal{H}}$; then the boundary CR operator of either X or X' is defined by setting $\overline{\partial}_b f = df|_{\mathcal{H}^{(0,1)}}$, for any C^∞ function f on X or X', respectively. Therefore, the boundary CR operator of X,

$$\overline{\partial}_b\colon C^\infty(X) \to C^\infty\left(X, \mathcal{H}^{(0,1)\vee}\right),$$

restricts to the corresponding operator of X'. It follows that there is a natural algebraic (non-isometric) inclusion of corresponding Hardy spaces, $H(X) \hookrightarrow H(X')$. The latter is an algebraic isomorphism if $\mathrm{codim}_{\mathbb{C}}(M \setminus M', M) \geq 2$.

The action μ^X plays the role of the structure circle action of $\mathbf{T}^1$ with respect to κ. Let $\widetilde{H}_k(X')$ be the kth isotype for the latter action. Condition (3) for $s \in L^2(X)$ to belong to $H_k^\mu(X)$ may be rewritten $s\left(e^{i\theta} \bullet x\right) = e^{ik\theta}\, s(x)$, for any $e^{i\theta} \in \mathbf{T}^1$ and $x \in X$. Therefore, the previous inclusion of Hardy spaces yields for every $k = 0, 1, 2, \ldots$ an algebraic inclusion $H_k^\mu(X) \hookrightarrow \widetilde{H}_k(X')$.

2.6 The line bundle on N'

Let B be the complex line bundle on N' associated with κ and the tautological action of $\mathbf{T}^1 = U(1)$ on $\mathbb{C}$, and let $B^\vee$ be its dual; thus, B (resp. $B^\vee$) is the quotient of $X' \times \mathbb{C}$ by the equivalence relation $(x, w) \sim \left(e^{i\theta} \bullet x, e^{i\theta} w\right)$ (resp. $(x, w) \sim \left(e^{i\theta} \bullet x, e^{-i\theta} w\right)$). We can embed $\jmath \colon X' \hookrightarrow B^\vee$ by $x \mapsto [x, 1]$. Then B and $B^\vee$ inherit natural Hermitian structures, that we shall denote by ℓ_B, uniquely determined by imposing $\jmath \colon X' \hookrightarrow B^\vee$ to embed as the unit circle bundle. We shall denote by $\widehat{\kappa} \colon B^\vee \to N'$ the projection, so that $\kappa = \widehat{\kappa} \circ \jmath$.

The connection form β on X determines a unique metric covariant derivative ∇_B on B, with curvature $\Theta_B = -2i\,\kappa^*(\eta)$. Since η is a Kähler form on N', there is a uniquely determined holomorphic structure on B, such that ∇_B is the only covariant derivative on B compatible with both the metric and the latter holomorphic structure. A local section σ of B is holomorphic for this structure if and only if the connection matrix with respect to σ is of type $(1, 0)$.

Lemma 31 *There is a natural biholomorphism $\overline{\Gamma} \colon B_0^\vee \cong A_0^{\vee\prime}$ of bundles over N'; when we view X' as a submanifold of $A_0^{\vee\prime}$ and $B_0^\vee$ in the natural manner, $\overline{\Gamma}$ restricts to the identity $X' \to X'$ (that is, $\overline{\Gamma}(\jmath(x)) = x$ for any $x \in X'$). Furthermore, $\overline{\Gamma}$ preserves the horizontal distributions, and maps biholomorphically the fibers of the bundle projection $\widehat{\kappa} \colon B^\vee \to N'$ onto the orbits of the action $\widetilde{\mu}^{A^\vee} \colon \mathbb{T}^1 \times A_0^\vee \to A_0^\vee$.*

Proof We have $B_0^\vee = X' \times \mathbb{C}^* / \sim$, where $(x, w) \sim \left(e^{i\theta} \bullet x, e^{-i\theta} w\right)$, for any $e^{i\theta} \in \mathbf{T}^1$. If $\Psi \colon X' \times \mathbb{C}^* \to B_0^\vee$ is the quotient map, $\Psi(x, w) = [x, w]$, for any $(x, w) \in X' \times \mathbb{C}^*$ the differential $d_{(x,w)}\Psi$ induces a $\mathbb{C}$-linear isomorphism

$$\mathcal{H}_x \oplus \mathbb{C} \subset T_{(x,w)}\left(X' \times \mathbb{C}^*\right) \cong T_{\Psi(x,w)} B_0^\vee,$$

which maps $\mathcal{H}_x \oplus (\mathbf{0})$ and $(\mathbf{0}) \oplus \mathbb{C}$, respectively, onto the horizontal and vertical tangent spaces of $B^\vee$ at $\Psi(x, w)$.

Let us consider the map

$$\Gamma \colon X' \times \mathbb{C}^* \to A_0^{\vee\prime}, \qquad (x, w) \mapsto w \bullet x.$$

Holomorphicity of the complexified action $\widetilde{\mu}$ implies that the differential $d_{(x,w)}\Gamma$ induces a $\mathbb{C}$-linear isomorphism

$$\mathcal{H}_x \oplus \mathbb{C} \subset T_{(x,w)}\left(X' \times \mathbb{C}^*\right) \cong T_{\Gamma(x,w)} A_0^\vee, \tag{48}$$

under which $\mathcal{H}_x \oplus \mathbb{C}$ maps onto the horizontal tangent space of $A^\vee$ and $(\mathbf{0}) \oplus \mathbb{C}$ onto the tangent space to the complex orbit $\widetilde{\mu}^{A^\vee}$ at $\Gamma(x, w)$.

On the other hand, for any $(x, w) \in X' \times \mathbb{C}^*$ and $e^{i\theta} \in \mathbf{T}^1$, we have

$$\Gamma(x, w) = w \bullet x = \left(w\, e^{-i\theta}\right) \bullet \left(e^{i\theta} x\right) = \Gamma\left(e^{i\theta} x, e^{-i\theta} w\right). \tag{49}$$

Therefore, Γ passes to the quotient under Ψ, that is, there exists a C^∞ map $\overline{\Gamma}\colon B_0^\vee \to A_0^{\vee\prime}$ such that $\Gamma = \overline{\Gamma} \circ \Psi$:

$$\overline{\Gamma}([x,w]) = \Gamma(x,w) \qquad ((x,w) \in X' \times \mathbb{C}^*);$$

the previous discussion implies that $\overline{\Gamma}$ is holomorphic.

Corollary 28 evidently implies that Γ is surjective, and therefore so is $\overline{\Gamma}$. To see that $\overline{\Gamma}$ is also injective, suppose that $\lambda_j = \Psi(x_j, w_j)$, $j = 1, 2$, satisfy $\overline{\Gamma}(\lambda_1) = \overline{\Gamma}(\lambda_2)$. Thus $w_1 \bullet x_1 = \Gamma(x_1, w_1) = \Gamma_2(x_2, w_2) = w_2 \bullet x_2$, whence $\left(w_2^{-1} w_1\right) \bullet x_1 = x_2$. This evidently implies $\mathcal{N}_A\left(\left(w_2^{-1} w_1\right) \bullet x_1\right) = \mathcal{N}_A(x_1) = \mathcal{N}_A(x_2) = 1$. Since by Corollary 27 the map $t \mapsto \mathcal{N}_A(t \bullet x_1)$ is strictly increasing, this forces $|w_1| = |w_2|$. If $w_2^{-1} w_1 = e^{i\theta}$, we then have $w_2 = e^{-i\theta} w_1$, $x_2 = e^{i\theta} \bullet x_1$; hence $(x_2, w_2) = \left(e^{i\theta} \bullet x_1, e^{-i\theta} \bullet w_1\right) \sim (x_1, w_1)$. Therefore, $\lambda_1 = \lambda_2$.

Finally, any $x \in X' \subset A_0^{\vee\prime}$ corresponds to $[x, 1] \in B_0^\vee$, and $\overline{\Gamma}([x, 1]) = 1 \bullet x = x$. Therefore, with the previous identification $\overline{\Gamma}$ induces the identity map on X'. □

Remark 32 $\overline{\Gamma}$ interwines fiberwise scalar multiplication $\cdot_B$ on $B_0^\vee$ and the map (45). In fact, if $b = [x, w] \in B_0^\vee$ and $z \in \mathbb{C}^*$, then $z \cdot_B b = [x, z\,w]$. Therefore,

$$\overline{\Gamma}(z \cdot_B b) = \overline{\Gamma}([x, z\,w]) = (z\,w) \bullet x = z \bullet (w \bullet x) = z \bullet \overline{\Gamma}(x).$$

Let $\mathcal{N}_B\colon A_0^\vee \to \mathbb{R}$ be the norm function associated with the Hermitian structure of $B^\vee$, viewed as a function on $A_0^\vee$ by means of the biholomorphism of Lemma 31. Then $\mathcal{N}_B(z \bullet \lambda) = |z|^2\, \|\lambda\|^2$.

Corollary 33 *$X' \subseteq A_0^\vee$ is a real-analytic submanifold, and the projection $\kappa\colon X' \to N'$ is real-analytic.*

Proof Since the Hermitian metric h on $A^\vee$ is real-analytic by assumption, the norm function $\mathcal{N}_A\colon A_0^\vee \to \mathbb{R}$ is a positive real-analytic function. Therefore, $X' = \mathcal{N}_A^{-1}(1) \cap A_0^{\vee\prime}$ is a real-analytic submanifold of $A_0^{\vee\prime}$ (see Section 2.7 of [18]). On the contrary, we have

$$\kappa = \widehat{\kappa} \circ \jmath = \widehat{\kappa} \circ \overline{\Gamma}^{-1}\Big|_X .$$

Thus κ is the restriction of a holomorphic map to a real-analytic submanifold, hence it is real-analytic. □

Proposition 34 *The Kähler form η on N' is real-analytic.*

Proof It suffices to prove that for any $n \in N'$ there is a real-analytic chart for N', defined on an open neighborhood $V \subseteq N'$ of n, such that the local expression of η in that chart is real-analytic. To this end, choose $x \in \kappa^{-1}(n)$ and let $m =: \pi(x) \in M'$. On some open neighborhood $U \subseteq M$ of m, we can find a local holomorphic frame φ on $A^\vee$ such that $\varphi(m) = x$. We can also suppose, without loss of generality, that φ is horizontal at m, that it, its differential at m maps isomorphically T_mM to the horizontal tangent space $\mathcal{H}_x$. If we assume, as we may, that U is the domain of a holomorphic local coordinate chart (z_j) centered at m, this means that $\mathcal{N}_A \circ \varphi = h(\varphi, \varphi) = 1 + O\left(\|z\|^2\right)$.

Let us write $\|\varphi\| =: \sqrt{h(\varphi, \varphi)}$, so that $\zeta =: \varphi/\|\varphi\| : U \to X'$ is a local unitary frame. Since h is real-analytic, ζ is real-analytic. The previous remark shows, in addition, that

$$d_m\zeta = d_m\varphi : T_mM \longrightarrow T_xX \subseteq T_xA^\vee;$$

therefore ζ is also horizontal at m, whence it is transverse at m to the μ^X-orbit through x, in view of (1) and the positivity of Φ. Since the latter orbit is the fiber through x of the projection $\kappa : X' \to N'$, this implies that the composition $\kappa \circ \zeta : U \to N'$ is a real-analytic local diffeomorphism at m; we have $\kappa \circ \zeta(m) = \kappa(x) = n$. Therefore, perhaps after replacing U with a smaller open neighborhood of m, we may assume that $\kappa \circ \zeta$ induces a real-analytic equivalence $U \cong V$, where $V =: \kappa \circ \zeta(U)$ is an open neighborhood of n (see Theorem 2.5.1 of [18]). Given the holomorphic chart on U, we may then interpret $\kappa \circ \zeta$ as a real-analytic chart for N' in the neighborhood of n.

Let $\theta_\zeta^\vee = i\zeta^*(\alpha)$ be the connection form of $A^\vee$ in the local frame ζ. Then under our assumptions, $\theta_\zeta^\vee$ is a real-analytic imaginary 1-form. The local expression of 2η in this chart, by (28), is

$$\begin{aligned}(\kappa \circ \zeta)^*(2\eta) &= \zeta^*\left(\kappa^*(2\eta)\right)\\ &= \zeta^*\left(\frac{2}{\Phi}\pi^*(\omega) - \frac{1}{\Phi^2}d\Phi \wedge \alpha\right)\\ &= \frac{2}{\Phi}\omega + \frac{i}{\Phi^2}d\Phi \wedge \theta_\zeta^\vee. \end{aligned} \tag{50}$$

In view of Lemma 12, we conclude that (50) is real-analytic, and this completes the proof. □

This also follows from

Lemma 35 *The Hermitian metric h on $B^\vee$ is real-analytic.*

Proof It suffices to show that the norm function $\mathcal{N}_B : B_0^\vee \to \mathbb{R}_+$ is real-analytic. To this end, it is equivalent to show that the composition $\mathcal{N}_B \circ \overline{\Gamma}^{-1} : A_0^\vee \to \mathbb{R}_+$ is real-analytic. Again, let us simplify our discussion by biholomorphically identifying $A_0^\vee$ with $B_0^\vee$, and leaving $\overline{\Gamma}^{-1}$ implicit. Then fiberwise scalar multiplication on $B_0^\vee$ corresponds to the map (45). Thus if $(t, x) \in \mathbb{R}_+ \times X'$, then $\mathcal{N}_B(t \bullet x) = t^2$.

Let $\Upsilon\colon \mathbb{R}_+ \times X' \to A_0^{\vee\prime}$ be as in Corollary 28; then Υ is a real-analytic equivalence by Remark 29, and the previous remark implies that

$$\mathcal{N}_B \circ \Upsilon\colon \mathbb{R}_+ \times X' \to \mathbb{R}_+$$

is real-analytic. Therefore, so is $\mathcal{N}_B = (\mathcal{N}_B \circ \Upsilon) \circ \Upsilon^{-1}$. □

We can consider the *equivariant distortion function* $K_k^\mu\colon M \to \mathbb{R}$ defined by setting

$$K_k^\mu(m) =: \Pi_k^\mu(x,x) = \sum_j \left|s_j^{(k)}(x)\right|^2, \tag{51}$$

for $m \in M$ and any choice of $x \in \pi^{-1}(m) \subseteq X$, where the $s_j^{(k)}$s are an orthonormal basis of $H_k^\mu(X)$ (see (4)). That K_k^μ is well-defined follows from the fact that μ^X and ν^X commute (Lemma 2.1 of [28]). For any $m \in M$ and $t \in \mathbf{T}^1$, given $x \in \pi^{-1}(m)$ by (3) we have

$$\begin{aligned} K_k^\mu\left(\mu_{t^{-1}}^M(m)\right) &= \Pi_k^\mu\left(\mu_{t^{-1}}^X(x), \mu_{t^{-1}}^X(x)\right) = \sum_j \left|s_j^{(k)}\left(\mu_{t^{-1}}^X(x)\right)\right|^2 \\ &= \sum_j \left|s_j^{(k)}(x)\right|^2 = \Pi_k^\mu(x,x) = K_k^\mu(m). \end{aligned}$$

Therefore, $K_k^\mu \in C^\infty(M)^\mu$; it may thus be regarded as a function on N' in a natural manner. We have in fact:

Lemma 36 *$K_k^\mu \in C^\varpi(M)^\mu$. As a function on N', $K_k^\mu \in C^\varpi(N')^\nu$.*

Proof By its very definition, $\Pi_k^\mu \in C^\infty(X \times X)$ restricts to a sesquiholomorphic complex function on $A_0^\vee \times A_0^\vee$, which is then *a fortiori* real-analytic. Since $X \times X$ is a real-analytic submanifold of $A_0^\vee \times A_0^\vee$ by Corollary 33, we have $\Pi_k^\mu \in C^\varpi(X \times X)$. If now φ is a local holomorphic frame on an open subset $U \subset M$, the unitarization $\varphi_u = \varphi/\|\varphi\|_A\colon U \to X$ is real-analytic, where $\|\varphi\|_A =: (\mathcal{N}_A \circ \varphi)^{1/2}$. Therefore,

$$K_k^\mu(m) = \Pi_k^\mu(\varphi_u(m), \varphi_u(m)) \qquad (m \in U)$$

is real-analytic on U. The second statement is proved similarly (Lemma 35). □

2.7 Asymptotics of sesquiholomorphic extensions

Every $s \in H_k^\mu(X)$ extends uniquely to a holomorphic function $\widetilde{s}\colon A_0^\vee \to \mathbb{C}$. Holomorphicity of the extended action $\widetilde{\mu}^{A^\vee}$ implies, in view of (3) and (45), that for every $(z,\lambda) \in \mathbb{C}^* \times A_0^\vee$ we have

$$\widetilde{s}(z \bullet \lambda) = z^k \, \widetilde{s}(\lambda). \tag{52}$$

Given this and (4), we see that $\Pi_k^\mu \colon X \times X \to \mathbb{C}$ extends uniquely to a sesquiholomorphic function $\mathcal{P}_k^\mu \colon A_0^\vee \times A_0^\vee \to \mathbb{C}$, given by

$$\mathcal{P}_k^\mu(\lambda, \lambda') = \sum_j \widetilde{s}_j^{(k)}(\lambda) \cdot \overline{\widetilde{s}_j^{(k)}(\lambda')} \qquad \left(\lambda, \lambda' \in A_0^\vee\right), \tag{53}$$

and satisfying, by (52),

$$\mathcal{P}_k^\mu(z \bullet \lambda, w \bullet \lambda') = z^k \, \overline{w}^k \, \mathcal{P}_k^\mu(\lambda, \lambda'), \tag{54}$$

for every $z, w \in \mathbb{C}^*$.

Let σ be a local holomorphic frame of $B^\vee$ on an open subset $V \subseteq N'$. Then

$$\mathcal{P}_k^\mu \circ (\sigma \times \sigma) \colon V \times V \to \mathbb{C}, \quad (n, n') \mapsto \mathcal{P}_k^\mu(\sigma(n), \sigma(n'))$$

is sesquiholomorphic. The unitarization $\sigma_u =: (1/\|\sigma\|_B) \bullet \sigma \colon V \to X'$ (see Remark 32) is a real-analytic section. Given (54), we have

$$\begin{aligned}
\Pi_k^\mu(\sigma_u(n), \sigma_u(n')) &= \mathcal{P}_k^\mu\left(\frac{1}{\|\sigma(n)\|_B} \bullet \sigma(n), \frac{1}{\|\sigma(n')\|_B} \bullet \sigma(n')\right) \\
&= \frac{1}{\|\sigma(n)\|_B^k} \, \frac{1}{\|\sigma(n')\|_B^k} \, \mathcal{P}_k^\mu(\sigma(n), \sigma(n')) \\
&= e^{-\frac{k}{2}\left(\Xi(n) + \Xi(n')\right)} \, \mathcal{P}_k^\mu(\sigma(n), \sigma(n')),
\end{aligned} \tag{55}$$

where we have set, for $n \in V$,

$$\Xi(n) = \ln\left(\|\sigma(n)\|_B^2\right) = \ln\left(\ell_B\big(\sigma(n), \sigma(n)\big)\right). \tag{56}$$

Then Ξ is real-analytic by Lemma 35, and furthermore $\partial_N \overline{\partial}_N \Xi = \Theta_B$, where $\Theta_B = -2i\,\eta \in \Omega^2(N')$ is the curvature form of B. In any given local coordinate chart (z_k) for N' this means that

$$\frac{\partial^2 \Xi}{\partial z_k \, \partial z_{\overline{l}}} = \Theta_{B k\overline{l}} = -2i\,\eta_{k\overline{l}} = 2\,h_{k\overline{l}},$$

where h is the Riemannian metric of (N', I, η). In other words, Ξ is a Kähler potential for $2\,h$.

Being real-analytic, Ξ has a unique sesquiholomorphic extension $\widetilde{\Xi}$ to an open neighborhood of the diagonal $\widetilde{V} \subseteq V \times V$. Similarly, by Lemma 36, K_k^μ also has a unique sesquiholomorphic extension $\widetilde{K_k^\mu}$ to an open neighborhood of the diagonal in $N' \times N'$.

Lemma 37 *Let $\widetilde{V} \subseteq V \times V$ be an appropriate open neighborhood of the diagonal. Then for every $(n, n') \in \widetilde{V}$, we have*

$$\mathcal{P}^{\mu}_{k}(\sigma(n), \sigma(n')) = e^{k\,\widetilde{\Xi}(n,n')}\,\widetilde{K^{\mu}_{k}}(n, n'). \tag{57}$$

Proof Both sides being sesquiholomorphic, it suffices to show that they have equal restrictions on the diagonal. If $n = n'$, by (55) we have

$$\begin{aligned}\mathcal{P}^{\mu}_{k}(\sigma(n), \sigma(n)) &= e^{k\,\Xi(n)}\,\Pi^{\mu}_{k}(\sigma_u(n), \sigma_u(n)) \\ &= e^{k\,\widetilde{\Xi}(n,n)}\,K^{\mu}_{k}(n) = e^{k\,\widetilde{\Xi}(n,n)}\,\widetilde{K^{\mu}_{k}}(n, n).\end{aligned} \tag{58}$$

□

Inserting (57) in (55), we obtain for $(n, n') \in \widetilde{V}$:

$$\Pi^{\mu}_{k}(\sigma_u(n), \sigma_u(n')) = e^{k\left[\widetilde{\Xi}(n,n') - \frac{1}{2}\,\Xi(n) - \frac{1}{2}\,\Xi(n')\right]}\,\widetilde{K^{\mu}_{k}}(n, n'). \tag{59}$$

As discussed in the Introduction, by [28] if $m \in M'$ and $x \in \pi^{-1}(m)$ there is an asymptotic expansion (5), smoothly varying on M' and uniform on compact subsets of M', with leading coefficient $S^{\mu}_{0} = \Phi^{-(d+1)}$. Since $\Pi^{\mu}_{k}(x, x)$ is μ^{M}-invariant, so is every S^{μ}_{j}. Therefore, viewing K^{μ}_{k} as defined on N', the expansion may naturally be interpreted as holding on N' (see (51) and Lemma 36)):

$$K^{\mu}_{k}(n) \sim \left(\frac{k}{\pi}\right)^{d} \sum_{j \geq 0} k^{-j}\,S^{\mu}_{j}(n), \tag{60}$$

where $S^{\mu}_{0} = \Phi^{-(d+1)}$. This suggest, heuristically, that $\widetilde{K^{\mu}_{k}}(n, n')$ should satisfy a similar expansion, with coefficients the sesquiholomorphic extensions of the S^{μ}_{j}s. This is indeed the case.

To see this, let us consider first the asymptotics of $\Pi^{\mu}_{k}(\sigma_u(n), \sigma_u(n'))$ for $(n, n') \in \widetilde{V}$. Let

$$\Pi(x, y) = \int_{0}^{+\infty} e^{it\,\psi(x,y)}\,s(x, y, t)\,dt \tag{61}$$

be the usual Fourier integral representation of the Szegö kernel of X determined in [6]; here we think of X as the unit circle bundle of $A^{\vee}$, with volume form dV_X. In particular, $\Im(\psi) \geq 0$, and s is a semiclassical symbol admitting an asymptotic expansion of the form

$$s(x, y, t) \sim \sum_{j \geq 0} t^{d-j}\,s_j(x, y) \tag{62}$$

(see also the discussion in [30] and [35]). For some $\epsilon > 0$, let $\varrho_1 \in C^{\infty}_{0}(-2\epsilon, 2\epsilon)$ be a bump function identically equal to 1 on $(-\epsilon, \epsilon)$. For some $C > 0$, let

$\varrho_2 \in C_0^\infty(1/(2C), 2C)$ be a bump function identically equal to 1 on $(1/C, C)$. Let us write $\mu^X_{-\vartheta}$ for $\mu^X_{e^{-i\vartheta}}$. Then, arguing as in the proof of Theorem 1 of [28],

$$\begin{aligned}\Pi^\mu_k(\sigma_u(n), \sigma_u(n')) &= \frac{1}{2\pi}\int_{-\pi}^{\pi} e^{-ik\vartheta}\,\Pi\left(\mu^X_{-\vartheta}(\sigma_u(n)), \sigma_u(n')\right) d\vartheta \\ &\sim \frac{1}{2\pi}\int_0^{+\infty}\int_{-\pi}^{\pi} e^{-ik\vartheta+it\,\psi\left(\mu^X_{-\vartheta}(\sigma_u(n)),\sigma_u(n')\right)}\, s\left(\mu^X_{-\vartheta}(\sigma_u(n)), \sigma_u(n'), t\right)\varrho_1(\vartheta)\,dt\,d\vartheta \\ &\sim \frac{k}{2\pi}\int_0^{+\infty}\int_{-\pi}^{\pi} e^{ik\,\Psi(n,n',t,\vartheta)}\, s\left(\mu^X_{-\vartheta}(\sigma_u(n)), \sigma_u(n'), kt\right)\varrho_1(\vartheta)\,\varrho_2(t)\,dt\,d\vartheta,\end{aligned} \tag{63}$$

where

$$\Psi(n, n', t, \vartheta) =: t\,\psi\left(\mu^X_{-\vartheta}(\sigma_u(n)), \sigma_u(n')\right) - \vartheta.$$

The last line of (63) is an oscillatory integral with phase $\Psi(n, n', t, \vartheta)$, and $\Im(\psi) \geq 0$ implies $\Im(\Psi) \geq 0$.

Suppose first $n = n'$. Then one can see by (a slight adaptation of) the argument in the proof of Theorem 1 of [28] that the phase $\Psi(n, n, t, \vartheta)$ has a unique stationary point $P(n, n) = (t_0, \vartheta_0) = (1/\Phi(n), 0)$, where as usual we think of the invariant function Φ as descended on N. Since $\psi(x, x) = 0$ identically, we have $\Psi(n, n, t_0, \vartheta_0) = 0$. Furthermore, the Hessian matrix at P_0 is

$$H_{P_0}(\Psi) = \begin{pmatrix} 0 & \Phi(n) \\ \Phi(n) & \partial^2_{\vartheta\vartheta}\Psi(P_0) \end{pmatrix}.$$

Therefore, P_0 is a non-degenerate critical point, and by applying the stationary phase lemma to it we obtain the asymptotic expansion (60).

By the theory of [27], the stationary point and the asymptotic expansion will deform smoothly with $(n, n') \in \widetilde{V}$, although the stationary point may cease to be real when $n \neq n'$ (and should then be regarded as the stationary point of an almost analytic extension of Ψ). More precisely, if $\widetilde{\Psi}\left(\widetilde{n}, \widetilde{n}', \widetilde{t}, \widetilde{\vartheta}\right)$ denotes an almost analytic extension of $\Psi(n, n', t, \vartheta)$, then the condition that $P(\widetilde{n}, \widetilde{n}') = \left(\widetilde{t}(\widetilde{n}, \widetilde{n}'), \widetilde{\vartheta}(\widetilde{n}, \widetilde{n}')\right)$ be a stationary point of $\widetilde{\Psi}(\widetilde{n}, \widetilde{n}', \cdot, \cdot)$ defines an almost analytic manifold $\left(\widetilde{t}, \widetilde{\vartheta}\right) = \left(\widetilde{t}(\widetilde{n}, \widetilde{n}'), \widetilde{\vartheta}(\widetilde{n}, \widetilde{n}')\right)$.

Applying to (63) the stationary phase lemma for complex phase functions from Section 2 of [27] for $(n, n') \in \widetilde{V}$ we obtain a smoothly varying asymptotic expansion

$$\Pi^\mu_k(\sigma_u(n), \sigma_u(n')) \sim \left(\frac{k}{\pi}\right)^d e^{ik\,\widetilde{\Psi}(n,n',P(n,n'))} \sum_{j\geq 0} k^{-j}\,S_j(n, n'), \tag{64}$$

for appropriate smooth functions $S_j(\cdot,\cdot)$ on $\widetilde{V} \subseteq V \times V$.

Given (55) and (64), we get

$$\mathcal{P}_k^\mu(\sigma(n),\sigma(n')) \tag{65}$$
$$\sim e^{k\left[\frac{1}{2}\left(\Xi(n)+\Xi(n')\right)+i\widetilde{\Psi}\left(n,n',P(n,n')\right)\right]}\left(\frac{k}{\pi}\right)^d \sum_{j\geq 0} k^{-j} S_j(n,n').$$

Since the expansion holds in C^j-norm for every j and the LHS is sesquiholomorphic, so is every term on the RHS. Therefore, each term

$$e^{k\left[\frac{1}{2}\left(\Xi(n)+\Xi(n')\right)+i\widetilde{\Psi}\left(n,n',P(n,n')\right)\right]} S_j(n,n')$$

is the sesquiholomorphic extension of its diagonal restriction.

On the contrary, on the diagonal (65) restricts to the uniquely determined asymptotic expansion for (58), and so we need to have $S_j(n,n) = S_j^\mu(n)$, whence $S_j(n,n') = \widetilde{S_j^\mu}(n,n')$. Furthermore, we see that

$$\frac{1}{2}\left(\Xi(n)+\Xi(n')\right)+i\widetilde{\Psi}(n,n',P(n,n')) = \widetilde{\Xi}(n,n').$$

Inserting this in (64), we obtain

$$\Pi_k^\mu(\sigma_u(n),\sigma_u(n')) \sim \left(\frac{k}{\pi}\right)^d e^{k\left[\widetilde{\Xi}(n,n')-\frac{1}{2}\left(\Xi(n)+\Xi(n')\right)\right]}$$
$$\cdot \sum_{j\geq 0} k^{-j}\widetilde{S_j^\mu}(n,n'). \tag{66}$$

Now (59) and (55) imply

$$\widetilde{K_k^\mu}(n,n') \sim \left(\frac{k}{\pi}\right)^d \sum_{j\geq 0} k^{-j}\widetilde{S_j^\mu}(n,n') \tag{67}$$

(see [17] and [35] for analogues in the standard case $\Phi = 1$).

Analogous considerations hold for Toeplitz operators; see Section 9.

2.8 The Laplacian on invariant functions

Let us now dwell on the relation between the Laplacian operators Δ_N and Δ_M of (M, J, ω) and (N', I, η) acting on invariant functions. Thus let $f \in C^\infty(M)^\mu$, so that f determines in a natural manner functions on X and N', respectively. It is convenient in the present argument to explicitly distinguish the domain of definition of the function in point, so we shall write $f = f_M$, and f_X and f_N to denote the induced functions on X and N', respectively. It is also notationally convenient to leave $\overline{\Gamma}$ implicit, and to identify $B_0^\vee$ with $A_0^{\vee\prime}$ (see Lemma 31). Thus we have holomorphic line bundle structures $\widehat{\pi}\colon A^\vee \to M$

and $\widehat{\kappa}\colon A^\vee \to N'$, where we write $\widehat{\kappa}$ for $\widehat{\kappa} \circ \overline{\Gamma}^{-1}$. The fibers of the latter are the orbits of the complexified action $\widetilde{\mu}^{A^\vee}$.

Suppose $m \in M'$, $x \in \pi^{-1}(m) \in X \subset A_0^\vee$, and set $n =: \kappa(x)$. Choose a local holomorphic frame φ for $A^\vee$ on an open neighborhood $U \subset M'$ of m, such that $\varphi(m) = x$ and which is horizontal at m, in the sense of the proof of Proposition 34. Then, as remarked in the same proof, $\varphi\colon U \to A^\vee$ is transverse at m to the orbit of μ^X through x. In fact, in view of (41), $\varphi\colon U \to A^\vee$ is transverse at m to the full orbit of $\widetilde{\mu}^{A^\vee}$ through x. Thus, the composition $\widehat{\kappa} \circ \varphi\colon U \to N'$ is holomorphic, has maximal rank at m, and satisfies $\widehat{\kappa} \circ \varphi(m) = n$. Therefore, there exists an open neighborhood $U \subseteq M'$ of m such that $V =: \widehat{\kappa} \circ \varphi(U)$ is open, and the induced map $\widehat{\kappa} \circ \varphi\colon U \to N'$ is a biholomorphism.

Let us set $Z =: \varphi(U) \subseteq A_0^\vee$. Then Z is a complex submanifold of $A_0^\vee$, and the restrictions of $\widehat{\pi}$ and $\widehat{\kappa}$ to Z determine biholomorphic maps $\pi_Z\colon Z \to U$ and $\kappa_Z\colon Z \to V$. The invariance hypothesis on f implies that $f_M \circ \pi_Z = f_N \circ \kappa_Z$; let us write f_Z for this function.

Furthermore, if K is the complex structure on Z then by holomorphicity we can pull back the Kähler structures (M, J, ω) and (N, I, η) under π_Z and κ_Z, respectively, to Kähler structures (Z, K, ω') and (Z, K, η'). Clearly

$$\Delta_M(f_M) \circ \pi_Z = \Delta_1(f_Z), \qquad \Delta_N(f_N) \circ \kappa_Z = \Delta_2(f_Z),$$

where Δ_1 and Δ_2 are the Laplacian operators in the Kähler structures $(Z, K, 2\omega')$ and $(Z, K, 2\eta')$, respectively. Therefore,

$$\Delta_M(f_M)(m) = \Delta_1(f_Z)(x), \qquad \Delta_N(f_N)(n) = \Delta_2(f_Z)(x). \tag{68}$$

Recall that $g(\cdot,\cdot) = \omega(\cdot, J(\cdot))$ and $h = \eta(\cdot, I(\cdot))$ are the Riemannian metrics on (M, J, ω) and (N, I, η), and by pull-back we view them as the Riemannian metrics of (Z, K, ω') and (Z, K, η'), respectively. Perhaps after restricting U to a smaller open neighborhood of m in M', we may assume without loss that on Z there is a global holomorphic coordinate chart (z_j). Let $g_{a\overline{b}} = g(\partial_a, \partial_{\overline{b}})$ and $h_{a\overline{b}} = h(\partial_a, \partial_{\overline{b}})$ be the respective covariant metric tensors, with associated contravariant tensors $(g^{\overline{b}a})$ and $(h^{\overline{b}a})$.

In particular, $(T_xZ, K_x, \omega'_x) = (\mathcal{H}_x, J_{\mathcal{H},x}, \omega_x)$, where ω_x is ω_m pulled back to $\mathcal{H}_x$ under $d_x\pi$. Similarly, with the same abuse of language, $(T_xZ, K_x, \eta'_x) = (\mathcal{H}_x, J_{\mathcal{H},x}, \eta_x)$. By horizontality, expression (28) for $\kappa^*(2\eta)$ implies that $\eta_x = \omega_x/\Phi(m)$. Hence $h_{a\overline{b}}(x) = g_{a\overline{b}}(x)/\Phi(m)$, and so $h^{\overline{b}a}(x) = \Phi(m)\, g^{\overline{b}a}(x)$. Thus we conclude that

$$\begin{aligned}\Delta_2(f_Z)(x) &= \frac{1}{2}\, h^{\overline{b}a}(x)\, \partial_a\, \partial_{\overline{b}} f_Z(x)\\ &= \frac{1}{2}\, \Phi(m)\, g^{\overline{b}a}(x)\, \partial_a\, \partial_{\overline{b}} f_Z(x) = \Phi(m)\, \Delta_1(f_Z)(x).\end{aligned} \tag{69}$$

Given (68) and (69), we conclude that

$$\Delta_N(f_N)(n) = \Phi(m)\,\Delta_M(f_M)(m). \tag{70}$$

Interpreting Δ_M and Δ_N as endomorphisms of $C^\infty(M')^\mu$, we can restate (70) by writing

$$\Delta_N = \Phi \cdot \Delta_M. \tag{71}$$

2.9 μ-Adapted Heisenberg local coordinates

As mentioned in the Introduction, HLC for X centered at some $x \in X$ were defined in [30]; it is in these local coordinates that near-diagonal Szegö kernel scaling asymptotics exhibit their universal nature. While we refer to [30] for a detailed discussion, let us recall that they consist of the choice of an adapted local coordinate chart for M centered at $m = \pi(x)$, intertwining the unitary structure on T_mM with the standard one on $\mathbb{C}^d$, and a preferred local frame of $A^\vee$ on a neighborhood of m, having a prescribed second-order jet at m.

Let $\mathfrak{r}\colon (-\pi,\pi)\times B_{2d}(\mathbf{0},\epsilon) \to X$, $\mathfrak{r}(\theta,\mathbf{v}) = x + (\theta,\mathbf{v})$, be a system of HLC centered at x. Then $\mathfrak{r}^*(dV_X)(\theta,\mathbf{0}) = (2\pi)^{-1}\,|d\theta|\,d\mathcal{L}(\mathbf{v})$, where $d\mathcal{L}(\mathbf{v})$ is the Lebesgue measure on $\mathbf{R}^{2d}$. For $\mathbf{v} \in B_{2d}(\mathbf{0},\epsilon)$, let us set $x + \mathbf{v} =: \mathfrak{r}(0,\mathbf{v})$.

It is natural here to modify the previous prescription so as to incorporate μ^X into an "equivariant" HLC system. Namely, let us define $\mathfrak{y}'\colon \mathbf{T}^1\times B_{2d}(\mathbf{0},\epsilon) \to X$ by letting

$$\mathfrak{y}'\left(e^{i\vartheta},\mathbf{w}\right) =: e^{i\vartheta} \bullet (x+\mathbf{w}). \tag{72}$$

Working in coordinates on $\mathbf{T}^1$, this yields a map $\mathfrak{y}\colon (-\pi,\pi)\times B_{2d}(\mathbf{0},\epsilon) \to X$ by setting

$$\mathfrak{y}(\vartheta,\mathbf{w}) =: e^{i\vartheta} \bullet (x+\mathbf{w}). \tag{73}$$

If $\mathbf{H}(m) \in \mathbf{R}^{2d}$ is the local coordinate expression of $\xi_M(m) \in T_mM$ (viewed as a column vector) then the local HLC expression of $\xi_X(x)$ is $(\mathbf{H}(m), -\Phi(m)) \in \mathbb{R}^{2d}\times\mathbb{R}$. If $(\theta,\mathbf{v}) \sim (0,\mathbf{0})$, then by (1)

$$\mathfrak{y}(\vartheta,\mathbf{w}) = x + (\mathbf{w} + \vartheta\,\mathbf{H}(m), -\vartheta\,\Phi(m)) + O\left(\|(\mathbf{w},\vartheta)\|^2\right). \tag{74}$$

The Jacobian matrix at the origin of $\mathfrak{r}^{-1}\circ\mathfrak{y}$ is then

$$\mathrm{Jac}_{(0,\mathbf{0})}\left(\mathfrak{r}^{-1}\circ\mathfrak{y}\right) = \begin{pmatrix} I_{2d} & \mathbf{H}_m \\ \mathbf{0}^t & -\Phi(m)\end{pmatrix}. \tag{75}$$

Since $\Phi > 0$, $\mathfrak{y}'$ is a local diffeomorphism at $(1,\mathbf{0}) \in \mathbf{T}^1 \times B_{2d}(\mathbf{0},\epsilon)$. Therefore, if $T_\delta =: \left\{e^{i\vartheta} :\ -\delta < \vartheta < \delta\right\} \subseteq \mathbf{T}^1$ then for all sufficiently small δ, $\epsilon > 0$, the restriction of $\mathfrak{y}'$ to $T_\delta \times B_{2d}(\mathbf{0},\epsilon)$ is a diffeomorphism onto its image.

Lemma 38 *Suppose $x \in X'$. Then, for all sufficiently small $\epsilon > 0$, the restriction $\mathfrak{y}' : \mathbf{T}^1 \times B_{2d}(\mathbf{0}, \epsilon) \to X$ is injective. Its image is a μ^X-invariant tubular neighborhood of the μ^X-orbit of x.*

Proof If not, there exists $\epsilon_j \to 0^+$ and for every j a choice of distinct pairs

$$(e^{i\vartheta_j}, \mathbf{w}_j), \quad (e^{i\vartheta'_j}, \mathbf{w}'_j) \in \mathbf{T}^1 \times B_{2d}(\mathbf{0}, \epsilon_j),$$

and such that, if $\lambda_j =: \vartheta'_j - \vartheta_j$,

$$e^{i\vartheta_j} \bullet (x + \mathbf{w}_j) = e^{i\vartheta'_j} \bullet (x + \mathbf{w}'_j) \quad \Longrightarrow \quad x + \mathbf{w}_j = e^{i\lambda_j} \bullet (x + \mathbf{w}'_j). \tag{76}$$

If $e^{i\lambda_j} \in T_\delta$, the previous considerations imply that $e^{i\lambda_j} = 1$, whence $e^{i\vartheta_j} = e^{i\vartheta'_j}$, and $\mathbf{w}_j = \mathbf{w}'_j$, against the assumptions. Therefore, it follows from (76) that $e^{i\lambda_j} \in \mathbf{T}^1 \setminus T_\delta$, a compact subset of $\mathbf{T}^1$. Perhaps after passing to a subsequence, we may therefore assume without loss that $e^{i\lambda_j} \to e^{i\lambda_\infty} \in \mathbf{T}^1 \setminus T_\delta$ as $j \to +\infty$. Since obviously $x + \mathbf{w}_j$, $x + \mathbf{w}'_j \to x$ as $j \to +\infty$, passing to the limit in (76) we obtain $e^{i\lambda_\infty} \bullet x = x$. But this is absurd by definition of X', given that $e^{i\lambda_\infty} \neq 1$ and $x \in X'$. □

It follows easily that if $x \in X'$ then $\mathfrak{y}' : \mathbf{T}^1 \times B_{2d}(\mathbf{0}, \epsilon) \to X$ is a diffeomorphism onto its image for all sufficiently small $\epsilon > 0$, and therefore that $\mathfrak{y} : (-\pi, \pi) \times B_{2d}(\mathbf{0}, \epsilon) \to X$ is a local coordinate chart. We shall say that η is a system of μ-adapted HLC.

In general, $\mathfrak{y}' : \mathbf{T}^1 \times B_{2d}(\mathbf{0}, \epsilon) \to X$ is an $l : 1$-covering, where $l = |T_m|$. To see this, let us consider the following generalization of Lemma 38:

Lemma 39 *Suppose $l = |T_m|$, where $m = \pi(x)$. Then, for all sufficiently small $\epsilon > 0$, the restriction $\mathfrak{y}' : \mathbf{T}^1 \times B_{2d}(\mathbf{0}, \epsilon) \to X$ is an $l : 1$-covering. Its image is a μ^X-invariant tubular neighborhood of the μ^X-orbit of x.*

Proof Suppose $x' = e^{i\vartheta_0} \bullet x \in \mathbf{T}^1 \cdot x$. Then, for any $g \in T_m$, we have that $\mathfrak{y}((e^{i\vartheta_0} g, \mathbf{0})) = x'$. Therefore, the inverse image $\mathfrak{y}'^{-1}(x')$ contains l distinct elements $(e^{i\vartheta_0} g, \mathbf{0})$ $(g \in T_m)$, and at each of these $\mathfrak{y}'$ is a local diffeomorphism. It follows that any x'' sufficiently close to the orbit $\mathbf{T}^1 \cdot x$ has at least l inverse images under $\mathfrak{y}'$, and that at each of these the latter is a local diffeomorphism.

I claim that in fact any x'' sufficiently close to the orbit $\mathbf{T}^1 \cdot x$ has exactly l inverse images under $\mathfrak{y}'$. If not, there exist $\epsilon_j \to 0^+$ and for every j distinct pairs

$$(g_j^{(a)}, \mathbf{v}_j^{(a)}) \in \mathbf{T}^1 \times B_{2d}(\mathbf{0}, \epsilon), \quad 1 \leq a \leq l + 1,$$

such that $g_j^{(a)} \bullet \mathbf{v}_j^{(a)} = g_j^{(b)} \bullet \mathbf{v}_j^{(b)}$, for every $1 \leq a, b \leq l+1$. Arguing as in the proof of Lemma 38, we conclude that $g_j^{(a)} g_j^{(b)^{-1}} \notin T_\delta$ for any $1 \leq b < a \leq l + 1$ and

$j \gg 0$. In particular, perhaps after passing to a subsequence, for $a = 2, \ldots, l+1$ we have $g_j^{(a)} g_j^{(1)^{-1}} \to \lambda_\infty^{(a)} \in T_m \setminus T_\delta$.

Suppose $\lambda_\infty^{(a)} = \lambda_\infty^{(b)}$ for $2 \leq a < b \leq l+1$. Then $g_j^{(a)} g_j^{(b)^{-1}} \to 1 \in T_\delta$ as $j \to +\infty$, absurd. Therefore, T_m contains the $l+1$ distinct elements $\{1, \lambda_\infty^{(2)}, \ldots, \lambda_\infty^{(l+1)}\}$, a contradiction. □

Lemma 40 *For any $\vartheta \in (-\pi, \pi)$, we have*

$$\mathfrak{y}^*(dV_X)(\vartheta, \mathbf{0}) = \frac{1}{2\pi}\,\Phi(m)\,|d\vartheta|\,d\mathcal{L}(\mathbf{w}),$$

where $d\mathcal{L}(\mathbf{w})$ is the Lebesgue measure on $\mathbf{R}^{2d}$.

Proof Let us write $\mathfrak{r}^*(dV_X) = \mathcal{V}(\theta, \mathbf{v})\,|d\theta|\,d\mathcal{L}(\mathbf{v})$, so that $\mathcal{V}(\theta, \mathbf{0}) = (2\pi)^{-1}$. Then

$$\begin{aligned}
\mathfrak{y}^*(dV_X) &= \left(\mathfrak{r} \circ \mathfrak{r}^{-1} \circ \mathfrak{y}\right)^*(dV_X) = \left(\mathfrak{r}^{-1} \circ \mathfrak{y}\right)^*\left(\mathfrak{r}^*(dV_X)\right)\\
&= \left(\mathfrak{r}^{-1} \circ \mathfrak{y}\right)^*\left(\mathcal{V}(\theta, \mathbf{v})\,d\theta\,d\mathcal{L}(\mathbf{v})\right)\\
&= \left(\mathcal{V} \circ \left(\mathfrak{r}^{-1} \circ \mathfrak{y}\right)\right) \cdot \left|\det\left(\mathrm{Jac}\left(\mathfrak{r}^{-1} \circ \mathfrak{y}\right)\right)\right|\,|d\vartheta|\,d\mathcal{L}(\mathbf{w}).
\end{aligned}$$

At $(0, \mathbf{0})$, in view of (75) and since $\Phi > 0$, we get

$$\mathfrak{y}^*(dV_X)(0, \mathbf{0}) = \frac{1}{2\pi}\,\Phi(m)\,|d\vartheta|\,d\mathcal{L}(\mathbf{w}). \tag{77}$$

This proves the claim at $(0, \mathbf{0})$. To prove it at $(\vartheta_0, \mathbf{0})$, we replace $\vartheta \sim \vartheta_0$ by $\vartheta + \vartheta_0$ with $\vartheta \sim 0$ and note that $e^{i(\vartheta+\vartheta_0)} \bullet (x + \mathbf{v}) = e^{i\vartheta} \bullet (e^{i\vartheta_0} \bullet (x + \mathbf{v}))$. Since $\mathfrak{r}_{\vartheta_0}(\theta, \mathbf{v}) = e^{i\theta} \cdot (e^{i\vartheta_0} \bullet (x + \mathbf{v}))$ is a system of HLC centered at $e^{i\vartheta_0} \bullet x$, one can argue as in the previous case. □

Corollary 41 *Under the assumptions of Lemma 39, if $\epsilon > 0$ is sufficiently small let $V_\epsilon = \mathfrak{y}'(\mathbf{T}^1 \times B_{2d}(\mathbf{0}, \epsilon))$. Then for any continuous function on X, we have*

$$\int_{V_\epsilon} f\,dV_X = \frac{1}{2\pi\,|T_m|}\int_{-\pi}^{\pi}\int_{B_{2d}(\mathbf{0},\epsilon)} f \circ \mathfrak{y} \cdot (\Phi(m) + A(\mathbf{w}))\,|d\vartheta|\,d\mathcal{L}(\mathbf{w}),$$

where $A(\mathbf{w}) = O(\|\mathbf{w}\|)$.

3 Proof of Theorem 1

Proof As the orthogonal projector $\Pi_k^\mu \colon L^2(X, dV_X) \to H_k^\mu(X)$, Π_k^μ is idempotent, then for every $x \in X$ the Schwartz kernel $\Pi_k^\mu \in C^\infty(X \times X)$ satisfies

$$\Pi_k^\mu(x,x) = \int_X \Pi_k^\mu(x,y)\,\Pi_k^\mu(y,x)\,dV_X(y). \tag{78}$$

Let us fix $x_0 \in X'$ and set $m_0 =: \pi(x_0) \in M'$, $n_0 =: \kappa(x_0) \in N'$, and apply (78) with $x = x_0$. Let σ be a local holomorphic frame of $B^\vee$ on an open neighborhood $V \subseteq N'$ of n_0; as usual we implicitly identify $B_0^\vee$ with $A_0^\vee$ by means of $\overline{\Gamma}$ (Lemma 31). We may assume without loss that $\sigma(n_0) = x_0$. Let $\|\sigma\|_B =: (\mathcal{N}_B \circ \sigma)^{1/2}$. Then $\|\sigma\|_B$ is a positive real-analytic function on V by Lemma 35. Therefore, the unitarization $\sigma_u =: (1/\|\sigma\|_B) \bullet \sigma\colon V \to X'$ (see Remark 32) is a real-analytic section and $\sigma_u(n_0) = x_0$.

There exists $\epsilon > 0$ such that $\mathrm{dist}_X(x_0, \mathbf{T}^1 \cdot y) \geq \delta$ for every $y \in X \setminus \kappa^{-1}(V)$. Therefore, by Theorem 1 of [28] we have $\Pi_k^\mu(x_0,\cdot) = O(k^{-\infty})$ uniformly on $X \setminus \kappa^{-1}(V)$. If $\sim$ stands for "has the same asymptotics as," we see from this and (78) for $x = x_0$ that

$$\Pi_k^\mu(x_0,x_0) \sim \int_{\kappa^{-1}(V)} \Pi_k^\mu(x_0,y)\,\Pi_k^\mu(y,x_0)\,dV_X(y). \tag{79}$$

We can parameterize the invariant open neighborhood $\kappa^{-1}(V) \subseteq X'$ by setting

$$\varrho\colon \mathbf{T}^1 \times V \to \kappa^{-1}(V), \quad \left(e^{i\vartheta}, n\right) \mapsto e^{i\vartheta} \bullet \sigma_u(n). \tag{80}$$

Then

$$\varrho^*(dW_X) = \frac{1}{2\pi}\,d\vartheta \wedge dV_N \tag{81}$$

where $dV_N = (1/d!)\,\eta^{\wedge d}$ (see Section 2.5). Now (3) means that $s(e^{i\vartheta} \bullet x) = e^{ik\vartheta}\,s(x)$, for every $e^{i\vartheta} \in \mathbf{T}^1$ and $x \in X$. Therefore, given (4), we have

$$\begin{aligned}
&\Pi_k^\mu\left(x_0, e^{i\vartheta} \bullet \sigma_u(n)\right)\,\Pi_k^\mu\left(e^{i\vartheta} \bullet \sigma_u(n), x\right)\\
&= \left[e^{-ik\vartheta}\,\Pi_k^\mu\left(x_0, \sigma_u(n)\right)\right]\left[e^{ik\vartheta}\,\Pi_k^\mu\left(\sigma_u(n), x_0\right)\right]\\
&= \Pi_k^\mu\left(x_0, \sigma_u(n)\right)\,\Pi_k^\mu\left(\sigma_u(n), x_0\right).
\end{aligned}$$

Inserting this and (47) into (79), we obtain

$$\begin{aligned}
&\Pi_k^\mu(x_0,x_0)\\
&\sim \frac{1}{2\pi}\int_{-\pi}^{\pi}\int_V \Pi_k^\mu\left(x_0,\sigma_u(n)\right)\,\Pi_k^\mu\left(\sigma_u(n),x_0\right)\,\Phi(n)^{d+1}\,d\vartheta\,dV_N(n)\\
&= \int_V \Pi_k^\mu\left(\sigma_u(n_0),\sigma_u(n)\right)\,\Pi_k^\mu\left(\sigma_u(n),\sigma_u(n_0)\right)\,\Phi(n)^{d+1}\,dV_N(n).
\end{aligned} \tag{82}$$

If we use (59) in (82) we get

$$\Pi_k^\mu(x_0, x_0) \sim \int_V e^{-k\,\mathcal{D}_N(n_0,n)}\, \widetilde{K_k^\mu}(n_0, n)\, \widetilde{K_k^\mu}(n, n_0)\, \Phi(n)^{d+1}\, dV_N(n), \tag{83}$$

where $\mathcal{D}$ is Calabi's diastasis function of $(N', I, 2\eta)$, defined in (7).

Let us set, for simplicity, $\eta' = 2\,\eta$. Also, suppose without loss that V is the domain of a holomorphic local coordinate chart (z_a) for N'. If $z_a + i\,y_a$, with x_a, y_a real-valued, then by (18) we have

$$\begin{aligned} dV_N &= \det\left([2\,\eta_{k\bar{l}}]\right) \cdot dx_1 \wedge \cdots dx_d \wedge dy_1 \cdots \wedge dy_d \\ &= \det\left([\eta'_{k\bar{l}}]\right) \cdot dx_1 \wedge \cdots dx_d \wedge dy_1 \cdots \wedge dy_d. \end{aligned} \tag{84}$$

In view of (67), we can thus rewrite (83) as follows:

$$\begin{aligned} &\Pi_k^\mu(x_0, x_0) \\ &\sim \left(\frac{k}{\pi}\right)^{2d} \sum_{j \geq 0} k^{-j} \int_B e^{-k\,\mathcal{D}_N(n_0,n)}\, Z_j(n_0, n)\, \det\left([\eta'_{k\bar{l}}]\right) dx\, dy, \end{aligned} \tag{85}$$

where now $B \subseteq \mathbb{C}^d$ is some open ball centered at the origin, and for every $j \geq 0$ we have

$$Z_j(n, n') =: \Phi(n')^{d+1} \sum_{a+b=j} \widetilde{S_a^\mu}(n, n')\, \widetilde{S_b^\mu}(n', n) \quad ((n, n') \in V \times V). \tag{86}$$

In particular, since $S_0^\mu = \Phi^{-(d+1)}$, for $j = 0$ we get from (86):

$$\begin{aligned} Z_0(n, n') &= \Phi(n')^{d+1}\, \widetilde{\Phi}(n, n')^{-(d+1)}\, \widetilde{\Phi}(n', n)^{-(d+1)} \\ &= \left(\frac{\Phi(n')}{\widetilde{\Phi}(n, n')\, \widetilde{\Phi}(n', n)}\right)^{d+1} = F_\Phi(n')^{d+1}, \end{aligned} \tag{87}$$

with the notation of Lemma 13, taking $p_0 = n$, and where $\widetilde{\Phi}$ is the sesqui-holomorphic extension of Φ (as a function on N') to some open neighborhood $\widetilde{N}$ of the diagonal (and we assume $V \times V \subseteq \widetilde{N}$). On the diagonal, $Z_0(n, n) = \Phi(n)^{-(d+1)}$.

On the contrary, for $j \geq 1$ we get

$$\begin{aligned} Z_j(n, n') = \Phi(n')^{d+1} &\left[\widetilde{\Phi}(n, n')^{-(d+1)}\, \widetilde{S_j^\mu}(n', n) + \widetilde{S_j^\mu}(n, n')\, \widetilde{\Phi}(n', n)^{-(d+1)}\right] \\ &+ \Phi(n')^{d+1} \sum_{0<a<j} \widetilde{S_a^\mu}(n, n')\, \widetilde{S_{j-a}^\mu}(n', n). \end{aligned} \tag{88}$$

On the diagonal,

$$Z_j(n, n) = 2\, S_j^\mu(n) + \Phi(n)^{d+1} \sum_{0<a<j} S_a^\mu(n)\, S_{j-a}^\mu(n). \tag{89}$$

Let us now consider the asymptotics of the jth summand in (85). Because $\mathcal{D}_N$ is the diastasis function of η', we can apply Theorem 3 of [13], and obtain an asymptotic expansion of the form

$$\int_B e^{-k\,\mathcal{D}_N(n_0,n)}\,Z_j(n_0,n)\,\det\left([\eta'_{k\bar{l}}]\right)dx\,dy \sim \left(\frac{\pi}{k}\right)^d \sum_{l\geq 0} k^{-l}\,R_l^N(Z_j(n_0,\cdot))\Big|_{n=n_0}, \tag{90}$$

where the R_j^Ns are Engliš operators for the Kähler manifold (N, I, η').

Using (90) within (85), we get

$$\begin{aligned}\Pi_k^\mu(x_0,x_0) &\sim \left(\frac{k}{\pi}\right)^d \sum_{j,l\geq 0} k^{-j-l}\,R_l^N(Z_j(n_0,\cdot))\Big|_{n=n_0}\\ &= \left(\frac{k}{\pi}\right)^d \sum_{j\geq 0} k^{-j} \sum_{a+b=j} R_a^N(Z_b(n_0,\cdot))\Big|_{n=n_0}.\end{aligned} \tag{91}$$

It follows from (5) and (91) that

$$\begin{aligned}S_j^\mu(n_0) &= \sum_{a+b=j} R_a^N(Z_b(n_0,\cdot))\Big|_{n=n_0}\\ &= Z_j(n_0,n_0) + \sum_{a=1}^{j} R_a^N(Z_{j-a}(n_0,\cdot))\Big|_{n=n_0}.\end{aligned} \tag{92}$$

Given (89), the latter relation may be rewritten

$$\begin{aligned}S_j^\mu(n_0) = 2\,S_j^\mu(n_0) + \Phi(n_0)^{d+1} \sum_{0<a<j} S_a^\mu(n_0)\,S_{j-a}^\mu(n_0)\\ + \sum_{a=1}^{j} R_a^N(Z_{j-a}(n_0,\cdot))\Big|_{n=n_0}.\end{aligned} \tag{93}$$

It follows that

$$\begin{aligned}S_j^\mu(n_0) = -\Phi(n_0)^{d+1} \sum_{0<a<j} S_a^\mu(n_0)\,S_{j-a}^\mu(n_0)\\ - \sum_{a=1}^{j} R_a^N(Z_{j-a}(n_0,\cdot))\Big|_{n=n_0},\end{aligned} \tag{94}$$

which determines S_j^μ for any $j \geq 1$ in terms of the S_k^μs with $0 \leq k < j$ and their sesquiholomorphic extensions. The proof is complete, for (94) is (11), with j in place of $j+1$. □

4 Proof of Corollary 2

Proof Let us apply (94) with $j = 1$. We get

$$S_1^\mu(n_0) = -\left. R_1^N(Z_0(n_0,\cdot))\right|_{n=n_0} = -\left.\left(\Delta_N - \frac{1}{2}\varrho_N\right)(Z_0(n_0,\cdot))\right|_{n=n_0}, \tag{95}$$

where Z_0 is defined by (87), and Δ_N and ϱ_N are defined by (20) and (19), respectively, with reference to the Kähler manifold $(P, K, \gamma) = (N, I, \eta')$, where $\eta' = 2\eta$.

We have, by (87),

$$Z_0(n_0, n) = \left(\frac{\Phi(n)}{\widetilde{\Phi}(n_0, n)\,\widetilde{\Phi}(n, n_0)}\right)^{d+1} = F_\Phi(n)^{d+1}, \tag{96}$$

where F_Φ is defined as in Lemma 13, with $f = \Phi$ and $n_0 = p_0$. Applying (23) with $l = d + 1$ and $f = F_\Phi$, we get

$$\begin{aligned}\left.\Delta_N(Z_0(n_0,\cdot))\right|_{n=n_0} &= \Delta_N\left(F_\Phi^{d+1}\right)(n_0)\\ &= (d+1)\,F_\Phi(n_0)^d \cdot \Delta_N(F_\Phi)(n_0) + \frac{d(d+1)}{2}\,F_\Phi(n_0)^{d-1}\left\|\mathrm{grad}_N(F_\Phi)(n_0)\right\|^2\\ &= (d+1)\,\Phi(n_0)^{-d}\cdot\Delta_N(F_\Phi)(n_0),\end{aligned} \tag{97}$$

where the gradient and the norm are taken with respect to the Riemannian metric $h' = 2h$, and in the last equation we have made use of Lemma 15.

Let us apply Lemma 13 with $(P, K, \gamma) = (N', I, \eta')$, $f = \Phi^{d+1} \in C^\varpi(N')$, and $p_0 = n_0$, so that in the statement we have $F = Z_0(n_0, \cdot)$. We obtain

$$\begin{aligned}&\left.\Delta_N(Z_0(n_0,\cdot))\right|_{n=n_0}\\ &= (d+1)\,\Phi(n_0)^{-(d+2)}\left[\Delta_P(\Phi)(n_0) - \frac{1}{2\,\Phi(n_0)}\left\|\mathrm{grad}_N(\Phi)(n_0)\right\|^2\right].\end{aligned} \tag{98}$$

Inserting (98) into (95),

$$\begin{aligned}S_1^\mu(n_0) &= \frac{1}{2}\varrho_N(n_0)\,\Phi(n_0)^{-(d+1)}\\ &\quad+(d+1)\,\Phi(n_0)^{-(d+2)}\left[\frac{1}{2\,\Phi(n_0)}\left\|\mathrm{grad}_N(\Phi)(n_0)\right\|^2 - \Delta_P(\Phi)(n_0)\right].\end{aligned}$$

□

5 Proof of Theorem 4

Proof Statements (1) and (2) follow quite straightforwardly by using the corresponding properties of Π_k^μ in Theorem 1 of [28] in the first line of (6).

To prove (3), we start from the relation

$$T_k^\mu(f)\left(x+\frac{\mathbf{v}}{\sqrt{k}},x+\frac{\mathbf{w}}{\sqrt{k}}\right) \tag{99}$$
$$=\int_X \Pi_k^\mu\left(x+\frac{\mathbf{v}}{\sqrt{k}},y\right) f(y)\,\Pi_k^\mu\left(y,x+\frac{\mathbf{w}}{\sqrt{k}}\right) dV_X(y).$$

If integration in $dV_X(y)$ in (99) is restricted to a given invariant tubular neighborhood V of the orbit $\mathbf{T}^1\cdot x$, only a negligible contribution to the asymptotics is lost. On the contrary, on V we can introduce μ-adapted HLC as in Section 2.9, so as to write $y=e^{i\theta}\bullet(x+\mathbf{u})$. Applying Corollary 41 (with $V=V_\epsilon$), we get

$$T_k^\mu(f)\left(x+\frac{\mathbf{v}}{\sqrt{k}},x+\frac{\mathbf{w}}{\sqrt{k}}\right) \tag{100}$$
$$\sim\frac{1}{2\pi\,|T_m|}\int_{-\pi}^{\pi}\int_{B_{2d}(\mathbf{0},\epsilon)}(\Phi(m)+A(\mathbf{u}))$$
$$\cdot\Pi_k^\mu\left(x+\frac{\mathbf{v}}{\sqrt{k}},e^{i\theta}\bullet(x+\mathbf{u})\right) f(m+\mathbf{u})\,\Pi_k^\mu\left(e^{i\theta}\bullet(x+\mathbf{u}),x+\frac{\mathbf{w}}{\sqrt{k}}\right)$$
$$\cdot|d\vartheta|\,d\mathcal{L}(\mathbf{u}),$$

where we used the fact that $f\in C^\infty(M)^\mu$.

Let $D_1,\,D_2>0$ be as in (13). Since $\|\mathbf{v}\|,\,\|\mathbf{w}\|\le C\,k^{1/9}$, we have

$$\mathrm{dist}_X\left(\mathbf{T}^1\cdot x,x+\frac{\mathbf{v}}{\sqrt{k}}\right)\le D_2\,C\,k^{-7/18}. \tag{101}$$

If $\mathrm{dist}_X\left(\mathbf{T}^1\cdot x,y\right)\ge 2\,D_2\,C\,k^{-7/18}$, then by (101) we have

$$\mathrm{dist}_X\left(\mathbf{T}^1\cdot y,x+\frac{\mathbf{v}}{\sqrt{k}}\right)\ge D_2\,C\,k^{-7/18},$$

and similarly for $\mathbf{w}$. It follows from this and statement (2). (with $\epsilon=1/9$) that the contribution to (99) and (100) coming from the locus where $\mathrm{dist}_X(\mathbf{T}^1\cdot x,y)\ge 2\,D_2\,C\,k^{-7/18}$ is rapidly decreasing. By (13), this means that in (100) the contribution of the locus where $\|\mathbf{u}\|\ge(2D_2/D_1)\,C\,k^{-7/18}$ is rapidly decreasing. Therefore, only a negligible contribution is lost in (100) if the integrand is multiplied by $\varrho(k^{7/18}\,\mathbf{w})$, where ϱ is an appropriate radial bump function, identically equal to 1 near the origin.

Furthermore, using (3) and (45), for any $x, x', x'' \in X$ and $e^{i\theta} \in \mathbf{T}^1$ we have

$$\Pi_k^\mu\left(x', e^{i\theta} \bullet x''\right) = e^{-ik\theta}\,\Pi_k^\mu(x', x'') = \overline{\Pi_k^\mu(e^{i\theta} \bullet x'', x')}.$$

Inserting this in (100), and applying the rescaling $\mathbf{u} \mapsto \mathbf{u}/\sqrt{k}$, we obtain

$$\begin{aligned} &T_k^\mu(f)\left(x + \frac{\mathbf{v}}{\sqrt{k}}, x + \frac{\mathbf{w}}{\sqrt{k}}\right) \qquad (102)\\ &\sim \frac{k^{-d}}{|T_m|}\int_{\mathbb{C}^d}\left(\Phi(m) + A\left(\frac{\mathbf{u}}{\sqrt{k}}\right)\right)\\ &\cdot \Pi_k^\mu\left(x + \frac{\mathbf{v}}{\sqrt{k}}, x + \frac{\mathbf{u}}{\sqrt{k}}\right) f\left(m + \frac{\mathbf{u}}{\sqrt{k}}\right)\Pi_k^\mu\left(x + \frac{\mathbf{u}}{\sqrt{k}}, x + \frac{\mathbf{w}}{\sqrt{k}}\right)\\ &\cdot \varrho\left(k^{-1/9}\,\mathbf{u}\right) d\mathcal{L}(\mathbf{u}); \end{aligned}$$

integration in $d\mathbf{u}$ is really over an expanding ball of radius $O(k^{1/9})$ in $\mathbb{C}^d$.

Now by (3) of Theorem 1 of [28] (and the remark immediately following the statement of that theorem) with $\upsilon_1 = (0, \mathbf{v})$ and $\upsilon_2 = (0, \mathbf{w})$, the sought expansion holds for Π_k^μ (that is, for $f = 1$). Thus

$$\begin{aligned} \Pi_k^\mu\left(x + \frac{\mathbf{v}}{\sqrt{k}}, x + \frac{\mathbf{u}}{\sqrt{k}}\right) &\sim \left(\frac{k}{\pi}\right)^d \cdot \sum_{t \in T_m} t^k\, e^{\psi_2\left(d_m\mu_{t^{-1}}^M(\mathbf{v}), \mathbf{u}\right)/\Phi(m)} \qquad (103)\\ &\cdot \left(\Phi(m)^{-(d+1)} + \sum_{j \geq 1} k^{-j/2} R_j\left(m, d_m\mu_{t^{-1}}^M(\mathbf{v}), \mathbf{u}\right)\right), \end{aligned}$$

where ψ_2 is as in (15), and $R_j(m, \mathbf{v}, \mathbf{u})$ is a polynomial function of $\mathbf{v}$ and $\mathbf{u}$. Clearly,

$$\frac{1}{\Phi(m)}\,\psi_2(\mathbf{v}, \mathbf{u}) = \psi_2\left(\frac{1}{\sqrt{\Phi(m)}}\,\mathbf{v}, \frac{1}{\sqrt{\Phi(m)}}\,\mathbf{u}\right) = \psi_2\left(\mathbf{v}', \mathbf{u}'\right),$$

where for any $\mathbf{p} \in \mathbb{C}^d$ we set $\mathbf{p}' = \mathbf{p}/\sqrt{\Phi(m)}$.

Using this and the Taylor expansion for $f(m + \mathbf{u}/\sqrt{k})$ at m, we get for (102) an asymptotic expansion in descending powers of $k^{1/2}$, whose leading term is given by

$$\begin{aligned} &\frac{k^{-d}}{|T_m|}\,\Phi(m)^{-2d-1}\, f(m)\left(\frac{k}{\pi}\right)^{2d} \qquad (104)\\ &\cdot \sum_{t,s \in T_m} (s\,t)^k \int_{\mathbb{C}^d} e^{\psi_2\left(d_m\mu_{t^{-1}}^M(\mathbf{v}'), \mathbf{u}'\right) + \psi_2\left(\mathbf{u}', d_m\mu_s^M(\mathbf{w}')\right)}\, d\mathcal{L}(\mathbf{u}). \end{aligned}$$

Applying the change of variable $\mathbf{u} = \sqrt{\Phi(m)}\,\mathbf{s}$, (104) becomes

$$\frac{1}{|T_m|}\,\Phi(m)^{-(d+1)}\,f(m)\,\frac{k^d}{\pi^{2d}} \tag{105}$$
$$\cdot \sum_{t,s\in T_m} (s\,t)^k \int_{\mathbb{C}^d} e^{\psi_2\left(d_m\mu^M_{t^{-1}}(\mathbf{v}'),\mathbf{s}\right)+\psi_2\left(\mathbf{s},d_m\mu^M_s(\mathbf{w}')\right)}\,d\mathcal{L}(\mathbf{s}).$$
$$= \frac{1}{|T_m|}\,\Phi(m)^{-(d+1)}\,f(m)\,\frac{k^d}{\pi^{2d}}\,\pi^d \sum_{t,s\in T_m} (s\,t)^k\, e^{\psi_2\left(d_m\mu^M_{t^{-1}}(\mathbf{v}'),d_m\mu^M_s(\mathbf{w}')\right)}$$
$$= \frac{1}{|T_m|}\,\Phi(m)^{-(d+1)}\,f(m)\left(\frac{k}{\pi}\right)^d \sum_{t,s\in T_m} (s\,t)^k\, e^{\psi_2\left(d_m\mu^M_{(st)^{-1}}(\mathbf{v}'),\mathbf{w}'\right)}$$
$$= \Phi(m)^{-(d+1)}\,f(m)\left(\frac{k}{\pi}\right)^d \sum_{t\in T_m} t^k\, e^{\psi_2\left(d_m\mu^M_{t^{-1}}(\mathbf{v}'),\mathbf{w}'\right)}$$
$$= \Phi(m)^{-(d+1)}\,f(m)\left(\frac{k}{\pi}\right)^d \sum_{t\in T_m} t^k\, e^{\psi_2\left(d_m\mu^M_{t^{-1}}(\mathbf{v}),\mathbf{w}\right)/\Phi(m)}.$$

We have used the fact that if $A:\mathbb{C}^d \to \mathbb{C}^d$ is unitary, then

$$\psi_2(\mathbf{u}, A\mathbf{t}) = \psi_2\left(A^{-1}\mathbf{u},\mathbf{t}\right)$$

for any $\mathbf{u}$, $\mathbf{t} \in \mathbb{C}^d$, and the relation

$$\int_{\mathbb{C}^d} e^{\psi_2(\mathbf{v},\mathbf{u})+\psi_2(\mathbf{u},\mathbf{w})}\,d\mathcal{L}(\mathbf{u}) = \pi^d\, e^{\psi_2(\mathbf{v},\mathbf{w})}.$$

Finally, when $\mathbf{v} = \mathbf{w} = \mathbf{0}$ the appearance of descending powers of k in the asymptotic expansion for (102) originates from Taylor expanding the integrand in $\mathbf{u}/\sqrt{k}$; half-integer powers of k are thus associated with odd homogeneous polynomials in $\mathbf{u}$, and therefore the corresponding contributions to the integral vanish by parity considerations. □

6 Proof of Theorem 6

Proof The proof of Theorem 6 is an adaptation of the proof of Theorem 1, so we'll be very sketchy. Adopting the same setup, rather than (78), (79), and (83) we now have

$$T^\mu_k[f](x_0,x_0) = \int_X \Pi^\mu_k(x_0,y)\,f(y)\,\Pi^\mu_k(y,x_0)\,dV_X(y) \tag{106}$$
$$\sim \int_{\kappa^{-1}(V)} \Pi^\mu_k(x_0,y)\,f(y)\,\Pi^\mu_k(y,x_0)\,dV_X(y)$$
$$= \int_V e^{-k\,\mathcal{D}_N(n_0,n)}\,\widetilde{K^\mu_k}(n_0,n)\,\widetilde{K^\mu_k}(n,n_0)\,f(n)\,\Phi(n)^{d+1}\,dV_N(n).$$

Therefore, we get in place of (85) and (91):

$$\begin{aligned} & T_k^{\mu}[f](x_0, x_0) \\ & \sim \left(\frac{k}{\pi}\right)^{2d} \sum_{j\geq 0} k^{-j} \int_B e^{-k\,\mathcal{D}_N(n_0,n)}\, Z_j(n_0,n)\, f(n)\, \det\left([\eta'_{k\bar{l}}]\right) dx\, dy \\ & \sim \left(\frac{k}{\pi}\right)^{d} \sum_{j\geq 0} k^{-j} \sum_{a+b=j} R_a^N\left(Z_b(n_0,\cdot)\, f(\cdot)\right)\Big|_{n=n_0}, \end{aligned} \tag{107}$$

which proves the claim (and reproves Corollary 5). □

7 Proof of Corollary 8

Proof Let us simplify notation in the following arguments by setting $f_j^{\mu} =: S_j^{\mu}[f]$. To begin with, we have from (88) that $Z_1(n_0) = 2\,S_1^{\mu}(n_0)$. We see from (107) that

$$\begin{aligned} f_1^{\mu}(n_0) &= R_0^N\left(Z_1(n_0,\cdot)\, f(\cdot)\right)\Big|_{n=n_0} + R_1^N\left(Z_0(n_0,\cdot)\, f(\cdot)\right)\Big|_{n=n_0} \\ &= Z_1(n_0,n_0)\, f(n_0) + \left(\Delta_N - \frac{1}{2}\,\varrho_N\right)\left(Z_0(n_0,\cdot)\, f(\cdot)\right)\Big|_{n=n_0} \\ &= \left[2\,S_1^{\mu}(n_0) - \frac{1}{2}\,\varrho_N(n_0)\,\Phi(n_0)^{-(d+1)}\right] f(n_0) \\ &\quad + \Delta_N\left(Z_0(n_0,\cdot)\, f(\cdot)\right)\Big|_{n=n_0}. \end{aligned} \tag{108}$$

Now in view of Lemma 15 we have

$$\begin{aligned} & \Delta_N\left(Z_0(n_0,\cdot)\, f(\cdot)\right)\Big|_{n=n_0} \\ & = \Delta_N\left(Z_0(n_0,\cdot)\right)\Big|_{n=n_0} f(n_0) + \Phi(n_0)^{-(d+1)}\ \Delta_N\left(f(\cdot)\right)\Big|_{n=n_0}. \end{aligned}$$

Inserting this in (108), and recalling (95), we obtain

$$\begin{aligned} f_1^{\mu}(n_0) &= \Phi(n_0)^{-(d+1)}\ \Delta_N\left(f(\cdot)\right)\Big|_{n=n_0} + 2\,S_1^{\mu}(n_0)\, f(n_0) \\ &\quad + \left(-\frac{1}{2}\,\varrho_N(n_0)\,\Phi(n_0)^{-(d+1)} + \Delta_N\left(Z_0(n_0,\cdot)\right)\Big|_{n=n_0}\right) f(n_0) \\ &= \Phi(n_0)^{-(d+1)}\ \Delta_N\left(f(\cdot)\right)\Big|_{n=n_0} + S_1^{\mu}(n_0)\, f(n_0). \end{aligned} \tag{109}$$

□

8 Proof of Corollary 10

Proof Notation being as in Definition 9 and the proof of Corollary 8, by Corollaries 2 and 8 we have on M':

$$\begin{aligned}\mathrm{Ber}^{\mu}_k[f] &= \frac{f^{\mu}_0 + k^{-1} f^{\mu}_1 + O\left(k^{-2}\right)}{S^{\mu}_0 + k^{-1} S^{\mu}_1 + O\left(k^{-2}\right)} = \frac{f^{\mu}_0}{S^{\mu}_0} \cdot \frac{1 + k^{-1} (f^{\mu}_1/f^{\mu}_0) + O\left(k^{-2}\right)}{1 + k^{-1} (S^{\mu}_1/S^{\mu}_0) + O\left(k^{-2}\right)} \\ &= f + k^{-1} f \cdot \left(\frac{f^{\mu}_1}{f^{\mu}_0} - \frac{S^{\mu}_1}{S^{\mu}_0}\right) + O\left(k^{-2}\right).\end{aligned}$$

Thus $B^{\mu}_0(f) = f$; furthermore, by Corollary 8 we have

$$\begin{aligned}B^{\mu}_1(f) &= f \cdot \left(\frac{f^{\mu}_1}{f^{\mu}_0} - \frac{S^{\mu}_1}{S^{\mu}_0}\right) = \Phi^{d+1} f^{\mu}_1 - f\,\Phi^{d+1} S^{\mu}_1 \\ &= \Delta_N(f) + \Phi^{d+1} S^{\mu}_1 \cdot f - f \cdot \Phi^{d+1} S^{\mu}_1 = \Delta_N(f).\end{aligned}$$

□

9 Proof of Theorem 11

Before tackling the proof, let us remark that considerations similar to those in Section 2.7 hold for Toeplitz operators. Namely, if $f \in C^{\infty}(M)^{\mu}$ let $T^{\mu}_k[f]\colon H^{\mu}_k(X) \to H^{\mu}_k(X)$ and $T^{\mu}_k[f] \in C^{\infty}(X \times X)$ denote both the induced operator and its Schwartz kernel, given by (6). The latter extends uniquely to a sesquiholomorphic function $\mathcal{T}^{\mu}_k[f]\colon A^{\vee}_0 \times A^{\vee}_0 \to \mathbb{C}$, which is the Toeplitz analogue of (53); explicitly, it is given by

$$\mathcal{T}^{\mu}_k[f]\,(\lambda, \lambda') = \sum_j \widetilde{T^{\mu}_k[f]\left(s^{(k)}_j\right)}(\lambda) \cdot \overline{\tilde{s}^{(k)}_j(\lambda')} \qquad \left(\lambda, \lambda' \in A^{\vee}_0\right), \tag{110}$$

and satisfies the equivariance law (54). Corresponding to (55) we now have

$$T^{\mu}_k[f](\sigma_u(n), \sigma_u(n')) = e^{-\frac{k}{2}\left(\Xi(n)+\Xi(n')\right)}\, \mathcal{T}^{\mu}_k[f]\,(\sigma(n), \sigma(n')). \tag{111}$$

Let us define $K^{\mu}_k[f]\colon X \to \mathbb{R}$, the Toeplitz analogue of (51), by setting

$$K^{\mu}_k[f](x) =: T^{\mu}_k[f](x, x) = \sum_j T^{\mu}_k[f]\left(s^{(k)}_j\right)(x)\, \overline{s^{(k)}_j(x)} \qquad (x \in X); \tag{112}$$

since f is real, $T^{\mu}_k[f]\colon H^{\mu}_k(X) \to H^{\mu}_k(X)$ is self-adjoint, and so

$$T^{\mu}_k[f](x, x) \in \mathbb{R}.$$

Then $K^{\mu}_k[f]$ descends to a ν-invariant C^{ϖ} function on N, by an obvious analogue of Lemma 36, and so we can consider its unique sesquiholomorphic extension $\widetilde{K_k^\mu[f]}$ to a neighborhood of the diagonal in $N \times N$. In place of (57) we now have that

$$\mathcal{T}_k^\mu[f]\,(\sigma(n), \sigma(n')) = e^{k\,\Xi(n,n')}\,\widetilde{K_k^\mu[f]}(n, n'). \tag{113}$$

Finally, a Toeplitz operator $T[f] = \Pi \circ M_f \circ \Pi$ is a zeroth-order FIO associated with the same almost complex Lagrangian relation as Π, and therefore also has a microlocal structure of the form (61), with an amplitude having an asymptotic expansion as in (62). Repeating the arguments following (63), therefore, leads to the Toeplitz generalization of the asymptotic expansion (66) and (67):

$$T_k^\mu[f](\sigma_u(n), \sigma_u(n')) \sim \left(\frac{k}{\pi}\right)^d e^{k\left[\Xi(n,n') - \frac{1}{2}\left(\Xi(n) + \Xi(n')\right)\right]} \cdot \sum_{j \geq 0} k^{-j}\,\widetilde{S_j^\mu[f]}(n, n'); \tag{114}$$

$$\widetilde{K_k^\mu}[f](n, n') \sim \left(\frac{k}{\pi}\right)^d \sum_{j \geq 0} k^{-j}\,\widetilde{S_j^\mu[f]}(n, n'). \tag{115}$$

Let us prove Theorem 11.

Proof We adopt the notation and setting of the proof of Theorem 1. Given (114), arguing as in the derivation of (85) we now obtain

$$\begin{aligned} E_k^\mu[f, g](x_0, x_0) &= \left(T_k^\mu[f] \circ T_k^\mu[g]\right)(x_0, x_0) \\ &\sim \left(\frac{k}{\pi}\right)^{2d} \sum_{j \geq 0} k^{-j} \int_B e^{-k\,\mathcal{D}_N(n_0, n)}\, Z_j[f, g](n_0, n)\, \det\left([\eta'_{k\bar{l}}]\right) dx\, dy, \end{aligned} \tag{116}$$

where

$$\begin{aligned} &Z_j[f, g](n, n') \\ &=: \Phi(n')^{d+1} \sum_{a+b=j} \widetilde{S_a^\mu[f]}(n, n')\, \widetilde{S_b^\mu[g]}(n', n) \qquad ((n, n') \in V \times V). \end{aligned} \tag{117}$$

Corresponding to (91), we have

$$\begin{aligned} E_k^\mu[f, g](x_0, x_0) &\sim \left(\frac{k}{\pi}\right)^d \sum_{j,l \geq 0} k^{-j-l}\, R_l^N\left(Z_j[f, g](n_0, \cdot)\right)\Big|_{n=n_0} \\ &= \left(\frac{k}{\pi}\right)^d \sum_{j \geq 0} k^{-j} \sum_{a+b=j} R_a^N\left(Z_j[f, g](n_0, \cdot)\right)\Big|_{n=n_0} \\ &= \left(\frac{k}{\pi}\right)^d \left\{Z_0[f, g](n_0, n_0) + k^{-1} A_1[f, g](n_0) + O\left(k^{-2}\right)\right\}, \end{aligned} \tag{118}$$

where

$$A_1[f,g](n_0) =: R_1^N\big(Z_0[f,g](n_0,\cdot)\big)\Big|_{n=n_0} + Z_1[f,g](n_0,n_0). \tag{119}$$

By (117), we have

$$\begin{aligned} Z_0[f,g](n_0,n_0) &= \Phi(n_0)^{d+1}\, S_a^\mu[f](n_0)\, S_b^\mu[g](n_0) \\ &= \Phi(n_0)^{-(d+1)}\, f(n_0)\, g(n_0) \\ &= Z_0[g,f](n_0,n_0); \end{aligned} \tag{120}$$

therefore,

$$\begin{aligned} &E_k^\mu[f,g](x_0,x_0) - E_k^\mu[g,f](x_0,x_0) \\ &= \left(\frac{k}{\pi}\right)^d \left[k^{-1}\big(A_1[f,g](n_0) - A_1[g,f](n_0)\big) + O\left(k^{-2}\right)\right]. \end{aligned} \tag{121}$$

Furthermore, by (117) we have

$$\begin{aligned} Z_1[f,g](n_0,n_0) &= \Phi(n_0)^{d+1}\left[S_0^\mu[f](n_0)\, S_1^\mu[g](n_0) + S_1^\mu[f](n_0)\, S_0^\mu[g](n_0)\right] \\ &= Z_1[g,f](n_0,n_0). \end{aligned} \tag{122}$$

We see from (119) and (122) that

$$\begin{aligned} &A_1[f,g](n_0) - A_1[g,f](n_0) \\ &= R_1^N\big(Z_0[f,g](n_0,\cdot)\big)\Big|_{n=n_0} - R_1^N\big(Z_0[g,f](n_0,\cdot)\big)\Big|_{n=n_0} \\ &= \Delta_N\big(Z_0[f,g](n_0,\cdot)\big)\Big|_{n=n_0} - \Delta_N\big(Z_0[g,f](n_0,\cdot)\big)\Big|_{n=n_0}; \end{aligned} \tag{123}$$

in the latter equality, we have used the fact that $R_1^N = \Delta_N - \varrho_N/2$ and (120).

To compute the latter commutator, let us remark that

$$\begin{aligned} Z_0[f,g](n_0,n) &= \Phi(n)^{d+1}\, \widetilde{S_0^\mu[f]}(n_0,n)\, \widetilde{S_0^\mu[g]}(n,n_0) \\ &= Z_0(n_0,n)\, \widetilde{f}(n_0,n)\, \widetilde{g}(n,n_0), \end{aligned} \tag{124}$$

where $Z_0(n_0,n)$ is as in (87).

It follows from (124) and Lemma 15 that

$$\begin{aligned} \Delta_N\big(Z_0[f,g](n_0,\cdot)\big)\Big|_{n=n_0} &= \Delta_N\big(Z_0(n_0,\cdot)\big)\Big|_{n=n_0} \cdot f(n_0)\, g(n_0) \\ &\quad + Z_0(n_0,n_0)\ \Delta_N\big(\widetilde{f}(n_0,\cdot)\,\widetilde{g}(\cdot,n_0)\big)\Big|_{n=n_0}. \end{aligned} \tag{125}$$

Let $(h'^{\bar{r}s})$ be the contravariant metric tensor of (N', I, η'), where $\eta' = 2\eta$ (thus $h'^{\bar{r}s} = h^{\bar{r}s}/2$). Since the former summand on the RHS of (125) is symmetric in f and g, we have

$$
\begin{aligned}
&A_1[f,g](n_0) - A_1[g,f](n_0) \\
&= \Phi(n_0)^{-(d+1)} \left[\Delta_N\big(\widetilde{f}(n_0,\cdot)\,\widetilde{g}(\cdot,n_0)\big)\Big|_{n=n_0} - \Delta_N\big(\widetilde{g}(n_0,\cdot)\,\widetilde{f}(\cdot,n_0)\big)\Big|_{n=n_0} \right] \\
&= \Phi(n_0)^{-(d+1)}\, h'^{\bar{r}s} \big(\partial_{\bar{r}} f(n_0)\, \partial_s g(n_0) - \partial_{\bar{r}} g(n_0)\, \partial_s f(n_0)\big) \\
&= -i\, \Phi(n_0)^{-(d+1)}\, \{f,g\}_N,
\end{aligned} \tag{126}
$$

where in the latter step we have used (24). The last equality in the statement now follows from (126) and Corollary 21. □

References

[1] Ali, S. T. and Engliš, M. Quantization methods: A guide for physicists and analysts. *Rev. Math. Phys.* **17**(4) (2005) 391–490.

[2] Berezin, F. A. General concept of quantization. *Comm. Math. Phys.* **40** (1975) 153–174.

[3] Bleher, P., Shiffman, B., and Zelditch, S. Universality and scaling of correlations between zeros on complex manifolds. *Invent. Math.* **142** (2000) 351–395.

[4] Bordemann, M., Meinrenken, E., and Schlichenmaier, M. Toeplitz quantization of Kähler manifolds and $gl(N)$, $N \to \infty$ limits. *Comm. Math. Phys.* **165**(2) (1994) 281–296.

[5] Boutet de Monvel, L. and Guillemin, V. The Spectral Theory of Toeplitz Operators. Annals of Mathematics Studies, Vol. 99. Princeton, NJ: Princeton University Press, 1981.

[6] Boutet de Monvel, L. and Sjöstrand, J. Sur la singularité des noyaux de Bergman et de Szegö. *Astérisque* **34/35** (1976) 123–164.

[7] Cahen, M., Gutt, S., and Rawnsley, J. Quantization of Kähler manifolds. I. Geometric interpretation of Berezin's quantization. *J. Geom. Phys.* **7** (1990) 45–62.

[8] Cahen, M., Gutt, S., and Rawnsley, J. Quantization of Kähler manifolds. II. *Trans. Amer. Math. Soc.* **337** (1993) 73–98.

[9] Calabi, E. Isometric imbedding of complex manifolds. *Ann. Math. (2)* **58** (1953) 123.

[10] Catlin, D. The Bergman kernel and a theorem of Tian. In *Analysis and Geometry in Several Complex Variables (Katata, 1997)*, pp. 1–23. Trends in Mathematics. Boston: Birkhäuser, 1999.

[11] Charles, L. Berezin–Toeplitz operators, a semi-classical approach. *Comm. Math. Phys.* **239**(1&2) (2003) 128.

[12] Engliš, M. Berezin quantization and reproducing kernels on complex domains. *Trans. Amer. Math. Soc.* **348**(2) (1996) 411–479.

[13] Engliš, M. The asymptotics of a Laplace integral on a Kähler manifold. *J. Reine Angew. Math.* **528** (2000) 139.

[14] Guillemin, V. Star products on compact pre-quantizable symplectic manifolds. *Lett. Math. Phys.* **35** (1995) 85–89.

[15] Guillemin, V., Ginzburg, V., and Karshon, Y. Moment Maps, Cobordisms, and Hamiltonian Group Actions. Appendix J by Maxim Braverman. Mathematical Surveys and Monographs, Vol. 98. Providence, RI: American Mathematical Society, 2002.

[16] Guillemin, V. and Sternberg, S. Geometric quantization and multiplicities of group representations. *Invent. Math.* **67**(3) (1982) 515–538.

[17] Karabegov, A. and Schlichenmaier, M. Identification of Berezin–Toeplitz deformation quantization. *J. Reine Angew. Math.* **540** (2001) 49–76.

[18] Krantz, S. and Parks, H. A Primer of Real Analytic Functions, 2nd edn. Boston: Birkhäuser, 2002.

[19] Loi, A. The Tian–Yau–Zelditch asymptotic expansion for real analytic Kähler metrics. *Int. J. Geom. Methods Mod. Phys.* **1**(3) (2004) 253–263.

[20] Loi, A. A Laplace integral on a Kähler manifold and Calabi's diastasis function. *Diff. Geom. Appl.* **23** (2005) 55–66.

[21] Loi, A. A Laplace integral, the T–Y–Z expansion, and Berezin's transform on a Kähler manifold. *Int. J. Geom. Methods Mod. Phys.* **2**(3) (2005) 359–371.

[22] Lu, Z. On the lower order terms of the asymptotic expansion of Tian–Yau–Zelditch. *Amer. J. Math.* **122**(2) (2000) 235–273.

[23] Ma, X. and Marinescu, G. *Holomorphic Morse Inequalities and Bergman Kernels.* Progress and in Mathematics, Vol. 254. Basel: Birkhäuser, 2007.

[24] Ma, X. and Marinescu, G. Toeplitz operators on symplectic manifolds. *J. Geom. Anal.* **18**(2) (2008) 565–611.

[25] Ma, X. and Marinescu, G. Berezin–Toeplitz quantization and its kernel expansion. In *Geometry and Quantization*, pp. 125–166. Travaux Mathématiques, Vol. 19. University of Luxembourg, Luxembourg, 2011.

[26] Ma, X. and Marinescu, G. Berezin–Toeplitz quantization on Kähler manifolds. *J. Reine Angew. Math.* **662** (2012) 1–56.

[27] Melin, A. and Sjöstrand, J. Fourier integral operators with complex-valued phase functions. In *Fourier Integral Operators and Partial Differential Equations* International Colloquium, University of Nice, Nice, 1974), pp. 120–223. Lecture Notes in Mathematics, Vol. 459. Berlin: Springer-Verlag, 1975.

[28] Paoletti, R. Asymptotics of Szegö kernels under Hamiltonian torus actions. *Israel J. Math.* **191** (2012) 363–403.

[29] Schlichenmaier, M. Berezin–Toeplitz quantization for compact Kähler manifolds. An introduction. In *Geometry and Quantization*, pp. 97–124. Travaux Mathématiques, Vol. 19. University of Luxenbourg, Luxembourg, 2011.

[30] Shiffman, B. and Zelditch, S. Asymptotics of almost holomorphic sections of ample line bundles on symplectic manifolds. *J. Reine Angew. Math.* **544** (2002) 181–222.

[31] Tian, G. On a set of polarized Kähler metrics on algebraic manifolds. *J. Diff. Geom.* **32** (1990) 99–130.

[32] Tian, G. Canonical metrics in Kähler geometry. Notes taken by Meike Akveld. *Lectures in Mathematics ETH Zürich.* Basel: Birkhäuser, 2000.

[33] Xu, H. An explicit formula for the Berezin star product. *Lett. Math. Phys.* **101**(3) (2012) 239–264.

[34] Zelditch, S. Index and dynamics of quantized contact transformations. *Ann. Inst. Fourier (Grenoble)* **47** (1997) 305–363.

[35] Zelditch, S. Szegö kernels and a theorem of Tian. *Int. Math. Res. Not.* **6** (1998) 317–331.

18

Gaussian maps and generic vanishing I: Subvarieties of abelian varieties

G. Pareschi
Università di Roma, Tor Vergata

Abstract

The aim of this paper is to present an approach to Green–Lazarsfeld's generic vanishing combining Gaussian maps and the Fourier–Mukai transform associated with the Poincaré line bundle. As an application, we prove the generic vanishing theorem for all normal Cohen–Macaulay subvarieties of abelian varieties over an algebraically closed field.

Dedicated to my teacher, Rob Lazarsfeld, on the occasion of his 60th birthday

1 Introduction

We work with irreducible projective varieties on an algebraically closed field of any characteristic, henceforth called *varieties*. The contents of this paper are:

(1) A general criterion expressing the vanishing of the higher cohomology of a line bundle on a Cohen–Macaulay variety in terms of certain first-order conditions on hyperplane sections (Theorem 2). Such conditions involve *Gaussian maps* and the criterion is a generalization of well-known results on hyperplane sections of K3 and abelian surfaces.
(2) Using a relative version of the above, we prove the vanishing of higher direct images of Poincaré line bundles of normal Cohen–Macaulay subvarieties of abelian varieties[1] (Theorem 5). As is well known, this is equivalent to Green–Lazarsfeld's *generic vanishing*, a condition satisfied by all irregular compact Kahler manifolds [5]. This implies in turn a Kodaira-type vanishing

[1] By the Poincaré line bundle of a subvariety X of an abelian variety A we mean the pull-back to $X \times \mathrm{Pic}^0 A$ of a Poincaré line bundle on $A \times \mathrm{Pic}^0 A$.

From *Recent Advances in Algebraic Geometry*, edited by Christopher Hacon, Mircea Mustaţă and Mihnea Popa

for line bundles which are restrictions to normal Cohen–Macaulay subvarieties of abelian varieties, of ample line bundles on the abelian variety (Corollary 6).

Concerning point (2), it should be mentioned that at present we are not able to extend this approach efficiently to the general generic vanishing theorem (GVT), i.e., for varieties *mapping to* abelian varieties, even for smooth projective varieties over the complex numbers (where it is well known by the work of Green and Lazarsfeld). This will be the object of further research. However, concerning possible extensions of the general GVT to singular varieties and/or to positive characteristic, one should keep in mind the work of Hacon and Kovacs [8] where – by exploiting the relation between GVT and the Grauert–Riemenschneider vanishing theorem – they show examples of failure of the GVT for mildly singular varieties (over $\mathbb{C}$) and even smooth varieties (in characteristic $p > 0$) of dimension ≥ 3, with a (separable) generically finite map to an abelian variety. This disproved an erroneous theorem of a previous preprint of the author.

Now we turn to a more detailed presentation of the above topics.

1.1 Motivation: Gaussian maps on curves and vanishing of the H^1 of line bundles on surfaces

We introduce part (1) starting from a particular case, where the essence of the story becomes apparent: the vanishing of the H^1 of a line bundle on a surface in terms of Gaussian maps on a sufficiently positive hyperplane section (Theorem 1 below).

To begin with, let us recall what Gaussian maps are. Given a curve C and a line bundle A on C, denote by M_A the kernel of the evaluation map of global sections of A:

$$0 \to M_A \to H^0(C, A) \otimes \mathcal{O}_C \to A.$$

This comes equipped with a natural $\mathcal{O}_C$-linear differentiation map

$$M_A \to \Omega^1_C \otimes A$$

defined as

$$M_A = p_*(\mathcal{I}_\Delta \otimes q^*A) \to p_*((\mathcal{I}_\Delta \otimes A)_{|\Delta}) = \Omega^1_C \otimes A,$$

where p, q, and Δ are the projections and the diagonal of the product $C \times C$. Twisting with another line bundle B and taking global sections, one gets the *Gaussian map* (or *Wahl map* [21]) of A and B:

$$\gamma_{A,B} : Rel(A,B) := H^0(C, M_A \otimes B) \to H^0(C, \Omega^1_C \otimes A \otimes B)\,.^2$$

In our treatment it is more natural to set $A = N \otimes P$ and $B = \omega_C \otimes P^\vee$ for suitable line bundles N and P on the curve C, and to consider the dual map

$$g_{N,P} : \mathrm{Ext}^1_C(\Omega^1_C \otimes N, O_C) \to \mathrm{Ext}^1_C(M_{N\otimes P}, P). \tag{1}$$

Note that $g_{N,P}$ can be defined directly (even if ω_C is not a line bundle) as $\mathrm{Ext}^1_C(\cdot\,, P)$ of the differentiation map of $M_{N\otimes P}$.

The relation with the vanishing of the H^1 of line bundles on surfaces lies in the following result, whose proof follows closely arguments contained in the papers of Beaville and Mérindol [2] and Colombo *et al.* [3]. Let X be a Cohen–Macaulay surface and Q a line bundle on X. Let L be a base-point-free line bundle on X such that $L \otimes Q$ is also base-point-free, and let C be a (reduced and irreducible) Cartier divisor in $|L|$, not contained in the singular locus of X. Let $N_C = L_{|C}$ be the normal bundle of C. We have the extension class

$$e \in \mathrm{Ext}^1_C(\Omega^1_C \otimes N_C, O_C)$$

of the normal sequence

$$0 \to N_C^\vee \to (\Omega^1_X)_{|C} \to \Omega^1_C \to 0.$$

We consider the (dual) Gaussian map

$$g_{N_C, Q_{|C}} : \mathrm{Ext}^1_C(\Omega^1_C \otimes N_C, O_C) \to \mathrm{Ext}^1_C(M_{N_C\otimes Q}, Q_{|C}). \tag{2}$$

Theorem 1 *(a) If $H^1(X,Q) = 0$, then $e \in \ker(g_{N_C,Q_{|C}})$.*

(b) If L is sufficiently positive,[3] then the converse also holds: if $e \in \ker(g_{N_C,Q_{|C}})$, then $H^1(X,Q) = 0$.

(Note that e is nonzero if L is sufficiently positive.) For example, if X is a smooth surface with trivial canonical bundle and $Q = O_X$, then (a) says that if X is a K3 then $e \in \ker(g_{K_C,O_C})$. This is a result of [2]. Conversely, (b) says that if X is abelian and C is sufficiently positive then $e \notin \ker(g_{K_C,O_C})$. This is a result of [3].

[2] The source is denoted $Rel(A,B)$, as it is the kernel of the multiplication of global sections of A and B.

[3] By this we mean that L is a sufficiently high multiple of a fixed ample line bundle on X.

1.2

The proof is a calculation with extension classes whose geometric motivation is as follows. Suppose that C is a curve in a surface X and that C is embedded in an ambient variety Z. From the cotangent sequence

$$0 \to \mathcal{I}/\mathcal{I}^2 \to (\Omega^1_Z)_{|C} \to \Omega^1_C \to 0$$

(where $\mathcal{I}$ is the ideal of C in Z), one gets the long cohomology sequence

$$\cdots \to \mathrm{Hom}_C(\mathcal{I}/\mathcal{I}^2, N_C^\vee) \overset{H_Z}{\to} \mathrm{Ext}^1_C(\Omega^1_C, N_C^\vee) \overset{G_Z}{\to} \mathrm{Ext}^1_C((\Omega^1_Z)_{|C}, N_C^\vee) \to \cdots \quad (3)$$

The problem of extending the embedding $C \hookrightarrow Z$ to the surface X has a natural first-order obstruction, namely the class e must belong to $\ker G_Z = Im\, H_Z$. Indeed, as is well known, if the divisor $2C$ on X, seen as a scheme, is embedded in Z, then it lives (as an embedded first-order deformation) in the Hom on the left.[4] The forgetful map H_Z, disregarding the embedding, takes it to the class of the normal sequence $e \in \mathrm{Ext}^1_C(\Omega^1_C, N_C^\vee)$.

Now we specialize this to the case where the ambient variety is a projective space, specifically:

$$Z = \mathbb{P}(H^0(C, N_C \otimes Q)^\vee) := \mathbb{P}_Q$$

(in this informal discussion we are assuming, for simplicity, that the line bundle $L \otimes Q$ is very ample). By the Euler sequence, the map $G_{\mathbb{P}_Q}$ is the (dual) Gaussian map $g_{N_C, Q_{|C}}$ of (2). Notice that in this case there is a special feature that our extension problem can be relaxed to the problem of extending the embedding of C in $\mathbb{P}_Q$ to an embedding of the surface X in a possibly bigger projective space $\mathbb{P}$, containing $\mathbb{P}_Q$ as a linear subspace. However, since the restriction to $\mathbb{P}_Q$ of the conormal sheaf of C in $\mathbb{P}$ splits, this has the same first-order obstruction, namely $e \in \ker(g_{N_C, Q_{|C}})$.

The relation of all that with the vanishing of the H^1 is classical: the embedding of C in $\mathbb{P}_Q$ can be extended (in the above relaxed sense) to an embedding of X if and only if the restriction map $\rho_X : H^0(X, L \otimes Q) \to H^0(C, N_C \otimes Q)$ is surjective. This is implied by the vanishing of $H^1(X, Q)$, so we get (a). The converse is a bit more complicated: by Serre vanishing, if L is sufficiently positive then the vanishing of $H^1(X, Q)$ is *equivalent* to the surjectivity of the restriction map ρ_X, and also to the surjectivity of the restriction map $\rho_{2C} : H^0(2C, (L \otimes Q)_{|2C}) \to H^0(C, N_C \otimes Q)$, hence to the fact that $2C$ "lives" in $\mathrm{Hom}_C(\mathcal{I}/\mathcal{I}^2, N_C^\vee)$. Now if e is in the kernel of $g_{N_C, Q_{|C}} = G_{\mathbb{P}_Q}$, then e comes from some embedded deformations in $\mathrm{Hom}_C(\mathcal{I}/\mathcal{I}^2, N_C^\vee)$. However,

[4] More precisely, the ideal of $2C$ in Z induces the morphism of $\mathcal{O}_Z$-modules $\mathcal{I}/\mathcal{I}^2 \to N_C$ whose kernel is $\mathcal{I}_{2C/Z}/\mathcal{I}^2$ (see, e.g., [1] or [4]).

these do not necessarily include $2C$. A more refined analysis proves that this is indeed the case as soon as L is sufficiently positive.

1.3 Gaussian maps on hyperplane sections and vanishing

The criterion of part (1) above is a generalization of the previous theorem to a higher dimension and to a relative (flat) setting. The relevant case deals with the vanishing of the H^n of a line bundle on a variety of dimension $n + 1$.[5,6] To this end, we consider "hybrid" Gaussian maps as follows: let C be a curve in an n-dimensional variety Y and let A_C be a line bundle on C. The *Lazarsfeld sheaf* (see [7]), denoted $F^Y_{A_C}$, is the kernel of the evaluation map of A_C, *seen as a sheaf on* Y:

$$0 \to F^Y_{A_C} \to H^0(A_C) \otimes \mathcal{O}_Y \to A_C$$

(note that $F^Y_{A_C}$ is never locally free if $\dim Y \geq 3$). As above, it comes equipped with a $\mathcal{O}_Y$-linear differentiation map

$$F^Y_{A_C} \to \Omega^1_Y \otimes A_C.$$

If B is a line bundle on Y, we define the *Gaussian map of* A_C *and* B as

$$\gamma^Y_{A_C,B} : Rel(A_C, B) = H^0(Y, F_{A_C} \otimes B) \to H^0(Y, \Omega^1_Y \otimes A_C \otimes B).$$

As above, we rather use the dual map

$$g^Y_{M_C,R} : \mathrm{Ext}^n_Y(\Omega^1_Y \otimes M_C, \mathcal{O}_Y) \to \mathrm{Ext}^n_Y(F_{M_C \otimes R}, R),$$

where M_C and R are line bundles respectively on C and Y such that $A_C = M_C \otimes R$ and $B = \omega_Y \otimes R^\vee$. Again, this map can be defined directly (even if ω_Y is not a line bundle) as $\mathrm{Ext}^1_C(\,\cdot\,, R)$ of the differentiation map of $F^Y_{M_C \otimes R}$. The case $n = 1$ is recovered by taking $Y = C$.

These maps can be extended to a relative flat setting. In this paper we consider only the simplest case, namely a family of line bundles on a fixed variety Y, as this is the only case needed in subsequent applications. In the notation above, let T be another projective CM variety (or scheme), and let $\mathcal{R}$ be a line bundle on $Y \times T$. Let ν and π denote the two projections, respectively

[5] In fact, for all positive k, with $k < n$, the vanishing of H^k can be reduced to this case, as it is equivalent (by Serre vanishing) to the vanishing of H^k of the restriction of the given line bundle to a sufficiently positive $(k + 1)$-dimensional hyperplane section.

[6] *Note*: One could think of using the equality $h^n(X, Q) = h^1(X, \omega_X \otimes Q^\vee)$ and then reducing, as in the previous footnote, to a surface. However, this is not possible in the relative case, since in general there is no Serre duality isomorphism of the direct images. Even in the non-relative case, the resulting criterion is usually more difficult to apply.

on Y and T. We can consider the *relative Lazarsfeld sheaf* $\mathcal{F}^Y_{M_C,\mathcal{R}}$, kernel of the relative evaluation map

$$0 \to \mathcal{F}^Y_{M_C,\mathcal{R}} \to \pi^*\pi_*(\mathcal{R} \otimes \nu^* M_C) \to \mathcal{R} \otimes \nu^* M_C$$

where, as above, we see M_C as a sheaf on Y. The $O_{Y\times T}$-module $\mathcal{F}^Y_{M_C,\mathcal{R}}$ is equipped with its $O_{Y\times T}$-linear differentiation map (see Section 2.1 below)

$$\mathcal{F}^Y_{M_C,\mathcal{R}} \to \nu^*(\Omega^1_Y \otimes M_C) \otimes \mathcal{R}. \tag{4}$$

Applying $\mathrm{Ext}^n_{Y\times T}(\ \cdot\ ,\mathcal{R})$ and restricting to the direct summand $\mathrm{Ext}^n_Y(\Omega^1_Y \otimes M_C, O_Y)$, we get the (dual) Gaussian map

$$g^Y_{M_C,\mathcal{R}} : \mathrm{Ext}^n_Y(\Omega^1_Y \otimes M_C, O_Y) \to \mathrm{Ext}^n_{Y\times T}(\mathcal{F}^Y_{M_C,\mathcal{R}}, \mathcal{R}).$$

The announced generalization of Theorem 1 is as follows. Let X be an $(n+1)$-dimensional Cohen–Macaulay variety, let T be a CM variety, and let Q be a line bundle on $X \times T$. In order to avoid heavy notation, we still denote by ν and π the two projections of $X \times T$ (however, see Notation 1 in Section 2.1). Let L be a line bundle on X, with n irreducible effective divisors $Y_1, \ldots, Y_n \in |L|$ such that their intersection is an integral curve C not contained in the singular locus of X. We assume also that the line bundle $Q \otimes \nu^* L^{\otimes n}$ is relatively base-point-free, namely the relative evaluation map $\pi^*\pi_*(Q \otimes \nu^* L^{\otimes n}) \to Q \otimes \nu^* L^{\otimes n}$ is surjective. We choose a divisor among $Y_1, \ldots, Y_n$, say $Y = Y_1$, such that C is not contained in the singular locus of Y. Let N_C denote the line bundle $L_{|C}$. We consider the "restricted normal sequence"

$$0 \to N_C^\vee \to (\Omega^1_X)_{|C} \to (\Omega^1_Y)_{|C} \to 0. \tag{5}$$

Via the canonical isomorphism

$$\mathrm{Ext}^1_C(\Omega^1_Y \otimes N_C, O_C) \cong \mathrm{Ext}^n_Y(\Omega^1_Y \otimes N_C^{\otimes n}, O_Y) \tag{6}$$

(see Section 2.1 below), we see that the class e of (5) belongs to $\mathrm{Ext}^n_Y(\Omega^1_Y \otimes N_C^{\otimes n}, O_Y)$. Finally, we consider the (dual) Gaussian map

$$g^Y_{N_C^{\otimes n}, Q_{|Y\times T}} : \mathrm{Ext}^n_Y\left(\Omega^1_Y \otimes N_C^{\otimes n}, O_Y\right) \to \mathrm{Ext}^n_{Y\times T}\left(\mathcal{F}^Y_{N_C^{\otimes n}, Q_{|Y\times T}}, Q_{|Y\times T}\right). \tag{7}$$

Then we have the following result, recovering part (b) of Theorem 1 as the case $n = 1$ and $T = \{point\}$:

Theorem 2 *If L is sufficiently positive and e is an element of* $\ker\left(g^Y_{N_C^{\otimes n}, Q_{|Y\times T}}\right)$, *then $R^n\pi_* Q = 0$.*

The following version is technically easier to apply:

Corollary 3 *Keeping the notation of Theorem 2, if the line bundle L is sufficiently positive, then the kernel of the map* $g^Y_{N_C^{\otimes n}, Q_{|Y\times T}}$ *is at most 1-dimensional (spanned by e). Therefore, if* $g^Y_{N_C^{\otimes n}, Q_{|Y\times T}}$ *is non-injective, then* $R^n\pi_* Q = 0$.

Concerning the other implication, what we can prove is:

Proposition 4 *(a) Assume that* $T = \{point\}$. *If* $H^n(X, Q) = 0$, *then* $e \in \ker(g_{N_C^{\otimes n}, Q_{|Y}})$.

(b) In general, assume that $R^i\pi_*(Q_{|Y\times T}) = 0$ *for* $i < n$. *If also* $R^n\pi_* Q = 0$, *then* $e \in \ker(g^Y_{N_C^{\otimes n}, Q_{|Y\times T}})$.

1.4

To motivate these statements, let us go back to the informal discussion of Section 1.2. We assume for simplicity that $T = \{point\}$. Let X be an $(n+1)$-dimensional variety and C a curve in X as above. It is easily seen, using the Koszul resolution of the ideal of C and Serre vanishing, that the vanishing of $H^n(X, Q)$ implies the surjectivity of the restriction map $\rho_X : H^0(X, L^{\otimes n} \otimes Q) \to H^0(C, N_C^{\otimes n} \otimes Q)$, and in fact the two conditions are equivalent as soon as L is sufficiently positive. Hence it is natural to look for first-order obstructions to extend to X an embedding of the curve C (a 1-dimensional complete intersection of linearly equivalent divisors of X) into

$$\mathbb{P}_Q := \mathbb{P}(H^0(C, N_C^{\otimes n} \otimes Q)^\vee) .$$

More generally, we can consider the same problem for any given ambient variety Z, rather than projective space.

To find a first-order obstruction one can no longer replace X by the first-order neighborhood of C in X. We rather have to pick a divisor in $|L|$ containing C, say $Y = Y_1$, and replace X by the scheme $2Y \cap Y_2 \cap \cdots \cap Y_n$. In analogy with the case of curves on surfaces, it is natural to consider the long cohomology sequence

$$\cdots \to \mathrm{Hom}_C\, \mathcal{I}_Y/\mathcal{I}_Y^2, N_C^\vee) \xrightarrow{H_Z^Y} \mathrm{Ext}^1_C((\Omega^1_Y)_{|C}, N_C^\vee) \xrightarrow{G_Z^Y} \mathrm{Ext}^1_C((\Omega^1_Z)_{|C}, N_C^\vee) \to \cdots \tag{8}$$

(where $\mathcal{I}_Y$ is the ideal of Y in Z). As above, a necessary condition for the lifting to X of the embedding of $C \hookrightarrow Z$ is that the "restricted normal class" e of (5) belongs to $\ker(G_Z^Y)$.

However, looking for *sufficient* conditions for lifting (in the relaxed sense, as in Section 1.2) the embedding $C \hookrightarrow \mathbb{P}_Q$ to X, one cannot assume that the divisor Y is already embedded in $\mathbb{P}_Q$. This is the reason why, differently from the case when X is a surface, the map $g_{N_C^{\otimes n}, Q_{|Y}}$ appearing in the statement of Theorem 2 and Corollary 3 is not the map G_Z^Y with $Z = \mathbb{P}_Q$, but rather a slightly more complicated "hybrid" version of a (dual) Gaussian map. After this modification, the geometric motivation for Theorem 2 is similar to that of Section 1.2.

1.5 Generic vanishing for subvarieties of abelian varieties

Although difficult – if not impossible – to use in most cases, the above results can be applied in some very special circumstances. For example, in analogy with the literature on curves sitting on K3 surfaces and Fano threefolds, Proposition 4 can supply nontrivial necessary conditions for an n-dimensional variety to sit in some very special $(n+1)$-dimensional varieties.

However, in this paper we rather focus on the sufficient condition for vanishing provided by Theorem 2 and Corollary 3, as it provides an approach to *generic vanishing*, a far-reaching concept introduced by Green and Lazarsfeld [5, 6]. Namely, we consider a variety X with a map to an abelian variety, generically finite onto its image

$$a : X \to A. \tag{9}$$

Denoting by $\mathrm{Pic}^0 A = \widehat{A}$ the dual variety, we consider the pull-back to $X \times \widehat{A}$ of a Poincaré line bundle $\mathcal{P}$ on $A \times \widehat{A}$:

$$Q = (a \times \mathrm{id}_{\widehat{A}})^* \mathcal{P}. \tag{10}$$

We keep the notation of the previous section. In particular, we denote by ν and π the projections of $X \times \widehat{A}$. A way of expressing generic vanishing is the vanishing of higher direct images

$$R^i \pi_* Q = 0 \quad \text{for} \quad i < \dim X. \tag{11}$$

For smooth varieties over the complex numbers, (11) was proved (as a particular case of a more general statement) by Hacon [7], settling a conjecture of Green and Lazarsfeld. Another way of expressing the generic vanishing condition involves the *cohomological support loci*

$$V_a^i(X) = \{\alpha \in \mathrm{Pic}^0 A \mid h^i(X, a^*\alpha) > 0\}.$$

Green and Lazarsfeld's theorem [5, 6] is that, if the map a is generically finite, then

$$\mathrm{codim}_{\widehat{A}} V_a^i(X) \geq \dim X - i.^7 \tag{12}$$

[7] In general, if the map a is not generically finite, Hacon's and Green and Lazarsfeld's theorems are respectively $R^i \pi_* Q = 0$ for $i < \dim a(X)$ and $\mathrm{codim}_{\mathrm{Pic}^0 A} V_a^i(X) \geq \dim a(X) - i$. However,

It is easy to see that (11) implies (12). Subsequently, it has been observed in [16, 17] that (11) is in fact *equivalent* to (12).[8] The heart of Hacon's proof of (11) consists of a clever reduction to Kodaira–Kawamata–Viehweg vanishing, while the argument of Green and Lazarsfeld for (12) uses Hodge theory. Both need characteristic 0, and that the variety X is smooth (or with rational singularities).

On the contrary, a characteristic-free example of both (11) and (12) is given by abelian varieties themselves [15, p. 127]. Here we extend this by proving that (11) (and, therefore, (12)) holds for normal Cohen–Macaulay subvarieties of abelian varieties on an algebraically closed field of any characteristic.

Theorem 5 *In the above notation, assume that X is normal Cohen–Macaulay and the morphism a is an embedding. Then $R^i\pi_*Q = 0$ for all $i < \dim X$.*

The strategy of the proof consists of applying Theorem 2 to the Poincaré line bundle Q. In order to do so we take a general complete intersection $C = Y \cap Y_2 \cap \cdots \cap Y_n$ of X, with $Y_i \in |L|$, where, as above, L is a sufficiently positive line bundle on X and $n + 1 = \dim X$. The main issue of the argument consists of comparing two spaces of first-order deformations: the first is the kernel of the (dual) Gaussian map $g_{N_C^{\otimes n}, Q_{Y\times\widehat{A}}}$. The second is the kernel of the map G_Z^Y of (8) with Z equal to the ambient abelian variety A[9] (by (6), the two maps have the same source). As in the discussion of Section 1.4, the variety $X \subset A$ induces naturally, via the restricted normal extension class e, a nontrivial element of $\ker G_A^Y$. Hence, to get the vanishing of $R^n\pi_*Q$, it would be enough to prove that $\ker G_A^Y$ is contained in $\ker g_{N_C^{\otimes n}, Q_{Y\times\widehat{A}}}$, or at least – in view of Corollary 3 – that the intersection of $\ker G_A^Y$ and $\ker g_{N_C^{\otimes n}, Q_{|Y\times\widehat{A}}}$ is nonzero. This analysis is accomplished by means of the Fourier–Mukai transform associated with the Poincaré line bundle.[10] In doing this we were inspired by the classical papers [10, 13] where the conceptually related problem of comparing the first-order embedded deformations of a curve in its Jacobian and the first-order deformations of the Picard bundle on the dual was solved.

The vanishing of $R^i\pi_*Q$ for $i < n$ follows from this step, after reducing to a sufficiently positive $(i + 1)$-dimensional hyperplane section.

Note that conditions (12) can be expressed dually as

$$\operatorname{codim}_{\operatorname{Pic}^0 A}\{\alpha \in \operatorname{Pic}^0 A \mid h^i(\omega_X \otimes \alpha) > 0\} \geq i \qquad \text{for all} \quad i > 0.$$

they can be reduced to the case of generically finite a by taking sufficiently positive hyperplane sections of dimension equal to the rank of a.

[8] In [17] this is stated only in the smooth case, but this hypothesis is unnecessary.

[9] This is simply the dual of the multiplication map $V \otimes H^0(N_C \otimes \omega_C) \to H^0(\Omega^1_Y \otimes N_C \otimes \omega_C)$, where V is the cotangent space of A at the origin.

[10] We remark, incidentally, that (11) for abelian varieties (Mumford's theorem) is the key point assuring that the Fourier–Mukai transform is an equivalence of categories.

According to the terminology of [17], this is stated by saying that the dualizing sheaf ω_X is a GV-sheaf. As a first application of Theorem 2 we note that, combined with Proposition 3.1 of [18] ("GV tensor $IT_0 \Rightarrow IT_0$"), we get the following Kodaira-type vanishing:

Corollary 6 *Let X be a normal Cohen–Macaulay subvariety of an abelian variety A, and let L be an ample line bundle on A. Then $H^i(X, \omega_X \otimes L) = 0$ for all $i > 0$.*

The rest of this paper is organized as follows: in Section 2 we prove Theorem 2 (and Proposition 4). Section 3 contains the proof of Corollary 3. In Sections 4 and 5 we establish the setup of the argument for Theorem 5. In particular, we interpret Gaussian maps in terms of the Fourier–Mukai transform. The conclusion of the proof of Theorem 5 takes up Section 6.

It seems possible that these methods can find application in wider generality.

2 Proof of Theorem 2 and Proposition 4

2.1 Preliminaries

The argument consists of a computation with extension classes. The geometric motivation is outlined in the Introduction (Sections 1.2 and 1.4). To get a first idea of the argument, it could be helpful to have a look at the proof of Lemma 3.1 of [3].

Notation 1 In the first place, some warning about the notation. We have the three varieties $C \subset Y \subset X$ (respectively of dimension 1, n, and $n+1$). The projections of $X \times T$ onto X and T are denoted respectively by ν and π. It will be different to consider the relative evaluation maps of a sheaf $\mathcal{A}$ on $C \times T$ seen as a sheaf on $Y \times T$, or on $X \times T$, or on $C \times T$ itself: their kernels are the various relative Lazarsfeld sheaves attached to $\mathcal{A}$ in different ambient varieties (see Section 1.3). Therefore, we denote

$$\pi_Y = \pi_{|Y \times T}, \qquad \pi_C = \pi_{|C \times T}.$$

For example, on $Y \times T$ we have

$$0 \to \mathcal{F}^Y_{A, Q_{|Y \times T}} \to \pi_Y^* \pi_*(Q \otimes \nu^* A) \to Q \otimes \nu^* A \tag{13}$$

while on $X \times T$

$$0 \to \mathcal{F}^X_{A,Q} \to \pi^* \pi_*(Q \otimes \nu^* A) \to Q \otimes \nu^* A. \tag{14}$$

Next, we clarify a few points appearing in the Introduction.

The differentiation map (4). We describe explicitly the differentiation map (4). We keep the notation there: M_C is a line bundle on the curve C while $\mathcal{R}$ is a line bundle on $Y \times T$. Now let p, q, and $\widetilde{\Delta}$ denote the two projections and the diagonal of the fibered product $(Y \times T) \times_T (Y \times T)$. Concerning the Lazarsfeld sheaf $\mathcal{F}^Y_{M_C,\mathcal{R}}$, we claim that there is a canonical isomorphism

$$\mathcal{F}^Y_{M_C,\mathcal{R}} \cong p_*(\mathcal{I}_{\widetilde{\Delta}} \otimes q^*(\mathcal{R} \otimes \nu^* M_C)). \tag{15}$$

Admitting the claim, the differentiation map (4) is defined as usual, as p_* of the restriction to $\widetilde{\Delta}$. The isomorphism (15): in the first place $p_*(\mathcal{I}_{\widetilde{\Delta}} \otimes q^*(\mathcal{R} \otimes \nu^* M_C))$ is the kernel of the map (p_* of the restriction map)

$$p_* q^*(\mathcal{R} \otimes \nu^* M_C) \to p_* q^*((\mathcal{R} \otimes \nu^* M_C)_{|\widetilde{\Delta}}) \cong \mathcal{R} \otimes \nu^* M_C \tag{16}$$

(it is easily seen that the sequence $0 \to \mathcal{I}_{\widetilde{\Delta}} \to \mathcal{O}_{Y \times_T Y} \to \mathcal{O}_{\widetilde{\Delta}} \to 0$ remains exact when restricted to $(Y \times T) \times_T (C \times T)$). To prove (15) we note that, by a flat base change,

$$\pi_Y^* \pi_*(\mathcal{R} \otimes \nu^* M_C) \cong p_* q^*(\mathcal{R} \otimes \nu^* M_C)$$

and, via such an isomorphism, the map (16) is identified with the relative evaluation map.

The isomorphism (6). This follows from the spectral sequence

$$\mathrm{Ext}^i_C(\Omega^1_Y \otimes N_C^{\otimes n}, \mathcal{E}xt^j_Y(\mathcal{O}_C, \mathcal{O}_Y)) \Rightarrow \mathrm{Ext}^{i+j}_Y(\Omega^1_Y \otimes N_C^{\otimes n}, \mathcal{O}_Y)$$

using the fact that, C being the complete intersection of $n-1$ divisors in the linear system $|L_{|Y}|$, we have $\mathcal{E}xt^j_Y(\mathcal{O}_C, \mathcal{O}_Y) = N_C^{\otimes n-1}$ if $j = n-1$ and zero otherwise. Seeing the elements of Ext-groups as higher extension classes with their natural multiplicative structure (Yoneda Exts; see, e.g., [12, Chapter III]), we denote by

$$\kappa \in \mathrm{Ext}^{n-1}_Y(\mathcal{O}_C, L_{|Y}^{\otimes -(n-1)}) \tag{17}$$

the extension class of the Koszul resolution of $\mathcal{O}_C$ as a $\mathcal{O}_Y$-module:

$$0 \to L_{|Y}^{\otimes -(n-1)} \to \cdots \to (L_{|Y}^{\otimes -1})^{\oplus n-1} \to \mathcal{O}_Y \to \mathcal{O}_C \to 0\ . \tag{18}$$

Then the multiplication by κ,

$$\mathrm{Ext}^1_C(\Omega^1_Y \otimes N_C, \mathcal{O}_C) \xrightarrow{\cdot\kappa} \mathrm{Ext}^n_Y(\Omega^1_Y \otimes N_C, L_{|Y}^{\otimes -(n-1)}) \cong \mathrm{Ext}^n_Y(\Omega^1_Y \otimes N_C^{\otimes n}, \mathcal{O}_Y),$$

is an isomorphism coinciding, up to scalar, with (6).

2.2 First step (statement)

Notation 2 From this point on we will adopt the hypotheses and notation of Theorem 2. We also adopt the following typographical abbreviations:

$$\mathcal{F}^Y = \mathcal{F}^Y_{N_C^{\otimes n}, Q_{|Y\times T}} \qquad \mathcal{F}^X = \mathcal{F}^X_{N_C^{\otimes n}, Q} \qquad g = g_{N_C^{\otimes n}, Q_{|Y\times T}}$$

The first, and most important, step of the proof of Theorem 2 and Proposition 4 consists of an explicit calculation of the class $g(e)$. This is the content of Lemma 7 below. The strategy is as follows. Applying $\mathrm{Ext}^n_{Y\times Y}(\,\cdot\,, Q_{|Y\times T})$ to the basic sequence

$$0 \to \mathcal{F}^Y \to \pi_Y^*\pi_*(Q \otimes \nu^* N_C^{\otimes n}) \to Q \otimes \nu^* N_C^{\otimes n} \to 0$$

(namely (13) for $A = N_C^{\otimes n}$ and $\mathcal{R} = Q_{|Y\times T}$ [11]), we get the following diagram with exact (in the middle) column

$$\begin{array}{ccc} & & \mathrm{Ext}^n_{Y\times T}(Q \otimes \nu^* N_C^{\otimes n}, Q_{|Y\times T}) \\ & & \downarrow h \\ & & \mathrm{Ext}^n_{Y\times T}(\pi_Y^*\pi_*(Q \otimes \nu^* N_C^{\otimes n}), Q_{|Y\times T}) \\ & & \downarrow f \\ \mathrm{Ext}^n_Y(\Omega^1_Y \otimes N_C^{\otimes n}, \mathcal{O}_Y) & \xrightarrow{\ g\ } & \mathrm{Ext}^n_{Y\times T}(\mathcal{F}^Y, Q_{|Y\times T}) \end{array} \tag{19}$$

In Definition 8 below we produce a certain class b in the source of f, namely

$$b \in \mathrm{Ext}^n_{Y\times T}(\pi_Y^*\pi_*(Q \otimes \nu^* N_C^{\otimes n}), Q_{|Y\times T}) \tag{20}$$

such that its coboundary map

$$\delta_b : \pi_*(Q \otimes \nu^* N_C^{\otimes n}) \to R^n\pi_*(Q_{|Y\times T})$$

is the composition

$$\begin{array}{ccc} \pi_*(Q \otimes \nu^* N_C^{\otimes n}) & \xrightarrow{\ \alpha\ } & R^n\pi_*(Q) \\ & \searrow^{\delta_b} & \downarrow \beta \\ & & R^n\pi_*(Q_{|Y\times T}) \end{array} \tag{21}$$

where the horizontal map α is the coboundary map of the natural extension of $\mathcal{O}_{X\times T}$-modules

$$0 \to Q \to \cdots \to (Q \otimes \nu^* L^{\otimes n-1})^{\oplus n} \to Q \otimes \nu^* L^{\otimes n} \to Q \otimes \nu^* N_C^{\otimes n} \to 0 \tag{22}$$

(ν^* of the Koszul resolution of $\mathcal{O}_C$ as an $\mathcal{O}_X$-module, twisted by $Q \otimes \nu^* L^{\otimes n}$) and the vertical map β is simply $R^n\pi_*$ of the restriction map $Q \to Q_{|Y\times T}$. The main lemma is:

[11] The surjectivity on the right follows from the hypotheses of Theorem 2.

Lemma 7 $f(b) = g(e)$.

Note that this will already prove Proposition 4. Indeed, if $T = \{point\}$ then the vector space of (20) is

$$\mathrm{Ext}^n_Y(H^0(Q \otimes \nu^* N_C^{\otimes n}) \otimes \mathcal{O}_Y, Q_{|Y}) \cong \mathrm{Hom}_k(H^0(Q \otimes \nu^* N_C^{\otimes n}), H^n(Y, Q_{|Y})). \quad (23)$$

Hence the class b coincides, up to scalar, with its coboundary map δ_b. From the description of δ_b we have that $\delta_b = 0$ if $H^n(X, Q) = 0$. If this is the case, then Lemma 7 says that $g(e) = 0$, proving Proposition 4 in this case. If $\dim T > 0$, we consider the spectral sequence

$$\mathrm{Ext}^i_T(\pi_*(Q \otimes N_C^{\otimes n}), R^j\pi_*(Q_{|Y\times T})) \Rightarrow \mathrm{Ext}^{i+j}_{Y\times T}(\pi_Y^*\pi_*(p^*(Q \otimes \nu^* N_C^{\otimes n})), Q_{|Y\times T})$$

coming from the isomorphism

$$\mathbf{R}\,\mathrm{Hom}_T(\pi_*(Q \otimes N_C^{\otimes n}), \mathbf{R}\pi_*(Q_{|Y\times T})) \cong \mathbf{R}\,\mathrm{Hom}_{Y\times T}(\pi_Y^*\pi_*(p^*(Q \otimes \nu^* N_C^{\otimes n})), Q_{|Y\times T}).$$

Since we are assuming that $R^i\pi_*(Q_{|Y\times T}) = 0$ for $i < n$, the spectral sequence degenerates, providing an isomorphism as (23), and Proposition 4 follows in the same way. □

Next, we give a definition of the class b of (20). In order to do so, we introduce some additional notation.

Notation 3 We denote by $\mathbf{K}^\bullet_{C,X}$ (resp. $\mathbf{H}^\bullet_{C,Y}$) the ν^* of the Koszul resolution of the ideal of C in X tensored with $Q \otimes \nu^* L^{\otimes n}$ (resp. the ν^* of the Koszul resolution of $\mathcal{O}_C$ as an $\mathcal{O}_Y$-module, tensored with $Q \otimes \nu^* L_{|Y}^{\otimes n-1}$):

$$\begin{array}{llllll} \mathbf{K}^\bullet_{C,X} & 0 \to & Q & \to \cdots \to (Q \otimes \nu^* L^{\otimes n-2})^{\oplus \binom{n}{2}} \to & (Q \otimes \nu^* L^{\otimes n-1})^{\oplus n} \\ \mathbf{H}^\bullet_{C,Y} & 0 \to & Q_{|Y\times T} & \to \cdots \to (Q \otimes \nu^* L_{|Y}^{\otimes n-2})^{\oplus n-1} \to & Q \otimes \nu^* L_{|Y}^{\otimes n-1} \end{array}$$

(note that they have the same length). For example, with this notation the exact complex of $\mathcal{O}_{X\times T}$-modules (22) is written as

$$\mathbf{K}^\bullet_{C,X} \to Q \otimes L^{\otimes n} \to Q \otimes \nu^* N_C^{\otimes n} \to 0. \quad (24)$$

Definition 8 (The class b of (20)) Composing (24) with the relative evaluation map of $Q \otimes \nu^* N_C^{\otimes n}$ (seen as a sheaf on $X \times T$),

$$\pi^*\pi_*(Q \otimes \nu^* N_C^{\otimes n}) \to Q \otimes \nu^* N_C^{\otimes n},$$

we get the commutative exact diagram

$$\begin{array}{ccccccc} \mathbf{K}^\bullet_{C,X} & \to & \mathcal{E} & \to & \pi^*\pi_*(Q\otimes \nu^* N_C^{\otimes n}) & \to 0 \\ \| & & \downarrow & & \downarrow & \\ \mathbf{K}^\bullet_{C,X} & \to & Q\otimes L^{\otimes n} & \to & Q\otimes \nu^* N_C^{\otimes n} & \to 0 \end{array} \tag{25}$$

where $\mathcal{E}$ is an $O_{X\times T}$-module. Since $tor^i_{X\times T}(\pi^*\pi_*(Q\otimes \nu^*N_C^{\otimes n}), \nu^*O_Y) = 0$ for $i > 0$, restricting the top row of (25) to $Y \times T$ we get an *exact* complex of $O_{Y\times T}$-modules

$$(\mathbf{K}^\bullet_{C,X})_{|Y\times T} \to \mathcal{E}_{|Y\times T} \to \pi_Y^*\pi_*(Q\otimes \nu^* N_C^{\otimes n}) \to 0. \tag{26}$$

We define the class $b \in \mathrm{Ext}^n_{Y\times T}(\pi_Y^*\pi_*(Q\otimes \nu^* N_C^{\otimes n}), Q_{|Y\times T})$ of (20) as the extension class of the exact complex (26). The assertion about its coboundary map follows from its definition.

We will need the following:

Lemma 9 *The row of the following diagram*

$$\begin{array}{ccccc} \mathbf{H}^\bullet_{C,Y} & \longrightarrow & \mathcal{E}_{|Y\times T} \to \pi_Y^*\pi_*(Q\otimes \nu^* N_C^{\otimes n}) \to 0 \\ & \searrow \quad\quad \nearrow & \\ & Q\otimes \nu^* N_C^{\otimes n-1} & \\ & \nearrow \quad\quad \searrow & \\ 0 & & 0 \end{array} \tag{27}$$

is an exact complex having the same extension class as (26), namely $b \in \mathrm{Ext}^n_{Y\times T}(\pi_Y^*\pi_*(Q\otimes \nu^* N_C^{\otimes n}), Q_{|Y\times T})$.

Proof For $n = 1$, i.e., $C = Y$, there is nothing to prove. For $n > 1$, recall that, by its definition, the top row of (25) is

$$\begin{array}{ccccc} \mathbf{K}^\bullet_{C,X} & \longrightarrow & \mathcal{E} \to \pi^*\pi_*(Q\otimes \nu^* N_C^{\otimes n}) \to 0 \\ & \searrow \quad\quad \nearrow & \\ & \mathcal{I}_{C/X}\otimes Q\otimes \nu^* L^{\otimes n} & \\ & \nearrow \quad\quad \searrow & \\ 0 & & 0 \end{array}$$

Recalling that the curve C is the complete intersection $Y_1 \cap \cdots \cap Y_n$, with $Y_i \in |L|$, and that $Y = Y_1$, restricting the ideal sheaf $\mathcal{I}_{C/X}$ to Y one gets $\mathcal{I}_{C/Y} \oplus N_C^{-1}$. Accordingly the Koszul resolution of $\mathcal{I}_{C/X}$, restricted to Y, splits as the direct sum of the Koszul resolution of $\mathcal{I}_{C/Y}$ and the Koszul resolution of O_C, as an O_Y-module, tensored with $L^{-1}_{|Y}$:

$$0 \to \begin{array}{c} 0 \\ \oplus \\ L^{-n}_{|Y} \end{array} \to \cdots \to \begin{array}{c} (L^{-2}_{|Y})^{\oplus\binom{n-1}{2}} \\ \oplus \\ (L^{-2}_{|Y})^{\oplus n-1} \end{array} \to \begin{array}{c} (L^{-1}_{|Y})^{\oplus n-1} \\ \oplus \\ L^{-1}_{|Y} \end{array} \left(\to \begin{array}{c} \mathcal{I}_{C/Y} \\ \oplus \\ N_C^{-1} \end{array} \to 0 \right).$$

Now, restricting the exact complex (25) to Y one gets the exact complex (26) whose "tail," namely the exact complex $(\mathbf{K}^\bullet_{C,X})_{|Y\times T}$, splits as above. Therefore, deleting the exact complex corresponding to the above upper row one gets the equivalent – as an extension – exact complex (27). This proves the claim. □

2.3 First step (proof)

In this section we prove Lemma 7. We first compute $g(e)$.[12] The exact sequences defining $\mathcal{F}^X$ and $\mathcal{F}^Y$ (see Notation 2) fit into the commutative diagram

$$\begin{array}{ccccccc}
0\to & \mathcal{F}^X & \to & \pi^*\pi_*(Q\otimes\nu^*N_C^{\otimes n}) & \to & Q\otimes\nu^*N_C^{\otimes n} & \to 0\\
 & \downarrow & & \downarrow & & \| & \\
0\to & \mathcal{F}^Y & \to & \pi_Y^*\pi_*(Q\otimes\nu^*N_C^{\otimes n}) & \to & Q\otimes\nu^*N_C^{\otimes n} & \to 0
\end{array}$$

yielding, after restricting the top row to $Y\times T$, the exact sequence

$$0\to Q\otimes\nu^*N_C^{\otimes n-1}\to(\mathcal{F}^X)_{|Y\times T}\to\mathcal{F}^Y\to 0 \tag{28}$$

where the sheaf on the left is $tor_1^{O_{X\times T}}(Q\otimes\nu^*N_C^{\otimes n},O_{Y\times T})$.

This sequence in turn fits into the commutative diagram with exact rows

$$\begin{array}{ccccccc}
0\to & Q\otimes\nu^*N_C^{\otimes n-1} & \to & (\mathcal{F}^X)_{|Y\times T} & \to & \mathcal{F}^Y & \to 0\\
 & \| & & \downarrow & & \downarrow & \\
0\to & Q\otimes\nu^*N_C^{\otimes n-1} & \to & Q\otimes\nu^*(\Omega^1_X\otimes N_C^{\otimes n}) & \to & Q\otimes\nu^*(\Omega^1_Y\otimes N_C^{\otimes n}) & \to 0
\end{array}$$

where the class of the bottow row is $\nu^*(e)\in\nu^*\operatorname{Ext}^1_C(\Omega^1_Y\otimes N_C,O_C)$. It follows that $g(e)$ (where now e is seen in $\operatorname{Ext}^n_Y(\Omega^1_Y\otimes N_C^{\otimes n},O_Y)$, see (6) and Section 2.1)) is the class of the sequence (28) with $\mathbf{H}^\bullet_{C,Y}$ attached on the left:

$$\mathbf{H}^\bullet_{C,Y}\to(\mathcal{F}^X)_{|Y\times T}\to\mathcal{F}^Y\to 0\,. \tag{29}$$

Next, we compute $f(b)$. The exact complex (25) is the middle row of the commutative exact diagram

$$\begin{array}{ccccccc}
 & & 0 & & 0 & & \\
 & & \downarrow & & \downarrow & & \\
 & & \mathcal{F}^X & = & \mathcal{F}^X & & \\
 & & \downarrow & & \downarrow & & \\
\mathbf{K}^\bullet_{C,X} & \to & \mathcal{E} & \to & \pi^*\pi_*(Q\otimes\nu^*N_C^{\otimes n}) & \to 0 & \\
\| & & \downarrow & & \downarrow & & \\
\mathbf{K}^\bullet_{C,X} & \to & Q\otimes\nu^*L^{\otimes n} & \to & Q\otimes\nu^*N_C^{\otimes n} & \to 0 & \\
 & & \downarrow & & \downarrow & & \\
 & & 0 & & 0 & &
\end{array} \tag{30}$$

[12] This argument follows [20, p. 252].

This provides us with the commutative exact diagram

$$\begin{array}{ccccc} & 0 & & 0 & \\ & \downarrow & & \downarrow & \\ \mathbf{H}^\bullet_{C,Y} \to & (\mathcal{F}^X)_{|Y\times T} & \to & \mathcal{F}^Y & \to 0 \\ & \downarrow & & \downarrow & \\ & \mathcal{E}_{|Y\times T} & \to & \pi_Y^*\pi_*(Q\otimes \nu^* N_C^{\otimes n}) & \to 0 \end{array}$$

where the long row is (29), whose class is $g(e)$. By Lemma 9, we can complete the above diagram as follows:

$$\begin{array}{ccccccc} \mathbf{H}^\bullet_{C,Y} & \to & (\mathcal{F}^X)_{|Y\times T} & \to & \mathcal{F}^Y & \to 0 \\ \| & & \downarrow & & \downarrow & \\ \mathbf{H}^\bullet_{C,Y} & \to & \mathcal{E}_{|Y\times T} & \to & \pi_Y^*\pi_*(Q\otimes \nu^* N_C^{\otimes n}) & \to 0 \end{array}$$

and the class of the bottow row is b. By definition, the class of the top row is $f(b)$, and it is equal to $g(e)$. This proves Lemma 7. □

2.4 Conclusion of the proof of Theorem 2

The last step is:

Lemma 10 *We keep the notation and setting of Lemma 7. Assume that the line bundle L on X is sufficiently positive. If $f(b) = 0$ then $b = 0$.*

Assuming this, Theorem 2 follows: if $g(e) = 0$ then, by Lemmas 7 and 10, it follows that $b = 0$, hence its coboundary map $\delta_b = \beta\circ\alpha$ is zero (see (21)). Taking L sufficiently positive, Serre vanishing yields that α is surjective and β is injective. Therefore the target of δ_b, namely $R^n\pi_*Q$, is zero. □

Proof (of Lemma 10) The proof is a somewhat tedious repeated application of Serre vanishing. Going back to diagram (19), we have that if $f(b) = 0$ then there is a $c \in \mathrm{Ext}^n_{Y\times T}(Q\otimes \nu^* N^{\otimes n}_{|C}, Q_{|Y\times T})$ such that

$$h(c) = b\ . \tag{31}$$

Now we consider the commutative diagram

$$\begin{array}{ccc} \mathrm{Ext}^n_{Y\times T}(Q\otimes \nu^* N_C^{\otimes n}, Q_{|Y\times T}) & \xrightarrow{\ r\ } & \mathrm{Ext}^n_{X\times T}(Q\otimes \nu^* L^{\otimes n}, Q_{|Y\times T}) \\ \downarrow h & & \downarrow h' \\ \mathrm{Ext}^n_{Y\times T}(\pi_Y^*\pi_*(Q\otimes \nu^* N_C^{\otimes n}), Q_{|Y\times T}) & \xrightarrow{\ s\ } & \mathrm{Ext}^n_{X\times T}(\pi^*\pi_*(Q\otimes \nu^* L^{\otimes n}), Q_{|Y\times T}) \\ \downarrow \mu & & \downarrow \mu' \\ \mathrm{Hom}(\pi_*(Q\otimes \nu^* N_C^{\otimes n}), R^n\pi_*Q_{|Y\times T}) & \xrightarrow{\ t\ } & \mathrm{Hom}(\pi_*(Q\otimes \nu^* L^{\otimes n}), R^n\pi_*Q_{|Y\times T}) \end{array} \tag{32}$$

where:
(a) h is as above and h' is the analogous map $\mathrm{Ext}^n_X(ev_X, Q_{|Y\times T})$, where ev_X is the relative evaluation map on $X\times T$: $\pi^*\pi_*(Q\otimes \nu^*L^{\otimes n}) \to Q\otimes \nu^* L^{\otimes n}$.
(b) μ is the map taking an extension to its coboundary map. Consequently, the map $\mu\circ h$ takes an extension class $e\in \mathrm{Ext}^n_{Y\times T}(Q\otimes \nu^*N_C^{\otimes n}, Q_{|Y\times T})$ to its coboundary map

$$\pi_*(Q\otimes \nu^* N_C^{\otimes n}) \to R^n\pi_*(Q_{|Y\times T}).$$

The map $\mu'\circ h'$ operates in the same way.
(c) Notice that the target of r is simply $H^n(\nu^* L^{\otimes -n}_{|Y\times T})$, i.e., $\mathrm{Ext}^n_{Y\times T}(\nu^* L^{\otimes n}_{|Y}, O_{Y\times T})$. Via this identification, the map r is defined as the natural map

$$\mathrm{Ext}^n_{Y\times T}(\nu^* L^{\otimes n}_{|C}, O_{Y\times T}) \to \mathrm{Ext}^n_{Y\times T}(\nu^* L^{\otimes n}_{|Y}, O_{Y\times T}).$$

(d) s and t are the natural maps.

We know that the coboundary map of the extension class b factorizes through the natural coboundary map $\alpha : \pi_*(Q\otimes \nu^* N_C^{\otimes n}) \to R^n\pi_*(Q)$. This implies that $(t\circ\mu)(b) = 0$. Therefore, by (31) and (32), we have that $(\mu'\circ h'\circ r)(c) = 0$. The lemma will follow from the fact that both r and $\mu'\circ h'$ are injective.

Injectivity of r: In the case $n = 1$, i.e. $Y = C$, the map r is just the identity (cf. (c) above). Assume that $n > 1$. Chasing in the Koszul resolution of O_C as an O_Y-module one finds that the injectivity of r holds as soon as $\mathrm{Ext}^{n-i}_{Y\times T}(\nu^* L^{\otimes n-i}_{|Y}, O_{Y\times T}) = 0$ for $i = 1,\dots,n-1$. But these are simply $H^{n-i}(Y\times T, L^{\otimes i-n}_{|Y} \boxtimes O_T)$ and the result follows easily from Künneth decomposition, Serre vanishing, and Serre duality.

Injectivity of $\mu'\circ h'$: We have that $\mathrm{Ext}^n_{X\times T}(Q\otimes \nu^* L^{\otimes n}, Q_{|Y\times T}) \cong$ $\cong H^n(Y\times T, L^{-n}_{|Y}\boxtimes O_T)$. If L is sufficiently positive, it follows as above that this is isomorphic to $H^n(Y, L^{-n}_{|Y})\otimes H^0(T, O_T)$. Therefore, the map $\mu'\circ h'$ is identified with H^0 of the following map of O_T-modules:

$$H^n(Y, L^{-n}_{|Y})\otimes O_T \to \mathcal{H}om_T(\pi_*(Q\otimes \nu^* L^{\otimes n}), R^n\pi_* Q_{|Y\times T}). \tag{33}$$

Hence the injectivity of $\mu'\circ h'$ holds as soon as (33) is injective at a general fiber. For a closed point $t\in T$, let $Q_t = Q_{|X\times\{t\}}$. By base change, the map (33) at a general fiber $X\times\{t\}$ is

$$H^n(Y, L^{\otimes -n}_{|Y}) \to H^0(X, Q_t\otimes L^{\otimes n})^\vee \otimes H^n(Y, Q_{t|Y}), \tag{34}$$

which is the Serre dual of the multiplication map of global sections

$$H^0(X, Q_t\otimes L^{\otimes n})\otimes H^0(Y, (\omega_X\otimes Q_t^{-1}\otimes L)_{|Y}) \to H^0(Y, (\omega_X\otimes L^{\otimes n+1})_{|Y}). \tag{35}$$

At this point a standard argument with Serre vanishing shows that (35) is surjective as soon as L is sufficiently positive.[13] This proves the injectivity of $\mu' \circ h'$ and concludes also the proof of the lemma. □

3 Proof of Corollary 3

The deduction of Corollary 3 from Theorem 2 is a standard argument with Serre vanishing. However, there are some complications due to the weakness of the assumptions on the singularities of the variety X.

A Gaussian map on the ambient variety $X \times T$. The argument makes use of a (dual) Gaussian map defined on the ambient variety $X \times T$ itself. Namely, for a line bundle A on X we define $\mathcal{M}^X_{A,Q}$ as the kernel of the relative evaluation map

$$\pi^*\pi_*(Q \otimes \nu^*A) \to Q \otimes \nu^*A.$$

As in (4) and Section 2.1, there is the isomorphism

$$\mathcal{M}^X_{A,Q} \cong p_{X*}(\mathcal{I}_{\tilde{\Delta}_X} \otimes q_X^*(Q \otimes \nu^*A)) \tag{36}$$

(where p_X, q_X, and $\tilde{\Delta}_X$ denote the projections and the diagonal of $(X \times Y) \times_T (X \times T)$). There is also the differentiation map $\mathcal{M}^X_{A,Q} \to Q \otimes \nu^*(\Omega^1_X \otimes A)$.

Now, taking as $A = L^{\otimes n}$ and taking $\mathrm{Ext}^{n+1}_{X\times T}(\,\cdot\,, Q \otimes \nu^*L^\vee)$, we get the desired dual Gaussian map on X:

$$g_X : \mathrm{Ext}^{n+1}_X(\Omega^1_X \otimes L^{\otimes n+1}, \mathcal{O}_X) \to \mathrm{Ext}^{n+1}_{X\times T}(\mathcal{M}_{L^{\otimes n},Q}, Q \otimes \nu^*L^\vee).$$

Note that there are natural maps $\mathcal{M}_{L^{\otimes n},Q} \to \mathcal{F}^X \to \mathcal{F}^Y$ (see Notation 2).

First step. We consider the commutative diagram

$$\begin{array}{ccc}
\mathrm{Ext}^n_Y(\Omega^1_Y \otimes N_C^{\otimes n}, \mathcal{O}_Y) & \xrightarrow{\;g\;} & \mathrm{Ext}^n_{Y\times T}(\mathcal{F}^Y_{N_C^{\otimes n},Q_{|Y\times T}}, Q_{|Y\times T}) \\
\downarrow{\scriptstyle \mu} & & \downarrow{\scriptstyle \eta} \\
\mathrm{Ext}^{n+1}_X(\Omega^1_X \otimes L^{\otimes n+1}, \mathcal{O}_X) & \xrightarrow{\;g_X\;} & \mathrm{Ext}^{n+1}_{X\times T}(\mathcal{M}^X_{L^{\otimes n},Q}, Q \otimes \nu^*L^\vee)
\end{array} \tag{37}$$

where, as in the previous section, g denotes the main character, namely the (dual) Gaussian map $g_{N_C^{\otimes n},Q_{Y\times T}}$. The maps μ and η are the natural ones, and the

[13] In brief, one shows that the desired surjectivity follows from the surjectivity of $H^0(X, Q_t \otimes L^{\otimes n}) \otimes H^0(X, \omega_X \otimes Q_t^{-1} \otimes L) \to H^0(X, \omega_X \otimes L^{\otimes n+1})$. This in turn is proved by interpreting such a multiplication map as the H^0 of a restriction-to-diagonal map of $\mathcal{O}_{X\times X}$-modules.

definition is left to the reader.[14] However such maps are more easily understood by considering the following commutative diagram, whose maps are the natural ones:

$$\begin{array}{ccc} H^d(\mathcal{M}^X_{L^{\otimes n},Q} \otimes Q^\vee \otimes ((\omega_X \otimes L) \boxtimes \omega_T))) & \xrightarrow{g'_X} & H^0(\Omega^1_X \otimes L^{\otimes n+1} \otimes \omega_X) \otimes H^d(\omega_T) \\ \downarrow \eta' & & \downarrow \mu' \\ H^d(\mathcal{F}^Y_{N_C^{\otimes n},Q} \otimes Q^\vee \otimes (\omega_Y \boxtimes \omega_T)) & \xrightarrow{g'} & H^0(\Omega^1_Y \otimes L^{\otimes n} \otimes \omega_Y) \otimes H^d(\omega_T) \end{array} \tag{38}$$

where $d = \dim T$. (Note that, since X is Cohen–Macaulay, adjunction formulas for dualizing sheaves do hold.) Notice also that, if X and T are Gorenstein, then (38) is the dual of diagram (37).

As is easy to see, after tensoring with $\omega_C \otimes N_C$ the restricted normal sequence (5) remains exact:

$$0 \to \omega_C \to \Omega^1_X \otimes L \otimes \omega_C \to \Omega^1_Y \otimes N_C \otimes \omega_C \to 0. \tag{39}$$

Therefore e defines naturally a linear functional on $H^0(\Omega^1_Y \otimes N_C \otimes \omega_C)$ (cf. (44) below), still denoted by e. We have

Claim 11 *If L is sufficiently positive, then the map g'_X is surjective, while cokerμ' is 1-dimensional, with $(\text{coker}\mu')^\vee$ spanned by e.*

Proof Serre vanishing ensures the surjectivity of the restriction

$$H^0(\Omega^1_X \otimes L^{\otimes n+1} \otimes \omega_X) \to H^0(\Omega^1_X \otimes L \otimes \omega_C).$$

Since the map μ' is the composition of the above map with H^0 of the right arrow of sequence (39), the claim for μ' follows.

Concerning the surjectivity of the map g'_X, we first note that by Serre vanishing,

$$R^i p_*(\mathcal{I}_{\widetilde{\Delta}_X} \otimes q_X^*(\nu^* L^{\otimes n} \otimes Q)) = \begin{cases} 0 & \text{for } i > 0, \\ \text{locally free} & \text{for } i = 0. \end{cases} \tag{40}$$

Now we project on T. A standard computation using (40), base change, Serre vanishing, Leray spectral sequence, and Künneth decomposition shows that the map g'_X is identified as

[14] For example, μ is defined by sending $\mathrm{Ext}^n_Y(\Omega^1_Y \otimes N_C^{\otimes n}, \mathcal{O}_Y)$ to $\mathrm{Ext}^n_X(\Omega^1_Y \otimes N_C^{\otimes n}, \mathcal{O}_Y)$ and then composing with the natural map $\Omega^1_X \otimes L^{\otimes n} \to \Omega^1_Y \otimes N_C^{\otimes n}$ on the left, and with the natural extension $0 \to L^\vee \to \mathcal{O}_X \to \mathcal{O}_Y \to 0$ on the right.

$$H^d(T, \pi_*(p_*(\mathcal{I}_{\widetilde{\Delta}_X} \otimes q^*(v^* L^{\otimes n} \otimes Q)) \otimes Q^\vee \otimes ((L \otimes \omega_X) \boxtimes \omega_T))) \\ \to H^d(T, H^0(X, \Omega^1_X \otimes L^{\otimes n+1} \otimes \omega_X) \otimes \omega_T). \quad (41)$$

This is the H^d of a map of coherent sheaves on the q-dimensional variety T. Hence the surjectivity of (41) is implied by the generic surjectivity of the map of sheaves itself. By base change, at a generic fiber $X \times t$ the map of sheaves is the Gaussian map

$$\gamma_t : H^0(X, p_*(\mathcal{I}_{\Delta_X} \otimes q^*(L^{\otimes n} \otimes Q_t)) \otimes L \otimes Q_t^\vee \otimes \omega_X) \\ \to H^0(X, \Omega^1_X \otimes L^{\otimes n+1} \otimes \omega_X).$$

The map γ_t is defined by restriction to the diagonal in the usual way. Once again it follows from relative Serre vanishing (on $(X \times T) \times_T (X \times T)$) that, as soon as L is sufficiently positive, γ_t is surjective for all t. This proves the surjectivity of the g'_X and concludes the proof of the claim. □

Last step. If C is Gorenstein, Claim 11 achieves the proof of Corollary 3. Indeed, diagram (37) is dual to diagram (38) and it follows that the kernel of our map $g = g^Y_{N_C^{\otimes n}, Q}$ is at most 1-dimensional, spanned by e. In the general case, Corollary 3 follows in the same way once we have proved the following:

Claim 12 *As soon as L is sufficiently positive, the maps g'_X and μ' are respectively Serre duals of the maps g_X and μ.*

Proof To prove this assertion for g'_X we note that, concerning its source, the sheaf $\mathcal{M}^X_{L^{\otimes n}, Q}$ is locally free by (40). Therefore,

$$\mathrm{Ext}^{n+1}_{X \times T}(\mathcal{M}^X_{L^{\otimes n}, Q}, Q \otimes v^* L^\vee) \cong H^{n+1}((\mathcal{M}^X_{L^{\otimes n}, Q})^\vee \otimes Q \otimes v^* L^\vee) \\ \cong H^d(\mathcal{M}^X_{L^{\otimes n}, Q} \otimes Q^\vee \otimes ((L \otimes \omega_X) \boxtimes \omega_T))^\vee. \quad (42)$$

Next, we show the Serre duality

$$H^0(X, \Omega^1_X \otimes \omega_X \otimes L^{\otimes n+1})^\vee \cong \mathrm{Ext}^{n+1}_X(\Omega^1_X \otimes L^{\otimes n}, L^\vee). \quad (43)$$

By definition of a dualizing complex (see, e.g., [9, Chapter V, Section 2, Proposition 2.1 on p. 258]), in the derived category of X we have that $\mathcal{O}_X = \mathbf{R}\mathcal{H}om(\omega_X, \omega_X)$. Therefore it follows that

$$\mathbf{R}\mathrm{Hom}_X(\Omega^1_X \otimes L^{\otimes n+1}, \mathcal{O}_X) = \mathbf{R}\mathrm{Hom}_X(\Omega^1_X \otimes L^{\otimes n+1}, \mathbf{R}\mathcal{H}om(\omega_X, \omega_X)) \\ = \mathbf{R}\mathrm{Hom}_X(\Omega^1_X \otimes \mathcal{O}_X \underline{\otimes}^{\mathbf{L}} \omega_X \otimes L^{\otimes n+1}, \omega_X).$$

By Serre–Grothendieck duality, this is isomorphic to

$$\mathbf{R}\mathrm{Hom}_k(\mathbf{R}\Gamma(X, \Omega^1_X \underline{\otimes}^{\mathbf{L}} \omega_X \otimes L^{\otimes n+1}[n+1]), k).$$

The spectral sequence computing $\mathbf{R}\Gamma(X, \Omega^1_X \underline{\otimes}^{\mathbf{L}} \omega_X \otimes L^{\otimes n+1})$ degenerates to the isomorphisms

$$H^i(X, \Omega^1_X \underline{\otimes}^{\mathbf{L}} \omega_X \otimes L^{\otimes n+1}) \cong \bigoplus_i H^0(X, tor_i^X(\Omega^1_X, \omega_X) \otimes L^{\otimes n+1})$$

(if L is sufficiently positive, by Serre vanishing there are only H^0s). Therefore, (43) follows. By (42) and (43) we have proved the part of the claim concerning g'_X.

Concerning μ', at this point it is enough to prove the Serre duality

$$\mathrm{Ext}^n_Y(\Omega^1_{X|C} \otimes L^n, O_Y) \cong \mathrm{Ext}^1_C(\Omega^1_{Y|C} \otimes L, O_C) \cong H^0(\Omega^1_C \otimes N \otimes \omega_C)^\vee \tag{44}$$

where the first isomorphism is (6). Arguing as above, it is enough to prove that the O_C-modules

$$tor^i_C((\Omega^1_Y)_{|C}, \omega_C) \otimes N_C$$

have vanishing higher cohomology for all i. For $i > 0$ this follows simply because they are supported on points. For $i = 0$ note that, by the exact sequence (39), it is enough to show that

$$H^1(\Omega^1_X \otimes N_C \otimes \omega_C) = 0. \tag{45}$$

To prove this, we tensor the Koszul resolution of O_C as an O_X-module with $\Omega^1_X \otimes \omega_X \otimes L^{\otimes n+1}$, getting a complex (exact at the last step on the right)

$$0 \to \Omega^1_X \otimes \omega_X \otimes L \to \cdots \to (\Omega^1_X \otimes \omega_X \otimes L^{\otimes n})^{\oplus n}$$
$$\to \Omega^1_X \otimes \omega_X \otimes L^{\otimes n+1} \to \Omega^1_X \otimes \omega_C \otimes N_C \to 0.$$

Since C is not contained in the singular locus of X, the cohomology sheaves are supported on points. Therefore the required vanishing (45) follows from Serre vanishing via a diagram chase. This concludes the proof of Claim 12 and of Corollary 3. □

4 Gaussian maps and the Fourier–Mukai transform

In this section we describe the setup of the proof of Theorem 5. We show that when the variety X is a subvariety of an abelian variety A, the parameter variety T is the dual abelian variety $\widehat{A}$, and the line bundle Q is the restriction to $X \times \widehat{A}$ of the Poincaré line bundle then the (dual) Gaussian map $g_{N_C^{\otimes n}, Q_{|Y \times T}}$ of the Introduction can naturally be interpreted as a piece of a (relative version of) the classical Fourier–Mukai transform associated with the Poincaré line bundle, applied to a certain space of morphisms.

Notation/Assumptions 1 We keep all the notation and hypotheses of the Introduction. Explicitly:

- Let X be an $(n+1)$-dimensional normal Cohen–Macaulay subvariety of a d-dimensional abelian variety A. As usual we choose an ample line bundle L on X such that we can find n irreducible divisors $Y = Y_1, Y_2, \dots, Y_n \in X$ such that their intersection is an irreducible curve C. We assume also that C is not contained in the singular loci of X and Y. The line bundle $L_{|C}$ is denoted N_C.
- Let $\mathcal{P}$ be a Poincaré line bundle on $A \times \widehat{A}$. We denote

$$Q = \mathcal{P}_{|X\times\widehat{A}} \qquad \text{and} \qquad \mathcal{R} = \mathcal{P}_{|Y\times\widehat{A}}.$$

- ν and π are the projections of $Y \times \widehat{A}$.
- We assume that the line bundle $\nu^*L^{\otimes n} \otimes Q$ is relatively base-point-free, namely the evaluation map $\pi^*\pi_*(\nu^*L^{\otimes n} \otimes Q) \to \nu^*L^{\otimes n} \otimes Q$ is surjective (here ν and π denote also the projection of $X \times \widehat{A} \to \widehat{A}$).
- p, q, and $\widetilde{\Delta}$ are the projections and the diagonal of $(Y \times \widehat{A}) \times_{\widehat{A}} (Y \times \widehat{A})$.
- The Gaussian map of the Introduction (see (7)) is

$$g = g_{N_C^{\otimes n},\mathcal{R}} : \mathrm{Ext}^n_Y(\Omega^1_Y \otimes N_C^{\otimes n}, \mathcal{O}_Y) \to \mathrm{Ext}^n_{Y\times\widehat{A}}(p_*(q^*(\mathcal{I}_{\widetilde{\Delta}} \otimes \mathcal{R} \otimes \nu^* N_C^{\otimes n})), \mathcal{R}) \quad (46)$$

obtained as (the restriction to the relevant Künneth direct summand of) $\mathrm{Ext}^n_{Y\times\widehat{A}}(\,\cdot\,, \mathcal{R})$ of the differentiation (i.e., restriction to the diagonal) map (see Section 2.1). We recall also the identification of the source:

$$\mathrm{Ext}^n_Y(\Omega^1_Y \otimes N_C^{\otimes n}, \mathcal{O}_Y) \cong \mathrm{Ext}^1_C(\Omega^1_Y \otimes N_C, \mathcal{O}_C)$$

(see (6) and Section 2.1).
- The projections of $Y \times A$ will be denoted p_1 and p_2.

Remark 1 Since the variety X is assumed to be smooth in codimension 1, and our arguments concern a sufficiently positive line bundle L, we could have assumed from the beginning that the curve C is smooth and the divisor Y is smooth along C. However, we preferred to assume the smoothness of C only where needed, namely at the end of the proof. See also Remarks 2 and 5 below.

Fourier–Mukai transform. Now we consider the trivial abelian scheme $Y \times A \to Y$ and its dual $Y \times \widehat{A} \to Y$. The Poincaré line bundle $\mathcal{P}$ induces naturally a Poincaré line bundle $\tilde{\mathcal{P}}$ on $(Y \times A) \times_Y (Y \times \widehat{A})$ (namely the pull-back of $\mathcal{P}$ to $Y \times A \times \widehat{A}$) and we consider the functors

$$\mathbf{R}\Phi : \mathbf{D}(Y \times A) \to \mathbf{D}(Y \times \widehat{A}) \qquad \text{and} \qquad \mathbf{R}\Psi : \mathbf{D}(Y \times \widehat{A}) \to \mathbf{D}(Y \times A)$$

defined respectively by $\mathbf{R}\pi_{Y\times\widehat{A}*}(\pi^*_{Y\times A}(\,\cdot\,)\otimes\tilde{P})$ and $\mathbf{R}\pi_{Y\times A*}(\pi^*_{Y\times\widehat{A}}(\,\cdot\,)\otimes\tilde{P})$. By Mukai's theorem [14, Theorem 1.1] they are equivalences of categories, more precisely

$$\mathbf{R}\Psi\circ\mathbf{R}\Phi\cong(-1)^*[-q]\qquad\text{and}\qquad\mathbf{R}\Phi\circ\mathbf{R}\Psi\cong(-1)^*[-q]. \tag{47}$$

In particular it follows that, given $\mathcal{O}_{Y\times A}$-modules $\mathcal{F}$ and $\mathcal{G}$, we have the functorial isomorphism

$$FM_i:\mathrm{Ext}^i_{Y\times A}(\mathcal{F},\mathcal{G})\xrightarrow{\cong}\mathrm{Ext}^i_{Y\times\widehat{A}}(\mathbf{R}\Phi(\mathcal{F}),\mathbf{R}\Phi(\mathcal{G})) \tag{48}$$

(note that the Ext-spaces on the right are usually hyperexts).

The Gaussian map. Now we focus on the target of the Gaussian map (46). Let $\Delta_Y\subset Y\times A$ be the graph of the embedding $Y\hookrightarrow A$. In other words, Δ_Y is the diagonal of $Y\times Y$, seen as a subscheme of $Y\times A$. It follows from the definitions that

$$\mathbf{R}\Phi(\mathcal{O}_{\Delta_Y})=\mathcal{P}_{|Y\times\widehat{A}}=\mathcal{R}. \tag{49}$$

Moreover, we have that

$$p_*(q^*(\mathcal{I}_{\widetilde{\Delta}}\otimes\mathcal{R}\otimes v^*N_C^{\otimes n}))\cong R^0\Phi(\mathcal{I}_{\Delta_Y}\otimes p_2^*N_C^{\otimes n}). \tag{50}$$

This is because of the natural isomorphisms

$$(Y\times Y)\times_Y(Y\times\widehat{A})\cong Y\times Y\times\widehat{A}\cong(Y\times\widehat{A})\times_{\widehat{A}}(Y\times\widehat{A})$$

yielding the identifications $\tilde{\mathcal{P}}_{|(Y\times Y)\times_Y(Y\times\widehat{A})}\cong q^*(\mathcal{P}_{|Y\times\widehat{A}})=q^*(\mathcal{R})$. Moreover, for any sheaf $\mathcal{F}$ supported on $Y\times C$ (as $\mathcal{I}_{\Delta_Y}\otimes p_2^*N_C^{\otimes n}$), we have that $R^i\Phi(\mathcal{F})=0$ for $i>1$. Therefore the fourth-quadrant spectral sequence

$$\mathrm{Ext}^p_{Y\times\widehat{A}}(R^q\Phi(\mathcal{F}),\mathcal{R})\Rightarrow\mathrm{Ext}^{p-q}(\mathbf{R}\Phi(\mathcal{F}),\mathcal{R})$$

is reduced to a long exact sequence

$$\begin{aligned}\cdots\to\mathrm{Ext}^{i-1}_{Y\times\widehat{A}}(R^0\Phi(\mathcal{F}),\mathcal{R})&\to\mathrm{Ext}^{i+1}_{Y\times\widehat{A}}(R^1\Phi(\mathcal{F}),\mathcal{R})\\&\to\mathrm{Ext}^i_{Y\times\widehat{A}}(\mathbf{R}\Phi(\mathcal{F}),\mathcal{R})\to\mathrm{Ext}^i_{Y\times\widehat{A}}(R^0\Phi(\mathcal{F}),\mathcal{R})\to\cdots\end{aligned} \tag{51}$$

Putting all that together we get the following diagram, with right column exact in the middle:

$$\begin{array}{ccc}
\mathrm{Ext}^n_Y(\Omega^1_Y \otimes N_C^{\otimes n}, \mathcal{O}_Y) & & \\
\downarrow = & & \\
\mathrm{Ext}^n_{\Delta_Y}((\mathcal{I}_{\Delta_Y} \otimes p_2^* N_C^{\otimes n})_{|\Delta_Y}, \mathcal{O}_{\Delta_Y}) & & \mathrm{Ext}^{n+1}_{Y\times\widehat{A}}(R^1\Phi_{\widetilde{\mathcal{P}}}(\mathcal{I}_{\Delta_Y} \otimes p_2^* N_C^{\otimes n}), \mathcal{R}) \\
\downarrow u & & \downarrow \alpha \\
\mathrm{Ext}^n_{Y\times A}(\mathcal{I}_{\Delta_Y} \otimes p_2^* N_C^{\otimes n}, \mathcal{O}_{\Delta_Y}) & \xrightarrow[\cong]{FM_n} & \mathrm{Ext}^n_{Y\times\widehat{A}}(\mathbf{R}\Phi_{\widetilde{\mathcal{P}}}(\mathcal{I}_{\Delta_Y} \otimes p_2^* N_C^{\otimes n}), \mathcal{R}) \\
 & & \downarrow \beta \\
 & & \mathrm{Ext}^n_{Y\times\widehat{A}}(R^0\Phi_{\widetilde{\mathcal{P}}}(\mathcal{I}_{\Delta_Y} \otimes p_2^* N_C^{\otimes n}), \mathcal{R})
\end{array} \tag{52}$$

where u is the natural map (see also (65) below). In conclusion, the kernel of the Gaussian map can be described as follows:

Lemma 13 *The Gaussian map* $g = g_{N_C^{\otimes n}, \mathcal{R}}$ *of (46) is the composition* $\beta \circ FM_n \circ u$. *Therefore*

$$\ker(g) \cong Im(FM_n \circ u) \cap Im(\alpha).$$

Proof The identification of the two maps follows using (50), simply because they are defined in the same way. □

5 Cohomological computations on $Y \times A$

In this section we describe the source of the Fourier–Mukai map FM_n of diagram (52) above, together with other related cohomology groups. We use the Grothendieck duality (or change of rings) spectral sequence

$$\mathrm{Ext}^i_{Y\times C}(\mathcal{F}, \mathcal{E}xt^j_{Y\times A}(\mathcal{O}_{Y\times C}, \mathcal{O}_{\Delta_Y})) \Rightarrow \mathrm{Ext}^{i+j}_{Y\times A}(\mathcal{F}, \mathcal{O}_{\Delta_Y}). \tag{53}$$

With this in mind, we compute the sheaves $\mathcal{E}xt^i_{Y\times A}(\mathcal{O}_{Y\times C}, \mathcal{O}_{\Delta_Y})$ in Proposition 14 below.

5.1 Preliminaries

The following standard identifications will be useful:

$$\bigoplus_i \mathcal{E}xt^i_{Y\times A}(\mathcal{O}_{\Delta_Y}, \mathcal{O}_{\Delta_Y}) \cong \bigoplus_i \delta_*(\Lambda^i T_{A,0} \otimes \mathcal{O}_Y) \tag{54}$$

(as graded algebras), where $T_{A,0}$ is the tangent space of A at 0 and δ denotes the diagonal embedding

$$\delta : Y \hookrightarrow Y \times A.$$

This holds because Δ_Y is the pre-image of 0 via the difference map $Y \times A \to A$, $(y, x) \mapsto y - x$ (which is flat), and $\mathcal{E}xt^{\bullet}_A(k(0), k(0))$ is $\Lambda^{\bullet} T_{A,0} \otimes k(0)$.

Moreover, letting $\Delta_C \subset Y \times C \subset Y \times A$ the diagonal of $C \times C$ (seen as a subscheme of $Y \times A$), we have

$$\bigoplus_i \mathcal{E}xt^i_{Y\times A}(\mathcal{O}_{\Delta_C}, \mathcal{O}_{\Delta_Y}) \cong \bigoplus_i \delta_*(\Lambda^{i-n+1} T_{A,0} \otimes N_C^{\otimes n-1}) \tag{55}$$

(as graded modules on the above algebra). This is seen as follows: since C is the complete intersection of $n - 1$ divisors of Y, all of them in $|L_{|Y}|$, then $\mathcal{E}xt^j_{\Delta_Y}(\mathcal{O}_{\Delta_C}, \mathcal{O}_{\Delta_Y}) = 0$ if $j \neq n - 1$ and $\mathcal{E}xt^{n-1}_{\Delta_Y}(\mathcal{O}_{\Delta_C}, \mathcal{O}_{\Delta_Y}) = \delta_* N_C^{\otimes n-1}$. Therefore (55) follows from (54) and the spectral sequence

$$\mathcal{E}xt^h_{\Delta_Y}(\mathcal{O}_{\Delta_C}, \mathcal{E}xt^j_{Y\times A}(\mathcal{O}_{\Delta_Y}, \mathcal{O}_{\Delta_Y})) \Rightarrow \mathcal{E}xt^{h+j}_{Y\times A}(\mathcal{O}_{\Delta_C}, \mathcal{O}_{\Delta_Y}).$$

5.2 The (equisingular) restricted normal sheaf

We consider the $\mathcal{O}_C$-module $\mathcal{N}'$ defined by the sequence

$$0 \to (\mathcal{T}_Y)_{|C} \to (\mathcal{T}_A)_{|C} \to \mathcal{N}' \to 0. \tag{56}$$

When $Y = C$ the sheaf $\mathcal{N}'$ is usually called the *equisingular normal sheaf* [19, Proposition 1.1.9]. Therefore we refer to $\mathcal{N}'$ as the restricted equisingular normal sheaf.

Remark 2 Note that, since X is non-singular in codimension 1, the curve C can be taken to be smooth and the divisor Y smooth along C so that $\mathcal{N}'$ is locally free and it is the restriction to C of the normal sheaf of Y. Eventually we will make this assumption in the last section. However, the computations of the present section work in the more general setting.

The sheaves $\mathcal{E}xt^j_{Y\times A}(\mathcal{O}_{Y\times C}, \mathcal{O}_{\Delta_Y})$ appearing in (53) are described as follows:

Proposition 14 *(a)* $\bigoplus_i \mathcal{E}xt^i_{Y\times A}(\mathcal{O}_{Y\times C}, \mathcal{O}_{\Delta_Y}) \cong \bigoplus_i \delta_*(\overset{i-n+1}{\Lambda} \mathcal{N}' \otimes N_C^{\otimes n-1})$ *(as graded modules on the algebra (54)). In particular, the LHS is zero for* $i < n - 1$.

(b) $\mathcal{E}xt^{d-1}_{Y\times A}(\mathcal{O}_{Y\times C}, \mathcal{O}_{\Delta_Y}) \cong \delta_* \omega_C$.

Proof (a) We apply $\mathcal{H}om_{Y\times A}(\,\cdot\,, \mathcal{O}_{\Delta_Y})$ to the basic exact sequence

$$0 \to \mathcal{I}_{\Delta_C/Y\times C} \to \mathcal{O}_{Y\times C} \to \mathcal{O}_{\Delta_C} \to 0 \tag{57}$$

where $\mathcal{I}_{\Delta_C/Y\times C}$ denotes the ideal of Δ_C in $Y \times C$. Since Δ_C is the intersection (in $Y \times A$) of Δ_Y and $Y \times C$, the resulting long exact sequence is chopped into short exact sequences (where we plug in the isomorphism (55))

$$0 \to \mathcal{E}xt^{i-1}_{Y\times A}(\mathcal{I}_{\Delta_C/Y\times C}, \mathcal{O}_{\Delta_Y}) \to \delta_*(\overset{i-n+1}{\Lambda} T_{A,0} \otimes N_C^{\otimes n-1}) \\ \to \mathcal{E}xt^{i}_{Y\times A}(\mathcal{O}_{Y\times C}, \mathcal{O}_{\Delta_Y}) \to 0. \quad (58)$$

This proves that

$$\mathcal{E}xt^{i}_{Y\times A}(\mathcal{O}_{Y\times C}, \mathcal{O}_{\Delta_Y}) = \begin{cases} 0 & \text{if } i < n-1, \\ \delta_* N_C^{\otimes n-1} & \text{if } i = n-1. \end{cases} \quad (59)$$

For $i = n$ it follows from (59) and the spectral sequence (53) applied to $\mathcal{I}_{\Delta_C/Y\times C}$ that

$$\mathcal{E}xt^{n-1}_{Y\times A}(\mathcal{I}_{\Delta_C/Y\times C}, \mathcal{O}_{\Delta_Y}) \\ \cong \mathcal{H}om_{Y\times C}(\mathcal{I}_{\Delta_C/Y\times C}, \mathcal{E}xt^{n-1}_{Y\times A}(\mathcal{O}_{Y\times C}, \Delta_Y)) \cong \delta_*(\mathcal{T}_Y \otimes N_C^{\otimes n-1})$$

and that δ_* identifies (58) with (56), tensored with $N_{|C}^{n-1}$, i.e.,

$$0 \to \mathcal{T}_Y \otimes N_C^{\otimes n-1} \to T_{A,0} \otimes N_C^{\otimes n-1} \to \mathcal{N}' \otimes N_C^{\otimes n-1} \to 0. \quad (60)$$

This proves the statement for $i = n$. For $i > n$, Proposition 14 follows by induction. Indeed $\mathcal{E}xt^{\bullet}_{Y\times A}(\mathcal{O}_{Y\times C}, \mathcal{O}_{\Delta_Y})$ is naturally a graded module over the exterior algebra $\mathcal{E}xt^{\bullet}_{Y\times A}(\mathcal{O}_{\Delta_Y}, \mathcal{O}_{\Delta_Y}) \cong \delta_*(\Lambda^{\bullet} T_{A,0} \otimes \mathcal{O}_Y)$ (see (54)). Assume that the statement of the present proposition holds for the positive integer $i-1$. Because of the action of the exterior algebra, sequences (58) and (60) yield that the kernel of the map

$$\delta_*(\overset{i-n+1}{\Lambda} T_{A,0} \otimes N_C^{\otimes n-1}) \to \mathcal{E}xt^{i}_{Y\times A}(\mathcal{O}_{Y\times C}, \mathcal{O}_{\Delta}) \to 0$$

is surjected (up to twisting with $N_C^{\otimes n-1}$) by $\delta_*(\Lambda^{i-n} T_{A,0} \otimes (\mathcal{T}_Y)_{|C})$. This presentation yields that $\mathcal{E}xt^{i}_{Y\times A}(\mathcal{O}_{Y\times C}, \mathcal{O}_{\Delta})$ is equal to $\delta_*((\overset{i-n+1}{\Lambda} \mathcal{N}') \otimes N_C^{\otimes n-1})$. This proves (a).

(b) If Y is smooth along C then $\mathcal{N}'$ is locally free (coinciding with the restricted normal bundle) (see Remark 2). In this case (b) follows at once from (a). In the general case the proof is as follows. We claim that for each i the LHS of (a) can alternatively be described as

$$\mathcal{E}xt^{i}_{Y\times A}(\mathcal{O}_{Y\times C}, \mathcal{O}_{\Delta_Y}) \cong \mathcal{T}or^{Y\times A}_{d-1-i}(p_2^*\omega_C, \mathcal{O}_{\Delta_Y}).$$

This is proved by means of the isomorphism of functors

$$\mathbf{R}\operatorname{Hom}_{Y\times A}(\mathcal{O}_{Y\times C}, \mathcal{O}_{\Delta_Y}) \cong \mathbf{R}\operatorname{Hom}_{Y\times A}(\mathcal{O}_{Y\times C}, \mathcal{O}_{Y\times A}) \underline{\otimes}^{\mathbf{L}}_{Y\times A} \mathcal{O}_{\Delta_Y}$$

and the corresponding spectral sequences. In fact, since C is Cohen–Macaulay, we have that $\mathcal{E}xt^i_{Y\times A}(\mathcal{O}_{Y\times C}, \mathcal{O}_{Y\times A}) = 0$ for $i \neq d-1$ and equal to $p_2^*\omega_C$ for $i = d-1$. Thus the spectral sequence computing the RHS degenerates, proving the claim. In particular, for $i = d-1$, we have $\mathcal{E}xt^{d-1}_{Y\times A}(\mathcal{O}_{Y\times C}, \mathcal{O}_{\Delta_Y}) \cong (p_2^*\omega_C)\otimes \mathcal{O}_{\Delta_Y} \cong \delta_*\omega_C$. □

5.3 Reduction of the statement of Theorem 5

As a first application of Proposition 14, we reduce the statement of Theorem 5 – in the equivalent formulation provided by Lemma 13 – to a simpler one. This will involve the issue of comparing two spaces of first-order deformations mentioned in the Introduction (Section 1.5), and it will be the content of Proposition 15 and Corollary 16 below.

Notation 4 We consider the first spectral sequence (53) applied to $\mathcal{F} = p_2^*N_C^{\otimes n}$, rather than to $\mathcal{I}_\Delta \otimes p_2^*N_C^{\otimes n}$. Plugging the identification provided by Lemma 14, we get

$$H^j(C, \Lambda^{i-n+1}\mathcal{N}' \otimes N_C^{-1}) \Rightarrow \mathrm{Ext}^{j+i}_{Y\times A}(p_2^*N_C^{\otimes n}, \mathcal{O}_{\Delta_Y}).$$

Since the H^is on the left are zero for $i \neq 0, 1$, the spectral sequence is reduced to short exact sequences

$$0 \to H^1(C, \Lambda^{i-n}\mathcal{N}' \otimes N_C^{-1}) \xrightarrow{v_i} \mathrm{Ext}^i_{Y\times A}(p_2^*N_C^{\otimes n}, \mathcal{O}_{\Delta_Y}) \\ \xrightarrow{w_i} H^0(C, \Lambda^{i-n+1}\mathcal{N}' \otimes N_C^{-1}) \to 0. \quad (61)$$

In particular, for $i = n$ we have the exact sequence

$$0 \to H^1(C, N_C^{-1}) \xrightarrow{v_n} \mathrm{Ext}^n_{Y\times A}(p_2^*N_C^{\otimes n}, \mathcal{O}_{\Delta_Y}) \xrightarrow{w_n} H^0(C, \mathcal{N}' \otimes N_C^{-1}) \to 0.$$

Combining with the exact sequence coming from the spectral sequence (51), applied to $\mathcal{F} = p_2^*N_C^{\otimes n-1}$ we get

$$\begin{array}{ccc} H^1(C, N_C^{-1}) & & \mathrm{Ext}^{n+1}_{Y\times\widehat{A}}(R^1\Phi(p_2^*N_C^{\otimes n}), \mathcal{R}) \\ \downarrow{\scriptstyle v_n} & & \downarrow{\scriptstyle a_n} \\ \mathrm{Ext}^n_{Y\times A}(p_2^*N_C^{\otimes n}, \mathcal{O}_{\Delta_Y}) & \xrightarrow[\cong]{FM_n} & \mathrm{Ext}^n_{Y\times\widehat{A}}(\mathbf{R}\Phi(p_2^*N_C^{\otimes n}), \mathcal{R}) \\ \downarrow{\scriptstyle w_n} & & \downarrow{\scriptstyle b_n} \\ H^0(C, \mathcal{N}' \otimes N_C^{-1}) & & \mathrm{Ext}^n_{Y\times\widehat{A}}(R^0\Phi(p_2^*N_C^{\otimes n}), \mathcal{R}) \end{array} \quad (62)$$

Remark 3 Note that, as shown by the exact sequence (56) defining the restricted equisingular normal sheaf, we get that

$$H^0(C, \mathcal{N}' \otimes N_C^\vee) = \ker(H^1(C, \mathcal{T}_Y \otimes N_C^\vee) \xrightarrow{G} H^1(C, \mathcal{T}_A \otimes N_C^\vee)).$$

This map G is the restriction to $H^1(C, \mathcal{T}_Y \otimes N_C^\vee)$ of the map G_A^Y of (8) in the Introduction (see also Remark 4 below).

Proposition 15 *In diagram (62), if the map a_n is nonzero then the map $w_n \circ FM_n^{-1} \circ a_n$ is nonzero and its image is contained in the kernel of the Gaussian map (46).*

Combining with Theorem 2, and noting that the assumptions in Notation/Assumptions 1 are certainly satisfied by a sufficiently positive line bundle L on the variety X, we get

Corollary 16 *If the map a_n is nonzero then $R^{n-1}\pi_* Q = 0$.*

Proof of Proposition 15 We apply $\mathrm{Ext}^n_{Y\times A}(\cdot, \mathcal{O}_{\Delta_Y})$ to the usual exact sequence

$$0 \to \mathcal{I}_{\Delta_C/Y\times C} \otimes p_2^* N_C^{\otimes n} \to p_2^* N_C^{\otimes n} \to \delta_* N_C^{\otimes n} \to 0. \tag{63}$$

Using the spectral sequence (53) and the isomorphisms provided by Proposition 14, we get the commutative exact diagram

$$\begin{array}{ccccc}
H^1(N_C^{-1}) \otimes \Lambda^0 T_{A,0} & \xrightarrow{=} & H^1(N_C^{-1}) \otimes \Lambda^0 T_{A,0} & & \\
\downarrow = & & \downarrow & & \\
H^1(N_C^{-1}) \otimes \Lambda^0 T_{A,0} & \overset{v_n}{\hookrightarrow} & \mathrm{Ext}^n_{Y\times A}(p_2^* N_C^{\otimes n}, \mathcal{O}_{\Delta_Y}) & \overset{w_n}{\twoheadrightarrow} & H^0(N' \otimes N_C^{-1}) \\
& & \downarrow f & & \downarrow \\
& & \mathrm{Ext}^n_{Y\times A}(\mathcal{I} \otimes p_2^* N_C^{\otimes n}, \mathcal{O}_{\Delta_Y}) & \overset{u}{\hookleftarrow} & H^1(\mathcal{T}_Y \otimes N_C^{-1}) \\
& & \downarrow & & \downarrow G_A^Y \\
& & H^1(N_C^{-1}) \otimes T_{A,0} & \xrightarrow{=} & H^1(N_C^{-1}) \otimes T_{A,0}
\end{array} \tag{64}$$

where:

- We have used (55) to compute

$$\mathrm{Ext}^i_{Y\times A}(\delta_{C*} N_C^{\otimes n}, \mathcal{O}_{\Delta_Y}) \cong H^1(C, N_C^{-1}) \otimes \Lambda^{i-n+1} T_A.$$

- For typographical brevity we have denoted

$$\mathcal{I} := \mathcal{I}_{\Delta_C/Y\times C}$$

and the map

$$u : H^1(\mathcal{T}_Y \otimes N_C^{-1}) \rightarrowtail \mathrm{Ext}^n_{Y\times A}(\mathcal{I} \otimes p_2^* N_C^{\otimes n}, \mathcal{O}_{\Delta_Y}) \tag{65}$$

is the composition of the natural inclusion

$$\begin{aligned} H^1(\mathcal{T}_Y \otimes N_C^{-1}) = H^1(\mathcal{H}om(\mathcal{I} \otimes p_2^* N_C\,,\, \mathcal{O}_{\Delta_C})) &\hookrightarrow \mathrm{Ext}^1_{Y\times C}(\mathcal{I}\,,\, \delta_{C*} N_C^{-1}) \\ &\cong \mathrm{Ext}^1_{Y\times C}(\mathcal{I} \otimes p_2^* N_C^{\otimes n}, \mathcal{E}xt^{n-1}_{Y\times A}(\mathcal{O}_{Y\times C}, \mathcal{O}_{\Delta_Y})) \end{aligned}$$

and of the natural injection, arising (by Proposition 14, as the last isomorphism) in the beginning of the spectral sequence (53),

$$\mathrm{Ext}^1_{Y\times C}(\mathcal{I}_{\Delta_C/Y\times C}\otimes p_2^*N_C^{\otimes n}, \mathcal{E}xt^{n-1}_{Y\times A}(\mathcal{O}_{Y\times C},\mathcal{O}_{\Delta_Y})) \\ \to \mathrm{Ext}^n_{Y\times A}(\mathcal{I}_{\Delta_C/Y\times C}\otimes p_2^*N_C^{\otimes n},\mathcal{O}_{\Delta_Y})\ .$$

Next, we look at the Fourier–Mukai image of the central column of (64). In order to do so, we first apply the Fourier–Mukai transform $\mathbf{R}\Phi$ to sequence (63). Then we apply $\mathbf{R}\mathrm{Hom}_{Y\times\widehat{A}}(\cdot\,,\mathcal{R})$ and the spectral sequence on the $Y\times\widehat{A}$ side, namely (51).

We claim that applying the Fourier–Mukai transform $\mathbf{R}\Phi$ to sequence (63), we get the exact sequence

$$0\to R^0\Phi(\mathcal{I}\otimes p_2^*N_C^{\otimes n})\to R^0\Phi(p_2^*N_C^{\otimes n})\to \nu^*(N_C^{\otimes n})\otimes\mathcal{R}\to 0 \qquad (66)$$

and the isomorphism

$$R^1\Phi(\mathcal{I}\otimes p_2^*N_C^{\otimes n})\ \tilde{\to}\ R^1\Phi(p_2^*N_C^{\otimes n}). \qquad (67)$$

Indeed we have that $R^i\Phi(\delta_*(N_C^{\otimes n}))=\nu^*(N_C^{\otimes n})\otimes\mathcal{R}$ for $i=0$ and zero otherwise. The map $R^0\Phi(p_2^*N_C^{\otimes n})\to\nu^*(N_C^{\otimes n})\otimes\mathcal{R}$ is nothing else but the relative evaluation map

$$\pi^*\pi_*(\nu^*(N_C^{\otimes n})\otimes\mathcal{R})\to\nu^*(N_C^{\otimes n})\otimes\mathcal{R}$$

and its surjectivity follows from the assumptions (see Notation/Assumptions 1). This proves what was claimed.

Eventually we get the following exact diagram, whose central column is the Fourier–Mukai transform of the central column of (64) and whose right column is (part of) the long cohomology sequence of $\mathbf{R}\mathrm{Hom}_{Y\times\widehat{A}}(\cdot\,,\mathcal{R})$ applied to the exact sequence (66):

$$\begin{array}{ccccc}
 & & H^1(N_C^\vee)\otimes H^0(\mathcal{O}_{\widehat{A}}) & \xrightarrow{=} & H^1(N_C^\vee)\otimes H^0(\mathcal{O}_{\widehat{A}}) \\
 & & \downarrow & & \downarrow \\
\mathrm{Ext}^{n+1}(R^1\Phi(p_2^*N_C^{\otimes n}),\mathcal{R}) & \xrightarrow{a_n} & \mathrm{Ext}^n(\mathbf{R}\Phi(p_2^*N_C^{\otimes n}),\mathcal{R}) & \xrightarrow{b_n} & \mathrm{Ext}^n(R^0\Phi(p_2^*N_C^{\otimes n}),\mathcal{R}) \\
\downarrow\cong & & \downarrow FM_n(f) & & \downarrow \\
\mathrm{Ext}^{n+1}(R^1\Phi(p_2^*N_C^{\otimes n}),\mathcal{R}) & \xrightarrow{\alpha} & \mathrm{Ext}^n(\mathbf{R}\Phi(\mathcal{I}\otimes p_2^*N_C^{\otimes n}),\mathcal{R}) & \xrightarrow{\beta} & \mathrm{Ext}^n(R^0\Phi(\mathcal{I}\otimes p_2^*N_C^{\otimes n}),\mathcal{R}) \\
 & & \downarrow & & \downarrow \\
 & & H^1(N_C^\vee)\otimes H^1(\mathcal{O}_{\widehat{A}}) & \xrightarrow{=} & H^1(N_C^\vee)\otimes H^1(\mathcal{O}_{\widehat{A}})
\end{array} \qquad (68)$$

For brevity, at the place on the left of the third row we have plugged the isomorphism (67). From (66) and (67) it follows, in particular, that the map $FM_n(f)$ induces the isomorphism of the images of a_n and α:

$$FM_n(f) : \mathrm{im}(a_n) \xrightarrow{\cong} \mathrm{im}(\alpha). \tag{69}$$

An easy diagram chase in (64) and (68) proves the first part of the proposition, namely that if the map a_n is nonzero then the map $w_n \circ FM_n^{-1} \circ a_n$ is nonzero. The second part follows at once from the first one, (69), and Lemma 13. □

6 Proof of Theorem 5

The strategy of proof of Theorem 5 is to see the two vertical exact sequences of diagram (62) as the first homogeneous pieces of two exact sequences of graded modules over the exterior algebra. Namely, for each $i \geq n$ we have

$$\begin{array}{ccc}
\bigoplus_i \mathrm{Ext}^1_C(N_C^{\otimes n}, \Lambda^{i-n}N' \otimes N_C^{\otimes n-1}) & & \bigoplus_i \mathrm{Ext}^{i+1}_{Y\times\widehat{A}}(R^1\Phi(p_2^*N_C^{\otimes n}), \mathcal{R}) \\
\downarrow v_i & & \downarrow a_i \\
\bigoplus_i \mathrm{Ext}^i_{Y\times A}(p_2^*N_C^{\otimes n}, O_{\Delta_Y}) & \xrightarrow[\cong]{FM_i} & \bigoplus_i \mathrm{Ext}^i_{Y\times\widehat{A}}(\mathbf{R}\Phi(p_2^*N_C^{\otimes n}), \mathcal{R}) \\
\downarrow w_i & & \downarrow b_i \\
\bigoplus_i \mathrm{Hom}_C(N_C^{\otimes n}, \Lambda^{i-n+1}N' \otimes N_C^{\otimes n-1}) & & \bigoplus_i \mathrm{Ext}^i_{Y\times\widehat{A}}(R^0\Phi(p_2^*N_C^{\otimes n}), \mathcal{R})
\end{array} \tag{70}$$

The exterior algebra acts on the LHS as $\Lambda^\bullet T_{A,0} \hookrightarrow \mathrm{Ext}^\bullet_{Y\times A}(O_{\Delta_Y}, O_{\Delta_Y})$ (see (54) and (55)). After the Fourier–Mukai transform, it acts on the RHS as $\Lambda^\bullet H^1(O_{\widehat{A}}) \hookrightarrow \mathrm{Ext}^\bullet_{Y\times\widehat{A}}(\mathcal{R}, \mathcal{R})$.

6.1 Computations in degree $d-1$

In this section we will make some explicit calculations in degree $d-1$, where we have the special feature that the Hom space at the bottom of the left column is naturally isomorphic to $\mathrm{Hom}(N_C^{\otimes n}, \omega_C)$ (Proposition 14, part (b)). The following proposition shows that what we want to prove in degree n, namely that the map $w_n \circ FM_n^{-1} \circ a_n$ is nonzero, is true, in strong form, in degree $d-1$.

Proposition 17 *The map w_{d-1} has a canonical (up to scalar) section σ and the injective map $(FM_{d-1})_{|Im(\sigma)}$ factorizes through a_{d-1}. Summarizing, in degree $i = d-1$ diagram (70) specializes to*

$$\begin{array}{ccc}
\mathrm{Ext}^1_C(N_C^{\otimes n}, \Lambda^{d-1-n}N' \otimes N_C^{\otimes n-1}) & \nearrow & \mathrm{Ext}^d_{Y\times\widehat{A}}(R^1\Phi(p_2^*N_C^{\otimes n}), \mathcal{R}) \\
\downarrow v_{d-1} & & \downarrow a_{d-1} \\
\mathrm{Ext}^{d-1}_{Y\times A}(p_2^*N_C^{\otimes n}, O_{\Delta_Y}) & \xleftrightarrow[\cong]{FM_{d-1}} & \mathrm{Ext}^{d-1}_{Y\times\widehat{A}}(\mathbf{R}\Phi(p_2^*N_C^{\otimes n}), \mathcal{R}) \\
\sigma \curvearrowleft \;\downarrow w_{d-1} & & \downarrow b_{d-1} \\
\mathrm{Hom}_C(N_C^{\otimes n}, \omega_C) & & \mathrm{Ext}^d_{Y\times\widehat{A}}(R^0\Phi(p_2^*N_C^{\otimes n}), \mathcal{R})
\end{array} \tag{71}$$

Proof The section σ is given (up to scalar) by the product map

$$\mathrm{Ext}^{d-1}_{Y\times A}(p_2^* N_C^{\otimes n}, \mathcal{O}_{Y\times A}) \otimes \mathrm{Hom}_{Y\times A}(\mathcal{O}_{Y\times A}, \mathcal{O}_{\Delta_Y}) \xrightarrow{\sigma} \mathrm{Ext}^{d-1}_{Y\times A}(p_2^* N_C^{\otimes n}, \mathcal{O}_{\Delta_Y}). \quad (72)$$

In fact, note that $\mathrm{Ext}^{d-1}_{Y\times A}(p_2^* N_C^{\otimes n}, \mathcal{O}_{Y\times A}) \cong p_1^* H^0(\mathcal{O}_Y) \otimes p_2^* \mathrm{Ext}^{d-1}_A(N_C^{\otimes n}, \mathcal{O}_A) \cong p_1^* H^0(\mathcal{O}_Y) \otimes p_2^* \mathrm{Hom}_C(N_C^{\otimes n}, \omega_C)$. The fact that s is a section of w_{q-1} is clear, as the latter is the natural map

$$\mathrm{Ext}^{d-1}_{Y\times A}(p_2^* N_C^{\otimes n}, \mathcal{O}_{\Delta_Y}) \to H^0(\mathcal{E}xt^{d-1}_{Y\times A}(p_2^* N_C^{\otimes n}), \mathcal{O}_{\Delta_Y}) \cong$$

$$\cong H^0(\mathcal{E}xt^{d-1}_{Y\times A}(p_2^* N_C^{\otimes n}, \mathcal{O}_{Y\times A}) \otimes \mathcal{O}_{\Delta_Y}) \cong \mathrm{Ext}^{d-1}_{Y\times A}(p_2^* N_C^{\otimes n}, \mathcal{O}_{Y\times A}) \otimes H^0(\mathcal{O}_{\Delta_Y}).$$

Next, we prove the second part of the statement. On the $Y \times \widehat{A}$ side, we consider the following product map:

$$\begin{array}{ccc}
\mathrm{Hom}_{Y\times\widehat{A}}(R^1\Phi(p_2^* N_C^{\otimes n}), \mathcal{O}_{Y\times\hat{0}}) \otimes \mathrm{Ext}^d_{Y\times\widehat{A}}(\mathcal{O}_{Y\times\hat{0}}, \mathcal{R}) & \to & \mathrm{Ext}^d_{Y\times\widehat{A}}(R^1\Phi(p_2^* N_C^{\otimes n}), \mathcal{R}) \\
\downarrow \cong & & \downarrow a_{d-1} \\
\mathrm{Ext}^{-1}_{Y\times\widehat{A}}(\mathbf{R}\Phi(p_2^* N_C^{\otimes n}), \mathcal{O}_{Y\times\hat{0}}) \otimes \mathrm{Ext}^d_{Y\times\widehat{A}}(\mathcal{O}_{Y\times\hat{0}}, \mathcal{R}) & \longrightarrow & \mathrm{Ext}^{d-1}_{Y\times\widehat{A}}(\mathbf{R}\Phi(p_2^* N_C^{\otimes n}), \mathcal{R})
\end{array} \quad (73)$$

where the vertical isomorphism comes from the usual spectral sequence (51). By (47) the inverse of the Fourier–Mukai transform is $(-1)_A^* \circ \mathbf{R}\Psi[q]$. By (49) we have that

$$(-1)_A^* \circ \mathbf{R}\Psi(\mathcal{O}_{Y\times\hat{0}}) = (-1)_A^* \circ R^0\Psi(\mathcal{O}_{Y\times\hat{0}}) = \mathcal{O}_{Y\times A},$$

$$(-1)_A^* \circ \mathbf{R}\Psi(\mathcal{R}) = (-1)_A^* \circ R^d\Psi(\mathcal{R})[-d] = \mathcal{O}_{\Delta_Y}.$$

Therefore, thanks to Mukai's inversion theorem (47), the Fourier–Mukai transform identifies – on the $Y\times A$ side – the sources of both rows in diagram (73) to

$$\mathrm{Ext}^{-1}_{Y\times A}(p_2^* N_C^{\otimes n}[-d], \mathcal{O}_{Y\times A}) \otimes \mathrm{Ext}^d(\mathcal{O}_{Y\times A}, \mathcal{O}_{\Delta_Y}[-d])$$

$$\cong \mathrm{Ext}^{d-1}_{Y\times A}(p_2^* N_C^{\otimes n}, \mathcal{O}_{Y\times A}) \otimes \mathrm{Hom}_{Y\times A}(\mathcal{O}_{Y\times A}, \mathcal{O}_{\Delta_Y}).$$

This concludes the proof of the proposition. □

6.2 Conclusion of the proof of Theorem 5

Notation 5 We introduce the following typographical abbreviations on diagram (70): the isomorphic (via the Fourier–Mukai transform) spaces of the central row of diagram (70) are identified to vector spaces E_i, and we denote by V_i, E_i, W_i the spaces appearing in the left column of diagram (70) (from top to bottom), and by A_i, E_i, B_i the spaces appearing in the right column (from top to bottom). We denote also by $\Lambda^\bullet T_{A,0}$ the acting exterior algebra. The structure of $\Lambda^\bullet T_{A,0}$-graded modules induces a natural map of diagrams (we focus on degrees n and $d-1$ as they are the relevant ones in our argument)

$$\begin{array}{ccccc} & & A_n & & \\ & & \downarrow{\scriptstyle a_n} & & \\ V_n & \xrightarrow{v_n} & E_n & \xrightarrow{w_n} & W_n \\ & & \downarrow{\scriptstyle b_n} & & \\ & & B_n & & \end{array}$$

$$\downarrow \phi \tag{74}$$

$$\begin{array}{ccccc} & & \overset{d-1-n}{\Lambda} T^\vee_{A,0} \otimes A_{d-1} & & \\ & & \downarrow{\scriptstyle \tilde{a}_{d-1}} & & \\ \overset{d-1-n}{\Lambda} T^\vee_{A,0} \otimes V_{d-1} & \xrightarrow{\tilde{v}_{d-1}} & \overset{d-1-n}{\Lambda} T^\vee_{A,0} \otimes E_{d-1} & \xrightarrow{\tilde{w}_{d-1}} & \overset{d-1-n}{\Lambda} T^\vee_{A,0} \otimes W_{d-1} \\ & & \downarrow{\scriptstyle \tilde{b}_{d-1}} & & \\ & & \overset{d-1-n}{\Lambda} T^\vee_{A,0} \otimes B_{d-1} & & \end{array}$$

where we have denoted $\tilde{v}_{d-1} = \mathrm{id} \otimes v_{d-1}$ and so on. We denote also

$$\phi_{A_n} : A_n \to \overset{d-1-n}{\Lambda} T^\vee_{A,0} \otimes A_{d-1}$$

and, similarly, ϕ_{B_n}, ϕ_{V_n}, ϕ_{W_n}, ϕ_{E_n}.

At this point we make the following assumption:
(*) *The extension class e of the restricted cotangent sequence*

$$0 \to N^\vee_C \to (\Omega^1_X)_{|C} \to (\Omega^1_Y)_{|C} \to 0 \tag{75}$$

belongs to the subspace $H^1(\mathcal{T}_Y \otimes N^\vee_C)$ *of* $\mathrm{Ext}^1_C(\Omega^1_Y \otimes N_C, \mathcal{O}_C)$.[15] Note that if C is smooth and Y is smooth along C this is obvious, since the two spaces coincide.

Remark 4 Note that, if (*) holds then e belongs to the subspace $H^0(\mathcal{N}' \otimes N^\vee_C)$ of $H^1(\mathcal{T}_Y \otimes N^\vee_C)$: as mentioned in Remark 3, from the exact sequence defining the restricted equisingular normal sheaf (56) we get that

$$H^0(C, \mathcal{N}' \otimes N^\vee_C) = \ker(\, H^1(C, \mathcal{T}_Y \otimes N^\vee_C) \xrightarrow{G} H^1(C, \mathcal{T}_A \otimes N^\vee_C)).$$

[15] These are the locally trivial first-order deformations.

The fact that e belongs to $H^0(\mathcal{N}' \otimes N_C^\vee)$ essentially follows from the deformation-theoretic interpretation of this map G (it is the restriction (to $H^1(C, \mathcal{T}_Y \otimes N_C^\vee)$) of the map G_Z^Y of (8) in the Introduction, with $Z = A$). More formally: the target of G is $\mathrm{Hom}_k(\Omega^1_{A,0}, H^1(C, N_C^\vee))$ and G takes an extension class f to the map $\Omega^1_{A,0} \to H^1(N_C^{-1})$ obtained by composing the coboundary map of f with the map $\Omega^1_{A,0} \to H^0((\Omega^1_Y)_{|C})$. If the extension class is (75) then this map factorizes through $H^0((\Omega^1_X)_{|C})$, hence $e \in \ker G$.

From diagram (74) we have the map

$$\phi_{W_n} : H^0(C, \mathcal{N}' \otimes N_C^\vee) = \ker(G) \to \mathrm{Hom}(\overset{d-1-n}{\Lambda} T_{A,0}, H^0(\omega_C \otimes N_C^{\otimes -n})).$$

Lemma 18 $\phi_{W_n}(e) \neq 0.$

Proof We make the identification $\overset{d-1-n}{\Lambda} T_{A,0} \cong \overset{n+1}{\Lambda} \Omega^1_{A,0}$. Accordingly $\phi_{W_n}(e)$ is identified with a map

$$\phi_{W_n}(e) : \overset{n+1}{\Lambda} \Omega^1_{A,0} \to H^0(\omega_C \otimes N_C^{\otimes -n}).$$

We consider the map

$$\overset{n+1}{\Lambda} \Omega^1_{A,0} \to H^0((\overset{n+1}{\Lambda} \Omega^1_X)_{|C}) \tag{76}$$

obtained as H^0 of Λ^{n+1} of the codifferential $\Omega^1_{A,0} \otimes \mathcal{O}_C \to (\Omega^1_X)_{|C}$. Since the codifferential is surjective, the map (76) is nonzero. If C is smooth and X and Y are smooth along C then the target of (76) is $H^0((\omega_X)_{|C}) = H^0(\omega_C \otimes N_C^{\otimes -n})$. Via the above identifications, the map $\phi_{W_n}(e)$ coincides, up to a scalar, with (76). The lemma follows in this case. Even if X is not smooth along C, $\phi_{W_n}(e)$ is the composition of the map (76) and the H^0 of the canonical map $\Lambda^{n+1}((\Omega^1_X)_{|C}) \to (\omega_X)_{|C} \cong \omega_C \otimes N_C^{\otimes -n}$. Such a composition is clearly nonzero and the lemma follows as above. □

At this point, the line of the argument is clear. The class $\phi_{W_n}(e)$ is nonzero, and, in the splitting of Proposition 17, it belongs to the direct summand (of E_n) $\sigma(W_n) \subset \mathrm{Im}\,(\bar{a}_{d-1})$. Therefore the projection of $Im\,(\phi_{E_n})$ onto $\sigma(W_n)$ is nonzero. This implies that $\mathrm{Im}\,(a_n)$ is nonzero, since otherwise E_n would be isomorphic to a subspace of B_n and $\mathrm{Im}\,(\phi_{E_n})$ would be contained in the direct summand complementary to $\sigma(W_n)$.

By Corollary 16 this proves that $R^n\pi_*Q = 0$.

To prove the vanishing of $R^i\pi_*Q$ for $0 < i < n$ one takes a sufficiently positive ample line bundle M on X and an $(i+1)$-dimensional complete intersection of divisors in $|M|$, say X'. It follows from relative Serre vanishing that

$R^i\pi_*(Q) = R^i\pi_*(Q_{|X'\times\widehat{A}})$. Therefore the desired vanishing follows by the previous step. The vanishing of $R^0\pi_*Q$ is standard: as it is a torsion-free sheaf, it is enough to show that its support is a proper subvariety of $\widehat{A}$. By a base change, this is contained in the locus of $\alpha \in \widehat{A}$ such that $h^0(X, \alpha_{|X}) > 0$, i.e., the kernel of the homomorphism $\mathrm{Pic}^0 A \to \mathrm{Pic} X$, which is easily seen to be a proper subvariety of $\mathrm{Pic}^0 A$.[16] This concludes the proof of Theorem 5. □

Remark 5 The hypothesis that X is smooth in codimension 1 is used to ensure that assumption *(*)* can be made.

Acknowledgments Thanks to Christopher Hacon and Sándor Kovács for valuable correspondence, and especially for showing me their counterexamples to a previous wrong statement of mine. Thanks also to Antonio Rapagnetta and Edoardo Sernesi for valuable discussions. I also thank Christopher Hacon, Mircea Mustata, and Mihnea Popa for allowing me to contribute to this volume, even if I couldn't participate in the Robfest. Above all my gratitude goes to Rob Lazarsfeld. Most of my understanding of the matters of this paper goes back to his teaching.

References

[1] Banica, C. and Forster, O. Multiplicity structures on space curves. In *The Lefschetz Centennial Conference, Proceedings on Algebraic Geometry*, AMS (1984), pp. 47–64.

[2] Beauville, A. and Mérindol, J. Y. Sections hyperplanes des surfaces K3. *Duke Math. J.* **55**(4) (1987) 873–878.

[3] Colombo, E., Frediani, P., and Pareschi, G. Hyperplane sections of abelian surfaces. *J. Alg. Geom.* **21** (2012) 183–200.

[4] Ferrand, D. Courbes gauches et fibrés de rang 2. *C.R.A.S.* **281** (1977) 345–347.

[5] Green, M. and Lazarsfeld, R. Deformation theory, generic vanishing theorems, and some conjectures of Enriques, Catanese and Beauville. *Invent. Math.* **90** (1987) 389–407.

[6] Green, M. and Lazarsfeld, R. Higher obstructions to deforming cohomology groups of line bundles. *J. Amer. Math. Soc.* **1**(4) (1991) 87–103.

[7] Hacon, C. A derived category approach to generic vanishing. *J. Reine Angew. Math.* **575** (2004) 173–187.

[8] Hacon, C. and Kovacs, S. Generic vanishing fails for singular varieties and in characteristic $p > 0$. arXiv:1212.5105.

[9] Hartshorne, R. *Residues and Duality*. Berlin: Springer-Verlag, 1966.

[16] For example, one can reduce to prove the same assertion for a general curve C complete intersection of n irreducible effective divisors in $|L|$ for a sufficiently positive line bundle L on X.

[10] Kempf, G. Toward the inversion of abelian integrals, I. *Ann. Math.* **110** (1979) 184–202.
[11] Lazarsfeld, R. Brill–Noether–Petri without degeneration. *J. Diff. Geom. (3)* **23** (1986) 299–307.
[12] MacLane, S. *Homology*. Berlin: Springer-Verlag, 1963.
[13] Mukai, S. Duality between $D(X)$ and $D(\widehat{X})$ with its application to Picard sheaves. *Nagoya Math. J.* **81** (1981) 153–175.
[14] Mukai, S. Fourier functor and its application to the moduli of bundles on an abelian variety. In: *Algebraic Geometry (Sendai 1985)*. Advanced Studies in Pure Mathematics, Vol. 10, 1987, pp. 515–550.
[15] Mumford, D. *Abelian Varieties*, 2nd edn. London: Oxford University Press, 1974.
[16] Pareschi, G. and Popa, M. Strong generic vanishing and a higher dimensional Castelnuovo–de Franchis inequality. *Duke Math. J.* **150** (2009) 269–285.
[17] Pareschi, G. and Popa, M. GV-sheaves, Fourier–Mukai transform, and generic vanishing. *Amer. J. Math.* **133** (2011) 235–271.
[18] Pareschi, G. and Popa, M. Regularity on abelian varieties, III. Relationship with generic vanishing and applications. In *Grassmannians, Moduli Spaces and Vector Bundles*. AMS, 2011, pp. 141–167.
[19] Sernesi, E. *Deformations of Algebraic Schemes*. Berlin: Springer-Verlag, 2006.
[20] Voisin, C. Sur l'application de Wahl des courbes satisfaisant la condition de Brill-Noether–Petri, *Acta Math.* **168** (1992) 249–272.
[21] Wahl, J. Introduction to Gaussian maps on an algebraic curve. In *Complex Projective Geometry*. London Mathematical Society Lecture Note Series, No. 179. Cambridge: Cambridge University Press, 1992, pp. 304–323.

19

Torsion points on cohomology support loci: From $\mathscr{D}$-modules to Simpson's theorem

C. Schnell
Stony Brook University

Abstract

We study cohomology support loci of regular holonomic $\mathscr{D}$-modules on complex abelian varieties, and obtain conditions under which each irreducible component of such a locus contains a torsion point. One case is that both the $\mathscr{D}$-module and the corresponding perverse sheaf are defined over a number field; another case is that the $\mathscr{D}$-module underlies a graded-polarizable mixed Hodge module with a $\mathbb{Z}$-structure. As a consequence, we obtain a new proof for Simpson's result that Green–Lazarsfeld sets are translates of subtori by torsion points.

1 Overview

1.1 Introduction

Let X be a projective complex manifold. In their two influential papers about the generic vanishing theorem [6, 7], Green and Lazarsfeld showed that the so-called *cohomology support loci*

$$\Sigma_m^{p,q}(X) = \{ L \in \mathrm{Pic}^0(X) \mid \dim H^q(X, \Omega_X^p \otimes L) \geq m \}$$

are finite unions of translates of subtori of $\mathrm{Pic}^0(X)$. Beauville and Catanese [2] conjectured that the translates are always by *torsion points*, and this was proved by Simpson [19] with the help of the Gelfond–Schneider theorem from transcendental number theory. There is also a proof using positive characteristic methods by Pink and Roessler [13].

From *Recent Advances in Algebraic Geometry*, edited by Christopher Hacon, Mircea Mustaţă and Mihnea Popa

Over the past 10 years, the results of Green and Lazarsfeld have been reinterpreted and generalized several times [8, 14, 17], and we now understand that they are consequences of a general theory of holonomic $\mathscr{D}$-modules on abelian varieties. The purpose of this paper is to investigate under what conditions the result about torsion points on cohomology support loci remains true in that setting. One application is a new proof for the conjecture by Beauville and Catanese that does not use transcendental number theory or reduction to positive characteristic.

Note: In a recent preprint [20], Wang extends Theorem 1.4 to polarizable Hodge modules on compact complex tori; as a corollary, he proves the conjecture of Beauville and Catanese for arbitrary compact Kähler manifolds.

1.2 Cohomology support loci for $\mathscr{D}$-modules

Let A be a complex abelian variety, and let $\mathcal{M}$ be a regular holonomic $\mathscr{D}_A$-module; recall that a $\mathscr{D}$-module is called *holonomic* if its characteristic variety is a union of Lagrangian subvarieties of the cotangent bundle. Denoting by $A^\natural$ the moduli space of line bundles with integrable connection on A, we define the *cohomology support loci* of $\mathcal{M}$ as

$$S^k_m(A, \mathcal{M}) = \left\{ (L, \nabla) \in A^\natural \;\middle|\; \dim \mathbf{H}^k\big(A, \mathrm{DR}_A(\mathcal{M} \otimes_{\mathscr{O}_A} (L, \nabla))\big) \geq m \right\}$$

for $k, m \in \mathbb{Z}$. It was shown in [17, Theorem 2.2] that $S^k_m(A, \mathcal{M})$ is always a finite union of linear subvarieties of $A^\natural$, in the following sense:

Definition 1.1 A *linear subvariety* of $A^\natural$ is any subset of the form

$$(L, \nabla) \otimes \operatorname{im}(f^\natural \colon B^\natural \to A^\natural),$$

for $f \colon A \to B$ a homomorphism of abelian varieties, and (L, ∇) a point of $A^\natural$. An *arithmetic subvariety* is a linear subvariety that contains a torsion point.

Moreover, the analogue of Simpson's theorem is true for semisimple regular holonomic $\mathscr{D}$-modules of geometric origin: for such $\mathcal{M}$, every irreducible component of $S^k_m(A, \mathcal{M})$ contains a torsion point. We shall generalize this result in two directions:

1. Suppose that $\mathcal{M}$ is regular holonomic, and that both $\mathcal{M}$ and the corresponding perverse sheaf $\mathrm{DR}_A(\mathcal{M})$ are defined over a number field. We shall prove that the cohomology support loci of $\mathcal{M}$ are finite unions of arithmetic subvarieties; this is also predicted by Simpson's standard conjecture.
2. Suppose that $\mathcal{M}$ underlies a graded-polarizable mixed Hodge module with $\mathbb{Z}$-structure; for example, $\mathcal{M}$ could be the intermediate extension of a

polarizable variation of Hodge structure with coefficients in $\mathbb{Z}$. We shall prove that the cohomology support loci of $\mathcal{M}$ are finite unions of arithmetic subvarieties.

1.3 Simpson's standard conjecture

In his article [18], Simpson proposed several conjectures about regular holonomic systems of differential equations whose monodromy representation is defined over a number field. The principal one is the so-called "standard conjecture." Restated in the language of regular holonomic $\mathscr{D}$-modules and perverse sheaves, it takes the following form:

Conjecture 1.2 *Let $\mathcal{M}$ be a regular holonomic $\mathscr{D}$-module on a smooth projective variety X, both defined over $\bar{\mathbb{Q}}$. If $\mathrm{DR}_X(\mathcal{M})$ is the complexification of a perverse sheaf with coefficients in $\bar{\mathbb{Q}}$, then $\mathcal{M}$ is of geometric origin.*

He points out that, "there is certainly no more reason to believe it is true than to believe the Hodge conjecture, and whether or not it is true, it is evidently impossible to prove with any methods which are now under consideration. However, it is an appropriate motivation for some easier particular examples, and it leads to some conjectures which might in some cases be more tractable" [18, p. 372].

In the particular example of abelian varieties, Conjecture 1.2 predicts that when $\mathcal{M}$ is a regular holonomic $\mathscr{D}$-module with the properties described in the conjecture, then the cohomology support loci of $\mathcal{M}$ should be finite unions of arithmetic subvarieties. Our first result – actually a simple consequence of [17] and an old theorem by Simpson [19] – is that this prediction is correct.

Theorem 1.3 *Let A be an abelian variety defined over $\bar{\mathbb{Q}}$, and let $\mathcal{M}$ be a regular holonomic $\mathscr{D}_A$-module. If $\mathcal{M}$ is defined over $\bar{\mathbb{Q}}$, and if $\mathrm{DR}_A(\mathcal{M})$ is the complexification of a perverse sheaf with coefficients in $\bar{\mathbb{Q}}$, then all cohomology support loci $S^k_m(A, \mathcal{M})$ are finite unions of arithmetic subvarieties of $A^\natural$.*

Proof Let $\mathrm{Char}(A) = \mathrm{Hom}(\pi_1(A), \mathbb{C}^*)$ be the space of rank 1 characters of the fundamental group; for a character $\rho \in \mathrm{Char}(A)$, we denote by $\mathbb{C}_\rho$ the corresponding local system of rank 1. We define the cohomology support loci of a constructible complex of $\mathbb{C}$-vector spaces $K \in \mathrm{D}^b_c(\mathbb{C}_A)$ to be the sets

$$S^k_m(A, K) = \left\{ \rho \in \mathrm{Char}(A) \;\middle|\; \dim \mathbf{H}^k(A, K \otimes_{\mathbb{C}} \mathbb{C}_\rho) \geq m \right\}.$$

The correspondence between local systems and vector bundles with integrable connection gives a biholomorphic mapping $\Phi \colon A^\natural \to \mathrm{Char}(A)$; it takes a point

(L, ∇) to the local system of flat sections of ∇. According to [17, Lemma 14.1], the cohomology support loci satisfy

$$\Phi(S_m^k(A, \mathcal{M})) = S_m^k(A, \mathrm{DR}_A(\mathcal{M})).$$

Note that $\mathrm{Char}(A)$ is an affine variety defined over $\mathbb{Q}$; in our situation, $A^\natural$ is moreover a quasi-projective variety defined over $\bar{\mathbb{Q}}$, because the same is true for A. The assumptions on the $\mathscr{D}$-module $\mathcal{M}$ imply that $S_m^k(A, \mathcal{M}) \subseteq A^\natural$ is defined over $\bar{\mathbb{Q}}$, and that $S_m^k(A, \mathrm{DR}_A(\mathcal{M})) \subseteq \mathrm{Char}(A)$ is defined over $\bar{\mathbb{Q}}$. We can now use [19, Theorem 3.3] to conclude that both must be finite unions of arithmetic subvarieties. □

1.4 Mixed Hodge modules with $\mathbb{Z}$-structure

We now consider a much larger class of regular holonomic $\mathscr{D}_A$-modules, namely those that come from mixed Hodge modules with $\mathbb{Z}$-structure. This class includes, for example, intermediate extensions of polarizable variations of Hodge structure defined over $\mathbb{Z}$; the exact definition can be found below (Definition 2.2).

We denote by $\mathrm{MHM}(A)$ the category of graded-polarizable mixed Hodge modules on the abelian variety A, and by $\mathrm{D}^b\,\mathrm{MHM}(A)$ its bounded derived category [16, Section 4]; because A is projective, every mixed Hodge module is automatically algebraic. Let

$$\mathrm{rat}\colon \mathrm{D}^b\,\mathrm{MHM}(A) \to \mathrm{D}_c^b(\mathbb{Q}_A)$$

be the functor that takes a complex of mixed Hodge modules to the underlying complex of constructible sheaves of $\mathbb{Q}$-vector spaces; then a $\mathbb{Z}$-structure on $M \in \mathrm{D}^b\,\mathrm{MHM}(A)$ is a constructible complex $E \in \mathrm{D}_c^b(\mathbb{Z}_A)$ with the property that $\mathrm{rat}\, M \simeq \mathbb{Q} \otimes_{\mathbb{Z}} E$.

To simplify the notation, we shall define the *cohomology support loci* of a complex of mixed Hodge modules $M \in \mathrm{D}^b\,\mathrm{MHM}(A)$ as

$$S_m^k(A, M) = \{\rho \in \mathrm{Char}(A) \mid \dim H^k(A, \mathrm{rat}\, M \otimes_{\mathbb{Q}} \mathbb{C}_\rho) \geq m\},$$

where $k \in \mathbb{Z}$ and $m \geq 1$. Our second result is the following structure theorem for these sets.

Theorem 1.4 *If a complex of mixed Hodge modules $M \in \mathrm{D}^b\,\mathrm{MHM}(A)$ admits a $\mathbb{Z}$-structure, then all of its cohomology support loci $S_m^k(A, M)$ are complete unions of arithmetic subvarieties of* $\mathrm{Char}(A)$.

Definition 1.5 A collection of arithmetic subvarieties of $\mathrm{Char}(A)$ is called *complete* if it is a finite union of subsets of the form

$$\{\rho^k \mid \gcd(k, n) = 1\} \cdot \mathrm{im}(\mathrm{Char}(f)\colon \mathrm{Char}(B) \to \mathrm{Char}(A)),$$

where $\rho \in \mathrm{Char}(A)$ is a point of finite order n, and $f: A \to B$ is a surjective morphism of abelian varieties with connected fibers.

The proof of Theorem 1.4 occupies the remainder of the paper; it is by induction on the dimension of the abelian variety. Since we already know that the cohomology support loci are finite unions of linear subvarieties, the issue is to prove that every irreducible component contains a torsion point. Four important ingredients are the Fourier–Mukai transform for $\mathscr{D}_A$-modules [11, 15]; results about Fourier–Mukai transforms of holonomic $\mathscr{D}_A$-modules [17]; the theory of perverse sheaves with integer coefficients [3]; and of course Saito's theory of mixed Hodge modules [16]. Roughly speaking, they make it possible to deduce the assertion about torsion points from the following elementary special case: if V is a graded-polarizable variation of mixed Hodge structure on A with coefficients in $\mathbb{Z}$, and if $\mathbb{C}_\rho$ is a direct factor of $V \otimes_{\mathbb{Z}} \mathbb{C}$ for some $\rho \in \mathrm{Char}(A)$, then ρ must be a torsion character. The completeness of the set of components follows from the fact that $S^k_m(A, M)$ is defined over $\mathbb{Q}$, hence stable under the natural $\mathrm{Gal}(\mathbb{C}/\mathbb{Q})$-action; note that the $\mathrm{Gal}(\mathbb{C}/\mathbb{Q})$-orbit of a character ρ of order n consists exactly of the characters ρ^k with $\gcd(k, n) = 1$.

1.5 The conjecture of Beauville and Catanese

Now let X be a projective complex manifold. As a consequence of Theorem 1.4, we obtain a purely analytic proof for the conjecture of Beauville and Catanese.

Theorem 1.6 *Each $\Sigma^{p,q}_m(X)$ is a finite union of subsets of the form $L \otimes T$, where $L \in \mathrm{Pic}^0(X)$ is a point of finite order, and $T \subseteq \mathrm{Pic}^0(X)$ is a subtorus.*

Proof Inside the group $\mathrm{Char}(X)$ of rank 1 characters of the fundamental group of X, let $\mathrm{Char}^0(X)$ denote the connected component of the trivial character. If $f: X \to A$ is the Albanese morphism (for some choice of base point on X), then $\mathrm{Char}(f): \mathrm{Char}(A) \to \mathrm{Char}^0(X)$ is an isomorphism. As above, we denote the local system corresponding to a character $\rho \in \mathrm{Char}(X)$ by the symbol $\mathbb{C}_\rho$. Define the auxiliary sets

$$\Sigma^k_m(X) = \{\rho \in \mathrm{Char}^0(X) \mid \dim \mathbf{H}^k(X, \mathbb{C}_\rho) \geq m\};$$

by the same argument as in [1, Theorem 3], it suffices to prove that each $\Sigma^k_m(X)$ is a finite union of arithmetic subvarieties of $\mathrm{Char}^0(X)$. But this follows easily from Theorem 1.4. To see why, consider the complex of mixed Hodge modules

$M = f_* \mathbb{Q}_X^H[\dim X] \in \mathrm{D}^b \operatorname{MHM}(A)$. The underlying constructible complex is $\operatorname{rat} M = \mathbf{R} f_* \mathbb{Q}[\dim X]$, and so

$$\Sigma_m^{k+\dim X}(X) = \operatorname{Char}(f)(S_m^k(A, M)).$$

Because $\mathbf{R} f_* \mathbb{Z}[\dim X]$ is a $\mathbb{Z}$-structure on M, the assertion is an immediate consequence of Theorem 1.4. □

For some time, I thought that each $\Sigma_m^{p,q}(X)$ might perhaps also be complete in the sense of Definition 1.5, meaning a finite union of subsets of the form

$$\{ L^{\otimes k} \mid \gcd(k, n) = 1 \} \otimes T,$$

where n is the order of L. Unfortunately, this is not the case.

Example 1.7 Here is an example of a surface X where certain cohomology support loci are not complete. Let A be an elliptic curve. Choose a nontrivial character $\rho \in \operatorname{Char}(A)$ of order 3, let $L = \mathbb{C}_\rho \otimes_{\mathbb{C}} \mathscr{O}_A$, and let $B \to A$ be the étale cover of degree 3 that trivializes ρ. The Galois group of this cover is $G = \mathbb{Z}/3\mathbb{Z}$, and if we view G as a quotient of $\pi_1(A, 0)$, then the three characters of G correspond exactly to $1, \rho, \rho^2$. Finally, let ω be a primitive third root of unity, and let E_ω be the elliptic curve with an automorphism of order 3. Now G acts diagonally on the product $E_\omega \times B$, and the quotient is an isotrivial family of elliptic curves $f: X \to A$. Let us consider the variation of Hodge structure on the first cohomology groups of the fibers. Setting $H = H^1(E_\omega, \mathbb{Z})$, the corresponding representation of the fundamental group of A factors as

$$\pi_1(A, 0) \to G \to \operatorname{Aut}(H),$$

and is induced by the G-action on E_ω. This representation is the direct sum of the two characters ρ and ρ^2, because G acts as multiplication by ω and ω^2 on $H^{1,0}(E_\omega)$ and $H^{0,1}(E_\omega)$, respectively. For the same reason, $f_* \omega_X \simeq L$ and $R^1 f_* \omega_X \simeq \mathscr{O}_A$. Since $f^*: \operatorname{Pic}^0(A) \to \operatorname{Pic}^0(X)$ is an isomorphism, the projection formula gives

$$\Sigma_1^{2,0}(X) = \{L^{-1}\} \quad \text{and} \quad \Sigma_1^{2,1}(X) = \{L^{-1}, \mathscr{O}_A\}.$$

We conclude that not all cohomology support loci of X are complete.

Note: Although he does not state his result in quite this form, Pareschi [12, Scholium 4.3] shows that the set of positive-dimensional irreducible components of

$$V^0(\omega_X) = \Sigma_1^{\dim X, 0}(X)$$

is complete, provided that X has maximal Albanese dimension.

2 Preparation for the proof

2.1 Variations of Hodge structure

In what follows, A will always denote a complex abelian variety, and $g = \dim A$ its dimension. To prove Theorem 1.4, we have to show that certain complex numbers are roots of unity; we shall do this with the help of Kronecker's theorem, which says that if all conjugates of an algebraic integer have absolute value 1, then it is a root of unity. To motivate what follows, let us consider the simplest instance of Theorem 1.4, namely a polarizable variation of Hodge structure with coefficients in $\mathbb{Z}$.

Lemma 2.1 *If a local system with coefficients in $\mathbb{Z}$ underlies a polarizable variation of Hodge structure on A, then it is a direct sum of torsion points of* $\mathrm{Char}(A)$.

Proof The associated monodromy representation $\mu\colon \pi_1(A) \to \mathrm{GL}_n(\mathbb{Z})$, tensored by $\mathbb{C}$, is semisimple [4, Section 4.2]; the existence of a polarization implies that it is isomorphic to a direct sum of unitary characters of $\pi_1(A)$. Since μ is defined over $\mathbb{Z}$, the collection of these characters is preserved by the action of $\mathrm{Gal}(\mathbb{C}/\mathbb{Q})$. This means that the values of each character, as well as all their conjugates, are algebraic integers of absolute value 1; by Kronecker's theorem, they must be roots of unity. It follows that μ is a direct sum of torsion characters. □

Corollary *Let V be a local system of $\mathbb{C}$-vector spaces on A. If V underlies a polarizable variation of Hodge structure with coefficients in $\mathbb{Z}$, all cohomology support loci of V are finite unions of arithmetic subvarieties.*

Proof By Lemma 2.1, we have $V \simeq \mathbb{C}_{\rho_1} \oplus \cdots \oplus \mathbb{C}_{\rho_n}$ for torsion points $\rho_1, \ldots, \rho_n \in \mathrm{Char}(A)$. All cohomology support loci of V are then obviously contained in the set

$$\{\rho_1^{-1}, \ldots, \rho_n^{-1}\},$$

and are therefore trivially finite unions of arithmetic subvarieties. □

2.2 Mixed Hodge modules with $\mathbb{Z}$-structure

We shall say that a mixed Hodge module has a $\mathbb{Z}$-structure if the underlying perverse sheaf, considered as a constructible complex with coefficients in $\mathbb{Q}$, can be obtained by extension of scalars from a constructible complex with coefficients in $\mathbb{Z}$. A typical example is the intermediate extension of a variation of Hodge structure with coefficients in $\mathbb{Z}$. To be precise, we make the following definition:

Definition 2.2 A $\mathbb{Z}$-*structure* on a complex of mixed Hodge modules

$$M \in \mathrm{D}^b \, \mathrm{MHM}(A)$$

is a constructible complex $E \in \mathrm{D}_c^b(\mathbb{Z}_A)$ such that $\operatorname{rat} M \simeq \mathbb{Q} \otimes_{\mathbb{Z}} E$.

The standard operations on complexes of mixed Hodge modules clearly respect $\mathbb{Z}$-structures. For instance, suppose that $M \in \mathrm{D}^b \, \mathrm{MHM}(A)$ has a $\mathbb{Z}$-structure, and that $f \colon A \to B$ is a homomorphism of abelian varieties; then $f_* M \in \mathrm{D}^b \, \mathrm{MHM}(B)$ again has a $\mathbb{Z}$-structure. The proof is straightforward:

$$\operatorname{rat}(f_* M) = \mathbf{R} f_*(\operatorname{rat} M) \simeq \mathbf{R} f_*(\mathbb{Q} \otimes_{\mathbb{Z}} E) \simeq \mathbb{Q} \otimes_{\mathbb{Z}} \mathbf{R} f_* E.$$

By [3, Section 3.3] and [9], there are two natural perverse t-structures on the category $\mathrm{D}_c^b(\mathbb{Z}_A)$; after tensoring by $\mathbb{Q}$, both become equal to the usual perverse t-structure on $\mathrm{D}_c^b(\mathbb{Q}_A)$. We shall use the one corresponding to the perversity p_+; concretely, it is defined as follows:

$$E \in {}^{p_+}\mathrm{D}_c^{\leq 0}(\mathbb{Z}_A) \Longleftrightarrow \begin{cases} \text{for any stratum } S, \text{ the local system} \\ \mathcal{H}^m i_S^* E \text{ is zero if } m > -\dim S + 1, \\ \text{and } \mathbb{Q} \otimes_{\mathbb{Z}} \mathcal{H}^{-\dim S + 1} i_S^* E = 0 \end{cases}$$

$$E \in {}^{p_+}\mathrm{D}_c^{\geq 0}(\mathbb{Z}_A) \Longleftrightarrow \begin{cases} \text{for any stratum } S, \text{ the local system} \\ \mathcal{H}^m i_S^! E \text{ is zero if } m < -\dim S, \\ \text{and } \mathcal{H}^{-\dim S} i_S^! E \text{ is torsion-free.} \end{cases}$$

We can use the resulting formalism of perverse sheaves with integer coefficients to show that $\mathbb{Z}$-structures are also preserved under taking cohomology.

Lemma 2.3 *If* $M \in \mathrm{D}^b \, \mathrm{MHM}(A)$ *admits a* $\mathbb{Z}$-*structure, then each cohomology module* $\mathcal{H}^k(M) \in \mathrm{MHM}(A)$ *also admits a* $\mathbb{Z}$-*structure.*

Proof Let ${}^{p_+}\mathcal{H}^k(E)$ denote the p_+-perverse cohomology sheaf in degree k of the constructible complex $E \in \mathrm{D}_c^b(\mathbb{Z}_A)$. With this notation, we have

$$\operatorname{rat} \mathcal{H}^k(M) = {}^p\mathcal{H}^k(\operatorname{rat} M) \simeq {}^p\mathcal{H}^k(\mathbb{Q} \otimes_{\mathbb{Z}} E) \simeq \mathbb{Q} \otimes_{\mathbb{Z}} {}^{p_+}\mathcal{H}^k(E),$$

which gives the desired $\mathbb{Z}$-structure on $\mathcal{H}^k(M)$. □

There is also a notion of intermediate extension for local systems with integer coefficients. If $i \colon X \hookrightarrow A$ is a subvariety of A, and $j \colon U \hookrightarrow X$ is a Zariski-open subset of the smooth locus of X, then for any local system V on U with coefficients in $\mathbb{Z}$, one has a canonically defined p_+-perverse sheaf

$$i_*(j_{!*} V[\dim X]) \in {}^{p_+}\mathrm{D}_c^{\leq 0}(\mathbb{Z}_A) \cap {}^{p_+}\mathrm{D}_c^{\geq 0}(\mathbb{Z}_A).$$

After tensoring by $\mathbb{Q}$, it becomes isomorphic to the usual intermediate extension of the local system $\mathbb{Q} \otimes_{\mathbb{Z}} V$. This has the following immediate consequence:

Lemma 2.4 *Let M be a polarizable Hodge module. Suppose that M is the intermediate extension of $\mathbb{Q} \otimes_{\mathbb{Z}} V$, where V is a polarizable variation of Hodge structure with coefficients in $\mathbb{Z}$. Then M admits a $\mathbb{Z}$-structure.*

Proof In fact, $E = i_*(j_{!*}V[\dim X])$ gives a $\mathbb{Z}$-structure on M. □

We conclude our discussion of $\mathbb{Z}$-structures by improving Lemma 2.1.

Lemma 2.5 *Let M be a mixed Hodge module with $\mathbb{Z}$-structure. Let $\rho \in \operatorname{Char}(A)$ be a character with the property that, for all $g \in \operatorname{Gal}(\mathbb{C}/\mathbb{Q})$, the local system $\mathbb{C}_{g\rho}[\dim A]$ is a subobject of $\mathbb{C} \otimes_{\mathbb{Q}} \operatorname{rat} M$. Then ρ is a torsion point of $\operatorname{Char}(A)$.*

Proof Let $j \colon U \hookrightarrow A$ be the maximal open subset with the property that $j^*M = V[\dim A]$ for a graded-polarizable variation of mixed Hodge structure V. Consequently, $j^*\mathbb{C}_\rho$ embeds into the complex variation of mixed Hodge structure $\mathbb{C} \otimes_{\mathbb{Q}} V$. Since the variation is graded-polarizable, and since $j_* \colon \pi_1(U) \to \pi_1(A)$ is surjective, it follows that ρ must be unitary [5, Section 1.12]. On the other hand, we have $\operatorname{rat} M \simeq \mathbb{Q} \otimes_{\mathbb{Z}} E$ for a constructible complex E with coefficients in $\mathbb{Z}$. Then $H^{-\dim A} j^*E$ is a local system with coefficients in $\mathbb{Z}$, and $j^*\mathbb{C}_\rho$ embeds into its complexification. The values of the character ρ are therefore algebraic integers of absolute value 1. We get the same conclusion for all their conjugates, by applying the argument above to the characters $g\rho$, for $g \in \operatorname{Gal}(\mathbb{C}/\mathbb{Q})$. Now Kronecker's theorem shows that ρ takes values in the roots of unity, and is therefore a torsion point of $\operatorname{Char}(A)$. □

2.3 The Galois action on the space of characters

In this section, we study the natural action of $\operatorname{Gal}(\mathbb{C}/\mathbb{Q})$ on the space of characters, and observe that the cohomology support loci of a regular holonomic $\mathscr{D}$-module with $\mathbb{Q}$-structure are stable under this action.

The space of characters $\operatorname{Char}(A)$ is an affine algebraic variety, and its coordinate ring is easy to describe. We have $A = V/\Lambda$, where V is a complex vector space of dimension g, and $\Lambda \subseteq V$ is a lattice of rank $2g$; note that Λ is canonically isomorphic to the fundamental group $\pi_1(A)$. For a field k, we denote by

$$k[\Lambda] = \bigoplus_{\lambda \in \Lambda} k e_\lambda$$

the group ring of Λ with coefficients in k; the product is determined by $e_\lambda e_\mu = e_{\lambda+\mu}$. As a complex algebraic variety, $\mathrm{Char}(A)$ is isomorphic to $\operatorname{Spec} \mathbb{C}[\Lambda]$; in particular, $\mathrm{Char}(A)$ can already be defined over $\operatorname{Spec} \mathbb{Q}$, and therefore carries in a natural way an action of the Galois group $\mathrm{Gal}(\mathbb{C}/\mathbb{Q})$.

Proposition 2.6 *Let $M \in \mathrm{D}^b \operatorname{MHM}(A)$ be a complex of mixed Hodge modules on a complex abelian variety A. Then all cohomology support loci of* $\operatorname{rat} M$ *are stable under the action of* $\mathrm{Gal}(\mathbb{C}/\mathbb{Q})$ *on* $\mathrm{Char}(A)$.

Proof The natural Λ-action on the group ring $k[\Lambda]$ gives rise to a local system of k-vector spaces $\mathcal{L}_{k[\Lambda]}$ on the abelian variety. The discussion in [17, Section 14] shows that the cohomology support loci of $\operatorname{rat} M$ are computed by the complex

$$\mathbf{R}p_*(\operatorname{rat} M \otimes_{\mathbb{Q}} \mathcal{L}_{\mathbb{C}[\Lambda]}) \in \mathrm{D}^b_{coh}(\mathbb{C}[\Lambda]),$$

where $p\colon A \to pt$ denotes the morphism to a point. In the case at hand,

$$\mathbf{R}p_*(\operatorname{rat} M \otimes_{\mathbb{Q}} \mathcal{L}_{\mathbb{C}[\Lambda]}) \simeq \mathbf{R}p_*(\operatorname{rat} M \otimes_{\mathbb{Q}} \mathcal{L}_{\mathbb{Q}[\Lambda]}) \otimes_{\mathbb{Q}[\Lambda]} \mathbb{C}[\Lambda]$$

is obtained by extension of scalars from a complex of $\mathbb{Q}[\Lambda]$-modules [17, Proposition 14.7]; this means that all cohomology support loci of M are defined over $\mathbb{Q}$, and therefore stable under the $\mathrm{Gal}(\mathbb{C}/\mathbb{Q})$-action on $\mathrm{Char}(A)$. □

2.4 The Fourier–Mukai transform

In this section, we review a few results about Fourier–Mukai transforms of holonomic $\mathscr{D}_A$-modules from [17]. The Fourier–Mukai transform, introduced by Laumon [11] and Rothstein [15], is an equivalence of categories

$$\mathrm{FM}_A\colon \mathrm{D}^b_{coh}(\mathscr{D}_A) \to \mathrm{D}^b_{coh}(\mathscr{O}_{A^\natural});$$

for a single coherent $\mathscr{D}_A$-module $\mathcal{M}$, it is defined by the formula

$$\mathrm{FM}_A(\mathcal{M}) = \mathbf{R}(p_2)_* \, \mathrm{DR}_{A\times A^\natural/A^\natural}(p_1^* \mathcal{M} \otimes (P^\natural, \nabla^\natural)),$$

where $(P^\natural, \nabla^\natural)$ is the universal line bundle with connection on $A \times A^\natural$.

The Fourier–Mukai transform satisfies several useful exchange formulas [11, Section 3.3]; recall that for $f\colon A \to B$ a homomorphism of abelian varieties,

$$f_+\colon \mathrm{D}^b_{coh}(\mathscr{D}_A) \to \mathrm{D}^b_{coh}(\mathscr{D}_B) \quad \text{and} \quad f^+\colon \mathrm{D}^b_{coh}(\mathscr{D}_B) \to \mathrm{D}^b_{coh}(\mathscr{D}_A)$$

denote, respectively, the direct image and the shifted inverse image functor, while $\mathbf{D}_A\colon \mathrm{D}^b_{coh}(\mathscr{D}_A) \to \mathrm{D}^b_{coh}(\mathscr{D}_A)^{opp}$ is the duality functor.

Theorem 2.7 *Let $\mathcal{M}, \mathcal{M}_1, \mathcal{M}_2 \in \mathrm{D}^b_{coh}(\mathscr{D}_A)$ and $\mathcal{N} \in \mathrm{D}^b_{coh}(\mathscr{D}_B)$.*

(a) For any homomorphism of abelian varieties $f: A \to B$, one has

$$\mathbf{L}(f^\natural)^* \mathrm{FM}_A(\mathcal{M}) \simeq \mathrm{FM}_B(f_+\mathcal{M}),$$
$$\mathbf{R}f^\natural_* \mathrm{FM}_B(\mathcal{N}) \simeq \mathrm{FM}_A(f^+\mathcal{N}).$$

(b) One has $\mathrm{FM}_A(\mathbf{D}_A\mathcal{M}) \simeq \langle -1_{A^\natural}\rangle^* \mathbf{R}\mathcal{H}om(\mathrm{FM}_A(\mathcal{M}), \mathscr{O}_{A^\natural})$.

(c) Let $m: A \times A \to A$ be the addition morphism. Then one has

$$\mathrm{FM}_A(m_+(\mathcal{M}_1 \boxtimes \mathcal{M}_2)) \simeq \mathrm{FM}_A(\mathcal{M}_1) \overset{\mathbf{L}}{\otimes}_{\mathscr{O}_{A^\natural}} \mathrm{FM}_A(\mathcal{M}_2).$$

Now let $\mathrm{D}^b_h(\mathscr{D}_A)$ be the full subcategory of $\mathrm{D}^b_{coh}(\mathscr{D}_A)$ consisting of cohomologically bounded and holonomic complexes. We already mentioned that the cohomology support loci $S^k_m(A, \mathcal{M})$ of a holonomic complex are finite unions of linear subvarieties; here is another result from [17] that will be used below.

Theorem 2.8 *Let $\mathcal{M}$ be a holonomic $\mathscr{D}_A$-module. Then $\mathrm{FM}_A(\mathcal{M}) \in \mathrm{D}^{\geq 0}_{coh}(\mathscr{O}_{A^\natural})$, and for any $\ell \geq 0$, one has $\operatorname{codim} \operatorname{Supp} \mathcal{H}^\ell \mathrm{FM}_A(\mathcal{M}) \geq 2\ell$.*

The precise relationship between the support of $\mathrm{FM}_A(\mathcal{M})$ and the cohomology support loci of $\mathcal{M}$ is given by the base-change theorem, which implies that, for every $n \in \mathbb{Z}$, one has

$$\bigcup_{k \geq n} \operatorname{Supp} \mathcal{H}^k \mathrm{FM}_A(\mathcal{M}) = \bigcup_{k \geq n} S^k_1(A, \mathcal{M}). \tag{1}$$

In particular, the support of the Fourier–Mukai transform $\mathrm{FM}_A(\mathcal{M})$ is equal to the union of all the cohomology support loci of $\mathcal{M}$.

3 Proof of the theorem

Consider a complex of mixed Hodge modules $M \in \mathrm{D}^b\mathrm{MHM}(A)$ that admits a $\mathbb{Z}$-structure, and denote by $\operatorname{rat} M \in \mathrm{D}^b_c(\mathbb{Q}_A)$ the underlying complex of constructible sheaves. To prove Theorem 1.4, we have to show that all cohomology support loci of M are complete unions of arithmetic subvarieties of $\mathrm{Char}(A)$.

3.1 Reduction steps

Our first task is to show that every $S^k_m(A, M)$ is a finite union of arithmetic subvarieties. The proof is by induction on the dimension of A; we may therefore assume that *the theorem is valid on every abelian variety of strictly smaller dimension.* This has several useful consequences.

Lemma 3.1 *Let $f: A \to B$ be a homomorphism from A to a lower-dimensional abelian variety B. Then every intersection*

$$S^k_m(A, M) \cap \operatorname{im} \operatorname{Char}(f)$$

is a finite union of arithmetic subvarieties.

Proof The complex $f_*M \in \mathrm{D}^b \operatorname{MHM}(B)$ again admits a $\mathbb{Z}$-structure. If we now tensor by points of $\operatorname{Char}(B)$ and take cohomology, we find that

$$\operatorname{Char}(f)^{-1} S^k_m(A, M) = S^k_m(B, f_*M).$$

By induction, we know that the right hand side is a finite union of arithmetic subvarieties of $\operatorname{Char}(B)$; consequently, the same is true for the intersection $S^k_m(A, M) \cap \operatorname{im} \operatorname{Char}(f)$. □

The inductive assumption lets us show that all positive-dimensional components of the cohomology support loci of M are arithmetic.

Lemma 3.2 *Let Z be an irreducible component of some $S^k_m(A, M)$. If $\dim Z \geq 1$, then Z is an arithmetic subvariety of* $\operatorname{Char}(A)$.

Proof Since Z is a linear subvariety, it suffices to prove that Z contains a torsion point. Now A is an abelian variety, and so we can find a surjective homomorphism $f: A \to B$ to an abelian variety of dimension $\dim A - \dim Z/2$, such that $Z \cap \operatorname{im} \operatorname{Char}(f)$ is a finite set of points. According to Lemma 3.1, the intersection is a finite union of arithmetic subvarieties, hence a finite set of torsion points. In particular, Z contains a torsion point, and is therefore an arithmetic subvariety of $\operatorname{Char}(A)$. □

Irreducible components that are already contained in a proper arithmetic subvariety of $\operatorname{Char}(A)$ can also be handled by induction.

Lemma 3.3 *Let Z be an irreducible component of $S^k_m(A, M)$. If Z is contained in a proper arithmetic subvariety of* $\operatorname{Char}(A)$, *then Z is itself an arithmetic subvariety.*

Proof It again suffices to show that Z contains a torsion point. For some $n \geq 1$, there is a torsion point of order n on the arithmetic subvariety that contains Z. After pushing forward by the multiplication-by-n morphism $\langle n \rangle : \operatorname{Char}(A) \to \operatorname{Char}(A)$, which corresponds to replacing M by its inverse image $\langle n \rangle^* M$ under $\langle n \rangle : A \to A$, we can assume that $Z \subseteq \operatorname{im} \operatorname{Char}(f)$, where $f: A \to B$ is a morphism to a lower-dimensional abelian variety. The assertion now follows from Lemma 3.1. □

The following result allows us to avoid cohomology in degree 0:

Lemma 3.4 *Let $M \in \mathrm{MHM}(A)$, and let Z be an irreducible component of some cohomology support locus of M. If $Z \neq \mathrm{Char}(A)$, then Z is contained in $S^k_m(A, M)$ for some $k \neq 0$ and some $m \geq 1$.*

Proof This follows easily from the fact that the Euler characteristic

$$\chi(A, \mathrm{rat}\, M \otimes_{\mathbb{Q}} \mathbb{C}_\rho) = \sum_{k \in \mathbb{Z}} (-1)^k \dim H^k(A, \mathrm{rat}\, M \otimes_{\mathbb{Q}} \mathbb{C}_\rho)$$

is independent of the point $\rho \in \mathrm{Char}(A)$. □

3.2 Torsion points on components

Let Z be an irreducible component of some cohomology support locus of M. If $\dim Z \geq 1$, Lemma 3.2 shows that Z is an arithmetic subvariety; we may therefore assume that $Z = \{\rho\}$ consists of a single point. We have to prove that ρ has finite order in $\mathrm{Char}(A)$. There are three steps.

Step 1 We begin by reducing the problem to the case where M is a single mixed Hodge module. Each of the individual cohomology modules $\mathcal{H}^q(M) \in \mathrm{MHM}(A)$ also admits a $\mathbb{Z}$-structure (by Lemma 2.3); we know by induction that all positive-dimensional irreducible components of its cohomology support loci are arithmetic subvarieties. If ρ is contained in such a component, Lemma 3.3 proves that ρ is a torsion point; we may therefore assume that whenever there is some $p \neq 0$ such that $H^p(A, \mathrm{rat}\, \mathcal{H}^q(M) \otimes_{\mathbb{Q}} \mathbb{C}_\rho)$ is nontrivial, ρ is an isolated point of the corresponding cohomology support locus. To exploit this fact, let us consider the spectral sequence

$$E_2^{p,q} = H^p(A, \mathrm{rat}\, \mathcal{H}^q(M) \otimes_{\mathbb{Q}} \mathbb{C}_\rho) \Longrightarrow H^{p+q}(A, \mathrm{rat}\, M \otimes_{\mathbb{Q}} \mathbb{C}_\rho).$$

If $E_2^{p,q} \neq 0$ for some $p \neq 0$, then ρ must be an isolated point in some cohomology support locus of $\mathcal{H}^q(M)$; in that case, we can replace M by the single mixed Hodge module $\mathcal{H}^q(M)$. If $E_2^{p,q} = 0$ for every $p \neq 0$, then the spectral sequence degenerates and

$$H^k(A, \mathrm{rat}\, M \otimes_{\mathbb{Q}} \mathbb{C}_\rho) \simeq H^0(A, \mathrm{rat}\, \mathcal{H}^k(M) \otimes_{\mathbb{Q}} \mathbb{C}_\rho).$$

But $\rho \in S^k_m(A, M)$ is an isolated point, and so by semicontinuity, it must also be an isolated point in $S^0_m(A, \mathcal{H}^k(M))$; again, we can replace M by the single mixed Hodge module $\mathcal{H}^k(M)$.

Step 2 We now construct *another* mixed Hodge module with $\mathbb{Z}$-structure, such that the union of all cohomology support loci contains ρ but is not equal to $\operatorname{Char}(A)$. We can then use the inductive hypothesis to reduce the problem to the case where $\mathbb{C}_{\rho^{-1}}[\dim A]$ is a direct factor of $\mathbb{C} \otimes_{\mathbb{Q}} \operatorname{rat} M$; because of Lemma 2.5, this will be sufficient to conclude that the character ρ has finite order.

The idea for the construction comes from a recent article by Krämer and Weissauer [10, Section 13]. Since $M \in \operatorname{MHM}(A)$ is a single mixed Hodge module, we can use Lemma 3.4 to arrange that $\rho \in S_m^k(A, M)$ is an isolated point for some $k \neq 0$ and some $m \geq 1$; to simplify the argument, we shall take the absolute value $|k|$ to be as large as possible. Now let $A \times \cdots \times A$ denote the d-fold product of A with itself, and let $m \colon A \times \cdots \times A \to A$ be the addition morphism. The d-fold exterior product $M \boxtimes \cdots \boxtimes M$ is a mixed Hodge module on $A \times \cdots \times A$, and clearly inherits a $\mathbb{Z}$-structure from M. Setting

$$M_d = m_*(M \boxtimes \cdots \boxtimes M) \in \mathrm{D}^b \operatorname{MHM}(A),$$

it is easy to see from our choice of k that

$$H^{kd}(A, \operatorname{rat} M_d \otimes_{\mathbb{Q}} \mathbb{C}_\rho) \simeq H^k(A, \operatorname{rat} M \otimes_{\mathbb{Q}} \mathbb{C}_\rho)^{\otimes d}.$$

The right hand side is nonzero, and so $\rho \in S_{md}^{kd}(A, M_d)$. By a similar spectral sequence argument as above, we must have

$$H^p(A, \operatorname{rat} \mathcal{H}^q(M_d) \otimes_{\mathbb{Q}} \mathbb{C}_\rho) \neq 0$$

for some $p, q \in \mathbb{Z}$ with $p + q = kd$ and $-g \leq p \leq g$. If we take $d > g$, this forces $q \neq 0$. In other words, we can find $q \neq 0$ such that ρ lies in some cohomology support locus of the mixed Hodge module $\mathcal{H}^q(M_d)$.

Lemma 3.5 *If $q \neq 0$, all nontrivial cohomology support loci of $\mathcal{H}^q(M_d)$ are properly contained in* $\operatorname{Char}(A)$.

Proof It suffices to prove this for the underlying regular holonomic $\mathscr{D}$-module $\mathcal{H}^q m_+(\mathcal{M} \boxtimes \cdots \boxtimes \mathcal{M})$. The properties of the Fourier–Mukai transform in Theorem 2.7 imply that

$$\operatorname{FM}_A(m_+(\mathcal{M} \boxtimes \cdots \boxtimes \mathcal{M})) \simeq \operatorname{FM}_A(\mathcal{M}) \overset{\mathbf{L}}{\otimes}_{\mathscr{O}_{A^\natural}} \cdots \overset{\mathbf{L}}{\otimes}_{\mathscr{O}_{A^\natural}} \operatorname{FM}_A(\mathcal{M}),$$

and all cohomology sheaves of this complex, except possibly in degree 0, are torsion sheaves (by Theorem 2.8). In the spectral sequence

$$E_2^{p,q} = \mathcal{H}^p \operatorname{FM}_A(\mathcal{H}^q m_+(\mathcal{M} \boxtimes \cdots \boxtimes \mathcal{M})) \Longrightarrow \mathcal{H}^{p+q} \operatorname{FM}_A(m_+(\mathcal{M} \boxtimes \cdots \boxtimes \mathcal{M})),$$

the sheaf $E_2^{p,q}$ is zero when $p < 0$, and torsion when $p > 0$, for the same reason. It follows that $E_2^{0,q}$ is also a torsion sheaf for $q \neq 0$, which proves the assertion. □

Step 3 Now we can easily finish the proof. The mixed Hodge module $\mathcal{H}^q(M_d)$ again admits a $\mathbb{Z}$-structure by Lemma 2.3; by induction, all positive-dimensional irreducible components of its cohomology support loci are proper arithmetic subvarieties of Char(A). If ρ is contained in one of them, we are done by Lemma 3.3. After replacing M by $\mathcal{H}^q(M_d)$, we can therefore assume that, whenever $H^k(A, \mathrm{rat}\, M \otimes_{\mathbb{Q}} \mathbb{C}_\rho)$ is nontrivial, ρ is an isolated point of the corresponding cohomology support locus. Note that we now have this for all values of $k \in \mathbb{Z}$, including $k = 0$.

Let $\mathcal{M}$ denote the regular holonomic $\mathscr{D}$-module underlying the mixed Hodge module M. If $(L, \nabla) \in A^\natural$ is the flat line bundle corresponding to our character ρ, the assumptions on M guarantee that (L, ∇) is an isolated point in the support of $\mathrm{FM}_A(\mathcal{M})$. This means that, in the derived category, $\mathrm{FM}_A(\mathcal{M})$ has a direct factor supported on the point (L, ∇). But the Fourier–Mukai transform is an equivalence of categories, and so $\mathcal{M} \simeq \mathcal{M}' \oplus \mathcal{M}''$, where $\mathcal{M}'$ is a regular holonomic $\mathscr{D}$-module whose Fourier–Mukai transform is supported on (L, ∇). It is well known that $\mathcal{M}'$ is the tensor product of $(L, \nabla)^{-1}$ and a unipotent flat vector bundle; in particular, $\mathcal{M}$ contains a sub-$\mathscr{D}$-module isomorphic to $(L, \nabla)^{-1}$. Equivalently, $\mathbb{C} \otimes_{\mathbb{Q}} \mathrm{rat}\, M$ has a subobject isomorphic to $\mathbb{C}_{\rho^{-1}}[\dim A]$. Because the cohomology support loci of M are stable under the Gal($\mathbb{C}/\mathbb{Q}$)-action on Char(A) (by Proposition 2.6), the same is true for every conjugate $g\rho$, where $g \in \mathrm{Gal}(\mathbb{C}/\mathbb{Q})$. We can now apply Lemma 2.5 to show that ρ must be a torsion point.

This concludes the proof that all cohomology support loci of M are finite unions of arithmetic subvarieties of Char(A).

3.3 Completeness of the set of components

We finish the proof of Theorem 1.4 by showing that each cohomology support locus of M is a *complete* union of arithmetic subvarieties of Char(A). The argument is based on the following simple criterion for completeness:

Lemma 3.6 *A finite union of arithmetic subvarieties of* Char(A) *is complete if and only if it is stable under the action of* Gal($\mathbb{C}/\mathbb{Q}$).

Proof For a point $\tau \in$ Char(A) of order n, the orbit under the group $G = \mathrm{Gal}(\mathbb{C}/\mathbb{Q})$ consists precisely of the characters τ^k with $\gcd(k, n) = 1$; consequently, a complete collection of arithmetic subvarieties is stable under the

G-action. To prove the converse, let Z be a finite union of arithmetic subvarieties stable under the action of G. Let τL be one of its components; here L is a linear subvariety and $\tau \in \mathrm{Char}(A)$ a point of order n, say. Let p be any prime number with $\gcd(n, p) = 1$, and denote by $L[p]$ the set of points of order p. For any character $\rho \in L[p]$, we have $\mathrm{ord}(\tau\rho) = np$; the G-orbit of the set $\tau L[p]$ is therefore equal to

$$(G\tau) \cdot L[p] = \{ \tau^k \mid \gcd(k, n) = 1 \} \cdot L[p].$$

Because the union of all the finite subsets $L[p]$ with $\gcd(n, p) = 1$ is dense in the linear subvariety L, it follows that

$$\{ \tau^k \mid \gcd(k, n) = 1 \} \cdot L \subseteq Z;$$

this proves that Z is complete. □

Theorem 3.7 *Let $M \in \mathrm{D}^b \mathrm{MHM}(A)$ be a complex of mixed Hodge modules that admits a $\mathbb{Z}$-structure. Then all cohomology support loci of M are complete collections of arithmetic subvarieties of* Char(A).

Proof We already know that each $S_m^k(A, M)$ is a finite union of arithmetic subvarieties of Char(A). By Proposition 2.6, it is stable under the Gal($\mathbb{C}/\mathbb{Q}$)-action on Char(A); we can now apply Lemma 3.6 to conclude that $S_m^k(A, M)$ is complete. □

Acknowledgments This work was supported by the World Premier International Research Center Initiative (WPI Initiative), MEXT, Japan, and by NSF grant DMS-1331641. I thank Nero Budur, François Charles, and Mihnea Popa for several useful conversations about the topic of this paper, and Botong Wang for suggesting to work with arbitrary complexes of mixed Hodge modules. I am indebted to an anonymous referee for suggestions about the exposition. I also take this opportunity to thank Rob Lazarsfeld for the great influence that his work with Mark Green and Lawrence Ein has had on my mathematical interests.

References

[1] Arapura, D. 1992. Higgs line bundles, Green–Lazarsfeld sets, and maps of Kähler manifolds to curves. *Bull. Amer. Math. Soc. (N.S.)*, **26**(2), 310–314.

[2] Beauville, A. 1992. Annulation du H^1 pour les fibrés en droites plats. In *Complex Algebraic Varieties (Bayreuth, 1990)*, pp. 1–15. Lecture Notes in Mathematics, Vol. 1507. Berlin: Springer-Verlag.

[3] Beĭlinson, A. A., Bernstein, J., and Deligne, P. 1982. Faisceaux pervers. In *Analysis and Topology on Singular Spaces, I (Luminy, 1981)* pp. 5–171. Astérisque, Vol. 100. Paris: Société Mathématique de France.

[4] Deligne, P. 1971. Théorie de Hodge. II. *Inst. Hautes Études Sci. Publ. Math.*, **40**, 5–57.

[5] Deligne, P. 1987. Un théorème de finitude pour la monodromie. *Discrete Groups in Geometry and Analysis (New Haven, CT, 1984)*, pp. 1–19.. Progress in Mathematics, Vol. 67. Boston, MA: Birkhäuser Boston.

[6] Green, M. and Lazarsfeld, R. 1987. Deformation theory, generic vanishing theorems, and some conjectures of Enriques, Catanese and Beauville. *Invent. Math.*, **90**(2), 389–407.

[7] Green, M. and Lazarsfeld, R. 1991. Higher obstructions to deforming cohomology groups of line bundles. *J. Amer. Math. Soc.*, **4**(1), 87–103.

[8] Hacon, C. D. 2004. A derived category approach to generic vanishing. *J. Reine Angew. Math.*, **575**, 173–187.

[9] Juteau, D. 2009. Decomposition numbers for perverse sheaves. *Ann. Inst. Fourier (Grenoble)*, **59**(3), 1177–1229.

[10] Krämer, T. and Weissauer, R. 2011. Vanishing theorems for constructible sheaves on abelian varieties. `arXiv:1111.4947`.

[11] Laumon, G. 1996. Transformation de Fourier généralisée. `arXiv:alg-geom/9603004`.

[12] Pareschi, G. 2012. Basic results on irregular varieties via Fourier–Mukai methods. In *Current Developments in Algebraic Geometry*, pp. 379–403. Mathematical Sciences Research Institute Publications, Vol. 59. Cambridge: Cambridge University Press.

[13] Pink, R. and Roessler, D. 2004. A conjecture of Beauville and Catanese revisited. *Math. Ann.*, **330**(2), 293–308.

[14] Popa, M. and Schnell, C. 2013. Generic vanishing theory via mixed Hodge modules. *Forum of Mathematics, Sigma*, **1**(e1), 1–60.

[15] Rothstein, M. 1996. Sheaves with connection on abelian varieties. *Duke Math. J.*, **84**(3), 565–598.

[16] Saito, M. 1990. Mixed Hodge modules. *Publ. Res. Inst. Math. Sci.*, **26**(2), 221–333.

[17] Schnell, C. 2014. Holonomic $\mathscr{D}$-modules on abelian varieties. *Inst. Hautes Études Sci. Publ. Math.*. `10.1007/s10240-014-0061-x`.

[18] Simpson, C. T. 1990. Transcendental aspects of the Riemann–Hilbert correspondence. *Illinois J. Math.*, **34**(2), 368–391.

[19] Simpson, C. T. 1993. Subspaces of moduli spaces of rank one local systems. *Ann. Sci. École Norm. Sup. (4)*, **26**(3), 361–401.

[20] Wang, B. 2013. Torsion points on the cohomology jump loci of compact Kähler manifolds. `arXiv:1312.6619`.

20

Rational equivalence of 0-cycles on $K3$ surfaces and conjectures of Huybrechts and O'Grady

C. Voisin
CNRS and École Polytechnique

Abstract

We give a new interpretation of O'Grady's filtration on the CH_0 group of a $K3$ surface X. In particular, we get a new characterization of the canonical 0-cycles kc_X: given $k \geq 0$, kc_X is the only 0-cycle of degree k on X whose orbit under rational equivalence is of dimension k. Using this, we extend the results of Huybrechts and O'Grady concerning Chern classes of simple vector bundles on $K3$ surfaces.

1 Introduction

Let X be a projective $K3$ surface. In [1], Beauville and the author proved that X carries a canonical 0-cycle c_X of degree 1, which is the class in $CH_0(X)$ of any point of X lying on a (possibly singular) rational curve on X. This cycle has the property that for any divisors D, D' on X, we have

$$D \cdot D' = \deg(D \cdot D')\, c_X \text{ in } CH_0(X).$$

In recent works of Huybrechts [5] and O'Grady [11], this 0-cycle appeared to have other characterizations. Huybrechts proves, for example, the following result (which is proved in [5] to have much more general consequences on spherical objects and autoequivalences of the derived category of X):

Theorem 1 (Huybrechts [5]) *Let X be a projective complex $K3$ surface. Let F be a simple vector bundle on X such that $H^1(\mathrm{End}\, F) = 0$ (such an F is*

From *Recent Advances in Algebraic Geometry*, edited by Christopher Hacon, Mircea Mustaţă and Mihnea Popa

called spherical in [5]). Then $c_2(F)$ is proportional to c_X in $CH_0(X)$ if one of the following conditions holds:

(1) The Picard number of X is at least 2.
(2) The Picard group of X is $\mathbf{Z}H$ and the determinant of F is equal to kH with $k = \pm 1 \bmod r := \mathrm{rank}\, F$.

This result is extended in the following way by O'Grady. In [11], he introduces the following increasing filtration of $CH_0(X)$:

$$S_0(X) \subset S_1(X) \subset \cdots \subset S_d(X) \subset \cdots \subset CH_0(X),$$

where $S_d(X)$ is defined as the set of classes of cycles of the form $z + z'$, with z effective of degree d and z' a multiple of c_X. It is also convenient to introduce $S_d^k(X)$, which will by definition be the set of degree-k 0-cycles on X which lie in $S_d(X)$. Thus by definition

$$S_d^k(X) = \{z \in CH_0(X),\ z = z' + (k - d)c_X\},$$

where z' is effective of degree d.

Consider a torsion-free or more generally a pure sheaf $\mathcal{F}$ on X which is H-stable with respect to a polarization H. Let $2d(v_{\mathcal{F}})$ be the dimension of the space of deformations of $\mathcal{F}$, where $v_{\mathcal{F}}$ is the Mukai vector of $\mathcal{F}$ (cf. [6]). We recall that $v_{\mathcal{F}} \in H^*(X, \mathbf{Z})$ is the triple

$$(r, l, s) \in H^0(X, \mathbf{Z}) \oplus H^2(X, \mathbf{Z}) \oplus H^4(X, \mathbf{Z}),$$

with $r = \mathrm{rank}\, \mathcal{F}$, $l = c_1^{top}(\det \mathcal{F})$, and $s \in H^4(X, \mathbf{Z})$ is defined as

$$v_{\mathcal{F}} = ch(\mathcal{F}) \sqrt{td(X)}. \tag{1}$$

With this notation we get, by the Riemann–Roch formula, that

$$\sum_i (-1)^i \dim Ext^i(\mathcal{F}, \mathcal{F}) = < v_{\mathcal{F}}, v_{\mathcal{F}}^* > = 2rs - l^2 = 2 - 2d(v_{\mathcal{F}}),$$

where $<,>$ is the intersection pairing on $H^*(X, \mathbf{Z})$, and $v^* = (r, -l, s)$ is the Mukai vector of $\mathcal{F}^*$ (if $\mathcal{F}$ is locally free).

In particular $d(v_{\mathcal{F}}) = 0$ if $\mathcal{F}$ satisfies $\mathrm{End}\, \mathcal{F} = \mathbb{C}$ and $\mathrm{Ext}^1(\mathcal{F}, \mathcal{F}) = 0$, so that $\mathcal{F}$ is spherical as in Huybrechts' theorem. Noticing that $S_0(X) = \mathbf{Z}c_X$, one can then rephrase Huybrechts' statement by saying that if $\mathcal{F}$ satisfies $\mathrm{End}\,(\mathcal{F}) = \mathbb{C}$, $d(v_{\mathcal{F}}) = 0$, then $c_2(\mathcal{F}) \in S_0(X)$, assuming the Picard number of X is at least 2.

O'Grady then extends Huybrechts' results as follows:

Theorem 2 (O'Grady [11]) *Assuming $\mathcal{F}$ is H-stable, one has $c_2(\mathcal{F}) \in S_{d(v_{\mathcal{F}})}(X)$, $v_{\mathcal{F}} = (r, l, s)$, if furthermore one of the following conditions holds:*

(1) $l = H$, l is primitive and $s \geq 0$.
(2) The Picard number of X is at least 2, r is coprime to the divisibility of l, and H is v-generic.
(3) $r \leq 2$ and moreover H is v-generic if $r = 2$.

In fact, O'Grady's result is stronger, as he also shows that $S^k_{d(v)}(X)$, $k = \deg c_2(v)$, is equal to the set of classes $c_2(\mathcal{G})$ with $\mathcal{G}$ a deformation of $\mathcal{F}$. O'Grady indeed proves, by a nice argument involving the rank of the Mukai holomorphic 2-form on the moduli space of deformations of $\mathcal{F}$, the following result:

Proposition 3 (O'Grady [11, Proposition 1.3]) *If there is a H-stable torsion-free sheaf $\mathcal{F}$ with $v = v(\mathcal{F})$, and the conclusion of Theorem 2 holds for the deformations of $\mathcal{F}$, then*

$$\{c_2(\mathcal{G}),\ \mathcal{G} \in \overline{\mathcal{M}^{st}}(X, H, v)\ \} = S^k_{d(v)}(X),\ k = \deg c_2(\mathcal{F}).$$

In this statement, $\overline{\mathcal{M}^{st}}(X, H, v)$ is any smooth completion of the moduli space of H-stable sheaves with Mukai vector v.

Our results in this paper are of two kinds: First of all we provide another description of $S^k_d(X)$ for any $d \geq 0$, $k \geq d$. In order to state this result, let us introduce the following notation: Given an integer $k \geq 0$, and a cycle $z \in CH_0(X)$ of degree k, the subset O_z of $X^{(k)}$ consisting of effective cycles $z' \in X^{(k)}$ which are rationally equivalent to z is a countable union of closed algebraic subsets of $X^{(k)}$ (see [13, Lemma 10.7]). This is the "effective orbit" of z under rational equivalence, and the analogue of $|D|$ for a divisor $D \in CH^1(W)$ on any variety W. We define $\dim O_z$ as the supremum of the dimensions of the components of O_z. This is the analogue of $r(D) = \dim |D|$ for a divisor $D \in CH^1(W)$ on any variety W. We will prove the following:

Theorem 4 *Let X be a projective K3 surface. Let $k \geq d \geq 0$. We have the following characterization of $S^k_d(X)$:*

$$S^k_d(X) = \{z \in CH_0(X),\ O_z \text{ non-empty},\ \dim O_z \geq k - d\}.$$

Remark 1 The inclusion $S^k_d(X) \subset \{z \in CH_0(X),\ O_z$ non-empty, $\dim O_z \geq k - d\}$ is easy since the cycle $(k - d)c_X$ has its orbit of dimension $\geq k - d$ (e.g., $C^{(k-d)} \subset X^{(k-d)}$, for any rational curve $C \subset X$, is contained in the orbit of $(k - d)c_X$). Hence any cycle of the form $z + (k - d)c_X$ with z effective of degree d has an orbit of dimension $\geq k - d$.

A particular case of the theorem above is the case where $d(v) = 0$. By definition, $S_0(X)$ is the subgroup $\mathbf{Z}c_X \subset CH_0(X)$. We thus have:

Corollary 5 *For $k > 0$, the cycle kc_X is the unique 0-cycle z of degree k on X such that* $\dim O_z \geq k$.

Remark 2 We have in fact $\dim O_z = k$, $z = kc_X$ since by Mumford's theorem [10], any component L of O_z is Lagrangian for the holomorphic symplectic form on $S^{(k)}_{reg}$, hence of dimension $\leq k$ if L intersects $S^{(k)}_{reg}$. If L is contained in the singular locus of $S^{(k)}$, we can consider the minimal multiplicity-stratum of $S^{(k)}$ containing L, which is determined by the multiplicities n_i of the general cycle $\sum_i n_i x_i$, x_i distinct, parametrized by L and apply the same argument.

Remark 3 We will give in Section 2 an alternative proof of Corollary 5, using the remark above, and the fact that any Lagrangian subvariety of X^k intersects a product $D_1 \times \ldots \times D_k$ of ample divisors on X.

Our main application of Theorem 4 is the following result, which generalizes O'Grady's and Huybrechts' Theorems 2, 1 in the case of simple vector bundles (instead of semistable torsion-free sheaves). We do not need any of the assumptions appearing in Theorems 2, 1, but our results, unlike those of O'Grady, are restricted to the locally free case.

Theorem 6 *Let X be a projective $K3$ surface. Let F be a simple vector bundle on X with Mukai vector $v = v(F)$. Then*

$$c_2(F) \in S_{d(v)}(X).$$

A particular case of this statement is the case where $d = 0$. The corollary below proves Huybrechts' Theorem 1 without any assumption on the Picard group of the $K3$ surface or on the determinant of F. It is conjectured in [5].

Corollary 7 *Let F be a simple rigid vector bundle on a $K3$ surface. Then the class $c_2(F)$ in $CH_0(X)$ is a multiple of c_X.*

We also deduce the following corollary, in the same spirit (and with essentially the same proof) as Proposition 3:

Corollary 8 *Let $v \in H^*(X, \mathbf{Z})$ be a Mukai vector, with $k = c_2(v)$. Assume there exists a simple vector bundle F on X with Mukai vector v. Then*

$$S^k_d(X) = \{c_2(G),\ G \text{ a simple vector bundle on } X,\ v_G = v\},$$

where $k = c_2(v) := c_2^{\text{top}}(F) = \deg c_2(F)$.

These results answer, for simple vector bundles on $K3$ surfaces, questions asked by O'Grady (see [11, Section 5]) for simple sheaves.

The paper is organized as follows: in Section 2, we prove Theorem 4. We also show a variant concerning a family of subschemes (rather than 0-cycles) of given length in a constant rational equivalence class. In Section 3, Theorem 6 and Corollary 8 are proved.

2 An alternative description of O'Grady's filtration

This section is devoted to the proof of Theorem 4, which we state in the following form:

Theorem 9 *Let $k \geq d$ and let $Z \subset X^{(k)}$ be a Zariski closed irreducible algebraic subset of dimension $k - d$. Assume that all cycles of X parameterized by Z are rationally equivalent in X. Then the class of these cycles belongs to $S^k_d(X)$.*

We will need for the proof the following simple lemma, which already appears in [12]:

Lemma 10 *Let X be a projective $K3$ surface and let $C \subset S$ be a (possibly singular) curve such that all points of C are rationally equivalent in X. Then any point of C is rationally equivalent to c_X.*

Proof Let L be an ample line bundle on X. Then $c_1(L)_{|C}$ is a 0-cycle on C and our assumptions imply that $j_*(c_1(L)_{|C}) = \deg(c_1(L)_{|C})\, c$, for any point c of C.

Furthermore, we have

$$j_*(c_1(L)_{|C}) = c_1(L) \cdot C \text{ in } CH_0(X)$$

and thus, by [1], $j_*(c_1(L)_{|C}) = \deg(c_1(L)_{|C})\, c_X$ in $CH_0(X)$. Hence we have

$$\deg(c_1(L)_{|C})\, c = \deg(c_1(L)_{|C})\, c_X \text{ in } CH_0(X).$$

This concludes the proof, since c is arbitrary, $\deg(c_1(L)_{|C}) \neq 0$, and $CH_0(X)$ has no torsion. □

Lemma 11 *The union of curves C satisfying the property stated in Lemma 10 is Zariski dense in X.*

Proof The 0-cycle c_X is represented by any point lying on a (singular) rational curve $C \subset X$ (see [1]), so the result is clear if one knows that there are infinitely many distinct rational curves contained in X. This result is to our knowledge known only for general $K3$ surfaces but not for all $K3$ surfaces (see however [2] for results in this direction). In any case, we can use the following argument which already appears in [8]. By [9], there is a 1-parameter family of (singular) elliptic curves E_t on X. Let C be a rational curve on X which meets the fibers E_t. For any integer N, and any point t, consider the points $y \in \widetilde{E}_t$ (the desingularization of E_t), which are rationally equivalent in $\widetilde{E}_t$ to the sum of a point $x_t \in E_t \cap C$ (hence rationally equivalent to c_X) and an N-torsion 0-cycle on $\widetilde{E}_t$.

As $CH_0(X)$ has no torsion, the images y_t of these points in X are all rationally equivalent to c_X in X. Their images are clearly parameterized for N large enough by a (maybe reducible) curve $C_N \subset X$. Finally, the union over all N of the points y_t above is Zariski dense in each $\widetilde{E}_t$, hence the union of the curves C_N is Zariski dense in X. □

Proof of Theorem 9 The proof is by induction on k, the case $k = 1, d = 0$ being Lemma 10 (the case $k = 1, d = 1$ is trivial). Let Z' be an irreducible component of the inverse image of Z in X^k. Let $p: Z' \to X$ be the first projection. We distinguish two cases and note that they exhaust all possibilities, up to replacing Z' by another component Z'' deduced from Z' by letting the symmetric group $\mathfrak{S}_k$ act.

Case 1. The morphism $p: Z' \to X$ is dominant. For a curve $C \subset X$ parameterizing points rationally equivalent to c_X, consider the hypersurface

$$Z'_C := p^{-1}(C) \subset Z'.$$

Let $q: Z' \to X^{k-1}$ be the projection on the product of the $k-1$ last factors. Assume first that $\dim q(Z'_C) = \dim Z'_C = k - d - 1$. Note that all cycles of X parameterized by $q(Z'_C)$ are rationally equivalent in X. Indeed, an element z of Z'_C is of the form (c, z') with $c \in C$ so that $c = c_X$ in $CH_0(X)$. So, the rational equivalence class of z' is equal to $z - c_X$ and is independent of $z' \in Z'_C$. Thus the induction assumption applies and the cycles of degree $k-1$ parameterized by $\operatorname{Im} q$ belong to $S^{k-1}_d(X)$. It follows in turn that the classes of the cycles parameterized by Z' (or Z) belong to $S^k_d(X)$. Indeed, as just mentioned above, a 0-cycle z parameterized by Z' is rationally equivalent to $z = c_X + z'$ where $z' \in S^{k-1}_d(X)$, so z' is rationally equivalent to $(k-d-1)c_X + z''$, $z'' \in X^{(d)}$. Hence z is rationally equivalent in X to $(k-d)c_X + z''$, for some $z'' \in X^{(d)}$. Thus $z \in S^k_d(X)$.

Assume to the contrary that $\dim q(Z'_C) < \dim Z'_C = k-d-1$ for any curve C as above. We use now the fact (see Lemma 11) that these curves C are Zariski dense in X. We can thus assume that there is a point $x \in Z'_C$ which is generic in Z', so that both Z' and Z'_C are smooth at x, of respective dimensions $k-d$ and $k-d-1$. The fact that $\dim q(Z'_C) < k-d-1$ implies that q is not of maximal rank $k-d$ at x and as x is generic in Z', we conclude that q is of rank $< k-d$ everywhere on Z'_{reg}, so that $\dim \operatorname{Im} q \leq k-d-1$.

Now recall that all 0-cycles parameterized by Z' are rationally equivalent. It follows that for any fiber F of q, all points in $p(F)$ are rationally equivalent in X. This implies that all these points are rationally equivalent to c_X by Lemma 10. This contradicts the fact that p is surjective.

Case 2. None of the projections pr_i, $i = 1, \ldots, k$, from X^k to its factors restricts to a dominant map $p_i : Z' \to X$. Let $C_i := \operatorname{Im} p_i \subset X$ if $\operatorname{Im} pr_i$ is a curve, and any curve containing $\operatorname{Im} p_i$ if $\operatorname{Im} p_i$ is a point. Thus Z' is contained in $C_1 \times \cdots \times C_k$.

Let C be a non-necessarily irreducible ample curve such that all points in C are rationally equivalent to c_X. Observe that the line bundle $pr_1^*\mathcal{O}_X(C) \otimes \cdots \otimes pr_k^*\mathcal{O}_X(C)$ on X^k has its restriction to $C_1 \times \cdots \times C_k$ ample and that its $(k-d)$th self-intersection on $C_1 \times \ldots \times C_k$ is a complete intersection of ample divisors and is equal to

$$W := (k-d)! \sum_{i_1 < \ldots < i_{k-d}} p_{i_1}^*\mathcal{O}_{C_1}(C) \cdot \ldots \cdot p_{i_{k-d}}^*\mathcal{O}_{C_{k-d}}(C) \tag{2}$$

in $CH^{k-d}(C_1 \times \ldots \times C_k)$, where the p_i are the projections from $\prod_i C_i$ to its factors.

The cycle W of (2) is also the restriction to $C_1 \times \ldots \times C_k$ of the effective cycle

$$W' := (k-d)! \sum_{i_1 < \ldots < i_{k-d}} pr_{i_1}^* C \cdot \ldots \cdot pr_{i_{k-d}}^* C. \tag{3}$$

As the $(k-d)$-dimensional subvariety Z' of $C_1 \times \cdots \times C_k$ has a nonzero intersection with W, it follows that the intersection number of Z' with W' is nonzero in X^k, hence that

$$Z' \cap pr_{i_1}^* C \cdot \ldots \cdot pr_{i_{k-d}}^* C \neq \emptyset$$

for some choice of indices $i_1 < \cdots < i_{k-d}$. This means that there exists a cycle in Z which is of the form

$$z = z' + z''$$

with $z' \in C^{(k-d)}$ and $z'' \in X^{(d)}$. As z' is supported on C, it is equal to $(k-d)c_X$ in $CH_0(X)$ and we conclude that $z \in S_d^k(X)$. □

Let us now prove the following variant of Theorem 9. Instead of a family of 0-cycles (that is, elements of $X^{(k)}$), we now consider families of 0-dimensional *subschemes* (that is, elements of $X^{[k]}$):

Variant 12 *Let $k \geq d$ and let $Z \subset X^{[k]}$ be a Zariski closed irreducible algebraic subset of dimension $k-d$. Assume that all cycles of X parameterized by Z are rationally equivalent in X. Then the class of these cycles belongs to $S_d^k(X)$.*

Proof Let $z \in Z$ be a general point. The cycle $c(z)$ of z, where $c\colon X^{[k]} \to X^{(k)}$ is the Hilbert–Chow morphism, is of the form $\sum_i k_i x_i$, with $\sum_i k_i = k$, where x_i are k' distinct points of X. We have of course

$$k' = k - \sum_i (k_i - 1). \tag{4}$$

The fiber of c over a cycle of the form $\sum_i k_i x_i$ as above is of dimension $\sum_i (k_i - 1)$ (see, e.g., [3]). It follows that the image Z_1 of Z in $X^{(k)}$ is of dimension $\geq k - d - \sum_i (k_i - 1)$. By definition, Z_1 is contained in a multiplicity-stratum of $X^{(k)}$ where the support of the considered cycles has cardinality $\leq k'$. Let $Z_1' \subset X^{k'}$ be the set of $(x_1, \ldots, x_{k'})$ such that $\sum_i k_i x_i \in c(Z)$. Then the morphism

$$Z_1' \to Z_1,\ (x_1, \ldots, x_{k'}) \mapsto \sum_i k_i x_i$$

is finite and surjective, so that

$$\dim Z_1' = \dim Z_1 \geq k - d - \sum_i (k_i - 1), \tag{5}$$

which by (4) can be rewritten as

$$\dim Z_1' = \dim Z_1 \geq k' - d.$$

Note that by construction, Z_1' parameterizes k'-uples $(x_1, \ldots, x_{k'})$ with the property that $\sum_i k_i x_i$ is rationally equivalent to a constant cycle.

The proof of Variant 12 then concludes with the following statement:

Proposition 13 *Let l be a positive integer, $k_1 > 0, \ldots, k_l > 0$ be positive multiplicities. Let Z be a closed algebraic subset of X^l. Assume that $\dim Z \geq l - d$ and the cycles $\sum_i k_i x_i$, $(x_1, \ldots, x_l) \in Z$, are all rationally equivalent in X. Then the class of the cycles $\sum_i k_i x_i$, $(x_1, \ldots, x_l) \in Z$, belongs to $S_d^k(X)$, where $k = \sum_i k_i$.*

□

For the proof of Proposition 13, we have to start with the following lemma:

Lemma 14 *Let $x_1, \ldots, x_d \in X$ and let $k_i \in \mathbf{Z}$. Then $\sum_i k_i x_i \in S_d^k(X)$, $k = \sum_i k_i$.*

Proof We use the following characterization of $S_d(X)$ given by O'Grady:

Proposition 15 (O'Grady [11]) *A cycle $z \in CH_0(X)$ belongs to $S_d(X)$ if and only if there exists a (possibly singular, possibly reducible) curve $j\colon C \subset X$, such that the genus of the desingularization of C (or the sum of the genera of its components if C is reducible) is not greater than d and z belongs to $\operatorname{Im}(j_*\colon CH_0(C) \to CH_0(X))$.*

Let now $x_1, \ldots, x_d$ be as above. There exists by [9] a curve $C \subset X$, whose desingularization has genus $\leq d$ and containing $x_1, \ldots, x_d$. Thus for any k_i, the cycle $\sum_i k_i x_i$ is supported on C, which proves the lemma by Proposition 15. □

Proof of Proposition 13 Proposition 13 is proved exactly as Theorem 9, by induction on l. In case 1 considered in the induction step, we apply the same argument as in that proof. In case 2 considered in the induction step, using the same notation as in that proof, we conclude that there is in Z an l-uple $(x_1, \ldots, x_l)$ satisfying (up to permutation of the indices)

$$x_{d+1} \ldots, x_l \in C,$$

and as any point of C is rationally equivalent to c_X, we find that

$$\sum_i k_i x_i = (\sum_{i>d} k_i) c_X + \sum_{l \leq i \leq d} k_i x_i.$$

By Lemma 14, $\sum_{1 \leq i \leq d} k_i x_i \in S_d(X)$, so that $\sum_i k_i x_i \in S_d(X)$. □

As mentioned in the Introduction, Theorem 9 in the case $d = 0$ provides the following characterization of the cycle kc_X, $k > 0$: it is the only degree-k 0-cycle z of X, whose orbit $O_z \subset X^{(k)}$ is k-dimensional (cf. Corollary 5). Let us give a slightly more direct proof in this case. We use the following Lemma 16. Let V be a 2-dimensional complex vector space. Let $\eta \in \bigwedge^2 V^*$ be a nonzero generator, and let $\omega \in \bigwedge_{\mathbf{R}}^{1,1}(V^*)$ be a positive real $(1,1)$-form on V.

Lemma 16 *Let $W \subset V^k$ be a k-dimensional complex vector subspace which is Lagrangian for the nondegenerate 2-form $\eta_k := \sum_i pr_i^* \eta$ on V^k, where the pr_is are the projections from V^k to V. Then $\prod_i pr_i^* \omega$ restricts to a volume form on W.*

Proof The proof is by induction on k. Let $\pi \colon W \to V^{k-1}$ be the projector on the product of the last $k-1$ summands. We can clearly assume, up to changing the order of factors, that $\dim \operatorname{Ker} \pi < 2$. As $\dim \operatorname{Ker} \pi \leq 1$, we can choose a linear form μ on V such that the $(k-1)$-dimensional vector space $W_\mu := \ker pr_1^* \mu_{|W}$ is sent injectively by π to a $(k-1)$-dimensional subspace W' of V^{k-1}. Furthermore, since W is Lagrangian for η_k, W' is Lagrangian for η_{k-1} because $W_\mu \subset \operatorname{Ker} \mu \times V^{k-1}$, and on $\operatorname{Ker} \mu \times V^{k-1}$, $\eta_k = \pi^* \eta_{k-1}$. By the induction hypothesis, the form $\prod_{i>1} pr_i^* \omega$ restricts to a volume form on W', where the projections here are considered as restricted to $0 \times V^{k-1}$, and it follows that

$$pr_1^*(\sqrt{-1}\mu \wedge \overline{\mu}) \wedge \prod_{i>1} pr_i^* \omega$$

restricts to a volume form on W. It follows immediately that $\prod_{i\geq 1} pr_i^*\omega$ restricts to a volume form on W since for a positive number α, we have

$$\omega \geq \alpha\sqrt{-1}\mu \wedge \overline{\mu}$$

as real $(1, 1)$-forms on V. □

Proof of Corollary 5 Let $z \in CH_0(X)$ be a cycle of degree k such that $\dim O_z \geq k$. Let $\Gamma \subset X^k$ be an irreducible component of the inverse image of a k-dimensional component of $O_z \subset X^{(k)}$ via the map $X^k \to X^{(k)}$. By Mumford's theorem [10], using the fact that all the 0-cycles parameterized by Γ are rationally equivalent in X, Γ is Lagrangian for the symplectic form $\sum_i pr_i^*\eta_X$ on X^k, where $\eta_X \in H^{2,0}(X)$ is a generator. Let L be an ample line bundle on X such that there is a curve $D \subset X$ in the linear system $|L|$, all of whose components are rational. We claim that

$$\Gamma \cap D^k \neq \emptyset.$$

Indeed, it suffices to prove that the intersection number

$$[\Gamma] \cdot [D^k] \tag{6}$$

is positive. Let $\omega_L \in H^{1,1}(X)$ be a positive representative of $c_1(L)$. Then (6) is equal to

$$\int_{\Gamma_{reg}} \prod_i pr_i^*\omega_L. \tag{7}$$

By Lemma 16, the form $\prod_i pr_i^*\omega_L$ restricts to a volume form on Γ at any smooth point of Γ and the integral (7) is thus positive. □

3 Second Chern class of simple vector bundles

This section is devoted to the proof of Theorem 6. Recall first from [11] that, in order to prove the result for a vector bundle F on X, it suffices to prove it for $F \otimes L$, where L is a line bundle on X. Choosing L sufficiently ample, we can thus assume that F is generated by global sections, and furthermore that

$$H^1(X, F^*) = 0. \tag{8}$$

Let $r = \operatorname{rank} F$. Choose a general $(r-1)$-dimensional subspace W of $H^0(X, F)$, and consider the evaluation morphism

$$e_W : W \otimes O_X \to F.$$

The following result is well known (cf. [6, 5.1]):

Lemma 17 *The morphism e_W is generically injective, and the locus $Z \subset X$ where its rank is $< r-1$ consists of k distinct reduced points, where $k = c_2^{top}(F)$.*

Proof Let $G = Grass(r-1, H^0(X, F))$ be the Grassmannian of $(r-1)$-dimension subspaces of $H^0(X, F)$. Consider the following universal subvariety of $G \times X$:

$$G_{deg} := \{(W, x) \in G \times X, \text{rank}\, e_{W,x} < r-1\}.$$

Since F is generated by sections, G_{deg} is a fibration over X, with fibers smooth away from the singular locus

$$G_{deg}^{sing} := \{(W, x) \in G \times X, \text{rank}\, e_{W,x} < r-2\}.$$

Furthermore, we have

$$\dim G_{deg} = \dim (G \times X) - 2 = \dim G$$

and $\dim G_{deg}^{sing} < \dim G$.

Consider the first projection: $p_1 : G_{deg} \to G$. It follows from the observations above and from Sard's theorem that for general $W \in G$, $p_1^{-1}(W)$ avoids G_{deg}^{sing} and consists of finitely many reduced points in X. The statement concerning the number k of points follows from [4, 14.3], or from the following argument that we will need later on. Given W such that the morphism e_W is generically injective, and the locus Z_W where its rank is $< r-1$ consists of k distinct reduced points, we have an exact sequence

$$0 \to W \otimes \mathcal{O}_X \to F \to \mathcal{I}_{Z_W} \otimes \mathcal{L} \to 0, \tag{9}$$

where $\mathcal{L} = \det F$. Hence $c_2(F) = c_2(\mathcal{I}_Z \otimes \mathcal{L}) = c_2(\mathcal{I}_Z) = Z$, and in particular $c_2^{top}(F) = \deg Z$. This proves the lemma. □

By Lemma 17, we have a rational map

$$\phi : G \dashrightarrow X^{(k)},\ W \mapsto c(Z_W),$$

where $c : X^{[k]} \to X^{(k)}$ is the Hilbert–Chow morphism.

Proposition 18 *If F is simple and satisfies assumption (8), the rational map ϕ is generically one-to-one on its image.*

Proof Let $G^0 \subset G$ be the Zariski open set parameterizing the subspaces $W \subset H^0(X, F)$ of dimension $r-1$ satisfying the conclusions of Lemma 17. Note that c is a local isomorphism at a point Z_W of $X^{[k]}$ consisting of k distinct points, so that the dimension of the image of ϕ is equal to the dimension of the image of the rational map $G \dashrightarrow X^{[k]}$, $W \mapsto Z_W$, which we will also denote by ϕ. This ϕ is a morphism on G^0 and it suffices to show that the map $\phi^0 := \phi_{|G^0}$

is injective. Let $W \in G^0$, $Z := \phi(W)$. For any $W' \in {\phi^0}^{-1}(Z)$, we have an exact sequence as in (9):

$$0 \to W' \otimes \mathcal{O}_X \to F \to \mathcal{I}_Z \otimes \mathcal{L} \to 0, \tag{10}$$

so that W' determines a morphism

$$t_{W'} : F \to \mathcal{I}_Z \otimes \mathcal{L},$$

and conversely, we recover W' from the data of $t_{W'}$ up to a scalar as the space of sections of $\mathrm{Ker}\, t_{W'} \subset F$. We thus have an injection of the fiber ${\phi^0}^{-1}(Z)$ into $\mathbf{P}(\mathrm{Hom}\,(F, \mathcal{I}_Z \otimes \mathcal{L}))$.

In order to compute $\mathrm{Hom}\,(F, \mathcal{I}_Z \otimes \mathcal{L})$, we tensor by F^* the exact sequence (9). We then get the long exact sequence

$$\cdots \to \mathrm{Hom}\,(F, F) \to \mathrm{Hom}\,(F, \mathcal{I}_Z \otimes \mathcal{L}) \to H^1(X, F^* \otimes W). \tag{11}$$

By the vanishing (8), we conclude that the map

$$\mathrm{Hom}\,(F, F) \to \mathrm{Hom}\,(F, \mathcal{I}_Z \otimes \mathcal{L})$$

is surjective. As F is simple, the LHS is generated by Id_F, so the RHS is generated by t_W. The fiber ${\phi^0}^{-1}(Z)$ thus consists of one point. □

Proof of Theorem 6 Let F be a simple nontrivial globally generated vector bundle of rank r, with $h^1(F) = 0$ and with Mukai vector

$$v = v_F = (r, l, s) \in H^*(X, \mathbf{Z}).$$

This means that $r = \mathrm{rank}\, F$, $l = c_1^{top}(F) \in H^2(X, \mathbf{Z})$, and

$$\chi(X, \mathrm{End}\, F) = < v, v^* > = 2rs - l^2. \tag{12}$$

The Riemann–Roch formula applied to $\mathrm{End}\, F$ gives

$$\chi(X, \mathrm{End}\, F) = 2r^2 + (r-1)l^2 - 2rc_2^{top}(F), \tag{13}$$

hence we get the formula (which can also be derived from the definition (1))

$$s = r + \frac{l^2}{2} - c_2^{top}(F). \tag{14}$$

We have by definition of $d(v)$

$$\chi(X, \mathrm{End}\, F) = 2 - 2d(v),$$

and thus by (12)

$$d(v) = 1 - rs + \frac{l^2}{2}. \tag{15}$$

The Riemann–Roch formula applied to F gives, furthermore,

$$\chi(X, F) = 2r + \frac{l^2}{2} - c_2^{top}(F) \tag{16}$$

which by (14) gives

$$\chi(X, F) = r + s. \tag{17}$$

As we assume $h^1(F) = 0$ and we have $h^2(F) = 0$, since F is nontrivial, generated by sections and simple, we thus get

$$h^0(X, F) = r + s. \tag{18}$$

With the notation introduced above, we conclude that

$$\dim G = (r - 1)(s + 1).$$

By Proposition 18, as all cycles parameterized by $\operatorname{Im} \phi$ are rationally equivalent in X, the orbit under rational equivalence of $c_2(F)$ in $X^{(k)}$, $k = c_2^{top}(F)$, has dimension greater than or equal to

$$(r - 1)(s + 1) = rs - s + r - 1.$$

But we have by (14) and (15):

$$k - d(v) = r - s + rs - 1.$$

By Theorem 9, we conclude that $c_2(F) \in S^k_{d(v)}(X)$. □

Remark 4 Instead of proving that the general Z_W is reduced and applying Theorem 9, we could as well apply Variant 12 directly to the family of subschemes Z_W.

For completeness, we conclude this section with the proof of Corollary 8, although a large part of it mimics the proof of Proposition 3 in [11].

We recall for convenience the statement:

Corollary 19 *Let $v \in H^*(X, \mathbf{Z})$ be a Mukai vector. Assume there exists a simple vector bundle F on X with Mukai vector v. Then*

$$S^k_d(X) = \{c_2(G),\ G \text{ a simple vector bundle on } X,\ v_G = v\},$$

where $d = d(v)$, $k = c_2(v) := c_2^{top}(F)$, $v_F = v$.

Proof The inclusion

$$\{c_2(G),\ G \text{ a simple vector bundle on } X,\ v_G = v\} \subset S^k_d(X) \tag{19}$$

is the content of Theorem 6.

For the reverse inclusion, we first prove that there exists a Zariski open set $U \subset X^{(d)}$ such that

$$cl(U) + (k - d(v))c_X \subset \{c_2(G),\ G \text{ a simple vector bundle on } X,\ v_G = v\} \tag{20}$$

where $cl\colon X^{(d)} \to CH_0(X)$ is the cycle map.

As F is simple, the local deformations of F are unobstructed. Hence there exist a smooth connected quasi-projective variety Y, a locally free sheaf $\mathcal{F}$ on $Y \times X$, and a point $y_0 \in Y$ such that $\mathcal{F}_{y_0} \cong F$ and the Kodaira–Spencer map

$$\rho\colon T_{Y,y_0} \to H^1(X, \mathrm{End}\, F)$$

is an isomorphism.

As $\mathcal{F}_{y_0}$ is simple, so is $\mathcal{F}_y$ for y in a dense Zariski open set of Y. Shrinking Y if necessary, $\mathcal{F}_y$ is simple for all $y \in Y$. By Theorem 6, we have $c_2(\mathcal{F}_y) \in S_{d(v)}(X)$ for all $y \in Y$.

Let $\Gamma := c_2(\mathcal{F}) \in CH^2(Y \times X)$. Consider the following set $R \subset Y \times X^{(d(v))}$

$$R = \{(y, z),\ \Gamma_*(y) = c_2(\mathcal{F}_y) = cl(z) + (k - d(v))c_X \text{ in } CH_0(X)\},$$

where $cl\colon X^{(d(v))} \to CH_0(X)$ is the cycle map and $k = c_2(v)$.

R is a countable union of closed algebraic subsets of $Y \times X^{(d)}$ and by the above inclusion (19), the first projection

$$R \to Y$$

is surjective. By a Baire category argument, it follows that for some component $R_0 \subset R$, the first projection is dominant.

We claim that the second projection $R_0 \to X^{(d(v))}$ is also dominant. This follows from the fact that by Mumford's theorem, the pull-backs to R_0 of the holomorphic 2-forms on Y and $X^{(d(v))}$ are equal. As the first projection is dominant and the Mukai form on Y has rank $2d(v)$, the same is true for its pull-back to R_0 (or rather its smooth locus). Hence the pull-back to R_0 of the symplectic form on $X^{(d(v))}$ by the second projection also has rank $2d(v)$. This implies that the second projection is dominant, and hence that its image contains a Zariski open set. Thus (20) is proved. The proof of Corollary 19 is then concluded with Lemma 20 below. □

Lemma 20 *Let X be a K3 surface and $d > 0$ an integer. Then for any open set (in the analytic or Zariski topology) $U \subset X^{(d)}$, we have*

$$cl(U) = cl(X^{(d)}) \subset CH_0(X).$$

Proof It clearly suffices to prove the result for $d = 1$. It is proved in [8] that for any point $x \in X$, the set of points $y \in X$ rationally equivalent to x in X is dense in X for the usual topology. This set thus meets U, so $x \in cl(U)$. □

Acknowledgments I thank Daniel Huybrechts and Kieran O'Grady for useful and interesting comments on a preliminary version of this paper.

J'ai grand plaisir à dédier cet article à Rob Lazarsfeld, avec toute mon estime et ma sympathie. Son merveilleux article [7] redémontrant un grand théorème classique sur les séries linéaires sur les courbes génériques a en particulier joué un rôle décisif dans l'étude des fibrés vectoriels et des 0*-cycles sur les surfaces* $K3$.

References

[1] Beauville, A. and Voisin, C. On the Chow ring of a $K3$ surface. *J. Alg. Geom.* **13** (2004) 417–426.

[2] Bogomolov, F., Hassett, B., and Tschinkel, Y. Constructing rational curves on K3 surfaces. *Duke Math. J.* **157**(3) (2011) 535–550.

[3] Briançon, J. Description de $Hilb^n C\{x, y\}$. *Invent. Math.* **41** (1977) 45–89.

[4] Fulton, W. Intersection Theory. Ergebnisse der Mathematik und ihrer Grenzgebiete (3) Vol. 2. Berlin: Springer-Verlag, 1984.

[5] Huybrechts, D. Chow groups of $K3$ surfaces and spherical objects. *JEMS* 12 (2010) 1533–1551.

[6] Huybrechts, D. and Lehn, M. *The Geometry of Moduli Spaces of Sheaves*, 2nd edn. Cambridge Mathematical Library. Cambridge: Cambridge University Press, 2010.

[7] Lazarsfeld, R. Brill–Noether–Petri without degenerations. *J. Diff. Geom.* **23**(3) (1986) 299–307.

[8] Maclean, C. Chow groups of surfaces with $h^{2,0} \leq 1$. *C. R. Math. Acad. Sci. Paris* **338** (2004) 55–58.

[9] Mori, S. and Mukai. S. Mumford's theorem on curves on $K3$ surfaces. *Algebraic Geometry (Tokyo/Kyoto 1982)*, pp. 351–352. Berlin: Springer-Verlag, 1983.

[10] D. Mumford. Rational equivalence of 0-cycles on surfaces. *J. Math. Kyoto Univ.* **9** (1968) 195–204.

[11] O'Grady, K. Moduli of sheaves and the Chow group of $K3$ surfaces. *J. Math. Pure Appl.* **100**(5) (2013) 701–718.

[12] Voisin, C. Chow rings and decomposition theorems for families of $K3$ surfaces and Calabi–Yau hypersurfaces. *Geom. Topol.* **16** (2012) 433–473.

[13] Voisin, C. *Hodge Theory and Complex Algebraic Geometry*, II. Cambridge Studies in Advanced Mathematics, No. 77. Cambridge: Cambridge University Press, 2003.

Printed in the United States
by Baker & Taylor Publisher Services